Managing Salt Tolerance in Plants

Managing Salt Tolerance in Plants

Edited by Walker Williams

SYRAWOOD
PUBLISHING HOUSE

New York

Published by Syrawood Publishing House,
750 Third Avenue, 9th Floor,
New York, NY 10017, USA
www.syrawoodpublishinghouse.com

Managing Salt Tolerance in Plants
Edited by Walker Williams

International Standard Book Number: 978-1-64740-419-2 (Hardback)

Trademark Notice: Registered trademark of products or corporate names are used only for explanation and identification without intent to infringe.

Cataloging-in-publication Data

Managing salt tolerance in plants / edited by Walker Williams.
    p. cm.
Includes bibliographical references and index.
ISBN 978-1-64740-419-2
1. Plants--Effect of salts on. 2. Salt-tolerant crops--Physiology. 3. Plant physiology.
4. Plants--Hardiness. I. Williams, Walker.
QK753.S3 S25 2023
581.133 54--dc23

TABLE OF CONTENTS

PREFACE

Soil salinity is a major abiotic stress, which limits the growth and productivity of plants and crops. There are two phases of salinity stress on plants, namely, the osmotic phase and the ionic phase. The osmotic phase happens immediately after salinity stress and leads to a rapid inhibition in the growth of the plant. On the other hand, ionic phase occurs after several days or weeks of salinity stress. In the ionic phase, ions are accumulated in highly toxic levels in the shoot, which adversely affects the shoot function. Plants have developed various physiological and biological mechanisms for tolerating and dealing with salinity. It can be bifurcated into three mechanisms which include osmotic tolerance, ion exclusion and tissue tolerance. Research has also been conducted for developing molecular and genomic approaches that can help crop plants develop tolerance towards salinity stress. This book provides significant information on this topic to help develop a good understanding of salinity tolerance in plants. It consists of contributions made by international experts. With state-of-the-art inputs by acclaimed experts of this field, this book targets students and professionals associated with the field of agriculture and plant science.

After months of intensive research and writing, this book is the end result of all who devoted their time and efforts in the initiation and progress of this book. It will surely be a source of reference in enhancing the required knowledge of the new developments in the area. During the course of developing this book, certain measures such as accuracy, authenticity and research focused analytical studies were given preference in order to produce a comprehensive book in the area of study.

This book would not have been possible without the efforts of the authors and the publisher. I extend my sincere thanks to them. Secondly, I express my gratitude to my family and well-wishers. And most importantly, I thank my students for constantly expressing their willingness and curiosity in enhancing their knowledge in the field, which encourages me to take up further research projects for the advancement of the area.

Editor

Comparison of Biochemical, Anatomical, Morphological and Physiological Responses to Salinity Stress in Wheat and Barley Genotypes Deferring in Salinity Tolerance

Muhammad Zeeshan [1], Meiqin Lu [3], Shafaque Sehar [1], Paul Holford [4] and Feibo Wu [1,2,*]

[1] Institute of Crop Science, Department of Agronomy, College of Agriculture and Biotechnology, Zijingang Campus, Zhejiang University, Hangzhou 310058, China; 11616102@zju.edu.cn (M.Z.); 21516206@zju.edu.cn (S.S.)

[2] Jiangsu Co-Innovation Center for Modern Production Technology of Grain Crops, Yangzhou University, Yangzhou 225009, China

[3] Australian Grain Technologies, Narrabri, NSW 2390, Australia; meiqin.lu@ausgraintech.com

[4] School of Science and Health, Western Sydney University, Penrith, NSW 2751, Australia; p.holford@westernsydney.edu.au

* Correspondence: wufeibo@zju.edu.cn

Abstract: A greenhouse hydroponic experiment was performed using salt-tolerant (cv. Suntop) and -sensitive (Sunmate) wheat cultivars and a salt-tolerant barley cv. CM72 to evaluate how cultivar and species differ in response to salinity stress. Results showed that wheat cv. Suntop performed high tolerance to salinity, being similar tolerance to salinity with CM72, compared with cv. Sunmate. Similar to CM72, Suntop recorded less salinity induced increase in malondialdehyde (MDA) accumulation and less reduction in plant height, net photosynthetic rate (Pn), chlorophyll content, and biomass than in sensitive wheat cv. Sunmate. Significant time-course and cultivar-dependent changes were observed in the activities of antioxidant enzymes such as superoxide dismutase (SOD), peroxidase (POD), catalase (CAT), ascorbate peroxidase (APX), and glutathione reductase (GR) in roots and leaves after salinity treatment. Higher activities were found in CM72 and Suntop compared to Sunmate. Furthermore, a clear modification was observed in leaf and root ultrastructure after NaCl treatment with more obvious changes in the sensitive wheat cv. Sunmate, rather than in CM72 and Suntop. Although differences were observed between CM72 and Suntop in the growth and biochemical traits assessed and modified by salt stress, the differences were negligible in comparison with the general response to the salt stress of sensitive wheat cv. Sunmate. In addition, salinity stress induced an increase in the Na^+ and Na^+/K^+ ratio but a reduction in K^+ concentrations, most prominently in Sunmate and followed by Suntop and CM72.

Keywords: antioxidants; ultrastructure; osmotic stress; salinity; wheat; barley

---

1. Introduction

Saline soils are a major problem in many countries with the Environment Program of United Nations estimating that of the 9–34% of the world's irrigated land is adversely affected by salinity [1]. Salinity can kill plants and other soil organisms and is referred to as a "silent killer" in some regions or as "white death" in others as it invokes images of a lifeless, shining land studded with dead trees. Approximately 32 million ha of dry lands [2] and 60 million ha of irrigated land [3] are affected by human-induced soil salinization, and it is well documented that salinity is one of the most severe environmental stresses hampering crop production [4,5]. At high electrical conductivity (EC) resulting from salinization, crop yields can decline drastically rendering crop cultivation no longer profitable and making soil amendments inevitable [6]. World agriculture needs to feed about 2.3 billion people globally by 2050 [7]; thus, it is imperative to understand the mechanisms associated with tolerance to salinity so that breeding programs and agronomic practices can be put in place that will allow production to meet this increasing demand [8].

Saline soils limit plant growth due to osmotic stress, ionic toxicity, and a reduced ability to take up essential minerals [9,10]. In extreme cases, root cells may lose water instead of absorbing it due to the hyperosmotic pressure of the soil solution. Water deficits affect a cascade of physical, signaling, gene expression, biochemical, and physiological pathways and processes, resulting in decreased cell elongation, wilting, and, ultimately, plant death; these harmful effects of salinity can be considered as water-deficit effects [3,11,12]. In saline soils, NaCl comprises 50–80% of the total soluble salts [13] causing elevated, and potentially toxic, concentrations of Na^+ and/or Cl^- in the plant. These ions affect many enzymes or cellular functions such as photosynthesis signaling systems [14–16]. In addition, because of their physicochemical similarities and a shared transport system, the Na^+ in the soil solution of saline soils competes for uptake with K^+ [17] and can lead to K^+ deficiency [18,19]. The induced K^+ deficiency inhibits growth because it plays a critical role in maintaining cell turgor, membrane potentials, and enzyme activities.

As a consequence of the primary effects of salinity described above, secondary stresses such as oxidative stress often occurs due to an overproduction of reactive oxygen species (ROS) [20]. These ROS cause lipid peroxidation leading to increased membrane fluidity and permeability [21,22], the denaturation of functional and structural proteins [23], and can affect nucleic acids through base modifications, induce inter- and intra-strand crosslinks, crosslinks with proteins as well as creating strand breaks [24]. However, plants have developed comprehensive internal resistance systems to combat the outcomes of ROS that are comprised of enzymatic as well as non-enzymatic antioxidants [25]. ROS-scavenging enzymes include those that are playing a direct role in the processing of ROS such as superoxide dismutase (SOD), peroxidase (POD), and catalase (CAT), and those glutathione reductase (GR) and ascorbate peroxidases (APX) that mediate in the reaction cycle of antioxidant chemicals such as glutathione (GSH) and ascorbic acid (AsA) [26–30]. The other half of the antioxidant machinery includes nonenzymatic antioxidants comprising of ascorbic acid, phenolic compounds (flavonoids, anthocyanins), α-tocopherol, carotenoids, and amino acid cum osmolyte proline. Besides the synthesis and modulation of osmolytes some phytohormones and regulatory molecules also play prominent role in triggering salinity tolerance effector molecules [31].

Barley and wheat have different salt tolerances capacities and are grown as major grain crops in both saline and non-saline soils [32]. Previous studies have focused on salinity stress in either barley or wheat alone, with little inter-specific comparison. Thus, this study is the first to compare the mechanisms that confer salinity tolerance in these two species. We aimed to explore the similarities or differences in their physiological mechanisms upon exposure to salt stress. We also hypothesized that there may be species-specific mechanisms that can be co-related with the salt sensitivity of wheat or the tolerance of barley. Thus, this research can enhance our understanding of holistic salinity tolerance mechanisms and will aid in the breeding of salt-tolerant crops.

2. Material and Methods

2.1. Plant Material and Growth Conditions

A shade house hydroponic experiment was carried out on the Zijingang Campus, Zhejiang University, China. Two wheat (*Triticum aestivum* L.) cv. Suntop (salt-tolerant) and Sunmate (salt-sensitive) and a salt-tolerant barley (*Hordeum vulgare* L.) cv. CM72, were used in the experiment. Suntop and Sunmate are two high yielding Australian Prime Hard varieties bred by Australian Grain Technologies (AGT, Narrabri, Australia. Although they were derived from the same cross, Suntop and Sunmate differ significantly in salt tolerance. Seeds of each cultivar were disinfected in 2% (*v/v*) H_2O_2 then washed thoroughly using double distilled water (ddH$_2$O). The seeds were germinated on filter paper in germination boxes in a plant growth chamber (23/18 °C, day/night) in darkness for 3 days and incubated for further 4 days in the light. The uniform 7-day-old seedlings of each cultivar were selected and transplanted into 5 L pots in a hydroponic solution containing 4.5 L basic nutrient solution (BNS) as described by [33,34], with continuous aeration using air pumps. Each container was covered with a polystyrene plate with 7 evenly spaced holes (2 plants per hole) and placed in a greenhouse with natural light and a temperature of 20 ± 2 °C/day and 15 ± 2 °C/night. At the two-leaf stage, plants were treated with 100 mM NaCl; non-NaCl treated plants were used as controls (BNS). The solution pH was adjusted to 5.8 with NaOH or HCl, as required. The experiment was arranged in a randomized block design with four replications. Plants were sampled at 1, 5, 10, and 15 days after treatment (DAT) for time-course analysis of the salt treatments. For morphological and physiological analyses, plants were harvested 25 DAT and either analyzed immediately or frozen in liquid nitrogen and stored at −80 °C for further analysis.

2.2. Measurement of Growth Traits and Mineral Concentrations

Shoot height, root length, and fresh weights were determined 25 DAT (days after treatment) and then the samples were separated into shoot and root and perfectly washed with ddH$_2$O to eliminate any foreign material. Samples from each treatment with 4 biological replicates were oven dried at 75 °C for three days and subsequently, the dry weight of the roots and shoots were determined in gram. Later each dried sample was weighed (about 0.2 g), ground, and made into ashes by heating the samples at 550 °C for half a day. Before dilution with ddH$_2$O, the ashes were digested in 30% HNO_3 and then Na^+ and K^+ concentration were quantified using flame atomic absorption spectrometry (SHIMADZU AA-6300, Columbia, Maryland, USA) [35].

2.3. Measurement of Photosynthesis Parameters, Chlorophyll Contents, and Chlorophyll Fluorescence

Intact, second fully-expanded leaves from the apex were used to measure relative chlorophyll content with the help of a handheld chlorophyll meter (Minolta SPAD-502, Tokyo, Japan) according to Wu et al. [36]. Three measurements were recorded from each leaf and averaged. The gas exchange parameters (i.e., photosynthetic rates (Pn), intercellular CO_2 concentrations (Ci), stomatal conductance (Gs), and transpiration rates (Tr)) were measured on a bright sunny day between 9 a.m. to 11 a.m. using a Li-Cor-6400 portable photosynthesis system (Li-Cor, Lincoln, NE, USA).

Chlorophyll fluorescence (*Fv/Fm*) was measured at 25 DAT according to Genty et al. [37]. Both treated and control plants were shifted to an experimental room, kept in the dark for 25 min, flag leaf were cut for the determination of chlorophyll fluorescence using a pulse-modulated chlorophyll fluorimeter using IMAGING-PAM (Walz; Effeltrich, Germany) image processing software. Fluorescence values observed comprised of *Fo*, initial/minimal fluorescence, *Fm*, the maximal fluorescence value, and *Fv/Fm*, the maximum quantum yield of PSII photochemistry. The data were noted at five different points at 40, 70, 120, 150, and 180 mm from leaf tips.

2.4. Lipid Peroxidation and Antioxidative Enzyme Activity Assay

The roots and second leaf from the apex were sampled at 1, 5, 10, and 15 DAT. Lipid peroxidation was measured in the tissues and expressed as malondialdehyde (MDA) content using the TBA (thiobarbituric acid) method according to the Wu et al. [34]. The activity-specific and non-specific absorbance was determined at 532 and 600 nm, respectively. Enzymatic antioxidants activity was determined as described by Leul and Zhou [38]. Briefly, 0.2 g of frozen leaf and root plant tissue were ground in a pre-chilled mortar and pestle and homogenized in 2 mL of 1 M Tris buffer (pH 8). Later, the samples were briefly centrifuged at 10,000× g at 4 °C for about 15 min and the supernatants were used for the following assays. The activity assay of superoxide dismutase (SOD, EC 1.15.1.1), peroxidase (POD, EC 1.11.1.7), and catalysis (CAT, EC 1.11.1.6) were recorded according to Wu et al. [34]. Ascorbate peroxidase (EC 1.11.1.11) activity was determined at 290 nm using ascorbate (AsA) as a substrate and 2.8 mM cm^{-1} as an extinction co-efficient [39], while Jiang and Zhang [40] methods were used to determine the activity of glutathione reductase (GR, EC. 1.6.4.2).

2.5. Cell Ultrastructure

For transmission electron-microscopy, fresh roots (about 2–3 mm in length) and leaf pieces (about 1 mm^2) without veins were hand sectioned and treated with 100 mM sodium phosphate buffer (PBS; pH 7.0) containing 2.5% (*v/v*) glutaraldehyde and placed overnight at 4 °C, then briefly washed; this step was performed 3 times with the same buffer. Each sample was treated for 60 min with 1% osmium tetroxide OsO_4 (*v/v*) followed by washing with PBS (sodium phosphate buffer) further 3 times. Thereafter, the leaf and root samples were dehydrated with a diluted series of ethanol (50%, 60%, 70%, 80%, 90%, 95%, and 100%) for about 20 min in each solution, later all the samples were dried for 20 min in concentrated acetone. Finally, ultrathin sections (80 nm) were cut and affixed to copper grids for study using transmission electron microscopy (TEM 1230EX, JEOL, Akishima, Tokyo, Japan) at 60 kV.

2.6. Statistical Analysis

Statistical analysis was performed with the Data Processing System statistical software package [41] using ANOVA followed by Duncan's Multiple Range Tests (DMRT) to evaluate significant treatment effect at a significance level of $p < 0.05$. Origin Pro (Version 8.0, Origin lab corporation, Wellesley Hills, Wellesley, MA, USA) was used to prepare graphs.

3. Results

3.1. Plant Growth Parameters

Salt inhibited the growth of the barley and wheat plants, with the treated plants showing wilting, necrosis and chlorosis (Figure 1A). Salt damage was most severe in the wheat cv., Sunmate, while in the other cultivars, the damage was less pronounced. Salinity stress significantly ($p < 0.05$) reduced plant biomass in the wheat and barley cultivars (Figure 1B). In comparison to the other two cultivars, the effects of salt stress on plant growth was much more noticeable in Sunmate; it had the least effect on shoot height and the biggest effect on shoot weight. Shoot height was reduced under salinity stress by 29%, 12%, and 13% in the Sunmate, Suntop and CM72 cultivars, respectively. The fresh shoot weight was reduced by 68%, 55%, and 59%, while shoot dry weight was reduced 68%, 53%, and 49% in Sunmate, Suntop and CM72 in salinity stress plants respectively. Similarly, compared to the control plants, the root length was reduced under salinity stress by 37%, 8%, and 24% in Sunmate, Suntop, and CM72, respectively, while the reductions in fresh root weight were 42%, 33%, and 11% and dry root weight were 65%, 39%, and 30% in Sunmate, Suntop, and CM72, respectively in salinity treated plants (Table S1).

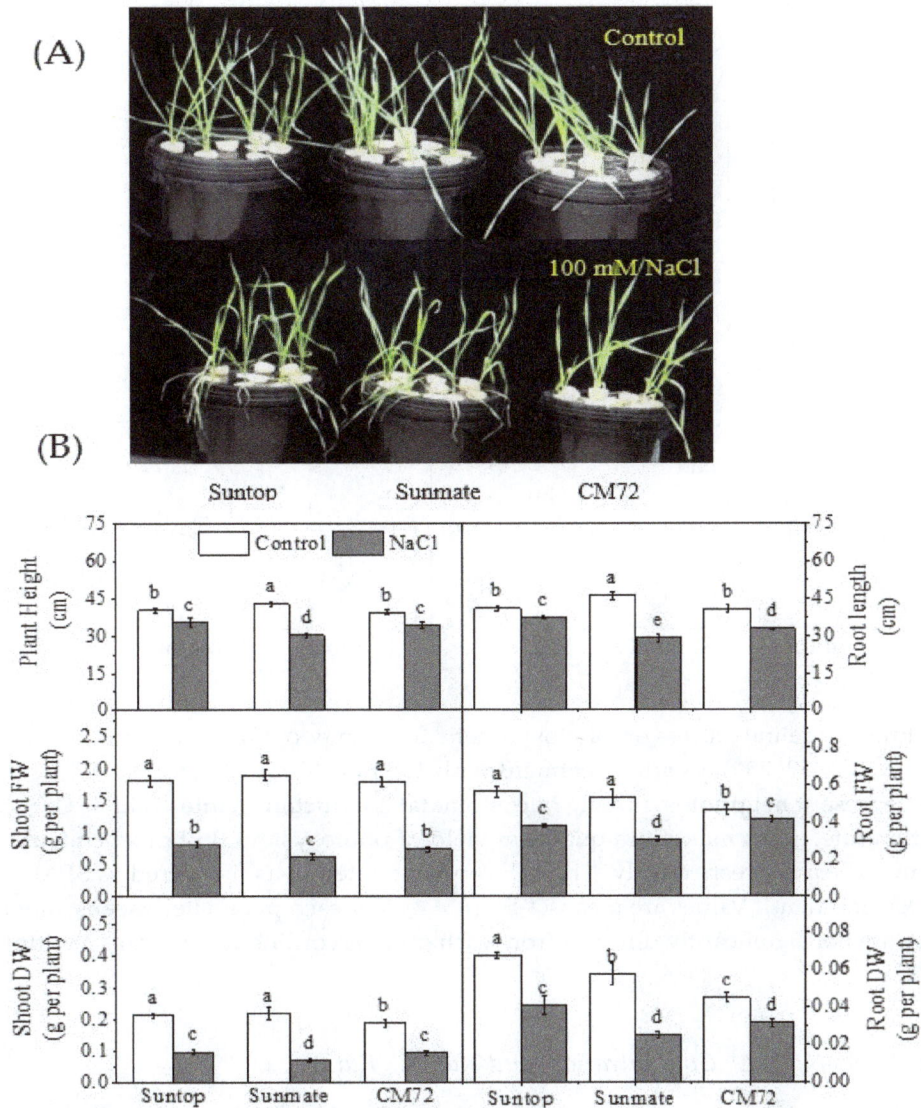

Figure 1. Morphology (**A**) and growth parameters (**B**) of seedlings of two wheat cv., Suntop and Sunmate, and one barley cv., CM72, 25 days after treatment with NaCl. Control and NaCl represent 0 and 100 mM NaCl, respectively. Values are means + SE ($n = 4$). For each parameter, means annotated with the same letter are not significantly different from each other according to Duncan's Multiple Range Tests at $p \leq 0.05$.

3.2. Chlorophyll and Photosynthetic Parameters

Gas exchange parameters were recorded 25 DAT (days after treatment) and significant ($p < 0.05$) decreases in net photosynthetic rate (Pn), stomatal conductance (Gs), intercellular CO_2 concentration (Ci), and transpiration rate (Tr) were detected in both wheat and barley in comparison to their respective controls (Figure 2A–F). No significant changes were observed in Gs, chlorophyll contents and Fv to Fm ratios, however, significant differences were observed in Pn, Ci, and Tr among all the cultivars in the salinity treated plants. Interestingly, the two-salt tolerant cultivars, Suntop and CM72, showed no significant difference in regard to the photosynthetic parameters, but differences were noted in Sunmate, which is salt-sensitive.

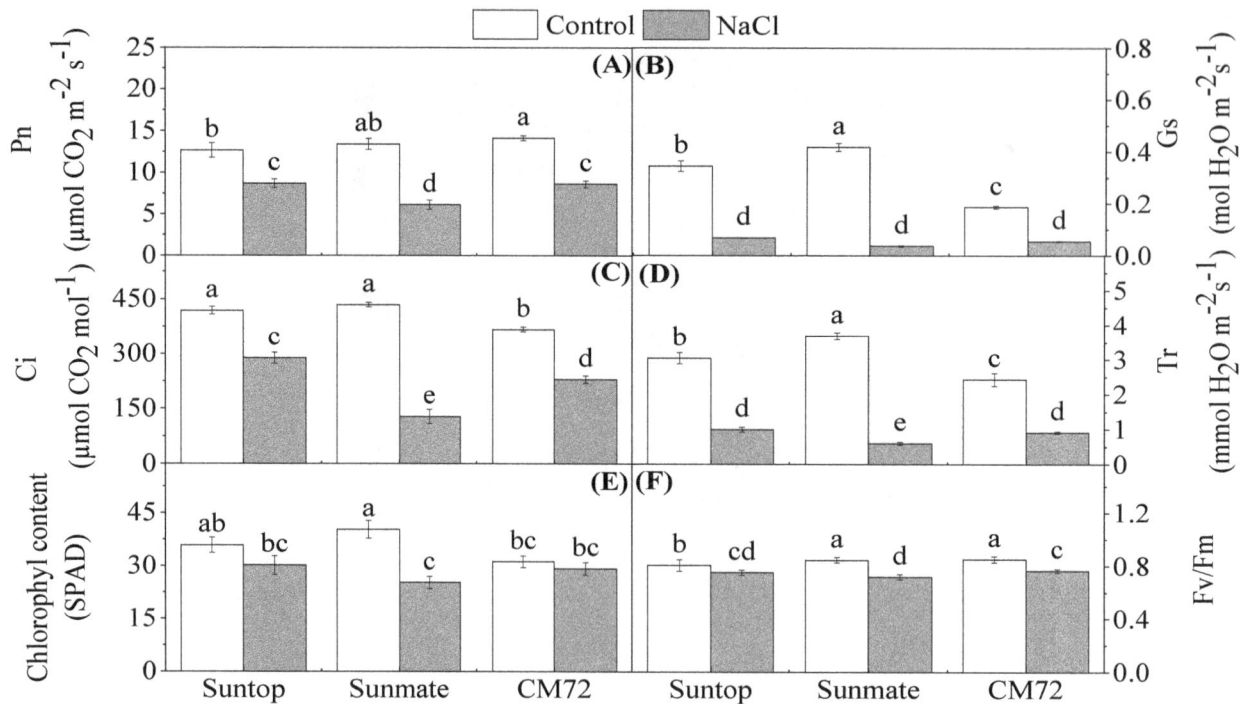

Figure 2. Effect of salinity stress on photosynthetic traits in two wheat cv., Suntop and Sunmate, and the barley cv., CM72, 25 days after treatment with 100 mM NaCl. Pn (**A**), Gs (**B**), Ci (**C**), Tr (**D**) and Fv/Fm (**F**), represent net photosynthetic rate, stomatal conductance, intercellular CO_2 concentration, transpiration rate, and a maximum quantum yield of photosystem II photochemistry of the second fully expanded leaves, respectively. The chlorophyll content was measured as SPAD (**E**) (Soil Plant analysis Development). Values are means + SE ($n = 4$). For each parameter, means annotated with the same letter are not significantly different from each other according to Duncan's Multiple Range Tests at $p \leq 0.05$.

3.3. Shoots and Roots Na^+, K^+ Concentration, and Na^+:K^+ Ratio

Internal Na^+ and K^+ concentrations were determined, salinity significantly ($p < 0.05$) increased Na^+, decreased K^+ content, and increased Na^+:K^+ ratio in both shoots and roots of all cultivars in the saline-treated plants relative to the control plants (Figure 3A–F). In general, roots contained more Na^+ and K^+ compared to shoots, regardless of the cultivar or treatment. With regard to the plants given the salt treatment, in both shoots and roots, the increase in the Na^+ content followed the trend CM72 < Suntop < Sunmate in both organs, while the K^+ content decreased in the following trend CM72 > Suntop > Sunmate. Interestingly, the increase in Na^+ content among the cultivars was inversely proportional to the decrease in K^+ content. As a consequence of the changes in both minerals, the Na^+:K^+ ratio increased under salt stress. The greatest Na^+:K^+ ratios were observed in Sunmate (0.339) and Suntop (0.127), while the smallest were observed in CM72 (0.075) in shoots. The same trend in Na^+:K^+ values were also observed in the roots (Table S2).

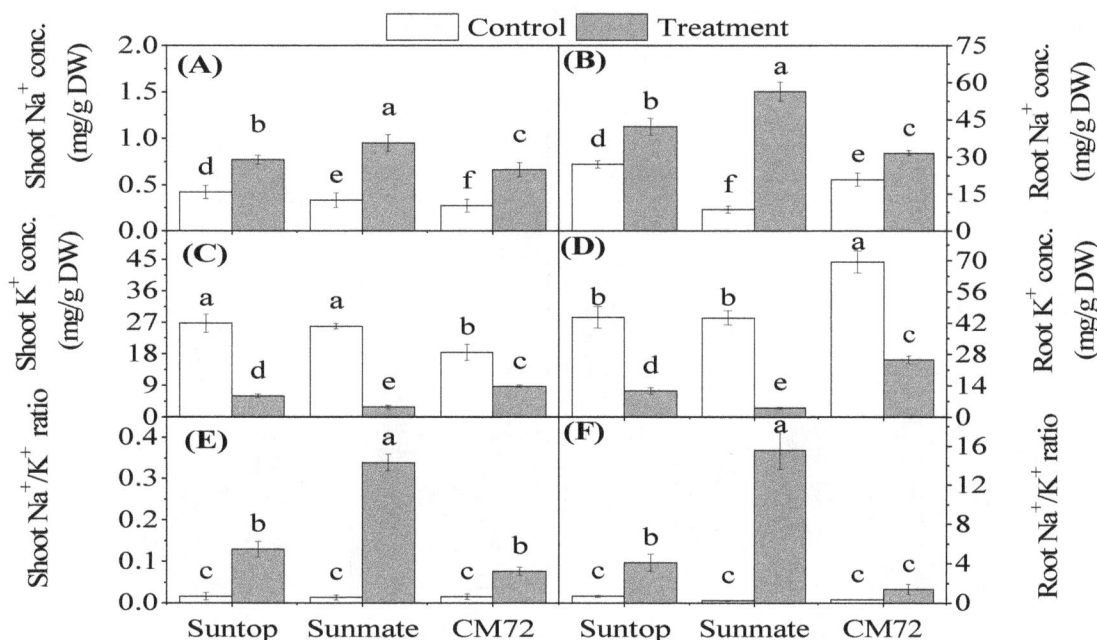

Figure 3. Effect of salinity stress on Na$^+$ and K$^+$ concentrations and Na$^+$:K$^+$ ratio in shoots (**A,C,E**) and roots (**B,D,F**) of two wheat cv., Suntop and Sunmate, and the barley cv., CM72, 25 days after treatment with 100 mM NaCl. Error bars represent SE (n = 4). Different letters indicate significant differences ($p \leq 0.05$) among the 3 cultivars.

3.4. Lipid Peroxidation Assay and Antioxidative Enzyme Activities

Lipid peroxidation measurements at 1, 5, 10, and 15 DAT showed that salt stress induced significant changes among the cultivars and treatments (Figure 4A,B and Supplementary Tables S3 and S4). Regardless of the cultivar, MDA contents were significantly increased by the salt treatment in both leaves and roots, indicating enhanced lipid peroxidation. In general, in plants given salinity treatments, the MDA content was highest in Sunmate followed by Suntop then CM72, with the highest increase observed 15 and 10 d after treatment in leaves and roots, respectively.

Significant differences ($p < 0.05$) were detected among the cultivars and between the treatment in both roots and leaves for all measured antioxidant enzymes (Supplementary Tables S3 and S4, and expression relative to control activities in Figure 5A–J). In general, in the leaves, the relative activities of all enzymes were highest in Suntop, followed by CM72 and then Sunmate; however, in the roots, there was little difference between CM72 and Suntop. For SOD in the leaves, the activities of this enzyme were similar on Days 1 and 5, rose on Day 10, and then dropped below those measured on Day 1. In the roots, SOD activities rose on Day 10 and remained high on Day 15. For POD in the leaves, activities rose on Day 5 and then declined during the remainder of the assessment period, whilst in the roots POD activities did not rise until Day 10 and then declined. For APX in the leaves, activities started to rise on Day 10 and were highest on Day 15 whilst in the root's activities rose on Day 10 and remained high. For CAT in leaves, activities were highest on Day 5 and then declined, except in Suntop, where a decline was observed on Day 10. In the roots, CAT activities tended to stay on similar levels throughout the treatment period. Finally, for GR for the two wheat cv., there was little change in activities in the leaves during the assessment period; however, for CM72, GR activities were highest on Days 5 and 10 and then declined. In the roots, for Suntop, GR activities increased on Day 5 and then remained high, for Sunmate, activities increased on Day 10, and then reduced and for CM72, GR activities were high throughout the experiment. Peaks of antioxidant enzyme activity were observed generally earlier in shoots than in roots, while the earliest peak was observed for CAT and the latest for APX.

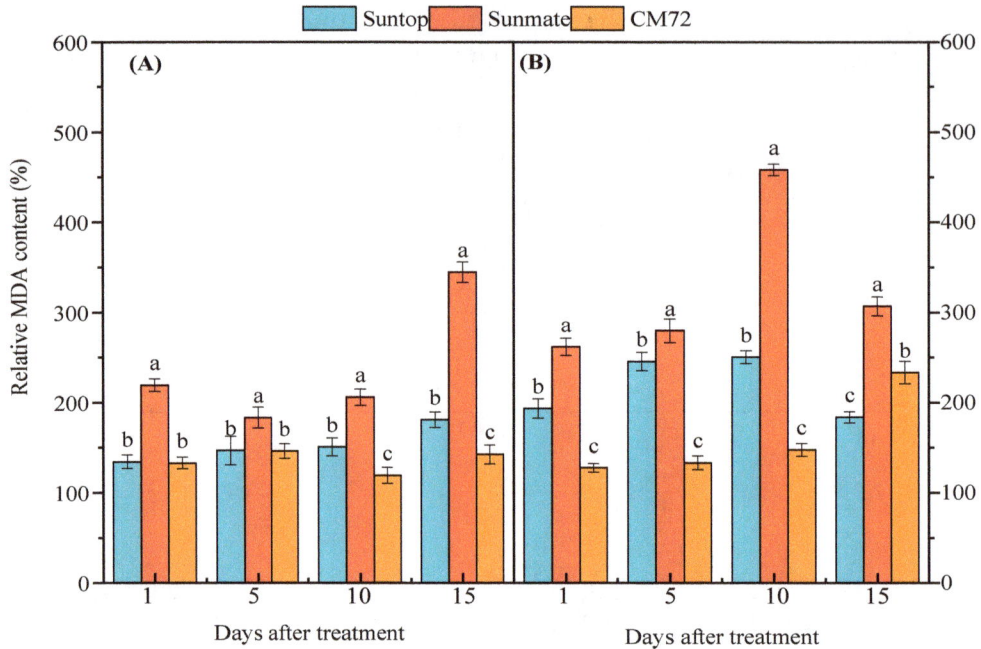

Figure 4. Effect of salinity stress on malondialdehyde contents (MDA, nmol⁻¹ FW) in leaves (**A**) and roots (**B**) of Suntop (Aqua), Sunmate (Red), and CM72 (Orange) 1, 5, 10, and 15 days after treatment with 100 mM NaCl. The data are expressed as a percentage of control content. Different letters indicate significant differences ($p \leq 0.05$) among the 3 cultivars within each sampling day. Error bars represent SE ($n = 4$).

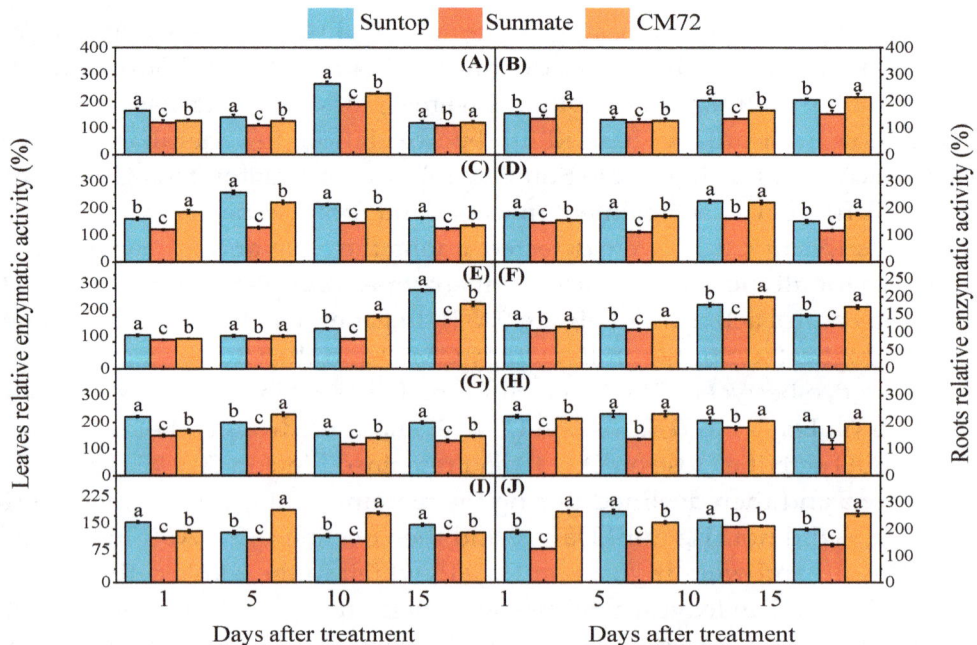

Figure 5. Effect of salinity stress on (**A,B**) superoxide dismutase (SOD) (U g⁻¹ FW); (**C,D**) peroxidase (POD) (OD 470 g⁻¹ min⁻¹); (**E,F**) ascorbate peroxidases (APX) (mmol g⁻¹ FW min⁻¹); (**G,H**) catalase (CAT) (mmol⁻¹ FW min⁻¹; (**I,J**) glutathione reductase (GR) (mmol g⁻¹ FW min⁻¹) activities in leaves (**A,C,E,G,I**) and roots (**B,D,F,H,J**) of Suntop (Aqua), Sunmate (Red), and CM72 (Orange) 1, 5, 10, and 15 days after treatment with 100 mM NaCl. The data are expressed as a percentage of control activities. Different letters indicate significant differences ($p \leq 0.05$) among the 3 cultivars within each sampling day. Error bars represent SE ($n = 4$).

3.5. Leaf and Root Ultrastructure

The chloroplast ultrastructure of Sunmate was more severely affected by salt stress relative to controls and also to Suntop and CM72. Under control conditions, the chloroplasts of Suntop mesophyll cells usually had normal morphology with distinct grana and stroma lamellae, large starch grains with numerous plastoglobuli and well-organized, round mitochondria (Figure 6A); after the salt treatment, there were fewer plastoglobuli, no starch grains were apparent, and the grana and stroma lamellae were diffuse (Figure 6B). In contrast, chloroplasts of Sunmate were severely damaged by salinity stress, i.e., the chloroplast envelope showed disintegration with reduced grana stacks and less distinct thylakoids membranes, swollen oval-shaped mitochondria and larger osmophilic plastoglobuli (Figure 6C,D). As with Suntop, the chloroplasts of CM72 remained relatively normal in response to the salt treatment except for the disappearance of starch grains and thinner lamellae (Figure 6E,F).

When viewed using transmission electron microscopy, the root cells of all cultivars grown without salt treatment (control) had dense cytoplasm and organelles, and organized and large nuclei and nucleoli (Suntop, Figure 7A; Sunmate, Figure 7C; CM72, Figure 7E). Treatment with salt induced a number of ultrastructural changes from mild to severe, with the most visible alteration being the disappearance of nucleoli and vacuoles in Sunmate (Suntop, Figure 7B; Sunmate, Figure 7D; CM72, Figure 7F). Suntop and CM72 had clear nucleoli and larger and several vacuoles in comparison with Sunmate. However, the size of the nucleoli increased in CM72, and to a lesser extent in Suntop, upon exposure to salt.

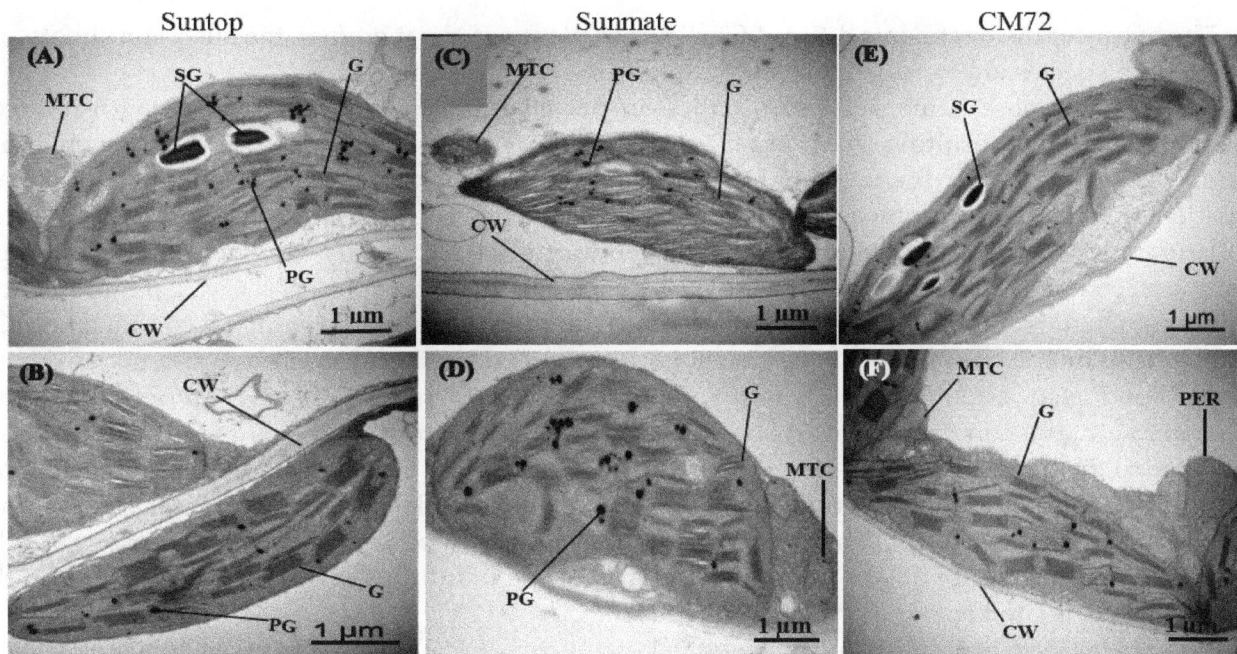

Figure 6. Transmission electron micrographs of chloroplasts of leaves of Suntop (**A,B**), Sunmate (**C,D**), and CM72 (**E,F**) under control (top panel) and 100 mM NaCl (bottom panel). CW, cell wall; G, grana; MTC, mitochondria; PG, plastoglobuli; SG, starch grains.

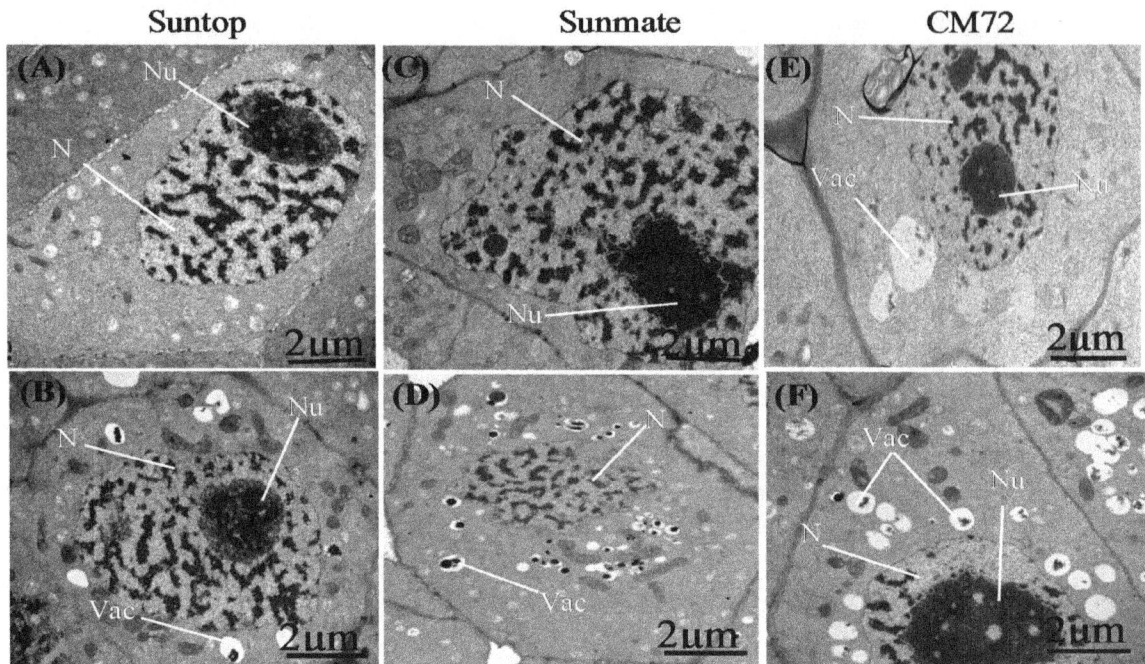

Figure 7. Electron micrographs of roots of Suntop (**A,B**), Sunmate (**C,D**), and CM72 (**E,F**) under control (top panel) and 100 mM NaCl (bottom panel). CW, cell wall; Nu, nucleolus; N, nucleus; Vac, vacuole.

4. Discussion

The effects of the treatments differed for different plant organs; therefore, the effects on shoots and roots of both species are considered separately.

Reduced biomass, a marked perturbation in photosynthetic parameters along with reduced chlorophyll contents resulting from salinity stress were observed in the wheat and barley cultivars. These effects are possibly due to either single or combined effects of reduced stomatal conductance, inhibition of metabolic phenomena, and increased ROS generation which can increase oxygen-induced cellular damage [42]. The reductions in stomatal conductance (Gs), photosynthesis rates (Pn), and leaf chlorophyll contents due to salinity were greater in Sunmate than in Suntop and CM72 (Figure 2). In a study conducted using sorghum, Netondo et al. [43] found that changes in stomatal conductance (Gs) and intercellular CO_2 concentration (Ci) were positively correlated under salt stress, concluding that stomatal conductance (Gs) was the key factor arresting net photosynthesis rates (Pn) under saline stress. The lower stomatal conductance (Gs) accompanied by low chlorophyll contents in Sunmate could contribute to the higher inhibition of net photosynthesis rates (Pn). Usually, plants close their stomata upon the onset of stressful conditions to save water, consequently reducing stomatal conductance (Gs) and photosynthesis [44]. The effect of salinity might be a secondary influence, arbitrated by the lower partial pressure of CO_2 in the green parts of the plant due to the stomata closure on the photosynthesis-related enzyme activities [45,46].

The *Fv/Fm* ratio reflects the photochemical efficiency of PSII [47]. Results in this study show that even a small but significant reduction in *Fv/Fm* with the greatest decline in Sunmate followed by CM72 then Suntop was recorded (Figure 2); these results are consistent with that presented by Ahmad et al. [48] and Ibrahim et al. [47]. NaCl stress can disturb the photosynthesis biochemistry, limiting the efficiency of two photosystems due to the disordering of chloroplast integrity [47]. In our study, salinity altered leaf chloroplast ultrastructure causing swelling of thylakoids, diffuse granular and stroma lamellae, a larger number of large-sized plastoglobuli and a reduction in leaf chlorophyll content in the sensitive wheat cultivar, Sunmate; these changes were not seen to the same extent in Suntop and CM72 (Figure 6). There may be several reasons for the disruptions to thylakoids and the chloroplast envelope. These include higher accumulation of lipids in chloroplasts, ion toxicity, or imbalance [49], and osmotic imbalance between chloroplast and stroma [50], which in turn cause a reduction of photosynthetic efficiency and the electron transport activity of chloroplasts [51].

Additionally, upon exposure to salinization, severe disruption of nuclei and nuclear membranes of roots were detected in Sunmate but to a lesser extent in Suntop and CM72 (Figure 7). Salinity largely affects roots because of their direct contact with the soil. Therefore, to protect the whole plant from the adverse effects of salinity, roots should better tolerate salinity stress [47]. Zhang and Blumwald [52] noted that in tolerant plants, Na^+ is kept away from the cytosol by compartmentalizing it into the vacuole, and due to the lack of this ability in sensitive plants, dehydration and ionic imbalance disturbed the metabolic process in sensitive plants [53]. We observed that the nucleolus disappeared in Sunmate. A common consequence of this type of alteration inside the nucleus would be a loss of function and/or even cell death [53].

It is important to determine Na^+ and K^+ concentrations and Na^+:K^+ ratios in shoots and roots to understand mechanisms of salinity tolerance [54]. In our study, under the salinity stress, significant differences were found in shoot Na^+ and K^+ along with Na^+:K^+ ratios in both species relative to controls, with the most severe effect in Sunmate (Figure 3A–F, Table S2). In general, CM72 accumulated less Na^+ and more K^+ in shoots followed by Suntop and then Sunmate. Hence, the low Na^+:K^+ ratios in CM72 and Suntop may explain the tolerance of these cultivars. Root to shoot Na^+ and K^+ translocation is limited, as all genotypes accumulated more Na^+ and K^+ in roots than in shoots. These results are consistent with the idea of differences in translocation restricting Na^+ movement to the shoot being one of the mechanisms of salinity tolerance. Na^+ and K^+ are interdependent under salinity stress. Previous studies have found a decrease in K^+ content in several plant species resulting from high salinity [35,55]. Increased Na^+ concentrations in root zones have an antagonistic effect on K^+ uptake. Consequently, a deficiency of K^+ has created stunting growth and reducing yields [56].

The most general consequence of salinization is the accumulation of hazardous substances in plant cells especially ROS such as singlet oxygen (O_2), superoxide radicals (O_2^-), and hydrogen peroxide (H_2O_2); these species cause damage to proteins, lipids, and nucleic acids thereby promoting rapid plant death [57]. Malondialdehyde (MDA), a product of polyunsaturated fatty acid peroxidation [58], is commonly considered as a sign of the extent of oxidation damage under stress [27,59]. The hostile influences of NaCl stress on lipid peroxidation have been reported in other plants, for example in *Brassica juncea* [60] and *Vicia faba* [61], and MDA has been widely recognized as a good salinity tolerance marker in plant species [62]. We found significantly lower MDA contents in the shoots and roots of CM72 and Suntop compared to the Sunmate (Figure 4A,B). These data suggest that CM72 and Suntop were better protected than Sunmate against oxidative damage under salinity stress.

After salt treatment, tolerant plants eventually develop an enhanced antioxidant enzyme system to handle the effects of ROS. In our study, significantly increased SOD, POD, APX, CAT, and GR activities were found in roots and leaf tissues of both species in the NaCl treated plants (Figure 5A–J). However, the relative activities of these enzymes were recorded higher in Suntop and CM72 than in cultivar Sunmate in both tissues. SOD provides the first line of defense against ROS and protects plants from severe damage generated by O_2^- and H_2O_2 in the presence of metal ions [63]. Many studies have found that salinity positively promotes SOD activity in tolerant cultivars in both roots [64] and leaves [65]. Subsequent reactions are required to convert the H_2O_2 produced by SOD to H_2O because H_2O_2 is still toxic to plants and reactions involving POD, CAT, and APX are important. Our research corroborates previous studies [47,48] had measured an enhanced activity of SOD, POD, and CAT in plants treated with a high NaCl dose and the activities of these enzymes were again higher in the two tolerant genotypes. Feki et al. [66] and Koca et al. [67] also demonstrated that tolerance to salinity in wheat and sesame genotypes was associated with lower MDA contents and higher activities of antioxidant enzymes. Thus, it is evident from our results and the results of others [10,68] that the higher POD, CAT, and APX activities coordinate with SOD activity to deal with the undesirable effect of O_2^- and H_2O_2 and the activities of these enzymes are strongly correlated with tolerance to salt-induced oxidative stress in wheat and barley.

The activity of GR in the leaves and roots was higher in CM72 compared to Suntop and Sunmate (Figure 5I,J). Other studies working with salt-sensitive and tolerant genotypes suggested that higher GR activities relate to salt tolerance [46,65]. The higher GR activity might be able to elevate $NADP^+$ concentrations to gain electrons from the photosynthetic electron transport chain thereby reducing the production of ROS [69]. Our results also suggest that the salt-tolerant cultivar may exhibit a more active ascorbate-glutathione cycle.

5. Conclusions

Although differences were observed between CM72 and Suntop in the growth and biochemical traits assessed and modified by salt stress, the differences are negligible in comparison with the response to the salt stress of sensitive wheat cv. Sunmate. The distinct differences between wheat and barley were lower MDA content, lower Na^+/K^+ ratio and a higher level of APX and GR content in the roots of barley cultivar CM72. These results lead us to infer that differences in response to salinity may be just as great within a species as between species. The most obvious mechanisms for salt tolerance in the tolerant barley and wheat cultivars are the increased activities of ROS-scavenging enzymes and a more balanced $Na^+:K^+$ ratio. Our results indicated that Suntop is highly tolerant against salinity, which is quite similar to barley CM72. Novel salt-tolerant related genes may be identified in Suntop for improving the salt tolerance of wheat cultivars, except for commercial application in saline-alkali soils.

Supplementary Materials:
Table S1: Effect of salinity on plant growth and biomass of Suntop, Sunmate and CM72, 25 days after treatment with 100 mM NaCl, Table S2: Shoot and root Na^+ and K^+ concentration and Na^+/K^+ ratio of two wheat cv. Suntop and Sunmate, and one barley cv. CM72, 25 days after treatment with 100 mM NaCl, Table S3: Effect of salinity stress on SOD, POD, CAT, APX and GR activities and MDA contents in the shoots of Suntop, Sunmate and CM72, after 1, 5, 10 and 15 days 100 mM NaCl treatment, Table S4: Effect of salinity stress on SOD, POD, CAT, APX, and GR activities and MDA contents in the roots of Suntop, Sunmate, and CM72, after 1, 5, 10, and 15 days 100 mM NaCl treatment.

Author Contributions: Conceptualization, F.W. and M.L.; data curation, M.Z. and F.W.; formal analysis, M.Z.; funding acquisition, F.W.; investigation, M.Z. and S.S.; methodology, F.W. and M.Z.; project administration, F.W.; supervision, F.W.; validation, M.Z., S.S., P.H., and F.W.B.; writing—original draft, M.Z.; writing—review and editing, M.Z., F.W., and P.H. All authors have read and agreed to the published version of the manuscript.

References

1. Ghassemi, F.; Jakeman, A.J.; Nix, H.A. *Salinisation of Land and Water Resources: Human Causes, Extent, Management and Case Studies*; CAB International Publishing: Wallingford, UK, 1995; p. 526.

2. FAO. Global Network on Integrated Soil Management for Sustainable Use of Salt-Affected Soils. 2000. Available online: http://www.fao.org/ag/AGL/agll/spush/intro.htm (accessed on 10 May 2004).

3. Zhang, H.X.; Hodson, J.N.; Williams, J.P.; Blumwald, E. Engineering salt-tolerant Brassica plants: characterization of yield and seed oil quality in transgenic plants with increased vacuolar sodium accumulation. *Proc. Natl. Acad. Sci. USA* **2001**, *98*, 12832–12836. [CrossRef] [PubMed]

4. Flowers, T.J. Improving crop salt tolerance. *J. Exp. Bot.* **2004**, *55*, 307–319. [CrossRef] [PubMed]

5. Munns, R.; Tester, M. Mechanisms of salinity tolerance. *Annu. Rev. Plant Biol.* **2008**, *59*, 651–681. [CrossRef] [PubMed]

6. Parida, A.K.; Das, A.B. Salt tolerance and salinity effects on plants: A review. *Ecotoxicol. Environ. Saf.* **2005**, *60*, 324–349. [CrossRef]

7. FAO. *The Special Challenge for Sub-Saharan Africa. How to Feed the World 2050*; FAO: Rome, Italy, 2009.

8. Ashraf, M.; Harris, P.J.C. Potential biochemical indicators of salinity tolerance in plants. *Plant Sci.* **2004**, *166*, 3–16. [CrossRef]

9. Ferguson, L.; Grattan, S.R. How salinity damages citrus: osmotic effects and specific ion toxicities. *Hort. Technol.* **2005**, *15*, 95–99. [CrossRef]

10. Munns, R.; James, R.A.; Lauchli, A. Approaches to increasing the salt tolerance of wheat and other cereals. *J. Exp. Bot.* **2006**, *57*, 1025–1043. [CrossRef]

11. Apse, M.P.; Blumwald, E. Engineering salt tolerance in plants. *Curr. Opin. Biotechol.* **2002**, *13*, 146–150. [CrossRef]

12. Machado, R.M.; Serralheiro, R.P. Soil salinity: effect on vegetable crop growth. Management practices to prevent and mitigate soil salinization. *Horticulturae* **2017**, *3*, 30. [CrossRef]

13. Rengasamy, P. World salinization with emphasis on Australia. *J. Exp. Bot.* **2006**, *57*, 1017–1023. [CrossRef]

14. Slabu, C.; Zörb, C.; Steffens, D.; Schubert, S. Is salt stress of faba bean (*Vicia faba*) caused by Na+ or Cl–toxicity? *J. Plant Nutr. Soil Sci.* **2009**, *172*, 644–651. [CrossRef]

15. Tavakkoli, E.; Rengasamy, P.; McDonald, G.K. High concentrations of Na^+ and Cl^- ions in soil solution have simultaneous detrimental effects on growth of faba bean under salinity stress. *J. Exp. Bot.* **2010**, *61*, 4449–4459. [CrossRef] [PubMed]

16. Cheeseman, J.M. The integration of activity in saline environments: Problems and perspectives. *Funct. Plant Biol.* **2013**, *40*, 759–774. [CrossRef]

17. Schachtman, D.P.; Liu, W.H. Molecular pieces to the puzzle of the interaction between potassium and sodium uptake in plants. *Trends Plant Sci.* **1999**, *4*, 281–287. [CrossRef]

18. Ball, M.C.; Chow, W.S.; Anderson, J.M. Salinity-induced potassium deficiency causes loss of functional photosystem II in leaves of the grey mangrove, *Avicennia marina*, through depletion of the atrazine-binding polypeptide. *Funct. Plant Biol.* **1987**, *14*, 351–361. [CrossRef]

19. Botella, M.A.; Martinez, V.; Pardines, J.; Cerda, A. Salinity induced potassium deficiency in maize plants. *J. Plant Physiol.* **1997**, *150*, 200–205. [CrossRef]

20. Pang, C.H.; Wang, B.S. Oxidative stress and salt tolerance in plants. In *Progress in Botany*; Lüttge, U., Beyschlag, W., Murata, J., Eds.; Springer: Berlin/Heidelberg, Germany, 2008; Volume 69, pp. 231–245.

21. Wong-Ekkabut, J.; Xu, Z.; Triampo, W.; Tang, I.M.; Tieleman, D.P.; Monticelli, L. Effect of lipid peroxidation on the properties of lipid bilayers: A molecular dynamics study. *Biophys. J.* **2007**, *93*, 4225–4236. [CrossRef]

22. Sharma, P.; Jha, A.B.; Dubey, R.S.; Pessarakli, M. Reactive oxygen species, oxidative damage, and antioxidative defense mechanism in plants under stressful conditions. *J. Bot.* **2012**, *2012*, 1–26. [CrossRef]

23. Smirnoff, N. Plant resistance to environmental stress. *Curr. Opin. Biotech.* **1998**, *9*, 214–219. [CrossRef]

24. Jena, N.R. DNA damage by reactive species: Mechanisms, mutation and repair. *J. Biosci.* **2012**, *37*, 503–517. [CrossRef]

25. Noctor, G.; Foyer, C.H. Ascorbate and glutathione: Keeping active oxygen under control. *Annu. Rev. Plant Biol.* **1998**, *49*, 249–279. [CrossRef] [PubMed]

26. Takahashi, M.; Asada, K. Superoxide production in aprotic interior of chloroplast thylakoids. *Arch. Biochem. Biophys.* **1988**, *267*, 714–722. [CrossRef]

27. Apel, K.; Hirt, H. Reactive oxygen species: Metabolism, oxidative stress, and signal transduction. *Annu. Rev. Plant Biol.* **2004**, *55*, 373–399. [CrossRef] [PubMed]

28. Mittler, R.; Vanderauwera, S.; Gollery, M.; Van Breusegem, F. Reactive oxygen gene network of plants. *Trends Plant Sci.* **2004**, *9*, 490–498. [CrossRef] [PubMed]

29. Dietz, K.J.; Jacob, S.; Oelze, M.L.; Laxa, M.; Tognetti, V.; de Miranda, S.M.; Baier, M.; Finkemeier, I. The function of peroxiredoxins in plant organelle redox metabolism. *J. Exp. Bot.* **2006**, *57*, 1697–1709. [CrossRef]

30. Türkan, I.; Demiral, T. Recent developments in understanding salinity tolerance. *Environ. Exp. Bot.* **2009**, *67*, 2–9. [CrossRef]

31. Gupta, B.; Huang, B. Mechanism of salinity tolerance in plants: Physiological, biochemical, and molecular characterization. *Int. J. Genom.* **2014**, *2014*, 1–18. [CrossRef]

32. Pessarakli, M. Dry matter yield, nitrogen-15 absorption, and water uptake by green bean under sodium chloride stress. *Crop Sci.* **1991**, *31*, 1633–1640. [CrossRef]

33. Wu, F.B.; Zhang, G.P.; Yu, J.S. Interaction of cadmium and four microelements for uptake and translocation in different barley genotypes. *Comm. Soil Sci. Plant Analysis* **2003**, *34*, 2003–2020. [CrossRef]

34. Wu, F.B.; Zhang, G.P.; Dominy, P. Four barley genotypes respond differently to cadmium: Lipid peroxidation and activities of antioxidant capacity. *Environ. Exp. Bot.* **2003**, *50*, 67–78. [CrossRef]

35. Chen, Z.; Zhou, M.; Newman, I.A.; Mendham, N.J.; Zhang, G.; Shabala, S. Potassium and sodium relations in salinized barley tissues as a basis of differential salt tolerance. *Funct. Plant Biol.* **2007**, *34*, 150–162. [CrossRef]

36. Wu, F.B.; Lianghuan, W.; Fuhua, X. Chlorophyll meter to predict nitrogen sidedress requirements for short-season cotton (*Gossypium hirsutum* L.). *Field Crop Res.* **1998**, *56*, 309–314.

37. Genty, B.; Briantais, J.M.; Baker, N.R. The relationship between the quantum yield of photosynthetic electron transport and quenching of chlorophyll fluorescence. *Biochim. Biophys. Acta-Gen. Subj.* **1989**, *990*, 87–92. [CrossRef]

38. Leul, M.; Zhou, W.J. Alleviation of waterlogging damage in winter rape by uniconazole application: Effects on enzyme activity, lipid peroxidation, and membrane integrity. *J. Plant Growth Regul.* **1999**, *18*, 9–14. [CrossRef]

39. Chen, F.; Wang, F.; Wu, F.B.; Mao, W.H.; Zhang, G.P.; Zhou, M.X. Modulation of exogenous glutathione in antioxidant defense system against Cd stress in the two barley genotypes differing in Cd tolerance. *Plant Physiol. Biochem.* **2010**, *48*, 663–672. [CrossRef]

40. Jiang, M.; Zhang, J. Effect of abscisic acid on active oxygen species, antioxidative defence system and oxidative damage in leaves of maize seedlings. *Plant Cell. Physiol.* **2001**, *42*, 1265–1273. [CrossRef]

41. Tang, Q.; Feng, M. *Practical Statistics and Its DPS Statistics Software Package*; China Agriculture Press: Beijing, China, 1997.

42. Neill, S.J.; Desikan, R.; Clarke, A.; Hurst, R.D.; Hancock, J.T. Hydrogen peroxide and nitric oxide as signaling molecules in plants. *J. Exp. Bot.* **2002**, *53*, 1237–1247. [CrossRef]

43. Netondo, G.W.; Onyango, J.C.; Beck, E. Sorghum and salinity. II. Gas exchange and chlorophyll fluorescence of sorghum under salt stress. *Crop Sci.* **2004**, *44*, 806–811.

44. Taiz, L.; Zeiger, E. *Plant Physiology*; Sinauer Associates Inc.: Sunderland, MA, USA, 2002.

45. Lawlor, D.W.; Cornic, G. Photosynthetic carbon assimilation and associated metabolism in relation to water deficits in higher plants. *Plant Cell Environ.* **2002**, *25*, 275–294. [CrossRef]

46. Meloni, D.A.; Oliva, M.A.; Martinez, C.A.; Cambraia, J. Photosynthesis and activity of superoxide dismutase, peroxidase and glutathione reductase in cotton under salt stress. *Environ. Exp. Bot.* **2003**, *49*, 69–76. [CrossRef]

47. Ibrahim, W.; Ahmed, I.M.; Chen, X.; Cao, F.; Zhu, S.; Wu, F. Genotypic differences in photosynthetic performance, antioxidant capacity, ultrastructure and nutrients in response to combined stress of salinity and Cd in cotton. *BioMetals* **2015**, *28*, 1063–1078. [CrossRef] [PubMed]

48. Ahmed, I.M.; Cao, F.; Zhang, M.; Chen, X.; Zhang, G.; Wu, F. Difference in yield and physiological features in response to drought and salinity combined stress during anthesis in Tibetan wild and cultivated barleys. *PLoS ONE* **2013**, *8*, e77869. [CrossRef] [PubMed]

49. Yamane, K.; Kawasaki, M.; Taniguchi, M.; Miyake, H. Differential effect of NaCl and polyethylene glycol on the ultrastructure of chloroplasts in rice seedlings. *J. Plant Physiol.* **2003**, *160*, 573–575. [CrossRef] [PubMed]

50. Naeem, M.S.; Warusawitharana, H.; Liu, H.; Liu, D.; Ahmad, R.; Waraich, E.A.; Xua, L.; Zhou, W.J. 5-aminolevulinic acid alleviates the salinity-induced changes in *Brassica napus* as revealed by the ultrastructural study of chloroplast. *Plant Physiol. Biochem.* **2012**, *57*, 84–92. [CrossRef] [PubMed]

51. Parida, A.K.; Das, A.B.; Mittra, B. Effects of NaCl stress on the structure, pigment complex composition, and photosynthetic activity of mangrove *Bruguiera parviflora* chloroplasts. *Photosynthetica* **2003**, *41*, 191–200. [CrossRef]

52. Zhang, H.X.; Blumwald, E. Transgenic salt-tolerant tomato plants accumulate salt in foliage but not in fruit. *Nat. Biotechnol.* **2001**, *19*, 765–768. [CrossRef] [PubMed]

53. Katsuhara, M.; Kawasaki, T. Salt stress induced nuclear and DNA degradation in meristematic cells of barley roots. *Plant Cell Physiol.* **1996**, *l37*, 169–173. [CrossRef]

54. Kader, M.A.; Lindberg, S. Uptake of sodium in protoplasts of salt-sensitive and salt-tolerant cultivars of rice, *Oryza sativa* L. determined by the fluorescent dye SBFI. *J. Exp. Bot.* **2005**, *56*, 3149–3158. [CrossRef]

55. Genc, Y.; McDonald, G.K.; Tester, M. Reassessment of tissue Na^+ concentration as a criterion for salinity tolerance in bread wheat. *Plant Cell. Environ.* **2007**, *30*, 1486–1498. [CrossRef]

56. Reuter, D.; Robinson, J.B. *Plant Analysis: An Interpretation Manual*, 2nd ed.; CSIRO Publishing: Melbourne, Australia, 1997.

57. Ozfidan-Konakci, C.; Yildiztugay, E.; Kucukoduk, M. Upregulation of antioxidant enzymes by exogenous gallic acid contributes to the amelioration in *Oryza sativa* roots exposed to salt and osmotic stress. *Environ. Sci. Pollut. Res. Int.* **2015**, *22*, 1487–1498. [CrossRef]

58. Del, R.D.; Stewart, A.J.; Pellegrini, N. A review of recent studies on malondialdehyde as toxic molecule and biological marker of oxidative stress. *Nutr. Metab. Cardiovasc Dis.* **2005**, *15*, 316–328. [CrossRef] [PubMed]

59. Davey, M.W.; Stals, E.; Panis, B.; Keulemans, J.; Swennen, R.L. High-throughput determination of malondialdehyde in plant tissues. *Anal. Biochem.* **2005**, *347*, 201–207. [CrossRef] [PubMed]

60. Ahmad, P.; Hakeem, K.R.; Kumar, A.; Ashraf, M.; Akram, N.A. Salt-induced changes in photosynthetic activity and oxidative defense system of three cultivars of mustard (*Brassica juncea* L.). *African J. Biotechnol.* **2012**, *11*, 2694–2703.

61. Azooz, M.M.; Youssef, A.M.; Ahmad, P. Evaluation of salicylic acid (SA) application on growth, osmotic solutes and antioxidant enzyme activities on broad bean seedlings grown under diluted seawater. *Int. J. Plant Physiol. Biochem.* **2011**, *3*, 253–264.

62. Katsuhara, M.; Otsuka, T.; Ezaki, B. Salt stress-induced lipid peroxidation is reduced by glutathione S-transferase, but this reduction of lipid peroxides is not enough for a recovery of root growth in Arabidopsis. *Plant Sci.* **2005**, *169*, 369–373. [CrossRef]

63. Bowler, C.; Montagu, M.V.; Inze, D. Superoxide dismutase and stress tolerance. *Annu. Rev. Plant Physiol. Plant Mol. Biol.* **1992**, *43*, 83–116. [CrossRef]

64. Shalata, A.; Mittova, V.; Volokita, M.; Guy, M.; Tal, M. Response of the cultivated tomato and its wild salt-tolerant relative *Lycopersicon pennellii* to salt-dependent oxidative stress: The root antioxidative system. *Physiol. Plant* **2001**, *112*, 487–494. [CrossRef]

65. Hernandez, J.A.; Jimenez, A.; Mullineaux, P.; Sevilla, F. Tolerance of pea (*Pisum sativum* L.) to long-term salt stress is associated with induction of antioxidant defences. *Plant Cell Environ.* **2000**, *23*, 853–862. [CrossRef]

66. Feki, K.; Tounsi, S.; Brini, F. Comparison of an antioxidant system in tolerant and susceptible wheat seedlings in response to salt stress. *Spanish J. Agric. Res.* **2017**, *15*, e0805. [CrossRef]

67. Koca, H.; Bor, M.; Özdemir, F.; Türkan, I. The effect of salt stress on lipid peroxidation, antioxidative enzymes and proline content of sesame cultivars. *Environ. Exp. Bot.* **2007**, *60*, 344–351. [CrossRef]

68. Temel, A.; Gozukirmizi, N. Physiological and molecular changes in barley and wheat under salinity. *Appl. Biochem. Biotechol.* **2015**, *175*, 2950–2960. [CrossRef] [PubMed]

69. Reddy, R.A.; Chaitanya, K.V.; Vivekanandan, M. Drought-induced responses of photosynthesis and antioxidant metabolism in higher plants. *J. Plant Physiol.* **2004**, *161*, 1189–1202. [CrossRef] [PubMed]

2

Germination and Growth of Spinach under Potassium Deficiency and Irrigation with High-Salinity Water

Kadir Uçgun [1]⬤, **Jorge F. S. Ferreira** [2,*], **Xuan Liu** [2], **Jaime Barros da Silva Filho** [3],
Donald L. Suarez [2], **Claudivan F. de Lacerda** [4] **and Devinder Sandhu** [2]⬤

[1]   Department of Plant and Animal Production, Technical Sciences Vocational School,
       Karamanoğlu Mehmetbey University, Karaman 70200, Turkey; kadirucgun@kmu.edu.tr
[2]   US Salinity Laboratory (USDA-ARS), 450 W. Big Springs Rd., Riverside, CA 92507, USA;
       xuan.liu@usda.gov (X.L.); donald.suarez@usda.gov (D.L.S.); devinder.sandhu@usda.gov (D.S.)
[3]   Departments of Microbiology and Plant Pathology, University of California Riverside, 900 University Ave.,
       Riverside, CA 92521, USA; jaimeba@ucr.edu
[4]   Department of Agricultural Engineering, Federal University of Ceará, Fortaleza-CE 60450-760, Brazil;
       cfeitosa@ufc.br
*   Correspondence: jorge.ferreira@usda.gov

Abstract: Information is scarce on the interaction of mineral deficiency and salinity. We evaluated two salt-tolerant spinach cultivars under potassium (K) doses (0.07, 0.15, 0.3, and 3.0 mmol$_c$ L^{-1}) and saline irrigation (5, 30, 60, 120, and 160 mmol$_c$ L^{-1} NaCl) during germination and growth. There was no interaction between salinity and K. Salinity decreased germination percent (GP), not always significantly, and drastically reduced seedling biomass. 'Raccoon' significantly increased GP at 60 mmol$_c$ L^{-1} while 'Gazelle' maintained GP up to 60 or 120 mmol$_c$ L^{-1}. After 50 days under saline irrigation, shoot biomass increased significantly at 30 and 60 mmol$_c$ L^{-1} at the lowest K dose but, in general, neither salinity nor K dose affected shoot biomass, suggesting that salinity supported plant growth at the most K-deficient dose. Salinity did not affect shoot N, P, or K but significantly reduced Ca, Mg, and S, although plants had no symptoms of salt toxicity or mineral deficiency. Although spinach seedlings are more sensitive to salt stress, plants adjusted to salinity with time. Potassium requirement for spinach growth was less than the current crop recommendation, allowing its cultivation with waters of moderate to high salinity without considerable reduction in yield, appearance, or mineral composition.

Keywords: *Spinacia oleracea*; potassium deficiency; high-salinity water; potassium–salinity interaction; salt-tolerant glycophyte

1. Introduction

The undesirable effects of salt stress on plant growth can be associated with (1) soil solution with low osmotic potential (water stress), (2) nutrient imbalance caused by excessive NaCl or by high soil pH, (3) specific ion effect (Na$^+$ and/or Cl$^-$), or (4) a combination of all three factors [1]. These factors cause adverse effects on plant growth and yield [2,3] and, when severe and prolonged, will lead to plant death. The most prevalent ions in salt-affected soils or saline irrigation water are Na$^+$ and Cl$^-$ and the great majority of salt stress is due to either one or both ions because they are highly soluble and become toxic to plants at high concentrations. These ions compete with the absorption of K$^+$, Ca^{2+}, and NO3$^-$ and often lead to pertinent mineral nutrient imbalance and deficiency [4]. Recent reports indicate that the effects of competition between different ions depend on their tissue concentration and may vary with plant species [5–7].

Depending on water or soil salinity, Na and Cl can accumulate in spinach shoots at levels as high as the macronutrients N and K [6,8], indicating the existence of an efficient vacuolar sequestration of potentially toxic Na^+ and Cl^- ions in this species. On the contrary, many glycophytes, including strawberry and Jerusalem artichoke, have mechanisms that prevent Na^+ toxicity by reducing Na^+ transport to leaves [5,9]. However, these mechanisms are not so efficient in controlling Cl^- absorption and translocation in most glycophytic plants, making Cl^- one of the main ions responsible for specific-ion toxicity. Thus, Cl alone, or associated with Na, may decrease plant performance after their excessive tissue accumulation [5,10].

Germination is highly influenced by environmental factors [11] and is a complex phenomenon involving many physiological and biochemical changes that are easily compromised by stress, including salt stress [12]. Salinity can affect the germination of seeds either by lowering soil osmotic potential and reducing water uptake by plant roots or by toxic effects caused by specific ions [13]. Salinity stress can change the activity of enzymes of the nucleic acid metabolism [14], alter protein metabolism [15,16], disturb hormonal balance [17], and reduce the utilization of seed reserves [18,19].

Data from several crops suggest that the level of salinity tolerance is highly dependent on the plant species [20,21], the cultivar within a species [9,22–24] or on the developmental stage of the plant [25]. Several researchers investigated the effect of salinity on the germination of plants like beans [26], quinoa [27,28], tomato [29], cabbage [30], cauliflower [30], canola [30,31], cowpea [32], onion [33], safflower [34], sunflower [35], eggplant [36] carrot [37], and spinach [20]. Salinity reduced germination rates in all of them.

Although it is known that salinity affects plant developmental stages differently, it is not known whether this would be associated with the duration of each stage or with the existence of specific mechanisms to maintain needed tissue levels of mineral nutrients for plant growth. For example, it is known that excess NaCl affects the absorption and accumulation of K in the leaves of many species, resulting in damage to cellular metabolism. In salt-tolerant species, however, Na may partially replace K to a great extent, but without damaging plant growth [38]. Thus, an important question arises: is this partial replacement of K by Na maintained and/or beneficial at all stages of the crop cycle? Furthermore, although it is known that salinity may affect the absorption of major mineral ions such as K^+ and NO_3^- in most glycophytes, there is no published data reporting the effect of moderate to high salinity of irrigation water, combined with deficient soil concentrations of either NO_3^- or K^+, on spinach germination and growth. Even data on ideal mineral (N, P, and K) nutrition for spinach without salinity was lacking until recently [39].

Considering that spinach is a moderately salt-tolerant glycophyte [6,8,40], we hypothesized that irrigation waters providing elevated NaCl-induced salinity combined with potassium deficiency would mimic similar conditions found in soil solutions, thus providing a model system to study spinach response to these stresses at both germination and late growth stage. Thus, the objective of this research was to examine the salt tolerance of two cultivars of spinach at different developmental stages (germination, seedling establishment, and vegetative growth) under cultivation with elevated salinity combined with K deficiency applied through the irrigation water.

2. Materials and Methods

2.1. Greenhouse Cultivation Details

Riverside, California, is situated at an altitude of 311 m with a latitude of 33.9°57′54″ N and a longitude of 117.3°20′13″ W. Although ideal temperatures for spinach growth are between 15 and 18 °C, the crop grows well in temperatures ranging from 5 to 30 °C [41]. Thus, greenhouse temperature control was set between 15 °C (night) and 28 °C (day). The greenhouse cultivation of both experiments spun from 19 October 2018 to 11 March 2019. During that time, spinach plants were grown under short days with 13–13.5 h dark/10.5 to 11.0 h of natural light, which is under the flowering inductive photoperiod (14 h day^{-1}) for spinach [42]. Minimum to maximum greenhouse average relative humidity, temperature, day length, and light intensity are shown in Table 1.

Table 1. Average (minimum/maximum) relative humidity (ARH), temperature, and light intensity as photosynthetic photon flux density (PPFD) per month in a greenhouse used for the germination and growth of spinach cv. Raccoon and Gazelle. Fall and winter in California were from 23 September–21 December 2018 and from 22 December 2018–19 March 2019.

Date	ARH (%)	Temp (°C)	PPFD * (μmol m^{-2} s^{-1})	Day Length ** (h:min:s)
October 19/2018 (min/max)	5.0/66.0	22.2/31.8	433.0/624.4	11:12:36
November/2018 (min/max)	3.0/73.4	19.8/30.1	224.4/517.6	10:04:37
December/2018 (min/max)	10.0/71.0	15.6/28.0	222.8/423.0	09:56:06
January/2019 (min/max)	8.0/69.0	14.6/27.8	200.0/458.6	10:31:53
February/2019 (min/max)	18.0/61.3	13.1/31.1	427.0/634.4	11:25:28
March 11/2019 (min/max)	30.0/66.0	17.0/30.7	351.6/771.2	11:48:34

* PPFD was estimated for 8.5–9.0 h day^{-1} (from 7:00 a.m. to 5:00 p.m.) with approximately half the hours averaged for minimum and half for maximum PPFD. Sensor was from Apogee Instruments, Inc. (Logan, UT, USA). ** Daylength data are given from 19 October 2018 to 11 March 2019, and for the last day of each month, except for October and March (daylength for 1st and last day of experiment only).

Two experiments were conducted to evaluate the effects of NaCl doses and K deficiency, separately and combined, under greenhouse conditions on two spinach (*Spinacia oleracea* L.) cultivars (Raccoon and Gazelle). One experiment focused on assessing germination and seedling establishment (4–6 true leaves). The other experiment was carried out to assess vegetative growth, when salinity was applied to plants with 6–8 true leaves and continued for seven weeks until harvest. The pots were arranged in a randomized complete block design with four replicates, each with 3 plants per pot. Both experiments had 20 treatments assigned to each cultivar (combination of 5 NaCl concentrations with 4 K doses: 3 deficient and 1 control) with a total of 80 pots per cultivar, in a factorial arrangement.

For Experiment 1 (germination and seedling establishment), five seeds were sown directly at a depth of 1.3 cm in 5.7-L pots (22.2 cm in diameter) on 07 February 2019. Pots were filled with non-washed, non-sterile, sand and saturated with the pertinent treatment solution. After sowing, pots were irrigated three times a week with each saline treatment solution. Data on seed gemination were recorded daily between 13 and 32 days after sowing for each cultivar. All Seedlings from germinated seeds (no thinning was performed in this experiment) were harvested on the 32nd day after sowing 11 March 2019) with shoots separated from roots and recorded for shoot fresh weight (SFW). The coefficient of velocity of germination (CVG in % day^{-1}) and the germination percentage (GP) at 13, 22, and 32 days were obtained according to [43], as follow:

CVG (% day^{-1}) = [$\sum$Ni/$\sum$(NiTi)] × 100, where N was the number of seeds germinated on day i, Ti is the number of days from sowing and GP (%) = No. of seeds germinated per cultivar at: 13 days for 'Gazelle', 22 days for 'Raccoon', and 32 days for both cultivars × 100.

For Experiment 2 (vegetative growth), 7 seeds were sown directly into pots at 1.3 cm depth. Seeds were sown on 19 October 2018 and harvested on 28 January 2019 (101 days). Pots had the same dimensions and growth medium as described above. After germination, plants were thinned to the most homogeneous three plants per pot. Plants were watered with 1/8-strength Hoagland's nutrition solution [44] made with Riverside municipal water with an electrical conductivity (EC_{iw}) of 0.7 dS m^{-1} three times a week until the growth stage of 6–8 true leaves. Salinity treatments were applied on 10 December 2018 and continued until 28 January 2019. Before treatments were applied, all pots were flushed four times with deionized water to remove any remaining fertilizer (including K) from the sand. After that, treatments combining Na (as NaCl) and K (substituting K for Na) began and continued for 7 weeks (50 days) with harvest taking place from 29 January to 1 February 2019. All solutions (control and salt treatments) contained all macronutrients and micronutrients needed for spinach growth, as previously determined [6]. Solution volumes used to water plants were determined to allow leaching of approximately 30% (water volume drained/volume applied) for both experiments. At harvest, all three plants in each pot were separated into shoots and roots. Unwashed shoots were weighed immediately upon harvest to obtain shoot fresh weight (SFW, g pot^{-1}). Then, shoots and roots were washed with tap water, twice with deionized water to remove remaining sand (roots) and mineral fertilizers from tissue surface (shoots and roots) before drying in a forced-air oven (70 °C) for at least 48 h to obtain constant shoot and root dry weight (SDW and RDW, respectively) in g pot^{-1} before grinding for mineral analysis.

2.2. Plant Mineral Analyses

In experiment 1 (germination and seedling establishment) plants were too small for mineral analysis and only the fresh weight of shoots was recorded. For Experiment 2, plants were harvested, separated into shoots and roots, washed with tap water, followed by deionized water and blotted dry with paper towels and bagged to be dried in a forced-air oven at 70 °C until stable dry weight. Dry weight was recorded for shoots and roots. Samples were then ground in a Wiley mill to pass a 20-mesh (0.84 mm) screen. Tissue mineral concentration was based on shoot or root dry weight. Chloride was determined from nitric/acetic acid extracts by amperometric titration. The concentrations of tissue Na, P, K, Ca, Mg, and total-S, and of the micronutrients Fe, Cu, Mn, Zn, and Mo were determined from nitric acid digestions (Milestone, Ethos EZ Microwave Digestion, Shelton, CT, USA) of the dried, ground, plant material by Inductively Coupled Plasma Optical Emission Spectrometry (ICP-OES, 3300DV, Perkin-Elmer Corp., Waltham, MA, USA). Nitrogen was determined by combustion in a Rapid N Exceed®analyzer (rNex, Elementar Americas Inc., Ronkonkoma, New York, NY, USA).

2.3. Saline Solutions

Saline solutions of the same salt composition were used as treatments in both experiments. In all, there were 20 treatments containing 5 nominal concentrations of Na (5, 30, 60, 120, and 160 mmol L^{-1}) and 4 concentration of K (0.07, 0.15, 0.3, and 3.0 mmol L^{-1}) prepared with deionized water (EC_w = 0.05 dS m^{-1}). NaCl and K (partially substitute for Na) were added to a basic, modified (no K), $\frac{1}{2}$-strength Hoagland's solution (Control) containing the same concentrations of macro- and micronutrients and 0.5-3.5 mmol L^{-1} of Na$^+$ (Table 2). The solutions were constructed and balanced using the free software ExtractChem V. 2.0 (by Suarez and Taber, Agricultural Water Use Efficiency and Salinity Research Unit in Riverside, or US Salinity Laboratory (USDA-ARS) and available at: https://www.ars.usda.gov/research/software/download/?softwareid=155&modecode=20-36-15-00.

The following salts were used to obtain target salinities in irrigation solution: NaCl, Ca(NO$_3$)$_2$·4H$_2$O, NaNO$_3$, KNO$_3$, KCl, NaH$_2$PO$_4$, MgSO$_4$·7H$_2$O (Table 1). NaFe-EDTA (0.05 mM), H$_3$BO$_3$ (0.023 mM), MnSO$_4$·H$_2$O (0.005 mM), ZnSO$_4$.7H$_2$O (0.0004 mM), CuSO$_4$·5H$_2$O (0.0003 mM), and (NH$_4$)$_6$Mo$_7$O$_{24}$·4H$_2$O (0.0001 mM) were added to every solution to meet plant micronutrient requirements. The concentrations of K$^+$ were achieved by substituting varying amounts of KNO$_3$ for NaNO$_3$ to achieve target K$^+$ levels. This resulted in slight decreases in Na$^+$ with increasing K+ that are insignificant, relative to the amounts of total Na+ applied, with real values given in Table 2. However, to simplify the presentation of results and discussion, we will refer to the Na doses with the nominal values of 5, 30, 60, 120, and 160 mmol$_c$ L^{-1}.

Table 2. Cation and anion concentrations of saline solutions used to irrigate seeds (germination experiment) for 32 days (up to 4–6 true-leaf stage) and young plants bearing 6–8 true leaves for seven weeks (growth experiment).

Treatment	K^+	Na^+	Cl^-	$H_2PO_4^-$	Ca^{2+}	Mg^{2+}	SO_4^{2-}	NO_3^-	TEC_{iw}	MEC_{iw}	pH
				(mmol$_c$ L^{-1})					(dS m^{-1})		
T1	0.07	3.5	0.07	0.5	5.0	2.0	2.0	8.0	1.0	1.3	6.14
T2	0.07	30.5	27.07	0.5	5.0	2.0	2.0	8.0	4.0	4.1	6.05
T3	0.07	60.5	57.07	0.5	5.0	2.0	2.0	8.0	7.2	7.2	6.15
T4	0.07	120.5	117.07	0.5	5.0	2.0	2.0	8.0	13.2	12.7	6.07
T5	0.07	160.5	157.07	0.5	5.0	2.0	2.0	8.0	17.1	16.8	6.09
T6	0.15	3.5	0.15	0.5	5.0	2.0	2.0	8.0	1.1	1.2	6.22
T7	0.15	30.5	27.15	0.5	5.0	2.0	2.0	8.0	4.0	4.0	6.17
T8	0.15	60.5	57.15	0.5	5.0	2.0	2.0	8.0	7.2	7.2	6.15
T9	0.15	120.5	117.15	0.5	5.0	2.0	2.0	8.0	13.2	13.2	6.18
T10	0.15	160.5	157.15	0.5	5.0	2.0	2.0	8.0	17.1	16.8	6.12
T11	0.3	3.5	0.30	0.5	5.0	2.0	2.0	8.0	1.1	1.3	6.19
T12	0.3	30.5	27.30	0.5	5.0	2.0	2.0	8.0	4.1	4.1	6.17
T13	0.3	60.5	57.30	0.5	5.0	2.0	2.0	8.0	7.2	7.4	6.21
T14	0.3	120.5	117.30	0.5	5.0	2.0	2.0	8.0	13.2	13.2	6.16
T15	0.3	160.5	157.30	0.5	5.0	2.0	2.0	8.0	17.1	16.9	6.09
T16	3.0	0.5	0.00	0.5	5.0	2.0	2.0	8.0	1.1	1.2	6.25
T17	3.0	27.5	27.00	0.5	5.0	2.0	2.0	8.0	4.1	4.8	6.17
T18	3.0	57.5	57.00	0.5	5.0	2.0	2.0	8.0	7.3	7.2	6.18
T19	3.0	117.5	117.00	0.5	5.0	2.0	2.0	8.0	13.2	13.2	6.18
T20	3.0	157.5	157.00	0.5	5.0	2.0	2.0	8.0	17.1	16.7	6.13

TEC_{iw}: Target electric conductivity of treatment irrigation water (EC_{iw}), MEC_{iw}: Measured EC_{iw} in treatment irrigation waters.

2.4. Statistical Analyses

Data for each cultivar was used for analyses of variance. Means were compared using Fisher's LSD test ($p < 0.05$). Statistical analysis was performed using SAEG version 9.1 [45] and the SigmaPlot/SigmaStat version 14.0 (Systat Software, San Jose, CA, USA, www.systatsoftware.com) software. Interactions were analyzed regardless of the significance of the F test. All interactions between Na and K doses were analyzed statistically independently of the significance of the F test. Analysis was also done for individual effects of either Na or K doses for all the parameters related to mineral tissue accumulation and biomass accumulation, except for germination percentage, shoot fresh weight, N, P, Ca, Mg, and S in roots and shoots of both cultivars. These parameters were analyzed taking the F test into consideration, also for individual effects of either NaCl or K doses. Because there was no significant effect of K dose, but there was of Na dose (salinity), the results are presented for salinity inside of each K dose.

3. Results

3.1. Effect of Combined Salinity and Potassium Doses on Germination and Seedling Establishment

There was no effect of K dose or significant interaction between Na and K doses but there was a significant effect of salinity on both germination and shoot fresh weight. During germination and seedling establishment, there was a significant decrease in the percentage of germination (GP) and in the coefficient of velocity of germination (CVG) with increased NaCl-induced salinity for both cultivars, regardless of the K dose. Thus, results were presented in terms of nominal NaCl doses (Figure 1), calling attention to the fact that Cl^- concentrations, after the control, increased in parallel with Na^+ in irrigation waters. At high NaCl doses, GP decreased for both cultivars. However, the GP of 'Raccoon' was significantly better than control (5 mmol$_c$ L^{-1}) at 60 mmol$_c$ L^{-1}, both at 22 and 32 days after seeding, while 'Gazelle' maintained GP up to 60 mmol$_c$ L^{-1} at 13 days and up to 120 mmol$_c$ L^{-1} at 32 days after germination (Figure 1). The CVG decreased significantly with NaCl dose for both cultivars, with the decrease more pronounced in 'Raccoon' than 'Gazelle', which maintained its CVG level up to 60 mmol$_c$ L^{-1}.

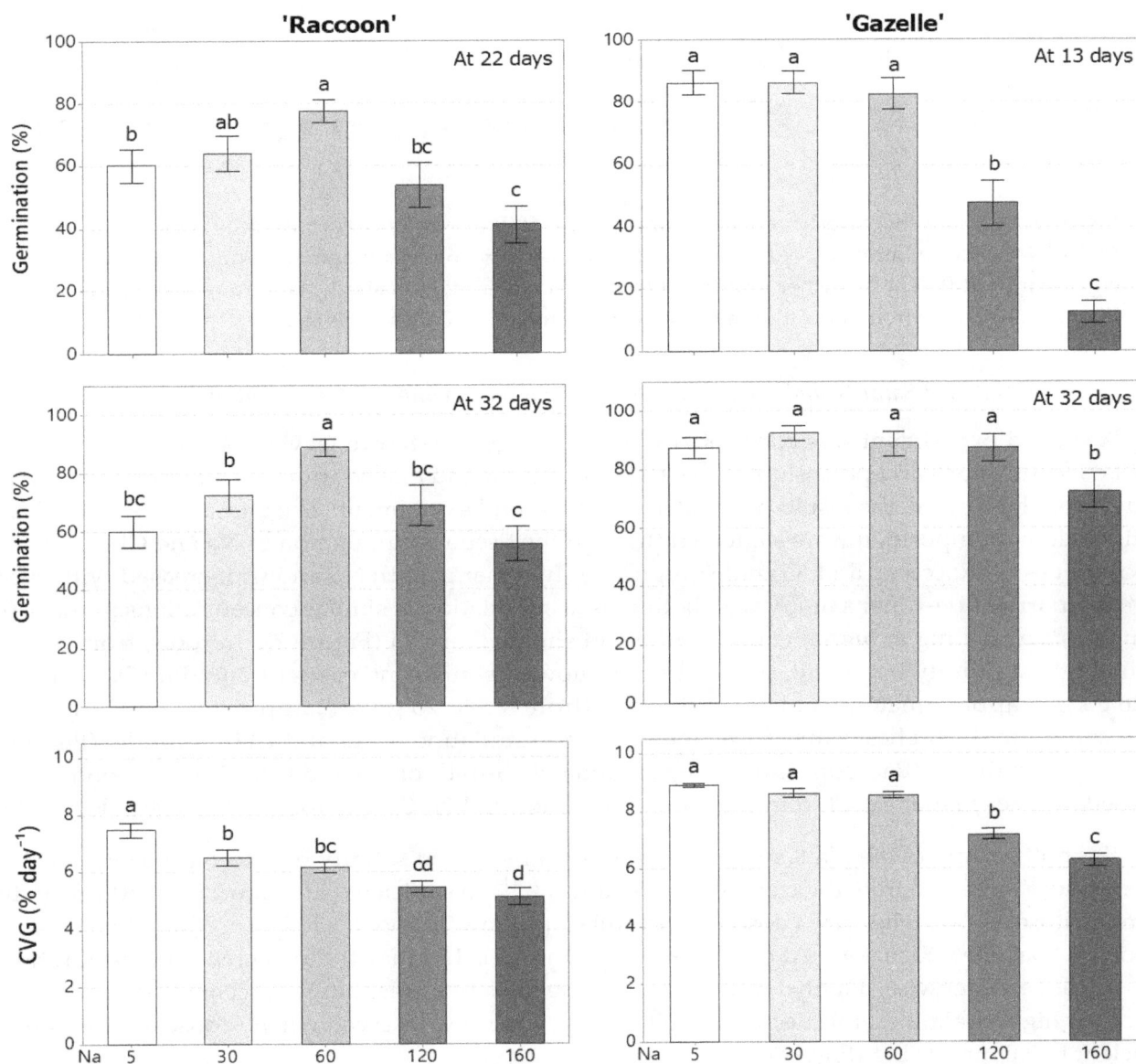

Figure 1. Germination percentage (%) for initial germination count and at 32 days and coefficient of velocity of germination (CVG) for 'Raccoon' and 'Gazelle' for different doses of NaCl (Na) in mmol$_c$ L^{-1}. Mean bars with different letters are significantly different by Fisher's LSD test ($p < 0.05$). Interval bars represent the standard errors of means. Lowercase letters: compare NaCl doses (Na) for germination parameters inside each cultivar. Because there was no effect of K doses, samples were grouped per salinity treatment, with $n = 16$.

Seedling shoot fresh weight decreased for both cultivars as salinity increased (Figure 2), regardless of K dose. However, after 32 days of irrigation with high-salinity water, seedlings had no visual signs of NaCl toxicity (Figure 2).

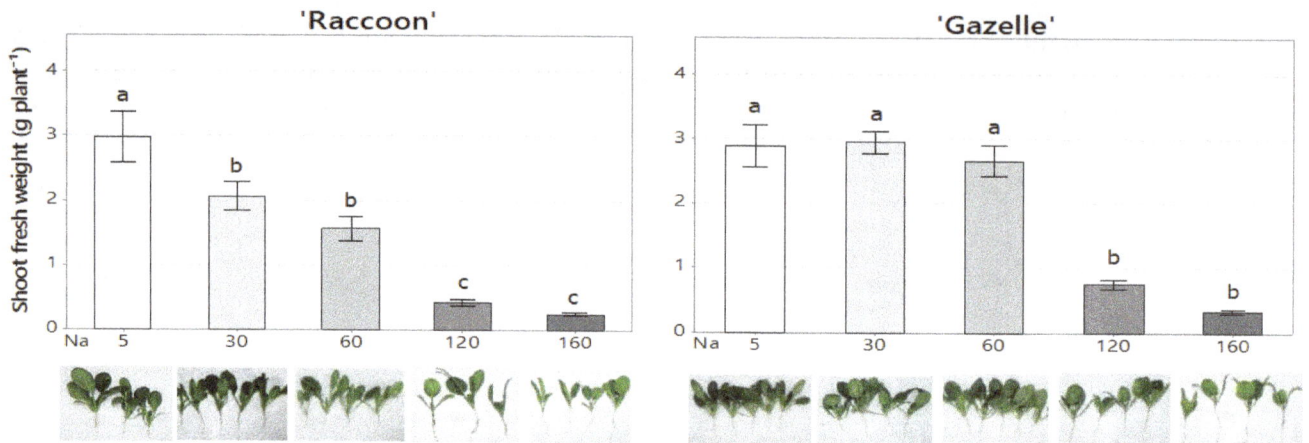

Figure 2. Seedling shoot fresh weight and appearance of 'Raccoon' and 'Gazelle' seedlings at different NaCl (Na) doses in $mmol_c$ L^{-1}, regardless of KCl dose ($n = 16$). Mean bars with different letters are significantly different by Fisher's LSD test ($p < 0.05$). Interval bars show standard errors of means. Plants pictured are from one random pot with 3.0 $mmol_c$ L^{-1} K at each Na dose.

3.2. Effect of Combined Salinity and Potassium Doses on Tissue Na and Cl Accumulation

A second experiment was conducted with seedlings of 6–8 true leaves submitted to the same NaCl concentrations and potassium doses (three deficient and one control) used in the germination study. From here on, all the results are related to this second experiment. The effects of increased NaCl on tissue ionic composition were evaluated through the tissue accumulation of Na and Cl in root and shoot tissues of 'Raccoon' and 'Gazelle' (Figure 3). In general, both Na and Cl increased with every increase in irrigation-water salinity with both ions accumulating at similar concentrations in roots but with Cl accumulating at higher concentrations in shoots than Na (Figure 3). In roots, from control to the highest salinity treatment, both cultivars showed similar increases of 2-fold for Na (from 13 to 26 g kg^{-1}, approximately) and 5.6-fold for Cl (from 4.5 to 25 g kg^{-1}, approximately). In shoots, from control to the highest salinity treatment, the accumulation of Na and Cl in 'Gazelle' (Figure 3) was similar to that of 'Raccoon' and was approximately 3-fold for Na and from 7.6 ('Raccoon') to 9.0 ('Gazelle') times higher for Cl from control to 160 $mmol_c$ L^{-1} NaCl, regardless of K doses (Figure 3).

For both cultivars the Na:K ratio increased with increased salinity but decreased significantly when K dose increased from deficient (0.07–0.3 $mmol_c$ L^{-1}) to sufficient (3.0 $mmol_c$ L^{-1}). Although the decrease lessened at higher NaCl doses, it was still significant for shoots of 'Raccoon' and for roots and shoots of 'Gazelle' (Figure 4). At control salinity (5 $mmol_c$ L^{-1} NaCl) the decrease in Na:K ratio in shoots was seen stepwise, mainly for 'Gazelle', despite the low increase in K concentration for deficient doses. At higher salinities, this decrease in Na:K ratio was only observed when K dose increased from deficient to 3.0 $mmol_c$ L^{-1} (Figure 4).

3.3. Effect of Combined Salinity and Potassium Doses on Root Mineral Composition

The effects of combined increased NaCl with potassium deficiency was evaluated on root and shoot composition of macronutrients (Table 3) and micronutrients (Supplementary Tables S1 and S2). Despite the significant increase in both Na and Cl, roots maintained their concentrations of K (14.3 to 26.8 g kg^{-1} for 'Raccoon' and 15.0 to 28.3 g kg^{-1} for 'Gazelle') (Figure 5) and of N (18.6 to 25.2 g kg^{-1} for both cultivars), even when K was applied at deficient doses (Table 3). However, Ca, Mg, and S decreased significantly with salinity by approximately 50%, 32%, and 54%, respectively, in roots of both cultivars (Table 3). Interestingly, P increased significantly with salinity, in each K dose, in roots of 'Gazelle'. However, the increase was not always significant in roots of 'Raccoon' (Table 3).

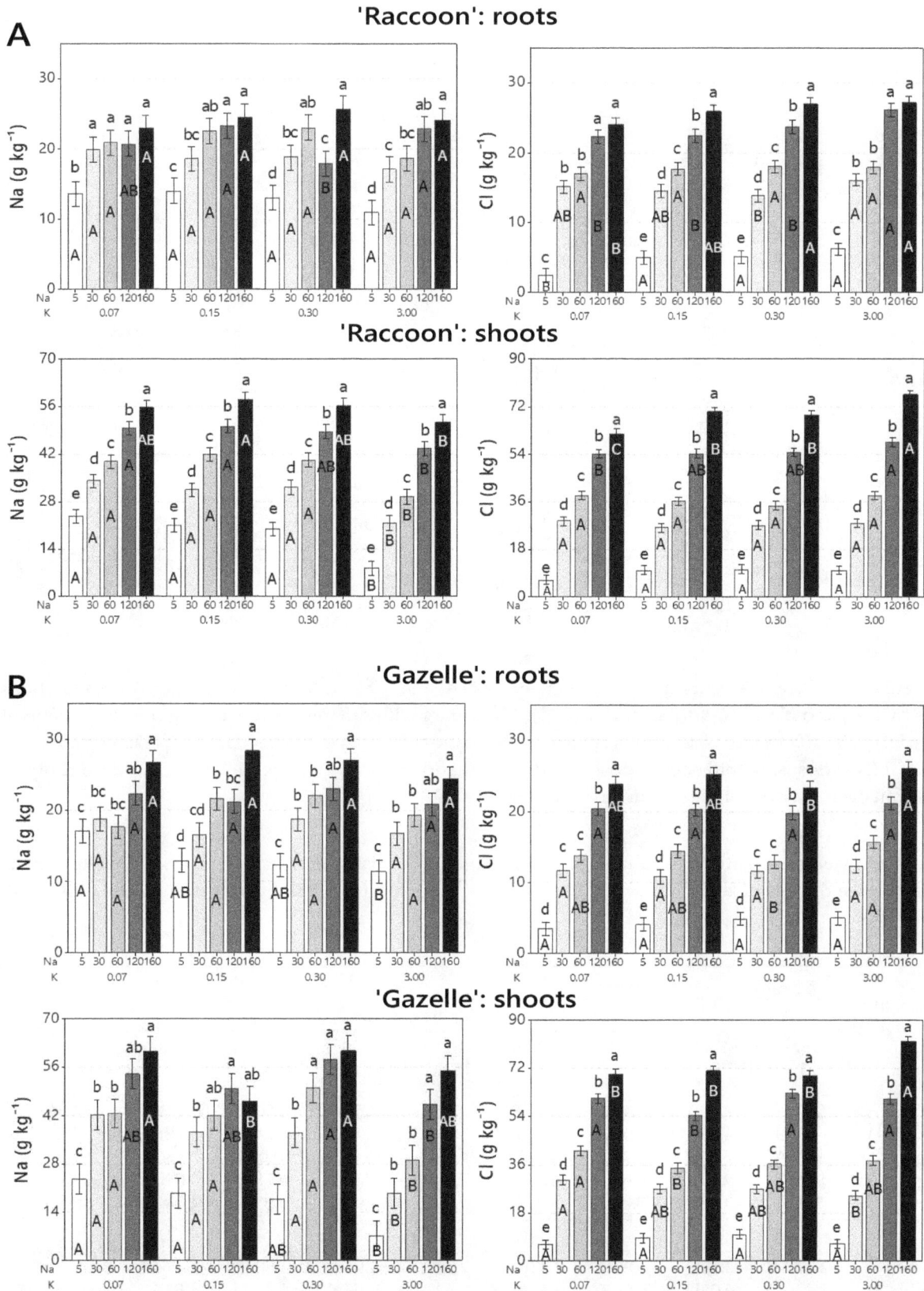

Figure 3. Concentrations of Na and Cl in roots and shoots of 'Raccoon' (**A**) and 'Gazelle' (**B**) for the respective combinations of NaCl (Na) and KCl (K) doses. Mean bars with different letters are significantly different by Fisher's LSD test ($p < 0.05$). Interval bars show standard errors of means. Uppercase letters compare K doses within each Na, inside each mineral. Lowercase letters compare NaCl doses (Na), within each K dose, for either Na or Cl.

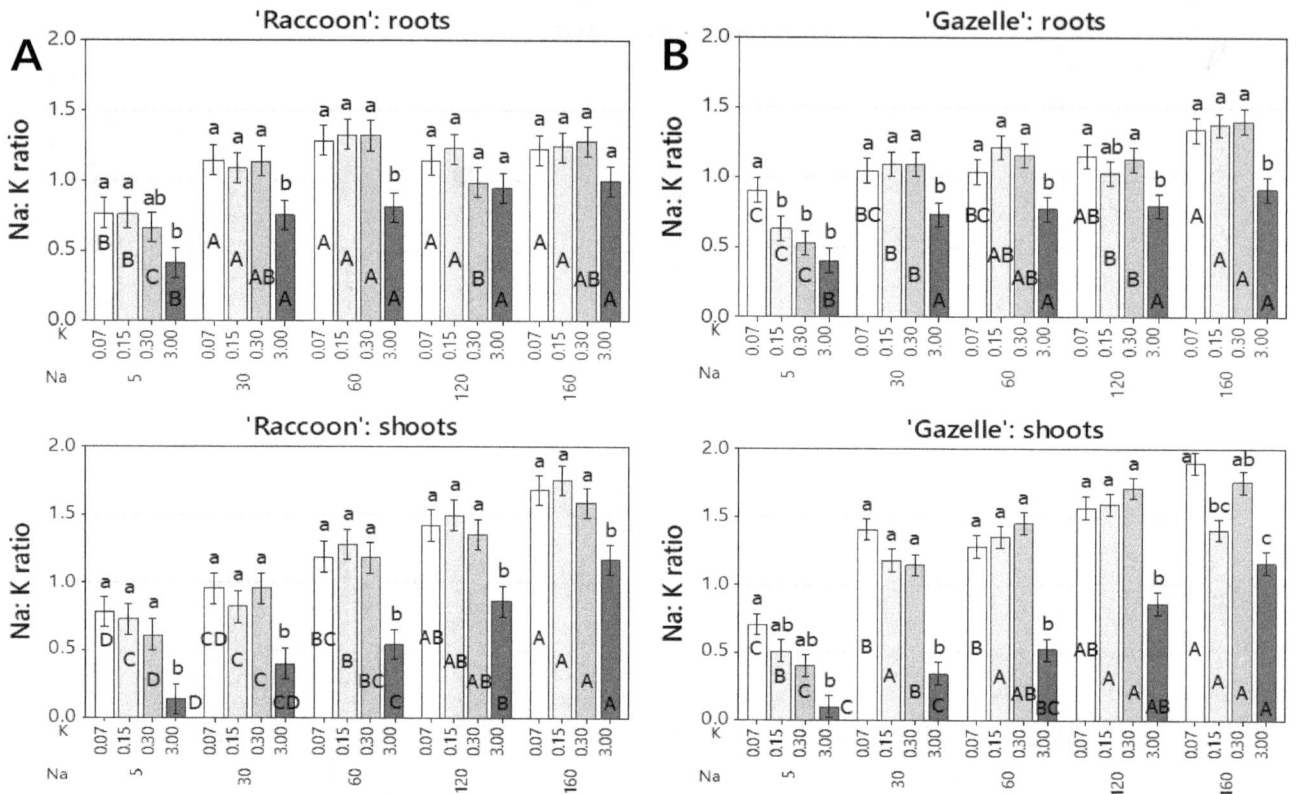

Figure 4. Na:K ratio in roots and shoots of 'Raccoon' (**A**) and 'Gazelle' (**B**) for the respective combinations of KCl (K) and NaCl (Na) doses (mmol$_c$ L^{-1}). Mean bars with different letters are significantly different by Fisher's LSD test ($p < 0.05$). Interval bars show standard errors of means. Uppercase letters compare NaCl (Na) doses, within each K dose, inside organ and cultivar. Lowercase letters compare K doses, within each Na dose, inside organ and cultivar.

Table 3. Mean concentrations of macronutrients in roots and shoots of 'Raccoon' and 'Gazelle' for the respective doses of NaCl (5.00 to 160.00 mmol$_c$ L^{-1}) pooled for all KCl doses due to lack of K effect ($n = 16$).

Nutrient	Roots					Shoots				
	NaCl Doses (mmol$_c$ L^{-1})									
	5.0	30.0	60.0	120.0	160.0	5.0	30.0	60.0	120.0	160.0
	'Raccoon'									
N (%)	2.42A	2.10B	2.19B	2.17B	2.35A	4.82A	4.50C	4.61BC	4.65ABC	4.70AB
P (g kg^{-1})	4.96B	4.86B	5.41AB	5.38B	5.95A	4.88AB	4.74B	4.88B	4.92AB	5.17A
Ca (g kg^{-1})	9.01A	7.33B	6.27C	4.28D	4.49D	15.21A	10.09B	7.56C	7.11C	7.03C
Mg (g kg^{-1})	11.50A	9.74B	8.65C	6.73D	7.59D	10.95A	9.78B	8.84C	7.88D	7.55D
S (g kg^{-1})	5.36A	3.78B	3.49B	2.11C	2.39C	4.98A	4.32B	4.06C	3.44D	3.24D
	'Gazelle'									
N (%)	2.27AB	2.02D	2.09CD	2.23BC	2.42A	3.94A	3.60A	3.47A	3.61A	3.97A
P (g kg^{-1})	5.32CD	5.08D	5.78BC	6.17AB	6.62A	4.57BC	4.31C	4.58B	4.60B	5.04A
Ca (g kg^{-1})	8.52A	6.68B	5.89C	4.52D	4.20D	10.90A	7.64B	6.30C	5.45D	5.04D
Mg (g kg^{-1})	10.62A	8.85B	8.29B	7.48C	7.57C	9.71A	8.32B	8.10B	7.32C	6.94C
S (g kg^{-1})	4.82A	3.59B	3.33B	2.60C	2.31C	4.40A	4.00B	3.79B	3.29C	3.11C

Means with different letters are significantly different by Fisher's LSD test ($p < 0.05$). Uppercase letters: comparisons between NaCl doses, within each cultivar.

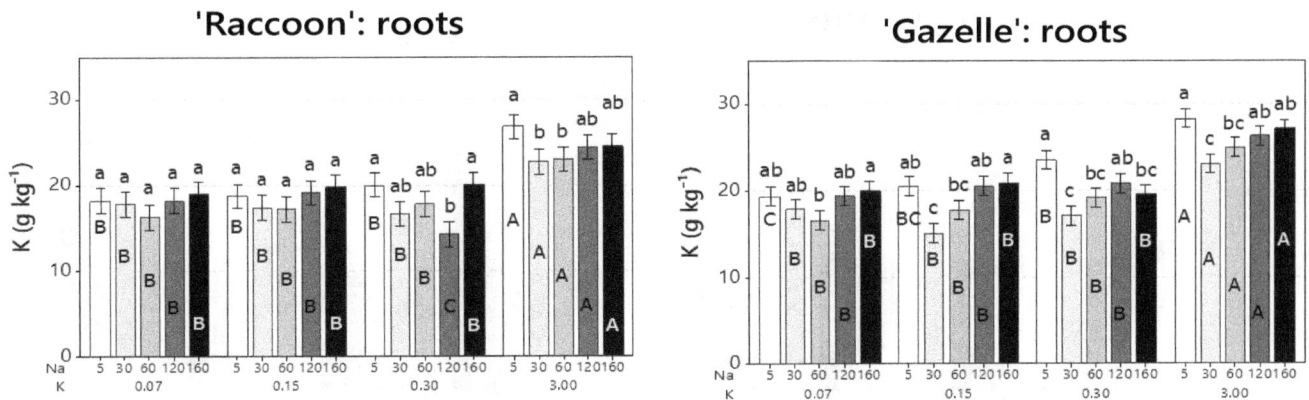

Figure 5. Concentrations of potassium in roots of 'Raccoon' (**left**) and 'Gazelle' (**right**) for the respective combinations of NaCl (Na) (from 5 to 160 $mmol_c$ L^{-1}), and KCl (K) doses (from 0.07 to 3.0 $mmol_c$ L^{-1}). Mean bars with different letters are significantly different by Fisher's LSD test ($p < 0.05$). Interval bars show standard errors of means. Lowercase letters compare Na doses within each K dose. Uppercase letters compare K doses within each Na dose.

Regarding root micronutrients, 'Raccoon' Fe was the highest, ranging from 820 (highest salinity) to 1350 mg kg^{-1} (control salinity), followed by Mn, which ranged from 254 (highest salinity) to 304 mg kg^{-1} (control salinity), with the remaining micronutrients (B, Cu, and Zn) ranging from 19 to 29.7 mg kg^{-1}. Salinity increase led to a significant reduction of Cu (35%), Fe (39%), and Mn (16.5%) but had no effect of the concentrations of B (18–24 mg kg^{-1}) or Zn (22 to 29.7 mg kg^{-1}) (Table S1). In 'Gazelle', Fe and Mn were also the highest micronutrients ranging, respectively, from 975 to 1730 mg kg^{-1} and from 250 to 452 mg kg^{-1}, while B ranged from 40–50 mg kg^{-1} and Zn from 19–25 mg kg^{-1}. Salinity led to significant reductions in Cu (50%), Fe (44%), Mn (43%), and Zn (24%) (Table S2).

3.4. Effect of Combined Salinity and Potassium Doses on Shoot Mineral Composition

Similar to roots, and regardless of the significant increases in Na and Cl, or K-deficient doses, shoot concentrations of K (Figure 6) and N (Table 3) were maintained and ranged from 31.9 (when K was provided in deficient doses) to 61.6 g kg^{-1} (when K was provided at 3.0 $mmol_c$ L^{-1}) for 'Raccoon' and from 30.5 (deficient K) to 73.5 g kg^{-1} (3.0 $mmol_c$ L^{-1} K) for 'Gazelle', while N ranged from 43.4 to 49.3 g kg^{-1} for 'Raccoon' (Figure 6, left, Table 3) and from 28.5 to 42.0 g kg^{-1} for 'Gazelle' (Figure 6, right, Table 3). As observed in roots, P was maintained or slightly increased (at high NaCl doses), while Ca, Mg, and S decreased with salinity in 54%, 31%, and 36%, respectively for 'Raccoon' and 53%, 29%, and 29.5%, respectively for 'Gazelle' (Table 3). When K was provided at the sufficient dose of 3.0 $mmol_c$ L^{-1}, there was a significant decrease in shoot K of both cultivars when salinity increased beyond control salinity and mainly from control to the highest salinity, with an average K decrease of 31% for 'Raccoon' and 34% for 'Gazelle' (Figure 6).

Like root micronutrient concentration, Fe and Mn were the highest micronutrients found in shoots of both cultivars without a trend in their concentration associated with salinity or K dose (Tables S1 and S2).

3.5. Effect of Combined Salinity and Potassium Doses on Plant Biomass Accumulation

Salinity effects were evaluated on plant vegetative growth and biomass accumulation through plant appearance (Figure 7), RDW, and SDW (Figure 8). Although salinity caused significant reductions in the concentrations of Ca, Mg, and S, these reductions were not enough to cause visual mineral deficiencies, regardless of K and NaCl doses, and allowed the development of healthy-looking spinach plants (Figure 7).

'Raccoon': shoots

'Gazelle': shoots

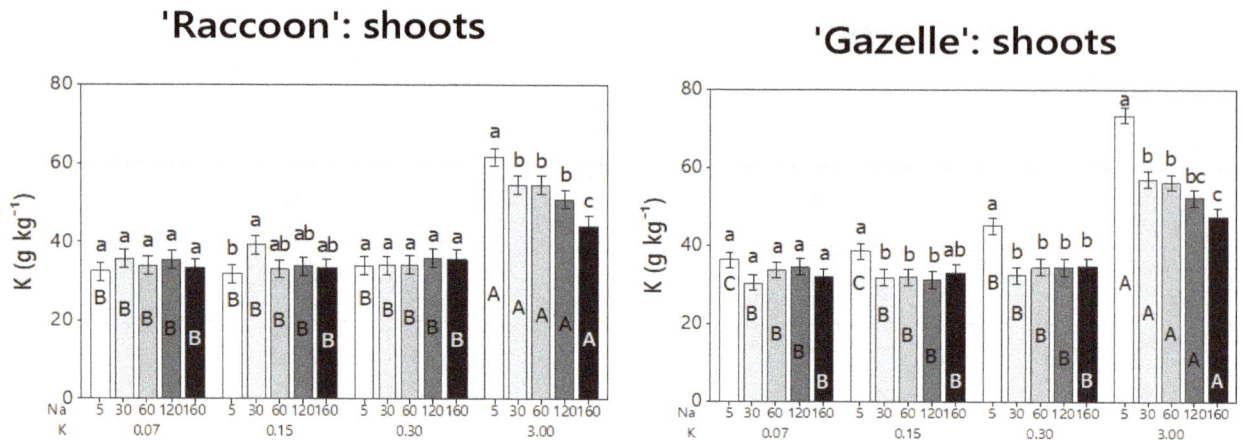

Figure 6. Concentrations of the mineral potassium (K, in g kg^{-1}) in shoots for 'Raccoon' (**left**) and 'Gazelle' (**right**) for the respective combinations of NaCl (Na) (from 5.0 to 160.0 mmol$_c$ L^{-1}) and KCl (K) doses (from 0.07 to 3.0 mmol$_c$ L^{-1}). Mean bars with different letters are significantly different by Fisher's LSD test ($p < 0.05$). Interval bars show standard errors of means. Uppercase letters compare K within each Na dose in shoots of each cultivar. Lowercase letters compare Na doses, within each K dose.

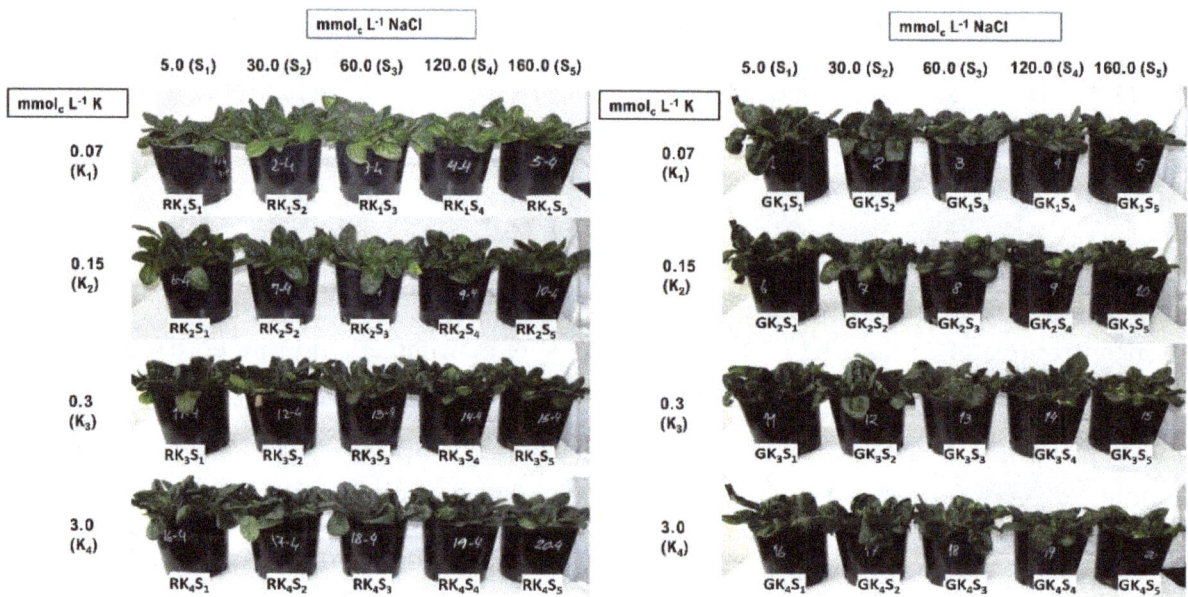

Figure 7. Appearance of spinach plants for 'Raccoon' (R, **left**) and' (G, **right**) seven weeks (50 days) after treatments with the respective combinations of NaCl (5.0 to 160.0 mmol$_c$ L^{-1}), or salinity treatments S_1 to S_5 and potassium, as potassium chloride (KCl) or K (0.07 to 3.0 mmol$_c$ L^{-1}) doses K_1 to K_4. Plants were photographed the same day the experiment was terminated.

Regarding K dose, 'Raccoon' plants had a significant increase in shoot biomass when K increased from 0.07 mmol$_c$ L^{-1} to higher K doses. 'Gazelle' had a similar response, although significant shoot biomass increase was only significant for the K doses of 0.3 and 3.0 mmol$_c$ L^{-1}. However, K doses higher than 0.07 mmol$_c$ L^{-1} had no effect on biomass of either cultivar when NaCl doses were higher than 30 mmol$_c$ L^{-1} (Figure 8). For SDW, at the lowest K dose of 0.07 mmol$_c$ L^{-1}, plants of both cultivars increased SDW significantly when Na concentration increased from 5.0 to 30 and 60 mmol$_c$ L^{-1}. The average SDW ranged from 7.9 to 10.4 g pot^{-1} for 'Raccoon' and from 8.2 to 10.6 g pot^{-1} for 'Gazelle' and, when there was a significant increase or decrease in SDW, the difference was approximately 2.0 g pot^{-1} (Figure 8). Although K doses had little or no effect on root dry weight (RDW) and shoot dry weight (SDW), RDW of 'Gazelle' increased significantly at 30 $mmol_c$ L^{-1} NaCl when K was sufficient (3.0 $mmol_c$ L^{-1}). The data showed a similar increase for 'Raccoon' at the same NaCl dose (although not significant) as if a small increase in salinity had favored root development (Figure 8).

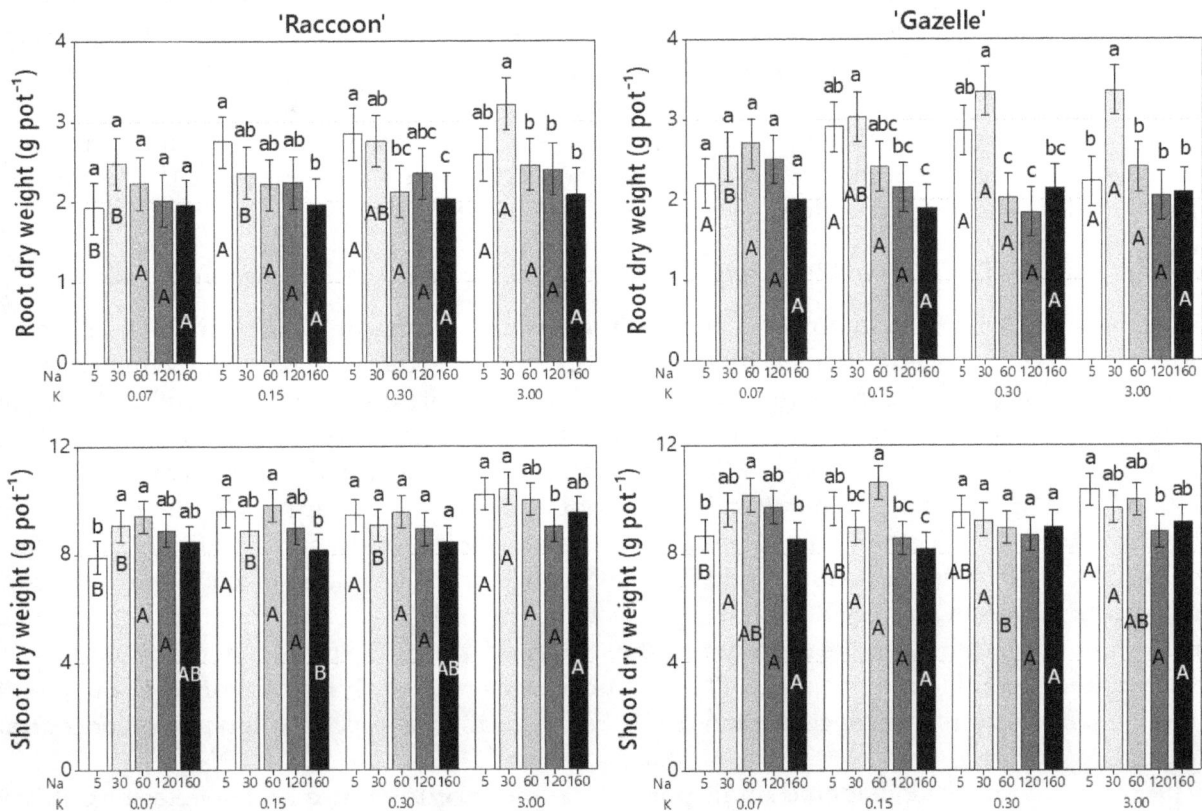

Figure 8. Dry biomass of roots and shoots for 'Raccoon' and 'Gazelle' for the respective combinations of NaCl (Na) and KCl (K) doses in $mmol_c$ L^{-1}. Mean bars with different letters are significantly different by Fisher's LSD test ($p < 0.05$). Interval bars show standard errors of means. Uppercase letters compare K doses within each Na dose, for roots or shoots inside each cultivar. Lowercase letters compare Na doses, within each K dose, for roots or shoots inside each cultivar.

4. Discussion

4.1. Effect of Combined Salinity and Potassium Doses on Germination and Seedling Establishment

While there was no effect of K dose on germination, 'Raccoon' had its germination improved by salinity while 'Gazelle' was more tolerant to salinity than 'Raccoon' during germination and seedling establishment. Different researchers cited in the introduction demonstrated that crop germination decreased with increased salinity of irrigation water. The salinity threshold of approximately 120 $mmol_c$ L^{-1} NaCl 32 days after germination for 'Gazelle' (Figure 1) was higher than that obtained for other cultivars of spinach such as 'Green Gold', 'Larisa', 'Mikado', and 'Ohio', in which the GP and relative germination rate were unaffected up to 50 mM (50 $mmol_c$ L^{-1}) NaCl but decreased significantly at 100 mM NaCl, and further at 200 mM NaCl [20]. Our germination data also indicate that GP improved from the first counting date for each cultivar (22 days for 'Raccoon' and 13 days for 'Gazelle') to the last counting at 32 days after seeding (Figure 1). Because most of 'Gazelle' seeds germinated before 'Raccoon's seeds, CVG values for Gazelle were higher than those for 'Raccoon'. Although not compared statistically, mean GP and CVG for 'Gazelle' appeared to be better than those of 'Raccoon'. Average seedling fresh weight (Figure 2) also indicates that 'Gazelle' has a higher salinity tolerance than 'Raccoon'. Our data suggest that irrigation water salinity below 7.0 dS m^{-1} (60 $mmol_c$ L^{-1}) would not affect either GP or seedling growth of Gazelle (Figure 2) if irrigation frequency is enough to maintain soil salinity at similar levels of irrigation-water salinity. Although 'Gazelle' showed a better recuperation in GP than 'Raccoon', seedling fresh weight data (Figure 2) clearly indicated that seedling growth was significantly affected by NaCl doses above 60 $mmol_c$ L^{-1} NaCl (7.0 dS m^{-1}). It has been previously shown that the lack of response to K fertilization at high levels of NaCl could be explained, at least in part, by the reduction in plant growth caused by NaCl ion toxicity, which limits the absorption of water and nutrients from the soil [46]. Although seedlings did not provide enough material for mineral analysis, they had no visual symptoms of either mineral deficiency or salt toxicity (Figure 2). Furthermore, because of 'Gazelle's CVG reflected a faster initial germination than that observed for 'Raccoon', seedling biomass observed for 'Gazelle' up to 60 $mmol_c$ L^{-1} may reflect the fact that 'Gazelle's seedlings had more time to adjust to salinity.

4.2. Effect of Combined Salinity and Potassium Doses on Tissue Na and Cl Accumulation

The significant increase in both Na and Cl in roots and shoots of both cultivars was expected with increasing NaCl concentrations. The steady and stepwise increase in tissue Cl, which surpassed the increase in tissue Na (Figure 3), indicates that spinach is more efficient in controlling Na than Cl. Increases in both tissue Na and Cl of similar magnitudes have been reported recently when irrigation water contained 90 and 120 $mmol_c$ L^{-1} of NaCl [8]. Interestingly, the use of saline water containing 120 and 160 $mmol_c$ L^{-1} in this study resulted in Na shoot accumulation of approximately 50–60 g kg^{-1}, smaller than the 60–70 g kg^{-1} of shoot Na reported previously [8] for the same cultivars under 90 and 120 $mmol_c$ L^{-1} of Na. The same can be said for Cl. Our data indicate that spinach plants have control mechanisms triggered by salinity that allows plants to control salt tissue accumulation once Na and Cl reach a certain level. However, the salinity threshold beyond which the plant will no longer be able to control its tissue accumulation of Na and Cl has not been determined. It is noteworthy to mention that in naturally-saline waters, the concentration of Cl^- can be as high to twice as high as the concentration of Na^+.

The ratio of Na:K is considered an important measure of tolerance of a species or cultivar to salinity. However, in the case of these two spinach cultivars, this ratio only increased with salinity because Na tissue levels increased significantly at every step while K tissue concentrations remained constant (Figures 5 and 6). However, the increase from 0.3 to 3.0 $mmol_c$ L^{-1} in K was enough to show that K tissue accumulation was antagonistic to Na tissue accumulation. These results agree with a previous report that concluded that a 20-fold increase in K from 0.25 to 5.0 $mmol_c$ L^{-1} decreased Na tissue accumulation [8].

4.3. Effect of Combined Salinity and Potassium Doses on Root Mineral Composition

Plants maintained adequate concentrations of N, P, and K in roots, regardless of salinity or K dose. Although Ca, Mg, and S decreased significantly with salinity, regardless of K dose, plants maintained a minimum concentration of these macronutrients needed for growth [8,47]. Root-P increase in both cultivars as salinity increased (Table 3) could have been an attempt for the plant to balance ions using HPO_4^{2-} to compensate for a constant concentration of NO_3^- in the leaf (interpreted through the constant tissue levels of mineral N), and in response to a steady increase of Na^+ in saline irrigation waters, as suggested by others [48]. Similar results were reported for 'Raccoon' and 'Gazelle' in previous salinity studies either when K was provided at sufficient doses [6] or when mineral composition was compared under salinity with K provided at both sufficient and deficient doses [8].

Regarding root micronutrients, Fe and Mn were the highest among other micronutrients tested. Although salinity significantly reduced Fe, Cu, and Mn (and Zn in 'Gazelle'), these micronutrients still were present at sufficient concentrations for root development. In a study evaluating the effect of combined salinity with N doses in spinach, the authors reported a significant decrease in Fe with increased salinity, in agreement with our results but, contrary to our results, they reported small, but significant, increases in Zn and Cu as salinity increased [49], which could be a cultivar-specific response. Although other micronutrients were present at similar concentration in roots and shoots,

Mn and Fe were found in roots in concentrations 6-fold or higher and 2.5-fold or higher in roots than in shoots, respectively. The data on mineral composition of spinach roots suggest that, instead of being discarded, they should be used as a rich source of Fe, Mn, and of macronutrients for human and animal diets.

4.4. Effect of Combined Salinity and Potassium Doses on Shoot Mineral Composition

Our shoot macronutrient data for both 'Raccoon' and Gazelle clearly established that concentrations of N, P, and K were maintained in shoots across salinity levels (Table 3; Figure 6). The increase in shoot P concentration, mainly observed in 'Gazelle' may have been an attempt for the plant to balance HPO_4^{2-} ions under an increased concentration of Na^+, as previously discussed under root mineral composition, and as suggested by others [48]. Similar results on the maintenance of N concentrations and increased P concentrations in spinach shoots, under similar salinity levels, were reported by others [49]. It has been established earlier that differences in salt tolerance of three brassica species could be related to K^+ retention in roots [50]. The high salt tolerance of 'Raccoon' and 'Gazelle' may be attributed to their ability to retain K^+ in both roots and shoots, regardless salinity increase in irrigation waters or the low availability of K in irrigation waters. This K homeostasis against its availability in soil or water was discussed at the molecular level elsewhere [8].

Interestingly, shoot micronutrients, in general, showed no significant differences in either cultivar as salinity increased. The concentrations of Fe and Mn were also the highest, as they were in roots, but at much lower magnitudes (Tables S1 and S2).

Homeostasis of N, P, and K was maintained in both spinach cultivars, regardless of the significant increase in Na and Cl and regardless the fact that three of the K doses were from 10 to 43 times lower than what was considered sufficient for spinach growth in previous works [6,40]. The concentrations of shoot Ca, Mg, and S, although with significant decreases, were still maintained at levels high enough for plant growth and shoot biomass accumulation, similar to our previous results comparing two doses of K (0.25 and 5.0 $mmol_c$ L^{-1}), also under elevated irrigation-water salinity [8].

4.5. Effect of Combined Salinity and Potassium Doses on Plant Biomass Accumulation

A similar increase in shoot biomass under low to mild salinity, as the one observed in this study, was reported for 'Raccoon' when irrigation salinity significantly increased total dry matter (roots + shoots) at 30 and 60 $mmol_c$ L^{-1} [8]. The SDW increase observed with 30 and 60 $mmol_c$ L^{-1} at the lowest K dose suggests that plants of both cultivars benefited from NaCl when K was deficient (Figure 8). Although salinity increase caused a decrease in RDW at the K doses of 0.15 and 0.30 $mmol_c$ L^{-1}, SDW remained mostly unchanged, regardless of K dose and salinity levels. However, a small, but significant, decrease in shoot biomass of both cultivars was observed at 0.15 $mmol_c$ L^{-1} K between control salinity and 160 $mmol_c$ L^{-1} NaCl. Thus, in general, neither K nor NaCl dose had a great impact on SDW of either cultivar.

Previous results with the same cultivars grown in soil under lower NaCl concentrations (from 5.0 to 120 $mmol_c$ L^{-1}) showed a significant decrease in dry shoot biomass for 'Raccoon' after 30 $mmol_c$ L^{-1} when K was 5.0 $mmol_c$ L^{-1}, while 'Raccoon' significantly increased SDW with 30–60 $mmol_c$ L^{-1} NaCl when K was 0.25 $mmol_c$ L^{-1} [8]. This significant increase in shoot biomass with 30–60 $mmol_c$ L^{-1} was also observed for both 'Raccoon' and 'Gazelle' when K dose was 0.07 $mmol_c$ L^{-1} in this experiment (Figure 8), 3.6-fold less than the deficient dose used by Ferreira and collaborators [8]. Plants in that experiment were cultivated in 1:1 (loamy sandy soil:sand) and submitted to salinity when they had 6–8 true leaves, as in this experiment. Because of the use of loamy sand soil and a leaching fraction of 0.25 (25% leaching) used by those authors, plants irrigated with 90 and 120 $mmol_c$ L^{-1} were exposed to soil salinities (EC_e) of 8.9 to 10.4 dS m^{-1}, respectively, at the end of the experiment. These soil–paste salinities correspond to irrigation-water salinities (EC_{iw}) of 19.6 and 22.9 dS m^{-1}, respectively, thus explaining the significant decrease in shoot biomass at these two salinities. In this study, plants were cultivated in an all-sand medium to reduce K to deficient levels for seven weeks under the salinity treatments, and a leaching fraction of 0.30 was applied. Thus, the low effect of salinity on SDW, even at 160 $mmol_c$ L^{-1} (EC_{iw} = 17 dS m^{-1}), can be partly explained by the fact that the sandy medium allowed a more efficient leaching of NaCl, preventing salinity build-up in the pot, and partly because spinach plants may have adapted to NaCl during the seven weeks (50 days) of cultivation, while in a previous experiment with 'Raccoon', plants were only allowed to grow under salinity for only 28 days [6].

Our results clearly indicate that K requirements of spinach are much lower relative to the ones previously recommended for the crop (63 to 138 kg ha^{-1} or 1.6 to 3.54 $mmol_c$ L^{-1}) [41]. Despite the fact that the lowest K level of 0.07 $mmol_c$ L^{-1} used (equivalent to 2.73 kg of K ha^{-1}) was from 23 to 50 times lower than recommended for spinach fertilization [41], plants maintained N, P, and K homeostasis in both roots and shoot, thus sustaining growth. Our results on K requirements confirm previously-reported data with spinach, in which plant biomass was similar under both 0.25 and 5.0 $mmol_c$ L^{-1} of K combined to a soil salinity of EC_e = 6.0 dS m^{-1} [8], equivalent to the EC_{iw} of 13.2 dS m^{-1} in this study. However, our results disagree with a previous study that reported that spinach plants benefitted from extra K when salinity increased from 50 to 250 mM NaCl [51]. In a salinity experiment with 'Raccoon' using sufficient K doses of 3, 5, and 7 $mmol_c$ L^{-1}, plants grew well up to 80 $mmol_c$ L^{-1} NaCl (EC_{iw} = 9.3–9.8 dS m^{-1}), regardless of the K doses and had no significant decrease in plant dry weight [6]. A study where spinach was cultivated in a loamy sand soil testing K doses of 0, 63, 85, 127, and 148 kg ha^{-1}, without salinity, reported that spinach shoot dry weight showed no difference among all the K doses tested [39]. However, those authors did not consider that the loamy soil used could have enough K for spinach to grow without any additional K needed in the control dose of 0.0 kg ha^{-1}. In our experiment, spinach shoot biomass was not generally affected by salinity up to 160 $mmol_c$ L^{-1}, despite significant shoot reductions in Ca, Mg, and S, and regardless of the very low doses of K provided. Thus, spinach plants behaved unlike what was expected for glycophytic plants irrigated with water electrical conductivity (EC_{iw}) up to 17 dS m^{-1} (160 $mmol_c$ L^{-1} NaCl) combined with deficient K in the growth medium.

Our results show clear differences in shoot biomass between seedlings and older spinach plants in response to the same water-salinity treatments. Plants exposed to salinity at the 6–8 true-leaf stage and allowed to grow for seven weeks into adult plants were more salt-tolerant than seedlings. Eggplant varieties subjected to increasing salinity during germination and seedling stages also responded differently depending on plant growth stage, with salinity tolerance of all cultivars increasing at the later growth stages [36]. Spinach plants, even under very low levels of K maintained optimal plant growth, suggesting that there are genetic mechanisms that are triggered under either low K, or both low K and high salinity, that are responsible for K and N, and maybe P, homeostasis in the plant as previously discussed [8,52,53]. Our results suggest that spinach plants can tolerate moderate to high salinity in sandy soils, and when applied a leaching fraction of approximately 0.30, having the ability to adjust to some degree of salinity as plants grow older. Thus, saline waters would be better suited for the production of "freezer spinach" (older plants) than of "baby spinach" as the former is harvested at a later developmental stage with mineral nutrient concentrations that can be twice that of baby spinach [41]. However, these results obtained in sand culture must be interpreted carefully and cannot be directly extrapolated to field cultivation under different leaching regimes before experimenting with the target soil to be used for spinach cultivation. Results clearly suggest that the salinity threshold of spinach, even under deficient levels of K in the soil solution, is higher than the previously reported EC_e of 2.0 dS m^{-1} (estimated EC_{iw} = 4.4 dS m^{-1}) [54–56]. Based on the EC_{iw} of 17 dS m^{-1} tested here, we can estimate an EC_e = 7.7 dS m^{-1} for 'Raccoon' and Gazelle, or 3.85-fold higher than the EC_e provided for spinach (no cultivar mentioned) by the previous authors.

5. Concluding Remarks

In general, spinach did not respond to K doses either during germination or vegetative growth and there was no interaction between salinity and K. However, salinity decreased germination parameters of Gazelle at the highest NaCl dose, and drastically reduced seedling biomass at every increase in NaCl for 'Raccoon' but only at 120 and 160 $mmol_c$ L^{-1} for Gazelle, indicating that Gazelle was more tolerant to salinity than 'Raccoon' during germination and seedling establishment, the most salt-susceptible stage. Interestingly, germination % of 'Raccoon' improved significantly at 60 $mmol_c$ L^{-1} NaCl, suggesting that moderate salinity can improve the germination of this cultivar. Our results also indicate that the responses of spinach to salinity depends on the plant growth stage and on the cultivar. At the later growth stage, 'Raccoon' and Gazelle were both tolerant to irrigation waters of moderate to high salinity.

Mature plants of both cultivars accumulated large concentrations of Na and Cl in both roots and shoots, and these ions could have contributed to osmotic balance to maintain cell turgor without causing symptoms of ion toxicity. Both cultivars maintained their macro and micronutrient at adequate, or close to adequate, tissue levels for growth. Even under severe K deficiency (0.07 $mmol_c$ L^{-1}, equivalent to a range of 23- to 50-fold less K than recommended for the crop), adult plants maintained K concentrations over the minimum required (20 g kg^{-1} SDW) for plant growth. Therefore, mature plants are more salt tolerant than young plants and saline waters would be better adapted to the production of "freezer spinach" (vegetatively mature plants) instead of "baby spinach" (young plants).

The increase in K from deficient to sufficient reduced the Na:K ratio in the shoots, regardless of the level of salinity but the beneficial effects of K on RDW and SDW were observed only at lowest level of salinity (5 $mmol_c$ L^{-1} NaCl). This demonstrates that the improvement in the nutritional status of the plant does not overcome the osmotic effects associated with the increase in the total concentration of salts in the root zone.

Plants from both cultivars seemed to benefit from salinity at the lowest K dose of 0.07 $mmol_c$ L^{-1} with shoot biomass increasing significantly at 60 $mmol_c$ L^{-1} of NaCl. Although not significant, a similar trend for increased root biomass was also observed for both cultivars. Although we assume that this salinity benefit was only caused by Na^+ (ions), as an osmoticum to maintain cell turgor but not as a substitute of K^+ in its physiological function, salinity treatments above control levels had similar concentrations of Cl^-. Regarding mineral tissue accumulation, Cl tissue accumulation was as substantial as that of Na. Interestingly, Cl is recognized to be essential to plant growth only at micronutrient levels (mg kg^{-1}), while Na is a needed nutrient for the growth of halophytes, but not of glycophytes, such as spinach. The spinach cultivars studied here have demonstrated to be salt tolerant and our results for Na:K ratio confirmed the results of a previous study with the same cultivars that reported that sufficient K decreases Na tissue accumulation.

Of course, these results cannot be directly extrapolated to field cultivation of spinach in general because salinity tolerance depends on the cultivar, type of soil, season, and irrigation frequency applied. What is clear is that K fertilization can be significantly decreased to increase farmer profits. These results also raise the question: Is K the only mineral that can be drastically reduced in spinach cultivation or would similar shoot biomass be achieved with N reduction? At a time that more food is needed to feed a growing world population and freshwater becomes scarcer and more expensive, or is unavailable, for irrigation in arid and semiarid areas of the planet, we must revisit mineral needs for crops cultivated under saline irrigation as a way to maintain food production, increase farmer's profits, while decreasing agriculture footprint and slowing down the salinization of our soils and water resources.

Spinach seems to be a salt-tolerant glycophyte in a family (Amaranthaceae) that has halophytes such as *Salicornia* spp., *Sarcocornia* spp., *Suaeda* spp., *Chenopodium* spp., and *Atriplex* spp., which have all adapted to saline and arid environments. The fact that germination percent improved with time and that plants showed no significant decrease in biomass at the adult stage, regardless of K dose and after 50 days of irrigation with moderate to high-salinity waters lead us to conclude that spinach can be successfully cultivated with saline recycled waters in arid and semiarid regions and requires much less K fertilization than previously recommended.

Author Contributions: Conceptualization, J.F.S.F.; data curation, X.L. and J.B.d.S.F.; formal analysis, K.U. and J.B.d.S.F.; investigation, J.F.S.F., D.S., D.L.S., and X.L.; methodology, J.F.S.F. and X.L.; project administration, J.F.S.F. and X.L.; visualization, J.F.S.F., K.U., J.B.d.S.F., and D.S.; writing, initial drafts, K.U. and J.F.S.F.; writing, review and editing, J.F.S.F., K.U., C.F.d.L., D.S., D.L.S., and X.L. All authors have read and agreed to the published version of the manuscript.

Acknowledgments: We acknowledge the financial support of a grant provided by TUBITAK (2219 International Postdoctoral Research Scholarship Program) for Kadir Uçgun's work on this project at the US Salinity Laboratory (USDA-ARS), Riverside, USA. We also acknowledge the help of Alysia Soria, Kimberly Wilkerson, and Noah Gangoso for their assistance with harvest, saline solutions and sample preparation and Pangki Xiong for the analyses of macro/micronutrients and of Na and Cl in tissues of spinach. We also acknowledge the help of Ray Anderson and Dennise Jenkins for the confirmation of greenhouse light intensity.

References

1. Ashraf, M. Breeding for salinity tolerance in plants. *Crit. Rev. Plant Sci.* **1994**, *13*, 17–42. [CrossRef]
2. Levitt, J. *Responses of Plants to Environmental Stresses, Water, Radiation, Salt and Other Stresses*; Academic Press: New York, NY, USA, 1980.
3. Munns, R. Comparative physiology of salt and water stress. *Plant Cell Environ.* **2002**, *25*, 239–250. [CrossRef] [PubMed]
4. Hu, Y.; Schmidhalter, U. Drought and salinity: A comparison of their effects on mineral nutrition of plants. *J. Plant Nutr. Soil Sci.* **2005**, *168*, 541–549. [CrossRef]
5. Dias, N.S.; Ferreira, J.F.S.; Liu, X.; Suarez, D.L. Jerusalem artichoke (*Helianthus tuberosus* L.) maintains high inulin, tuber yield, and antioxidant capacity under moderately-saline irrigation waters. *Ind. Crops Prod.* **2016**, *94*, 1009–1024. [CrossRef]
6. Ferreira, J.F.S.; Sandhu, D.; Liu, X.; Halvorson, J.J. Spinach (*Spinacea oleracea* L.) response to salinity: Nutritional value, physiological parameters, antioxidant capacity, and gene expression. *Agriculture* **2018**, *8*, 163. [CrossRef]
7. Ferreira, J.F.S.; Cornacchione, M.; Liu, X.; Suarez, D. Nutrient Composition, Forage Parameters, and Antioxidant Capacity of Alfalfa (*Medicago sativa*, L.) in Response to Saline Irrigation Water. *Agriculture* **2015**, *5*, 577–597. [CrossRef]
8. Ferreira, J.F.S.; da Silva, J.B.; Liu, X.; Sandhu, D. Spinach plants favor the absorption of K^+ over Na^+ regardless of salinity, and may benefit from Na^+ when K^+ is deficient in the soil. *Plants* **2020**, *9*, 507. [CrossRef]
9. Ferreira, J.F.S.; Liu, X.; Suarez, D.L. Fruit yield and survival of five commercial strawberry cultivars under field cultivation and salinity stress. *Sci. Hortic.* **2019**, *243*, 401–410. [CrossRef]
10. Suarez, D.L.; Grieve, C.M. Growth, Yield, and Ion Relations of Strawberry in Response To Irrigation With Chloride-Dominated Waters. *J. Plant Nutr.* **2013**, *36*, 1963–1981. [CrossRef]
11. Chinnusamy, V.; Jagendorf, A.; Zhu, J.K. Understanding and improving salt tolerance in plants. *Crop Sci.* **2005**, *45*, 437–448. [CrossRef]
12. Song, J.; Fan, H.; Zhao, Y.; Jia, Y.; Du, X.; Wang, B. Effect of salinity on germination, seedling emergence, seedling growth and ion accumulation of a euhalophyte *Suaeda salsa* in an intertidal zone and on saline inland. *Aquat. Bot.* **2008**, *88*, 331–337. [CrossRef]
13. Zhou, D.; Xiao, M. Specific ion effects on the seed germination of sunflower. *J. Plant Nutr.* **2010**, *33*, 255–266. [CrossRef]
14. Gomes-Filho, E.; Lima, C.R.F.M.; Costa, J.H.; Da Silva, A.C.M.; Da Guia Silva Lima, M.; De Lacerda, C.F.; Prisco, J.T. Cowpea ribonuclease: Properties and effect of NaCl-salinity on its activation during seed germination and seedling establishment. *Plant Cell Rep.* **2008**, *27*, 147–157. [CrossRef] [PubMed]
15. Yupsanis, T.; Moustakas, M.; Eleftheriou, P.; Damianidou, K. Protein phosphorylation-dephosphorylation in alfalfa seeds germinating under salt stress. *J. Plant Physiol.* **1994**, *143*, 234–240. [CrossRef]
16. Dantas, B.F.; De Sá Ribeiro, L.; Aragão, C.A. Germination, initial growth and cotyledon protein content of bean cultivars under salinity stress. *Rev. Bras. Sementes* **2007**, *29*, 106–110. [CrossRef]
17. Khan, M.A.; Rizvi, Y. Effect of salinity, temperature, and growth regulators on the germination and early seedling growth of *Atriplex griffithii* var. *stocksii*. *Can. J. Bot.* **1994**, *72*, 475–479. [CrossRef]
18. Promila, K.; Kumar, S. Vigna radiata seed germination under salinity. *Biol. Plant.* **2000**, *43*, 423–426. [CrossRef]

19. Tawaha, A.M.; Othman, Y.; Al-Karaki, G.; Al-Tawaha, A.R.; Al-Horani, A. Variation in germination and ion uptake in barley genotypes under salinity conditions. *World J. Agric. Sci.* **2006**, *2*, 11–15.

20. Turhan, A.; Kuşcu, H.; Şeniz, V. Effects of different salt concentrations (NaCl) on germination of some spinach cultivars. *Uludağ Üniv. Ziraat Fak. Derg.* **2011**, *25*, 65–77. [CrossRef]

21. Borsai, O.; Al-Hassan, M.; Boscaiu, M.; Sestras, R.E.; Vicente, O. Effects of salt and drought stress on seed germination and seedling growth in Portulaca. *Rom. Biotechnol. Lett.* **2018**, *23*, 13340–13349.

22. Sandhu, D.; Cornacchione, M.V.; Ferreira, J.F.S.; Suarez, D.L. Variable salinity responses of 12 alfalfa genotypes and comparative expression analyses of salt-response genes. *Sci. Rep.* **2017**, *7*. [CrossRef] [PubMed]

23. Sandhu, D.; Pudussery, M.V.; Ferreira, J.F.S.; Liu, X.; Pallete, A.; Grover, K.K.; Hummer, K. Variable salinity responses and comparative gene expression in woodland strawberry genotypes. *Sci. Hortic.* **2019**, *254*, 61–69. [CrossRef]

24. Zrig, A.; Ferreira, J.F.S.; Serrano, M.; Valero, D.; Tounekti, T.; Khemira, H. Polyamines and other secondary metabolites of green-leaf and red-leaf almond rootstocks trigger in response to salinity. *Pak. J. Bot.* **2018**, *50*, 1273–1279.

25. Abdul Jaleel, C.; Gopi, R.; Sankar, B.; Manivannan, P.; Kishorekumar, A.; Sridharan, R.; Panneerselvam, R. Studies on germination, seedling vigour, lipid peroxidation and proline metabolism in *Catharanthus roseus* seedlings under salt stress. *S. Afr. J. Bot.* **2007**, *73*, 190–195. [CrossRef]

26. Kaymakanova, M. Effect of salinity on germination and seed physiology in bean (*Phaseolus vulgaris* L.). *Biotechnol. Biotechnol. Equip.* **2009**, *23*, 326–329. [CrossRef]

27. Prado, F.E.; Boero, C.; Gallardo, M.; González, J.A. Effect of NaCl on germination, growth, and soluble sugar content in *Chenopodium quinoa* Willd. seeds. *Bot. Bull. Acad. Sin.* **2000**, *41*, 27–34. [CrossRef]

28. Brakez, M.; Harrouni, M.C.; Tachbibi, N.; Daoud, S. Comparative effect of NaCl and seawater on germination of quinoa seed (*Chenopodium quinoa* willd). *Emir. J. Food Agric.* **2014**, *26*, 1091–1096. [CrossRef]

29. Al-Harbi, A.R.; Wahb-Allah, M.A.; Abu-Muriefah, S.S. Salinity and nitrogen level affects germination, emergence, and seedling growth of tomato. *Int. J. Veg. Sci.* **2008**, *14*, 380–392. [CrossRef]

30. Jamil, M.; Lee, D.B.A.E.; Jung, K.Y.; Ashraf, M.; Chun, S.; Rha, E.U.I.S. Effect of salt (Nacl) stress on germination and early seedling growth of four vegetables species. *J. Cent. Eur. Agric.* **2006**, *7*, 273–282. [CrossRef]

31. Janagard, M.S.; Tobeh, A.; Esmailpour, B. Evaluation of salinity tolerance of three canola cultivars at germination and early seedling growth stage. *J. Food Agric. Environ.* **2008**, *6*, 272–275.

32. Abdel-Haleem, A.; El-Shaieny, H. Seed germination percentage and early seedling establishment of five (*Vigna unguiculata* L. (Walp) genotypes under salt stress. *Eur. J. Exp. Biol.* **2015**, *5*, 22–32.

33. Alemzadeh, N.; Khaleghi, A.S. Germination and emergence response of some onion cultivars of Southern Iran to salinity stress. *Seed Sci. Biotechnol.* **2009**, *3*, 21–23.

34. Kaya, M.D.; Ipek, A.; Öztürk, A. Effects of different soil salinity levels on germination and seedling growth of safflower (*Carthamus tinctorius* L.). *Turk. J. Agric. For.* **2003**, *27*, 221–227. [CrossRef]

35. Wu, G.Q.; Jiao, Q.; Shui, Q.Z. Effect of salinity on seed germination, seedling growth, and inorganic and organic solutes accumulation in sunflower (*Helianthus annuus* L.). *Plant Soil Environ.* **2015**, *61*, 220–226. [CrossRef]

36. Akinci, I.E.; Akinci, S.; Yilmaz, K.; Dikici, H. Response of eggplant varieties (*Solanum melongena*) to salinity in germination and seedling stages. *N. Z. J. Crop Hortic. Sci.* **2004**, *32*, 193–200. [CrossRef]

37. Kahouli, B.; Borgi, Z.; Hannachi, C. Effect of sodium chloride on the germination of the seeds of a collection of carrot accessions (*Daucus carota* L.) cultivated in the region of Sidi Bouzid. *J. Stress Physiol. Biochem.* **2014**, *10*, 28–36.

38. Subbarao, G.V.; Wheeler, R.M.; Stutte, G.W.; Levine, L.H. How-far can sodium substitute for potassium in red beet? *J. Plant Nutr.* **1999**, *22*, 1745–1761. [CrossRef]

39. Nemadodzi, L.E.; Araya, H.; Nkomo, M.; Ngezimana, W.; Mudau, N.F. Nitrogen, phosphorus, and potassium effects on the physiology and biomass yield of baby spinach (*Spinacia oleracea* L.). *J. Plant Nutr.* **2017**, *40*, 2033–2044. [CrossRef]

40. Ors, S.; Suarez, D.L. Salt tolerance of spinach as related to seasonal climate. *Hortic. Sci.* **2016**, *43*, 33–41. [CrossRef]

41. Koike, S.T.; Cahn, M.; Fennimore, S.; Lestrange, M.; Natwick, E.; Smith, R.F.; Takele, E. *Vegetable Production*

Series in Spinach Production in California; Publication 7212; University of California Agricultural and Natural Resources: Davis, CA, USA, 2011; 6p.

42. Karege, F.; Penel, C.; Greppin, H. Floral induction in spinach leaves by light, temperature and gibberellic acid: Use of the photocontrol of basic peroxidase activity as biochemical marker. *Z. Pflanzenphysiol.* **1982**, *107*, 357–365. [CrossRef]

43. Kader, M.A. A Comparison of seed germination calculation formulae and the associated interpretation of resulting data. *J. Proc. R. Soc. N. S. W.* **2005**, *138*, 65–75.

44. Hoagland, D.R.; Arnon, D.I. *The Water-Culture Method for Growing Plants without Soil*; California Agricultural Experiment Station: Berkeley, CA, USA, 1950; Volume Circular 347, 32p.

45. Ribeiro Junior, J.I.; Melo, A.L.P. *Guia Prático Para Utilização do Saeg*; Folha Artes Graficas Ltd.: Viçosa, Brazil, 2008.

46. Lacerda, C.F.; Ferreira, J.F.S.; Liu, X.; Suarez, D.L. Evapotranspiration as a criterion to estimate nitrogen requirement of maize under salt stress. *J. Agron. Crop Sci.* **2016**, *202*, 192–202. [CrossRef]

47. Morgan, K.T. (Ed.) *Nutrient Management of Vegetable and Row Crops Handbook (SP500)*; University of Florida: Gainesville, FL, USA, 2015.

48. Schröppel-Meier, G.; Kaiser, W.M. Ion homeostasis in chloroplasts under salinity and mineral deficiency. *Plant Physiol.* **1988**, *87*, 828–832. [CrossRef] [PubMed]

49. Sheikhi, J.; Ronaghi, A. Growth and macro and micronutrients concentration in spinach (*Spinacia oleracea* L.) as influenced by salinity and nitrogen rates. *Int. Res. J. Appl. Basic Sci.* **2012**, *3*, 770–777.

50. Chakraborty, K.; Bose, J.; Shabala, L.; Shabala, S. Difference in root K^+ retention ability and reduced sensitivity of K^+-permeable channels to reactive oxygen species confer differential salt tolerance in three Brassica species. *J. Exp. Bot.* **2016**, *67*, 4611–4625. [CrossRef] [PubMed]

51. Chow, W.; Ball, M.; Anderson, J. Growth and photosynthetic responses of spinach to salinity: Implications of K+ nutrition for salt tolerance. *Funct. Plant Biol.* **1990**, *17*, 563. [CrossRef]

52. Raddatz, N.; Morales de los Ríos, L.; Lindahl, M.; Quintero, F.J.; Pardo, J.M. Coordinated transport of nitrate, potassium, and sodium. *Front. Plant Sci.* **2020**, *11*, 247. [CrossRef]

53. Ragel, P.; Raddatz, N.; Leidi, E.O.; Quintero, F.J.; Pardo, J.M. Regulation of K^+ nutrition in plants. *Front. Plant Sci.* **2019**, *10*. [CrossRef]

54. Grieve, C.; Grattan, S.; Maas, E. Plant salt tolerance. In *Agricultural Salinity Assessment and Management*; Wallender, W.W., Tanji, K.K., Eds.; ASCE Press: Reston, VA, USA, 2012; pp. 405–459.

55. Maas, E.V.; Grattan, S.R. Crop Yields as Affected by Salinity. In *Agricultural Drainage*; Skaggs, R.W., Schilfgaarde, J.V., Eds.; ASA/CSSA/SSSA: Madison, WI, USA, 1999; pp. 55–108.

56. Maas, E.V.; Hoffman, G.J. Crop salt tolerance-current assessment. *J. Irrig. Drain. Div.* **1977**, *103*, 115–134.

3

A Genome-Wide Association Study Reveals Candidate Genes Related to Salt Tolerance in Rice (*Oryza sativa*) at the Germination Stage

Jie Yu [1,2,†], Weiguo Zhao [1,3,†], Wei Tong [1,2], Qiang He [1,4], Min-Young Yoon [1,5], Feng-Peng Li [1,6], Buung Choi [1,7], Eun-Beom Heo [1,8], Kyu-Won Kim [9,*] and Yong-Jin Park [1,9,*]

[1] Department of Plant Resources, College of Industrial Sciences, Kongju National University, Yesan 32439, Korea; agnesyu121@ahau.edu.cn (J.Y.); wgzsri@126.com (W.Z.); wtong@ahau.edu.cn (W.T.); qiangh06@gmail.com (Q.H.); myyoon0721@gmail.com (M.-Y.Y.); lifengpeng2013@gmail.com (F.-P.L.); pckorea1587@gmail.com (B.C.); hueunbum@gmail.com (E.-B.H.)

[2] State Key Laboratory of Tea Plant Biology and Utilization, Anhui Agricultural University, Hefei 230036, China

[3] School of Biotechnology, Jiangsu University of Science and Technology, Sibaidu, Zhenjiang, Jiangsu 212018, China

[4] National Key Facility for Crop Resources and Genetic Improvement, Institute of Crop Science, Chinese Academy of Agricultural Sciences, Beijing 100081, China

[5] Leader of Eco. Energy & Bio (LEEBCOR), 190-26 Hwangyeonggongwon-ro, Asan-si, Chungcheongnam-do 31529, Korea

[6] Suzhou GENEWIZ Biotechnology Co. LTD, C3 218 Xinghu Road Suzhou Industrial Park, Suzhou 215123, China

[7] Chemical Safety Division, National Institute of Agricultural Sciences (NIAS), Wanju 55365, Korea

[8] Breeding & Research Institute, Koregon Co. LTD, Anseong Center 60-34, Gokcheon-gil, Bogae-Myeon, Anseong-Si, Gyeonggi-Do 17509, Korea

[9] Center of Crop Breeding on Omics and Artificial Intelligence, Kongju National University, Yesan 32439, Korea

* Correspondence: sh.kyuwon@gmail.com (K.-W.K.); yjpark@kongju.ac.kr (Y.-J.P.)

† These authors contributed equally to this work.

Abstract: Salt toxicity is the major factor limiting crop productivity in saline soils. In this paper, 295 accessions including a heuristic core set (137 accessions) and 158 bred varieties were re-sequenced and ~1.65 million SNPs/indels were used to perform a genome-wide association study (GWAS) of salt-tolerance-related phenotypes in rice during the germination stage. A total of 12 associated peaks distributed on seven chromosomes using a compressed mixed linear model were detected. Determined by linkage disequilibrium (LD) blocks analysis, we finally obtained a total of 79 candidate genes. By detecting the highly associated variations located inside the genic region that overlapped with the results of LD block analysis, we characterized 17 genes that may contribute to salt tolerance during the seed germination stage. At the same time, we conducted a haplotype analysis of the genes with functional variations together with phenotypic correlation and orthologous sequence analyses. Among these genes, *OsMADS31*, which is a MADS-box family transcription factor, had a down-regulated expression under the salt condition and it was predicted to be involved in the salt tolerance at the rice germination stage. Our study revealed some novel candidate genes and their substantial natural variations in the rice genome at the germination stage. The GWAS in rice at the germination stage would provide important resources for molecular breeding and functional analysis of the salt tolerance during rice germination.

Keywords: rice; genome-wide association study; salt stress; germination; natural variation

1. Introduction

Salt stress undesirably affects plant growth during all developmental stages. Therefore, it is a major threat to crop productivity [1] and this situation has lasted in some parts of the world for over 3000 years with growth every year. As a monocotyledonous model plants, rice feeds more than one half of the world's population [2]. However, rice is also sensitive to salt stress and is currently listed as the most salt-sensitive cereal crop, which results in most cultivated varieties having a salinity threshold of 3 dSm^{-1} [3].

Seed germination is usually a very important stage in the seedling stable stand establishment and determines the success of crop production [4]. The effects of salt stress on seed germination are extremely complex involving various physical and biochemical cues. Generally, salt stress is negatively correlated with seed germination and seedling growth [5] in most plants such as *Oryza sativa* [6], *Zea mays* [7], *Helianthus annuus* [8], and *Brassica* spp. [9]. The impact of salt stress on seed germination is attributed to seed water uptake and ion toxic effect. During the seed germination, salinity alters the imbibition of water by reducing the osmotic potential of the germination medium [10], damages the ultrastructure of cells, tissue, and organs [11], changes the activity of enzymes [12], disturbs hormonal balance [13], alters protein metabolism [14], and reduces the use of seed reserves [15]. However, various environmental (external) and plant physiological (internal) factors affect seed germination under saline conditions including temperature, light, water and gasses, seed age, seed dormancy, nature of seed coat, seed morphology, and seedling vigor [16].

Recently, improving rice salt tolerance during the germination stage became more important because salinity may rapidly reduce the germination rate and percentage, which, in turn, may lead to a reduction of crop yields [17]. Many efforts have been made to improve seed germination and seedling vigor by optimizing the non-genetic factors [6]. However, success in improving salt tolerance in rice is made by identifying a major quantitative trait locus (QTL), which contributes to salt tolerance in rice. QTL analysis of seed germination have been reported in rice [18], soybean [19], wheat [20], *Arabidopsis* [21], and *Brassica rapa* [22]. However, it is difficult to develop rice elite varieties with a high level of salt tolerance due to a lack of understanding the mechanisms of salt tolerance during the seed germination stage. Moreover, QTLs conferring salt stress tolerance in rice were identified mainly at the seedling stage [23], but there are few reports on rice seed germination [18,24,25]. Wang et al. [18] detected 16 QTLs for the imbibition rate and germination percentage under control and salt stress with the recombinant inbred line (RIL) population derived from IR26/Jiucaiqing. Abe et al. [26] identified a candidate gene, *OsGA20ox1*, for a major QTL controlling seedling vigor in rice. Zheng et al. [27] identified 11 QTLs for salt tolerance at the germination and early seedling stage in *japonica* rice.

Genome-wide association study (GWAS) is an efficient method for detecting valuable natural variations in trait-associated loci as well as allelic variations in candidate genes underlying quantitative and complex traits [28,29]. Instead of SSR (simple sequence repeat) markers, which are commonly used for association mapping [30,31], SNPs (single-nucleotide polymorphisms) have become more popular for GWASs with the rapid development of NGS (next-generation sequencing) and the existence of high-density SNP markers by re-sequencing [32]. In rice, there are some successful reports to dissect genetic architecture of complex traits through GWAS [28,29,32,33]. However, limited studies have been carried out in rice to identify genes/QTLs for salt tolerance using GWAS. Kumar et al. [34] identified total 64 SNPs significantly associated with Na$^+$/K$^+$ ratio and other traits for reproductive stage salinity tolerance using GWAS. Yu et al. [35] identified 93 candidate genes significantly associated with salt tolerance at the rice seedling stage. Shi et al. [36] identified 22 significant salt tolerance associated SNPs based on the stress-susceptibility indices (SSIs) of vigor index (VI) and the mean germination time (MGT). Naveed et al. [37] identified 20 QTN for salinity tolerance at the germination and seedling stages in rice. Seed germination plays an important role in the cycle of plant growth. To our knowledge, there is little research until now on the identification of genes particularly for the germination stage salinity tolerance using GWAS in rice. Here, we applied GWAS mapping using ~1.65 million SNPs/indels covering all 12 rice chromosomes in a diverse rice collection to identify candidate genes and natural variation that may contribute to salt tolerance during the rice germination stage with the aim to guide breeding of salt-tolerant rice varieties.

2. Results

2.1. Phenotypic Screening and Evaluation

Individual value plots for GP with 0, 200 and 300 mM NaCl from a screening experiment using 12 randomly selected samples are shown in Figure 1a. According to the screening result, 200 mM NaCl fully exhibited their phenotypic variance, which resulted in the most diverse phenotypic distribution and facilitated discrimination of accessions with different salt tolerance levels. Thus, the treatment of 200 mM NaCl was chosen as the target salinity level for determining the salt tolerance of all accessions.

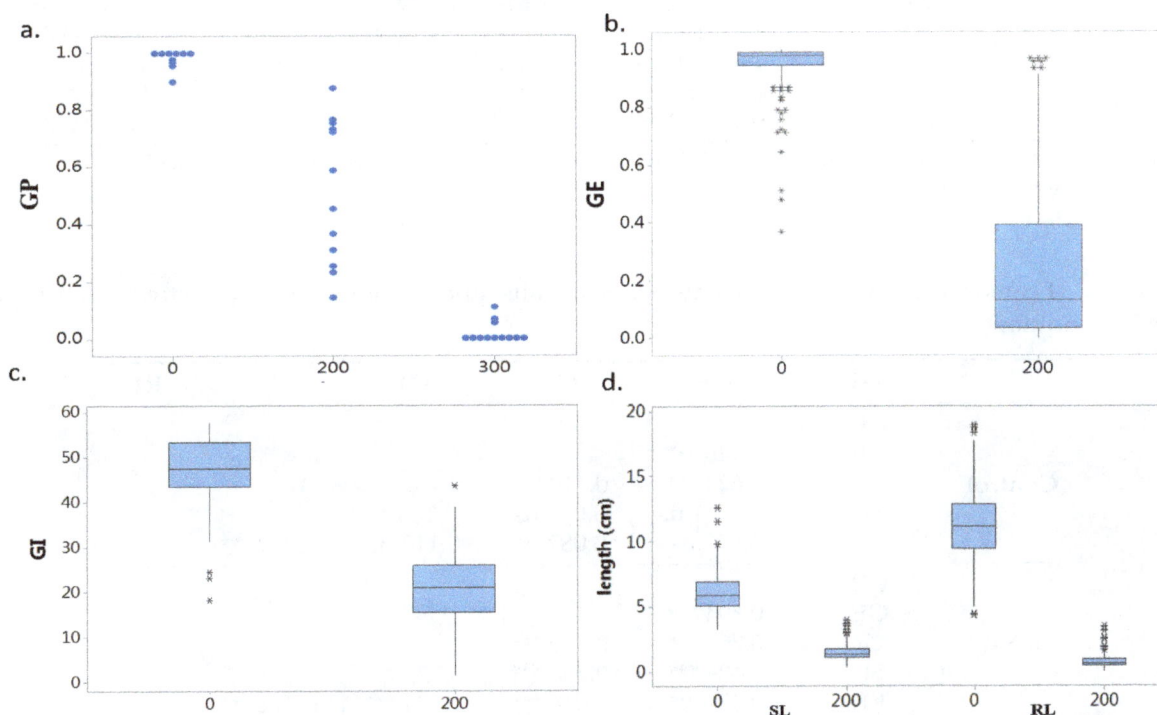

Figure 1. Determination of the optimum NaCl concentration and main phenotypes under salt stress and control conditions. (**a**) Individual value plot of germination percentage in the presence of 0, 200, and 300 mM NaCl (each dot represents an individual). (**b–d**) Box plots for phenotypic values in the presence of 0 and 200 mM NaCl (the asterisks are extreme outliers). GP: germination percentage, GE: germination energy, GI: Germination index, SL: shoot length, RL: root length.

The following traits: GP, GE, GI, SL, and RL were examined under 0 and 200 mM NaCl salt stress during the rice germination stage. Descriptive statistics of the phenotypes related to salt tolerance during the germination stage of the current collection were presented in Table 1. Box plots of phenotypes including GE, GI, SL, and RL in the presence of 0 and 200 mM NaCl are shown in Figure 1b–d. The data suggested that seed germination traits were negatively influenced by salt stress. Salt stress inhibits shoot and root elongation dramatically but GE and RL were more affected than SL [18]. These findings indicated that most germination parameters under salt stress exhibited lower performance than under control conditions, which may restrict plant growth.

The correlation coefficients of phenotypes under control and salt-stress conditions were also evaluated (Table 2). RL was significantly and positively correlated with SL only under control conditions. Excepting RL and SL was also significant positively correlated with GI. GP, GE, and GI were all significantly and positively correlated with each other under control conditions while all phenotypes were significantly and positively correlated with each other in salt stress conditions.

These results suggested that all phenotypes evaluated in this study could be used for GWAS and some overlapped results could be found among the phenotypes.

Table 1. Descriptive statistics for the traits in the control and salt-treated (200 mM NaCl) rice accessions.

Trait	Salinity Level (NaCl/mM)	Mean ± SD [a]	Range	Median	IQR [b]
GP	0	0.97 ± 0.06	0.47–1.00	0.99	0.97–1.00
	200	0.87 ± 0.18	0.14–1.00	0.94	0.83–0.98
GE	0	0.95 ± 0.08	0.37–1.00	0.98	0.94–0.99
	200	0.25 ± 0.29	0–0.97	0.13	0.03–0.39
GI	0	47.66 ± 6.71	18.33–57.87	47.55	43.40–53.44
	200	20.66 ± 8.53	1.51–43.72	21.05	15.48–26.03
SL	0	6.07 ± 1.38	3.25–12.62	5.82	5.08–6.90
	200	1.45 ± 0.63	0.38–3.87	1.29	1.03–1.76
RL	0	11.28 ± 2.89	4.28–18.89	11.11	9.46–12.81
	200	0.69 ± 0.49	0.013–3.44	0.55	0.39–0.87

[a] Standard deviation. [b] Interquartile range. GP: Germination percentage. GE: Germination energy. GI: Germination index. SL: shoot length. RL: root length.

Table 2. Pearson correlation coefficients among traits under control and salt stress (200 mM NaCl) conditions.

	Trait	GP	GE	GI	SL	RL
Control	GP					
	GE	0.946 ***				
	GI	0.624 ***	0.710 ***			
	SL	−0.001 ns	0.037 ns	0.168 **		
	RL	0.073 ns	0.087 ns	0.110 ns	0.254 ***	
200 mM	GP					
	GE	0.394 ***				
	GI	0.758 ***	0.849 ***			
	SL	0.439 ***	0.738 ***	0.725 ***		
	RL	0.357 ***	0.618 ***	0.594 ***	0.712 ***	

GP: germination percentage. GE: germination energy. GI: germination index. SL: shoot length. RL: root length.
*, **, ***, ns: significant at the 0.05, 0.01, and 0.001 probability level and not significant, respectively.

2.2. Principal Components Analysis (PCA)

PCA was performed with the 1.65 million high-quality SNPs/indels to mine the population structure in all rice accessions. Two components were suggested by the scree plot (Supplementary file 2: Figure S1a). Clear subpopulation structures were observed based on the first two PCs (PC1 and PC2), which resulted in two subpopulations, *indica* and *japonica*, with the admixture accessions located between the two groups (Figure 2a).

For the PCA using the phenotypes, we examined correlations between subspecies in salt tolerance levels using four main phenotypes that drove the differences among accessions. We used TASSEL to perform a PCA of R-GE, R-GI, R-RL, and R-SL in the rice collection. Most of the phenotypic variation (>91%) in the collection was explained by the first two PCs (Supplementary file 1: Figure S1b). Thus, we generated a PCA plot using PC1 and PC2. However, rice accessions in our study were not clustered into clearly defined groups (such as *indica* or *japonica*) based on the above four phenotypes (Figure 2b). This indicates that salt tolerance levels in rice (*O. sativa*) are not strongly correlated with the *indica* or *japonica* subgroups.

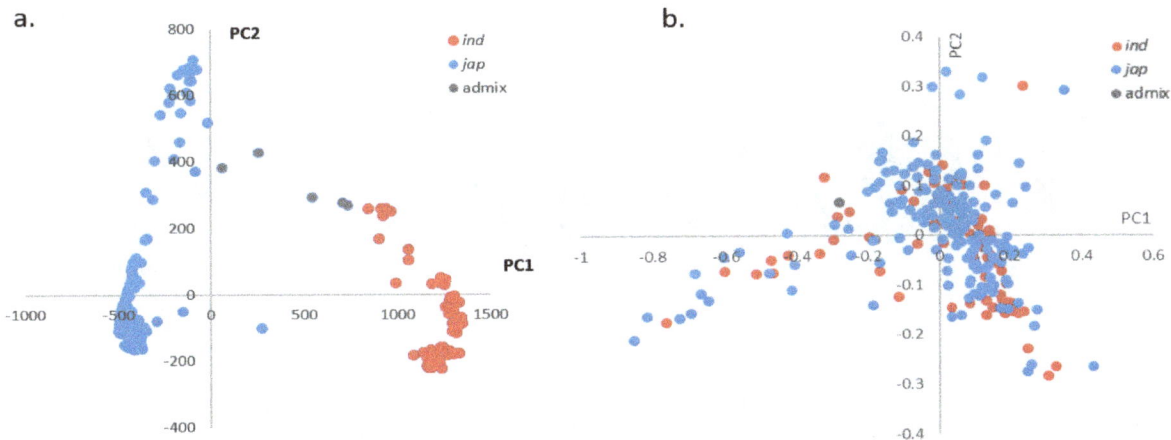

Figure 2. Principal components analysis (PCA) using genotype and phenotype data. (**a**). For genotype data, 295 accessions were divided into ind (*indica*) and jap (*japonica*) based on PC1 and PC2 along with an admixture group. (**b**). For phenotype data, no clear grouping was observed.

2.3. GWAS and Candidate Gene Identification

To generate the genotype dataset for GWAS, more than ~1.65 million SNPs/indels were identified across the accessions and subjected to GWAS applied with the CMLM [38]. The GWAS results were shown on Figure 3 and Supplementary file 1: Figure S2. We took associations held by the peaks with $-\log10\ (p)$ value > 5 and adjusted p-value (FDR, false discovery rate) < 0.05 for further analysis since the cutoff of $-\log10\ (p)$ value was five when the FDR ≤ 0.05. Under the salt stress condition, the significant signals were detected for GE, RL, SL, R/S, and R-R/S. In total, 10 SNPs were found significant for these traits and only one SNP (chr12_1628276) were found common for RL, R/S, and R-R/S traits. Excluding the common one, only 3, 2, 2, 1, and 1 SNPs were found uniquely associated for R-R/S, GE, RL, SL, and R/S, respectively. In addition, we also found two SNPs (chr02_1090174 and chr05_20164893) were the two strongest, significantly associated for GE in all observed traits ($p < 10^{-9}$).

We further conducted a genome-wide LD analysis of the candidate peak regions and determined LD blocks harboring significant SNPs/indels that characterized in the last step as regions containing putative candidate genes. LD block analysis was detected in a 400 kb range centered on the highest $-\log10\ (p)$ value (Figure 3c,d). Annotation of SNPs/indels from the 200 kb up-stream and down-stream ranges, together with the LD block analysis, resulted in the identification of 79 genes included in these peaks and some candidate genes have been reported previously to contribute to salt tolerance (Supplementary file 2: Table S2). Among the known genes, seven were associated with chromosome 1 (all in RL), 22 with chromosome 2 (7 in GE, 6 in R/S and 9 in R-R/S), 10 with chromosome 3 with R-R/S, 15 with chromosome 4 (3 in SL and 12 in RL), 6 with chromosome 5 (all in GE), 6 with chromosome 11 (all in R-R/S), and 13 with chromosome 12 (the peak region was identical in RL, R/S, and R-R/S). In the LD block analysis, most highly associated SNPs/indels were located in small or large LD blocks, which indicated that they were in significant linkage disequilibrium. Thus, these candidate genes may contribute to salt stress independently or co-operatively with other variations in other genes harboring these SNPs/indels. Simultaneously, we screened candidate genes containing many highly associated SNPs/indels in the genic region as well as some highly associated signals not located in known genes but suggesting that these unknown genes may also be related to salt tolerance (Supplementary file 4: Table S3). Some of those SNPs/indels were located in the coding region of the unknown genes rather than in the surrounding 200 kb regions. These genes could also be important determinants of salt tolerance in rice.

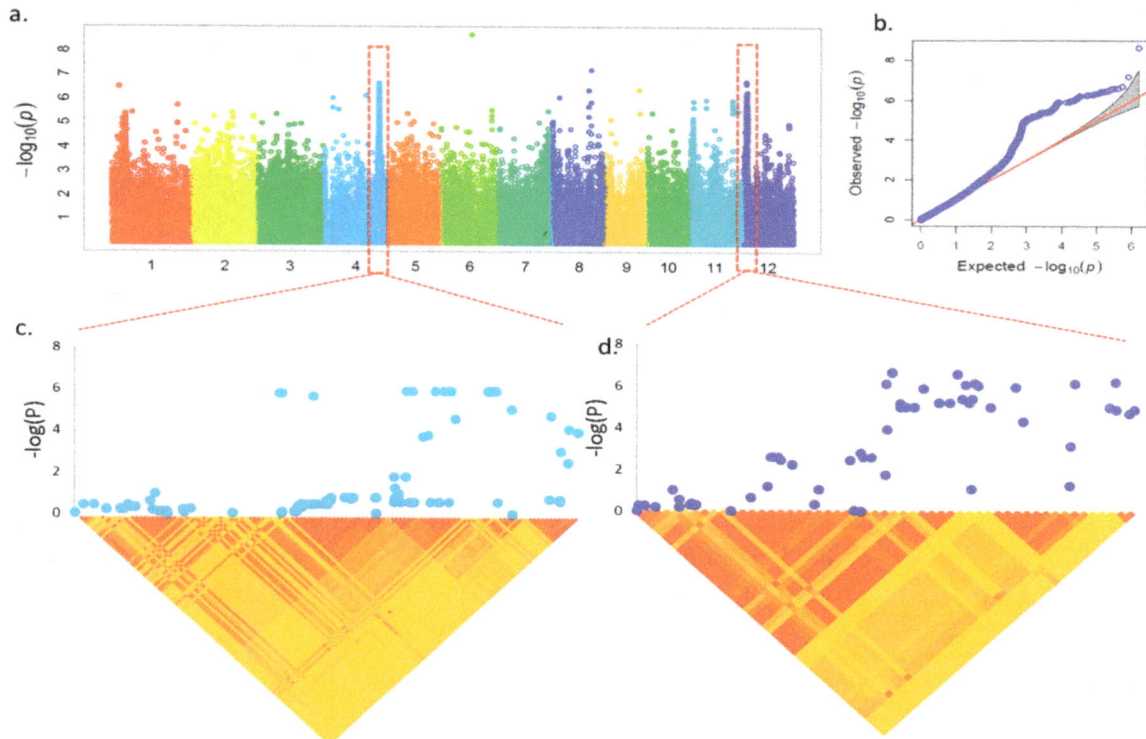

Figure 3. Genome-wide association mapping and LD block analysis for root length (RL) under salt stress (200 mM) conditions. (**a**) Manhattan plot from association mapping using the CMLM. (**b**) QQ plot of expected and observed P values. (**c**) The peak region on chromosome 4 along with the LD blocks. (**d**) The peak region on chromosome 12 along with the LD blocks. In (**c,d**) pair-wise LD between SNPs is indicated as D′ values: red indicates a value of 1 and yellow indicates 0. The LD region was 200 kb upstream and downstream of the top −log (*p*) value in the peak range.

2.4. Natural Variations in Candidate Genes and Sequence Analysis

Based on the associated peaks identified in the GWAS and by determining the LD blocks test, we identified many candidate genes associated with salt tolerance during the rice germination stage. To mine functional and novel candidate genes, we investigated 17 final candidate genes (Table 3) that contained highly associated SNPs/indels within the coding region. Many of these SNPs/indels have been reported to play a role in the salt stress in rice such as *OsAGO2* (Os04g0615700) [39], *OsZIFL13* (Os12g0133300) [40], and *OsHAK11* (Os04g0613900) [41], which are related to salt stress in rice. These genes are involved in the salt tolerance in rice by different pathways [42].

Natural variations of these 17 genes were mined and then functional variations were screened after checking the positions of the variations in genes and the corresponding amino acid change. Among the genes, *OsMADS31*, which is involved in floral organ specification and implicated in plant growth and development, was identified and predicted to be involved in salt tolerance [43]. As shown in Figure 4, one natural SNP substitution (T/A) was detected and caused an F/L amino acid change, which is presented by type 1 (reference sequence) and type 2 (variation) (Figure 4a). Furthermore, we generated a haplotype network of the whole collection, which was dominated by two common haplotypes including primarily the *japonica* type (type 1) and the *indica* type (type 2), respectively (Figure 4e). A phenotypic difference was observed in type 1 with 236 accessions and an average RL of 0.6293 and type 2 with 58 accessions and an RL of 0.9327 (Figure 4b). We conducted further orthologue alignment of *OsMADS31* in several rice groups and other species (Figure 4c). Type 2 (candidate SNP) showed an F/L amino acid change compared to other rice groups and species (type 1, *Oryza brachyantha, Oryza rufipogon, Oryza punctate, Hordeum vulgare, Triticum aestivum, Aegilops tauschii,* and *Triticum urartu*). However, type 2 shared this F/L with three other rice species (*Oryza glaberrima, Oryza barthii* and *Oryza glumaepatula*). *Oryza glaberrima* and *Oryza barthii* are African rice and its wild type have higher salt tolerance than *Oryza sativa* species. *Oryza glumaepatula* is a wild rice found in South America usually in deep and sometimes flowing water, which may also have salt tolerance characteristics based on the presence of related genes. Four salt tolerant accessions with type 2 haplotype and four salt sensitive accessions without the haplotype were used for the real-time expression analysis. Generally, the relative RNA expression level of OsMADS31 was higher in type 1 than in type 2, which indicates that the gene expression is down-regulated in salt conditions when compared to the control (Figure 4d).

Table 3. Candidate genes with highly associated signals in the coding region that overlapped with the GWAS and LD analysis.

Chr_Pos [a]	Trait	*p*-Value	FDR [b]	Gene ID	Description
chr02_19605493	R/S	1.72×10^{-7}	0.00779	Os02g0532500	Germin family protein, Germin-like protein 2-4
				Os02g0532900	Glycoside hydrolase family 17 protein
				Os02g0533300	Carbonic anhydrase, CAH1-like domain, containing protein
				Os02g0533800	Similar to ATPase inhibitor
chr04_31168058	RL	2.42×10^{-7}	0.00859	Os04g0612900	Vacuolar ATPase assembly integral membrane protein VMA21-like domain-containing protein
				Os04g0613900	Similar to Potassium transporter 18, OsHAK11
				Os04g0614000	Similar to Peroxisomal 2,4-dienoyl-CoA reductase
				Os04g0614100	MADS-box domain-containing protein, OsMADS31
				Os04g0614600	Similar to Viroid RNA-binding protein, aminotransferase
				Os04g0614500	Pyridoxal phosphate-dependent transferase, major region, subdomain 1 domain-containing protein
				Os04g0615100	Similar to Lecithine cholesterol acyltransferase-like protein
				Os04g0615700	Protein argonaute 2, OsAGO2
chr12_1628276	RL, R/S, R-R/S	2.02×10^{-7}	0.00859	Os12g0133100	Major facilitator superfamily protein, OsZIFL12
				Os12g0133300	zinc-induced facilitator-like 13, OsZIFL13
				Os12g0133400	4'-phosphopantetheinyl transferase domain-containing protein
				Os12g0133700	Stress-activated protein kinase pathway-regulating phosphatase 1
				Os12g0133800	Similar to Auxin efflux carrier protein, *OsPIN1d*

[a] The position was based on the annotation data on Os-Nipponbare-Reference-IRGSP-1.0 (RAP-DB, http://rapdb. dna.affrc.go.jp/). [b] FDR: False discovery rate. FDR Adjusted *p* values were calculated by GAPIT applying the Benjamini-Hochberg (1995) FDR-controlling procedure. RL: root length. R/S: root/shoot ratio. R-R/S: relative root/shoot ratio.

We also found several other functional SNPs/indels in the 17 candidate genes that were correlated with a phenotypic difference (Supplementary file 5: Table S4). These candidate genes may be related to rice salt tolerance, according to both previous reports and the natural variation mining in the current study. Novel polymorphisms of those genes may also contribute to salt tolerance that make the rice resistant to salinity.

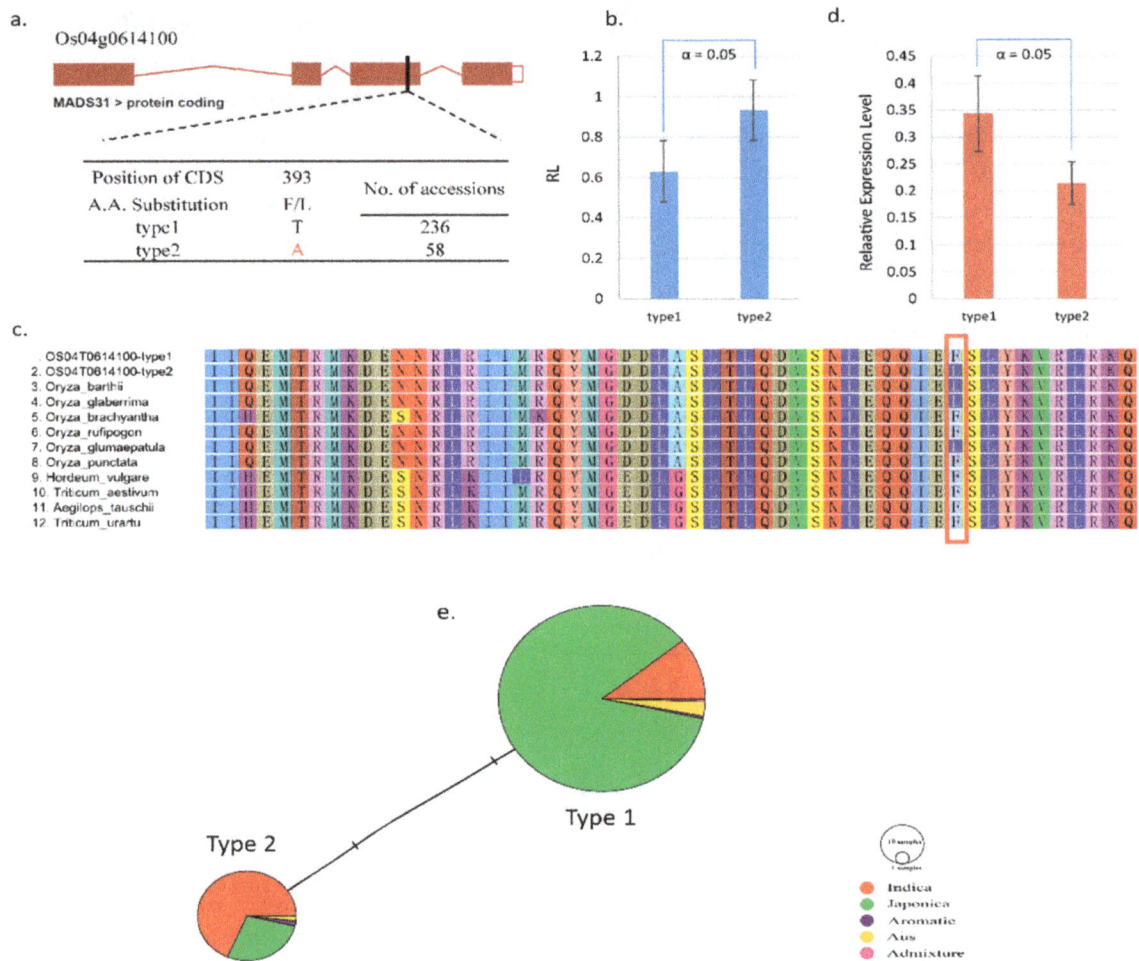

Figure 4. Haplotyping and sequence analysis of Os04g0614100, which was correlated with a phenotypic difference. (**a**) One functional SNP of the candidate gene in the CDS region. Type 1 is the reference and type 2 is the SNP. (**b**) The phenotypic difference based of the functional SNP. (**c**) Amino acid sequence alignment using several orthologues in various rice subgroups and species. Red box indicates the target amino acid change caused by the functional SNP. (**d**) RNA expression levels in rice accessions with type 1 and type 2. (**e**) Haplotype network analysis. Circle size is proportional to the number of samples within a given haplotype. Lines between haplotypes represent mutational steps between alleles. Colors denote rice designation.

3. Discussion

3.1. Salt Tolerance at Rice Germination Stage

The seed germination is one of the most critical steps in the life cycle of a crop. Seed germination begins with water uptake while salinity prevents water imbibition, which inhibits seed germination [15]. Experiments have shown that increased salinity delays the initiation of germination, which leads to a reduced germination percentage. However, salt tolerance during the early growth stages is not always correlated during subsequent growth stages [44,45]. The seeds of crops in different genotypes may germinate adequately under salt stress. However, the seedling may not become fully established later. We observed differential inhibition of the root length and shoot length in our study, which suggested that salinity can influence the germination quality of the seed.

By using the optimized salinity (200 mM NaCl) for discriminating accessions with different salt tolerance levels, we characterized the salt-tolerance-related phenotypes in a collection comprising 295 rice varieties. Phenotypic differences between the control and 200 mM NaCl salinity conditions suggested that rice growth during the seed germination stage can be markedly inhibited by salt stress, which results in very low germination energy and index (GE and GI) as well as reduced root and shoot lengths. This may suppress rice seed germination especially in some direct-sowing areas and decreases plant density and yield markedly. Therefore, development of rice varieties with salt-tolerant seeds would prevent salinity-mediated plant and yield loss during the early growth stage.

3.2. Salt Tolerance Is Not Strongly Correlated with Rice Subgroups

According to Lee et al. [46], the salt tolerance of *indica* rice was higher than that of *japonica* rice at the seedling stage, which was determined by measuring shoot Na^+ and K^+ absorption. However, as revealed in a recent study of the salt tolerance of 115 *O. sativa* and *O. glaberrima* accessions, salt tolerance was not strongly correlated with *O. sativa* cultivar groups [35]. Most of the *japonica* types were salt sensitive, but accessions from the *indica* group and *O. glaberrima* showed a wide range of sensitivities [47]. In our study, we performed a PCA of all rice germplasm using both genotype and phenotype data. Inconsistent with the genotypic PCA, which separated the collection into clear groups, phenotypic PCA using germination-related phenotypes showed no clear grouping (Figure 2b). This indicated that salt tolerance levels during the seed germination stage are not well correlated with the rice (*O. sativa*) subgroup.

3.3. GWAS and Candidate Gene Identification

In some direct-sowing areas, salt tolerance in rice during the seed germination stage is particularly important. To improve rice productivity in such areas, novel genes and alleles associated with complex quantitative salinity tolerance traits must be identified in diverse rice accessions and salt-tolerant varieties bred. An alternate and complementary approach is GWAS, which takes advantage of historical recombination events and, thus, enables a high-resolution genome wide mapping for the identification of target genomic regions in response to complex quantitative traits in rice [29]. In this study, we used a core set of rice collections and multiple bred varieties to investigate candidate loci and genes that regulate important phenotypes under salt stress in rice at the germination stage. Twelve GWAS peaks representing new QTLs on chromosomes 1, 2, 3, 4, 5, 11, and 12 during the rice germination stage were identified. The current association mapping can serve as source of novel salt tolerance genes and alleles. Thus, we found abundant candidate regions with high association peaks in five traits and were distributed on seven chromosomes. Now many QTL analysis of rice salt tolerance have been reported, but it is difficult to directly compare the chromosomal location of marker–trait associations detected in this study with previously reported QTLs because different materials at different stages, descriptive traits, and molecular maps have been used. Wang et al. [18] detected 16 QTLs for the imbibition rate and the germination percentage. Kumar et al. [34] identified 64 SNPs (loci) significantly associated with salt stress-related traits by GWAS. Leon et al. [48] identified 85 additive QTLs for seedling salinity tolerance by GBS. Yu et al. [35] identified 25 SNPs (loci) significantly associated with salt stress-related traits by GWAS. Shi et al. [36] identified 22 SNPs based on SSIs of VI and MGT by GWAS. In this study, we also found that some SNPs associated with salt-tolerance traits overlapped or located in similar or proximal regions such as a significant SNP (chr04_31168058) near *qRTL4.10* identified by Leon et al. [48] and SNPs (chr04_31164404) identified by Yu et al. (2017). This SNP is also located near the SNPs (chr04_34164920 and chr04_34292214) identified by Kumar et al. [34] associated with Na^+/K^+ ratio. Additionally, two QTLs for salt tolerance and potassium concentration were mapped just prior to this region, respectively, by Lin et al. [49] and Cai and Morishima [50]. The above results also indicated that chromosome 4 including many candidate genes in this region was found to be important for salt tolerance.

So far, about 70 salt tolerance QTLs had been located in rice using biparental mapping populations, but fine mapping and narrowing down reports are limited [34]. Driven by LD blocks to define the genomic regions for searching candidate genes has advantages over the fixed-window approach in which a certain distance from a significant SNP is considered to be the region containing candidate genes [51] by eliminating falsely included or excluded genes [52]. The wide candidate regions ranged from <1 kb to >1 Mb depending on the chromosomal position, which suggests that the resolution of the association mapping is highly dependent on the LD of the neighboring regions of the significant SNPs [34]. Since some of the LD blocks harboring significant SNPs did not contain an annotated gene, this method might have produced some false negatives or the identified region may have contained important DNA-binding or gene regulation sites, in which case, the causal gene was not detected in the LD block [53]. In this work, we used a 400 kb range of the strongest signal to locate the candidate genes, which is in line with previous studies [54]. From an LD block analysis, we obtained 79 candidate genes that had significant SNP/indel associations in LD block regions. Therefore, these regions and candidate genes have a statistically and genetically supported background and, therefore, may be important for the salt tolerance of rice during the germination stage. Apart from SNPs that had an association with previous known QTLs for salinity tolerance, there were a few SNPs, which hit specific genes that were known or functionally characterized for salt stress. Among 79 candidate genes including seven protein kinases (PK) (1 Serine/threonine protein kinase and 1 *OsCDPK26*), six ion exchanger and transporter related genes, five transcription factors (TFs), two electron carrier (peroxidase, Os01g0172600, and oxidoreductase, Os05g0411200), and two major facilitator superfamily proteins (Os12g0133100, *OsZIFL12*; Os12g0133300, *OsZIFL13*). In addition, we also found one stress-associated protein 18 (SAPs) (*OsSAP18*, Os02g0121600), one vacuolar ATPase assembly integral membrane protein (Os04g0612900), two argonaute family proteins (AGOs) (*OsAGO2*, Os04g0615700, and *OsAGO3*, Os04g0615800), one chloroplast precursor (*Ferritin1*, Os11g0106700), one calmodulin-like protein 3 (*OsCML3*, Os12g0132300), one Auxin efflux carrier protein (*OsPIN1d*, Os12g0133800), one Glycoside hydrolase (Os02g0532900), and one Glycosyltransferase (*ALG3*, Os01g0172000). The above results indicated that the candidate genes may play an essential role in salt tolerance mechanisms [34], which also indicated that salt tolerance genes are involved in ion pumps, calcium, the salt overly sensitive (SOS) pathway, mitogen-activated protein kinases (MAPK), glycine betaine, proline, and the reactive oxygen species pathways in a high salinity environment [42].

Furthermore, 17 candidate genes with high $-\log10$ (p) value-associated signals inside the coding region were also mined and may play an important role in salt tolerance. *OsHAKs* are candidates for high-affinity K^+ uptake transporters in the rice root. The transcription of *OsHAK11* (Os04g0613900) is significantly induced by salt stress and K^+ starvation, respectively [55]. *AGOs* (Os04g0615700) play important roles in the regulation of development and stress responses, antiviral immune response, transposons, and the regulation of chromatin structure and can affect the growth and development as well as the response to abiotic and biotic stress [56]. *OsPIN* (Os12g0133800), which encodes a member of the auxin efflux carrier proteins, is involved in the root elongation growth and lateral root formation patterns via the regulation of auxin distribution in rice [57]. The Germin family protein (Os02g0532500) had been revealed to be connected with a plant cell defense and diseases and to be highly resistant to sodium dodecyl sulfate (SDS) and proteases and important for early plant development and germination in plants [58]. *SAP* (Os12g0133700) is the A20/AN1 zinc-finger containing proteins, which can regulate the stress signaling in plants [59]. The Zinc-induced facilitator-like (ZIFL) family genes (Os12g0133100, *OsZIFL12*, Os12g0133300, *OsZIFL13*) are up-regulated under stress conditions [40].

Based on these regions and candidate genes, it may be possible to mine the natural variations of rice in response to salt stress in some tolerant accessions and apply those alleles to sensitive accessions via breeding methods. To the best of our knowledge, this is the first large-scale GWAS focusing on salt stress during the rice germination stage. These candidate regions and genes will facilitate the development of salt-tolerant rice varieties.

3.4. Novel Natural Variations of Candidate Genes

Investigation of new natural variations in focal traits can extend the tolerant varieties' functional alleles to other non-tolerant varieties. Breeding methods can then be used to transfer them to elite lines to produce tolerant varieties. Using the results of the GWAS and LD analysis, the haplotypes of candidate genes can be targeted and the functional alleles involved in responses can be identified.

New alleles in rice have been reported [60,61] and provide insight for researchers and breeders into the underlying mechanisms, which facilitates the breeding of improved varieties. According to Arora et al. [42], *OsMADS31* expression was low and not markedly affected by salt and cold stress. However, the expression was relatively down-regulated in seedlings under a salt condition. Nevertheless, *OsMADS31* expression was higher in seeds than during the panicle stage. In this study, we found that *OsMADS31* was associated with salt tolerance in rice at the seed germination stage with a down-regulated expression in the salt condition (Figure 4d). The contribution of MADS-box genes to flower organ specification is well developed in eudicots, but not very well in rice. Therefore, the roles of MADS genes and other candidate genes identified here using GWAS at the seed germination stage in response to salt stress should be investigated further. Moreover, by adapting functional studies (such as those performed using TALEN and CRISPR/Cas 9), the functions of genes and gene variations can be determined. Natural variations that have functional signals could be a good starting point for the exploration of gene-based assays of phenotypically different individuals such as salt tolerant vs. sensitive, drought resistant vs. susceptible, and more.

Overall, we investigated the genetics architecture of natural variation in rice salt-tolerance-related traits at the germination stage by GWAS mapping in 295 rice accessions. A total of 79 candidate genes were determined by LD blocks analysis. In addition, by detecting the highly associated variations located inside the genic region that overlapped with the results of LD block analysis, we finally characterized 17 genes that may contribute to salt tolerance during the seed germination stage. The salt tolerance related novel candidate genes would provide important resources for molecular breeding and functional analysis of the salt tolerance during the rice germination.

4. Materials and Methods

4.1. Materials

A core set of 137 rice accessions and 158 bred varieties from the National Gene Bank of the Rural Development Administration (RDA-Genebank, Korea) [62,63] was re-sequenced in the current study (Supplementary file 1: Table S1). We conducted a field experiment during the rice-growing season at the Kongju National University experimental farm and young leaves from a single plant were collected and immediately kept at $-80\ °C$ prior to genomic DNA extraction using the DNeasy Plant Mini Kit (Qiagen, Hilden, Germany). Qualified DNA was sent for the whole genome re-sequencing.

4.2. Whole Genome Re-Sequencing and Variation Detection

The genomes of all 295 rice accessions were sequenced with an average coverage of approximately $7.8\times$ on an Illumina HiSeq 2000 or 2500 Sequencing Systems Platform (Illumina Inc., San Diego, CA, USA). Raw reads were aligned against the rice reference genome (IRGSP 1.0) [64] for genotypes calling and only SNPs/indels without the missing value and a minor allele frequency (MAF) > 0.05 and containing genotype calls for all 295 accessions that were used. Lastly, ~1.65 million high-quality SNPs/Indels were obtained and used for the further GWAS [65].

4.3. Evaluation of Salt Stress and Phenotyping

We first carried out the pre-screening experiment using 12 randomly selected samples to determine the optimum level of NaCl concentration for the evaluation of salt stress during the germination stage. Seed germination were initially screened by germinating 30 seeds per genotype in petri dishes with two layers of filter papers soaked in two different NaCl concentrations: 200 and 300 mM NaCl. The germination percentage was recorded daily for 10 days. At the concentration of 300 mM NaCl, seeds hardly germinated and the seedlings did not grow out enough to be able to measure root and shoot length. Therefore, in this study, we used 0 mM NaCl (non-stress) and 200 mM NaCl (salt stress) for phenotyping all 295 accessions.

The following experiments were performed in petri dishes containing two-layered filter paper. Thirty seeds were first washed in water, then sterilized in 1% sodium hypochlorite solution for 10 min, and washed three times in deionized distilled water. Thereafter, seeds of each accession were soaked in petri dishes and then incubated at 30 °C with 40% relative humidity. Petri dishes were randomized in an incubator and three replicates of each accession under control and salt conditions (200 mM) were adopted. The solution was replaced every two days to maintain the NaCl concentration and the distilled water volume, respectively. The daily germination seed was measured and filter papers were replaced as necessary. Plumule emergence was taken as an index of germination. The length reached about 2 mm. At the end of day 10, we measured the root length (RL) and shoot length (SL) of the seedlings and the R/S (root/shoot ratio) was also calculated. Based on these experiments, several germination stage-related phenotypes (list below) were calculated and subjected to a GWAS. The mean value of the three biological replicates was calculated and used in the further analysis.

Germination Percentage (GP)

GP was recorded daily for 10 days and was calculated using the formula below.
GP = Number of germinated seeds at 10 days/Total number of seeds tested $\times$ 100%

Germination energy (GE)

GE was recorded daily for four days and was calculated using the formula below.
GE = Number of germinated seeds at four days/Total number of seeds tested $\times$ 100%

Germination index (GI)

GI was calculated using the formula below.
GI=Σ(Gt/t), where Gt is the number of seeds that germinated on day t (Alvarado et al. 1987, and Ruan et al. 2002).

Relative germination energy (R-GE)

R-GE was calculated by using the formula below.
R-GE = $GE_{200}/GE_{control}$.

Relative germination index (R-GI)

R-GI was calculated by using the formula below.
R-GI = $GI_{200}/GI_{control}$.

Relative root length (R-RL)

R-RLwas calculated by using the formula below R-RL = $RL_{200}/RL_{control}$.

Relative shoot length (R-SL)

R-SLwas calculated by using the formula below R-SL = $SL_{200}/SL_{control}$.

Relative R/S (R-R/S)

R-R/S was calculated by using the formula below R-R/S = $R/S_{200}/R/S_{control}$.

4.4. Principal Components and GWAS Analysis

Principal components analysis (PCA) of the genotype with ~1.65 million SNPs/indels and four main salt-tolerance-related phenotypes: R-GE, R-GI, R-RL, and R-SL was conducted using GAPIT and Trait Analysis by Association, Evolution and Linkage (TASSEL) 5.0 [66]. Principal component analyses (PCA) in genotypic and phenotypic were also performed using GAPIT and TASSEL 5 [66].

GWAS was performed in the GAPIT package (Genome Association and Prediction Integrated Tool) in which an advanced kinship clustering algorithm was implemented [38]. Only SNPs with adjusted p-values < 0.05 were considered significantly associated. Gene loci containing the SNPs with significantly associated peaks in the Manhattan plot of the GWAS result were considered to be candidate genes related to salt tolerance.

4.5. Linkage Disequilibrium (LD) Block, Haplotype Analysis, and Expression Analysis

LD analysis. LD analysis was calculated using TASSEL 5 [66] based on the high-quality variations (with neither missing genotype calls over all accessions nor MAF < 0.05) in a 400 kb range determined by the most closely associated SNP/indel. An LD block was recognized when the top 95% confidence intervals of the D' value exceeded 0.98 and the lower bounds exceeded 0.70 [67]. Loci with significant variations harbored by LD blocks were then defined as the candidate genes.

Haplotype analysis. With about 7.3× depth of genome coverage, we constructed the haplotyping of the identified candidate genes. Nucleotide polymorphisms on the target genes were captured according to the rice reference genome (IRGSP 1.0). The orthologous genes of the target candidate genes in several other plants were provided by Ensembl Plants (http://plants.ensembl.org). Alignments of orthologous gene sequences were conducted using Geneious (http://www.geneious.com) [68]. In addition, the TCS [69] haplotype network was conducted by PopART v 1.7 [70].

Gene expression analysis by qRT-PCR. Germinated seeds with shoot and root after 10 days in control (H_2O) and salt (200 mM NaCl) conditions were collected and used for expression analysis. Total RNA was prepared using an RNA extraction kit (Qiagen, Hilden, Germany). cDNA was synthesized according to the manufacturer's instructions using the PrimeScript™ RT reagent Kit (TaKaRa, Shiga, Japan). Real-time PCR was carried out using the SYBR Green method with the primers of OsMADS31 (MADS-F: TGGCTTCACTGACTCTGCAA, MADS-R: TACATACCCGGCTGTGCATC). Relative expression levels were calculated using the $2^{-\Delta\Delta CT}$ method [71] with *Ubiquitin 5* (*UBQ 5*) as the internal control [72] under three replicated tests.

Author Contributions: Conceptualization, Y.-J.P. Formal analysis, Q.H. Funding acquisition, Y.-J.P. Investigation, M.-Y.Y. and F.L. Methodology, M.-Y.Y. and B.C. Project administration, K.-W.K. and Y.-J.P. Software, W.T. and E.-B.H. Writing–original draft, J.Y. and W.Z.

Acknowledgments: This work was supported by the National Research Foundation of Korea (NRF) grant funded by the Korea government (MSIT) (NRF-2017R1A2B3011208). This work was carried out with the support of "Cooperative Research Program for Agriculture Science and Technology Development (Project No. PJ013405)" Rural Development Administration, Republic of Korea.

Abbreviations

GWAS	genome-wide association study
LD	linkage disequilibrium
NGS	next-generation sequencing
SNP	single-nucleotide polymorphism
INDEL	insertion and deletion
MAF	minor allele frequency
RL	root length
SL	shoot length
RS	root/shoot ratio
GP	Germination percentage
GE	Germination energy
GI	Germination index
CMLM	compressed mixed linear model
PCA	Principal component analysis

References

1. Sakadevan, K.; Nguyen, M.L. Extent, impact, and response to soil and water salinity in arid and semiarid regions. *Adv. Agron.* **2010**, *109*, 55.

2. Mather, K.A.; Caicedo, A.L.; Polato, N.R.; Olsen, K.M.; McCouch, S.; Purugganan, M.D. The extent of linkage disequilibrium in rice (*Oryza sativa* L.). *Genetics* **2007**, *177*, 2223–2232. [CrossRef] [PubMed]

3. Wang, Z.; Chen, Z.; Cheng, J.; Lai, Y.; Wang, J.; Bao, Y.; Huang, J.; Zhang, H. QTL analysis of Na^+ and K^+ concentrations in roots and shoots under different levels of NaCl stress in rice (*Oryza sativa* L.). *PLoS ONE* **2012**, *7*, e51202. [CrossRef] [PubMed]

4. Almansouri, M.; Kinet, J.M.; Lutts, S. Effect of salt and osmotic stresses on germination in durum wheat (*Triticum durum* Desf.). *Plant and Soil* **2001**, *231*, 243–254. [CrossRef]

5. Rehman, S.; Harris, P.J.; Bourne, W.F.; Wilkin, J. The relationship between ions, vigour and salinity tolerance of acacia seeds. *Plant Soil* **2000**, *220*, 229–233. [CrossRef]

6. Xu, S.; Hu, B.; He, Z.; Ma, F.; Feng, J.; Shen, W.; Yang, J. Enhancement of salinity tolerance during rice seed germination by presoaking with hemoglobin. *Int. J. Mol. Sci.* **2011**, *12*, 2488–2501. [CrossRef] [PubMed]

7. Khodarahmpour, Z.; Ifar, M.; Motamedi, M. Effects of NaCl salinity on maize (*Zea mays* L.) at germination and early seedling stage. *Afr. J. Biotechnol.* **2014**, *11*, 298–304. [CrossRef]

8. Mutlu, F.; Bozcuk, S. Salinity-induced changes of free and bound polyamine levels in sunflower (*Helianthus annuus* L.) roots differing in salt tolerance. *Pak. J. Bot.* **2007**, *39*, 1097–1102.

9. Ulfat, M.; Athar, H.U.R.; Ashraf, M.; Akram, N.A.; Jamil, A. Appraisal of physiological and biochemical selection criteria for evaluation of salt tolerance in canola (*Brassica napus* L.). *Pak. J. Bot.* **2007**, *39*, 1593–1608.

10. Khan, M.A.; Weber, D.J. *Ecophysiology of High Salinity Tolerant Plants*; Springer Science & Business Media: New York, NY, USA, 2006.

11. Koyro, H.W. Ultrastructural Effects of Salinity in Higher Plants. In *Salinity, Environment-Plants-Molecules*; Springer: New York, NY, USA, 2002; ISBN 978-1-4020-0492-6.

12. Gomes-Filho, E.; Lima, C.R.F.M.; Costa, J.H.; da Silva, A.C.M.; Lima, M.d.G.S.; de Lacerda, C.F.; Prisco, J.T. Cowpea ribonuclease, properties and effect of NaCl-salinity on its activation during seed germination and seedling establishment. *Plant Cell Rep.* **2008**, *27*, 147–157. [CrossRef] [PubMed]

13. Khan, M.A.; Rizvi, Y. Effect of salinity, temperature, and growth regulators on the germination and early seedling growth of *Atriplex griffithii* var. stocksii. *Can. J. Bot.* **1994**, *72*, 475–479. [CrossRef]

14. Yupsanis, T.; Moustakas, M.; Eleftheriou, P.; Damianidou, K. Protein phosphorylation-dephosphorylation in alfalfa seeds germinating under salt stress. *J. Plant Physiol.* **1994**, *143*, 234–240. [CrossRef]

15. Othman, Y.; Al-Karaki, G.; Al-Tawaha, A.; Al-Horani, A. Variation in germination and ion uptake in barley genotypes under salinity conditions. *World J. Agric. Sci.* **2006**, *2*, 11–15.

16. Wahid, A.; Rasul, E.; Rao, A.U.R. *Germination of Seeds and Propagules under Salt Stress*; Handbook of Plant and Crop Stress: Boca Raton, FL, USA, 1999; Volume 2, pp. 153–167.

17. Foolad, M.; Hyman, J.; Lin, G. Relationships between cold-and salt-tolerance during seed germination in tomato, Analysis of response and correlated response to selection. *Plant Breed.* **1999**, *118*, 49–52. [CrossRef]

18. Wang, Z.F.; Wang, J.F.; Bao, Y.M.; Wu, Y.Y.; Zhang, H.S. Quantitative trait loci controlling rice seed germination under salt stress. *Euphytica* **2011**, *178*, 297–307. [CrossRef]

19. Csanádi, G.; Vollmann, J.; Stift, G.; Lelley, T. Seed quality QTLs identified in a molecular map of early maturing soybean. *Theor. Appl. Genet.* **2001**, *103*, 912–919. [CrossRef]

20. Bai, C.; Liang, Y.; Hawkesford, M.J. Identification of QTLs associated with seedling root traits and their correlation with plant height in wheat. *J. Exp. Bot.* **2013**, *64*, 1745–1753. [CrossRef] [PubMed]

21. DeRose-Wilson, L.; Gaut, B.S. Mapping salinity tolerance during *Arabidopsis thaliana* germination and seedling growth. *PLoS ONE* **2011**, *6*, e22832. [CrossRef] [PubMed]

22. Basnet, R.K.; Duwal, A.; Tiwari, D.N.; Xiao, D.; Monakhos, S.; Bucher, J.; Visser, R.G.F.; Groot, S.P.C.; Bonnema, G.; Maliepaard, C. Quantitative trait locus analysis of seed germination and seedling vigor in *Brassica rapa* reveals QTL hotspots and epistatic interactions. *Front. Plant Sci.* **2015**, *6*, 1032. [CrossRef] [PubMed]

23. Hoang, T.M.L.; Tran, T.N.; Nguyen, T.K.T.; Williams, B.; Wurm, P.; Bellairs, S.; Mundree, S. Improvement of salinity stress tolerance in rice, challenges and opportunities. *Agronomy* **2016**, *6*, 54. [CrossRef]

24. Wang, Z.; Wang, J.; Bao, Y.; Wang, F.; Zhang, H. Quantitative trait loci analysis for rice seed vigor during the germination stage. *Zhejiang Univ. Sci. B (Biomed. & Biotechnol.)* **2010**, *11*, 958–964.

25. Cheng, J.; He, Y.; Yang, B.; Lai, Y.; Wang, Z.; Zhang, H. Association mapping of seed germination and seedling growth at three conditions in *indica* rice (*Oryza sativa* L.). *Euphytica* **2015**, *206*, 103–115. [CrossRef]

26. Abe, A.; Takagi, H.; Fujibe, T.; Aya, K.; Kojima, M.; Sakakibara, H. OsGA20ox1.; a candidate gene for a major QTL controlling seedling vigor in rice. *Theor. Appl. Genet.* **2012**, *125*, 647–657. [CrossRef] [PubMed]

27. Zheng, H.; Liu, B.; Zhao, H.; Wang, J.; Liu, H.; Sun, J.; Xing, J.; Zou, D. Identification of QTLs for salt tolerance at the germination and early seedling stage using linkage and association analysis in *japonica* rice. *Chin. J. Rice Sci.* **2014**, *28*, 358–366.

28. Huang, X.; Wei, X.; Sang, T.; Zhao, Q.; Feng, Q.; Zhao, Y.; Li, C.; Zhu, C.; Lu, T.; Zhang, Z.; et al. Genome-wide association studies of 14 agronomic traits in rice landraces. *Nat. Genet.* **2010**, *42*, 961–967. [CrossRef] [PubMed]

29. Zhao, K.; Tung, C.W.; Eizenga, G.C.; Wright, M.H.; Ali, M.L.; Price, A.H.; Norton, G.J.; Islam, M.R.; Reynolds, A.; Mezey, J.; et al. Genome-wide association mapping reveals a rich genetic architecture of complex traits in *Oryza sativa*. *Nat. Commun.* **2011**, *2*, 467. [CrossRef] [PubMed]

30. Zhao, W.; Park, E.J.; Chung, J.W.; Park, Y.J.; Chung, I.M.; Ahn, J.K.; Kim, G.H. Association analysis of the amino acid contents in rice. *J. Integr. Plant Biol.* **2009**, *51*, 1126–1137. [CrossRef] [PubMed]

31. Li, G.; Na, Y.W.; Kwon, S.W.; Park, Y.J. Association analysis of seed longevity in rice under conventional and high-temperature germination conditions. *Plant Syst. Evol.* **2014**, *300*, 389–402. [CrossRef]

32. Huang, X.; Zhao, Y.; Wei, X.; Li, C.; Wang, A.; Zhao, Q.; Li, W.; Guo, Y.; Deng, L.; Zhu, C.; et al. Genome-wide association study of flowering time and grain yield traits in a worldwide collection of rice germplasm. *Nat. Genet.* **2012**, *44*, 32–39. [CrossRef] [PubMed]

33. Ma, X.; Feng, F.; Wei, H.; Mei, H.; Xu, K.; Chen, S.; Li, T.; Liang, X.; Liu, H.; Luo, L. Genome-wide association study for plant height and grain yield in rice under contrasting moisture regimes. *Front. Plant Sci.* **2016**, *7*, 1801. [CrossRef] [PubMed]

34. Kumar, V.; Singh, A.; Mithra, S.V.; Krishnamurthy, S.L.; Parida, S.K.; Jain, S.; Tiwari, K.K.; Kumar, P.; Rao, A.R.; Sharma, S.K.; et al. Genome wide association mapping of salinity tolerance in rice (*Oryza sativa*). *DNA Res.* **2015**, *22*, 133–145. [CrossRef] [PubMed]

35. Yu, J.; Zhao, W.; He, Q.; Kim, T.S.; Park, Y.J. Genome-wide association study and gene set analysis for understanding candidate genes involved in salt tolerance at the rice seedling stage. *Mol. Genet. Genomics* **2017**, *292*, 1391–1403. [CrossRef] [PubMed]

36. Shi, Y.; Gao, L.; Wu, Z.; Zhang, X.; Wang, M.; Zhang, C.; Zhang, F.; Zhou, Y.; Li, Z. Genome-wide association study of salt tolerance at the seed germination stage in rice. *BMC Plant Biol.* **2017**, *17*, 92. [CrossRef] [PubMed]

37. Naveed, S.A.; Zhang, F.; Zhang, J.; Zheng, T.Q.; Meng, L.J.; Pang, Y.L.; Xu, J.L.; Li, Z.K. Identification of QTN and candidate genes for salinity tolerance at the germination and seedling stages in rice by genome-wide association analyses. *Sci. Rep.* **2018**, *8*, 6505. [CrossRef] [PubMed]

38. Lipka, A.E.; Tian, F.; Wang, Q.; Peiffer, J.; Li, M.; Bradbury, P.J.; Gore, M.A.; Buckler, E.S.; Zhang, Z. GAPIT, genome association and prediction integrated tool. *Bioinformatics* **2012**, *28*, 2397–2399. [CrossRef] [PubMed]

39. Kapoor, M.; Arora, R.; Lama, T.; Nijhawan, A.; Khurana, J.P.; Tyagi, A.K.; Kapoor, S. Genome-wide identification, organization and phylogenetic analysis of Dicer-like, argonaute and RNA-dependent RNA polymerase gene families and their expression analysis during reproductive development and stress in rice. *BMC Genom.* **2008**, *9*, 451. [CrossRef] [PubMed]

40. Ricachenevsky, F.K.; Sperotto, R.A.; Menguer, P.K.; Sperb, E.R.; Lopes, K.L.; Fett, J.P. ZINC-INDUCED FACILITATOR-LIKE family in plants, lineage-specific expansion in monocotyledons and conserved genomic and expression features among rice (*Oryza sativa*) paralogs. *BMC Plant Biol.* **2011**, *11*, 20. [CrossRef] [PubMed]

41. Okada, T.; Nakayama, H.; Shinmyo, A.; Yoshida, K. Expression of OsHAK genes encoding potassium ion transporters in rice. *Plant Biotechnol.* **2008**, *25*, 241–245. [CrossRef]

42. Wang, J.; Chen, L.; Wang, Y.; Zhang, J.; Liang, Y.; Xu, D. A computational systems biology study for understanding salt tolerance mechanism in rice. *PLoS ONE* **2013**, *8*, e64929. [CrossRef] [PubMed]

43. Arora, R.; Agarwal, P.; Ray, S.; Singh, A.K.; Singh, V.P.; Tyagi, A.K.; Kapoor, S. MADS-box gene family in rice, genome-wide identification, organization and expression profiling during reproductive development and stress. *BMC Genom.* **2007**, *8*, 242. [CrossRef] [PubMed]

44. Zeng, L.; Shannon, M.; Grieve, C. Evaluation of salt tolerance in rice genotypes by multiple agronomic parameters. *Euphytica* **2002**, *127*, 235–245. [CrossRef]

45. Ferdose, J.; Kawasaki, M.; Taniguchi, M.; Miyake, H. Differential sensitivity of rice cultivars to salinity and its relation to ion accumulation and root tip structure. *Plant Prod. Sci.* **2009**, *12*, 453–461. [CrossRef]

46. Lee, K.S.; Choi, W.Y.; Ko, J.C.; Kim, T.S.; Gregorio, G.B. Salinity tolerance of *japonica* and *indica* rice (*Oryza sativa* L.) at the seedling stage. *Planta.* **2003**, *216*, 1043–1046. [CrossRef] [PubMed]

47. Platten, J.D.; Egdane, J.A.; Ismail, A.M. Salinity tolerance.; Na$^+$ exclusion and allele mining of HKT1;5 in *Oryza sativa* and *O. glaberrima*, many sources.; many genes.; one mechanism? *BMC Plant Biol.* **2013**, *13*, 32. [CrossRef] [PubMed]

48. Leon, T.B.; Steven, L.; Prasanta, K.S. Molecular dissection of seedling salinity tolerance in rice (*Oryza sativa* L.) using a high-density GBS-based SNP linkage map. *Rice* **2016**, *9*, 52. [CrossRef] [PubMed]

49. Lin, H.X.; Yanagihara, S.; Zhuang, J.Y.; Senboku, T.; Zheng, K.L.; Yashima, S. Identification of QTLs for salt tolerance in rice via molecular markers. *Chin. J. Rice Sci.* **1998**, *12*, 72–78.

50. Cai, H.W.; Morishima, H. QTL clusters reflect character associations in wild and cultivated rice. *Theor. Appl. Genet.* **2002**, *104*, 1217–1228. [PubMed]

51. Courtois, B.; Audebert, A.; Dardou, A.; Roques, S.; Ghneim-Herrera, T.; Droc, G.; Frouin, J.; Rouan, L.; Goze, E.; Kilian, A.; et al. Genome-wide association mapping of root traits in a *japonica* rice panel. *PLoS ONE* **2013**, *8*, e78037. [CrossRef] [PubMed]

52. Chen, C.; DeClerck, G.; Tian, F.; Spooner, W.; McCouch, S.; Buckler, E. PICARA.; an analytical pipeline providing probabilistic inference about a priori candidates genes underlying genome-wide association QTL in plants. *PLoS ONE* **2012**, *7*, e46596. [CrossRef] [PubMed]

53. Sur, I.; Tuupanen, S.; Whitington, T.; Aaltonen, L.A.; Taipale, J. Lessons from functional analysis of genome-wide association studies. *Cancer Res.* **2013**, *73*, 4180–4184. [CrossRef] [PubMed]

54. Xu, X.; Liu, X.; Ge, S.; Jensen, J.D.; Hu, F.; Li, X.; Dong, Y.; Gutenkunst, R.N.; Fang, L.; Huang, L.; et al. Resequencing 50 accessions of cultivated and wild rice yields markers for identifying agronomically important genes. *Nat. Biotechnol.* **2012**, *30*, 105–111. [CrossRef] [PubMed]

55. Reddy, I.N.B.L.; Kim, B.K.; Yoon, I.S.; Kim, K.H.; Kwon, T.R. Salt tolerance in rice, focus on mechanisms and approaches. *Rice Sci.* **2017**, *24*, 123–144. [CrossRef]

56. Yang, Y.; Zhong, J.; Ouyang, Y.D.; Yao, J. The integrative expression and co-expression analysis of the AGO gene family in rice. *Gene* **2013**, *528*, 221–235. [CrossRef] [PubMed]

57. Inahashi, H.; Shelley, I.J.; Yamauchi, T.; Nishiuchi, S.; Takahashi-Nosaka, M.; Matsunami, M.; Ogawa, A.; Noda, Y.; Inukai, Y. OsPIN2, which encodes a member of the auxin efflux carrier proteins, is involved in root elongation growth and lateral root formation patterns via the regulation of auxin distribution in rice. *Physiol. Plantarum* **2018**. [CrossRef] [PubMed]

58. Rebecca, M.D.; Patrick, A.R.; Patricia, M.M.; Jan, E.L. Germins, a diverse protein family important for crop improvement. *Plant Sci.* **2009**, *177*, 499–510.

59. Kothari, K.S.; Dansana, P.K.; Giri, J.; Tyagi, A.K. Rice stress associated protein 1 (OsSAP1) interacts with aminotransferase (OsAMTR1) and pathogenesis-related 1a protein (OsSCP) and regulates abiotic stress responses. *Front. Plant Sci.* **2016**, *7*, 1057. [CrossRef] [PubMed]

60. Konishi, S.; Izawa, T.; Lin, SY.; Ebana, K.; Fukuta, Y.; Sasaki, T.; Yano, M. An SNP caused loss of seed shattering during rice domestication. *Science* **2006**, *312*, 1392–1396. [CrossRef] [PubMed]

61. Hu, B.; Wang, W.; Ou, S.; Tang, J.; Li, H.; Che, R.; Zhang, Z.; Chai, X.; Wang, H.; Wang, Y. Variation in NRT1.1B contributes to nitrate-use divergence between rice subspecies. *Nat. Genet.* **2015**, *47*, 834–838. [CrossRef] [PubMed]

62. Kim, K.W.; Chung, H.K.; Cho, G.T.; Ma, K.H.; Chandrabalan, D.; Gwag, J.G.; Kim, T.S.; Cho, E.G.; Park, Y.J. PowerCore, a program applying the advanced M strategy with a heuristic search for establishing core sets. *Bioinformatics* **2007**, *23*, 2155–2162. [CrossRef] [PubMed]

63. Zhao, W.; Cho, G.T.; Ma, K.H.; Chung, J.W.; Gwag, J.G.; Park, Y.J. Development of an allele-mining set in rice using a heuristic algorithm and SSR genotype data with least redundancy for the post-genomic era. *Mol. Breed.* **2010**, *26*, 639–651. [CrossRef]

64. Kawahara, Y.; de la Bastide, M.; Hamilton, J.P.; Kanamori, H.; McCombie, W.R.; Ouyang, S.; Schwartz, D.C.; Tanaka, T.; Wu, J.; Zhou, S.; et al. Improvement of the *Oryza sativa* Nipponbare reference genome using next generation sequence and optical map data. *Rice* **2013**, *6*, 4. [CrossRef] [PubMed]

65. Kim, T.S.; He, Q.; Kim, K.W.; Yoon, M.Y.; Ra, W.H.; Li, F.P.; Tong, W.; Yu, J.; Oo, W.H.; Choi, B.; et al. Genome-wide resequencing of KRICE_CORE reveals their potential for future breeding.; as well as functional and evolutionary studies in the post-genomic era. *BMC Genom.* **2016**, *17*, 408. [CrossRef] [PubMed]

66. Bradbury, P.J.; Zhang, Z.; Kroon, D.E.; Casstevens, T.M.; Ramdoss, Y.; Buckler, E.S. TASSEL, software for association mapping of complex traits in diverse samples. *Bioinformatics* **2007**, *23*, 2633–2635. [CrossRef] [PubMed]

67. Gabriel, S.B.; Schaffner, S.F.; Nguyen, H.; Moore, J.M.; Roy, J.; Blumenstiel, B.; Higgins, J.; DeFelice, M.; Lochner, A.; Faggart, M. The structure of haplotype blocks in the human genome. *Science* **2002**, *296*, 2225–2229. [CrossRef] [PubMed]

68. Kearse, M.; Moir, R.; Wilson, A.; Stones-Havas, S.; Cheung, M.; Sturrock, S.; Buxton, S.; Cooper, A.; Markowitz, S.; Duran, C. Geneious basic, an integrated and extendable desktop software platform for the organization and analysis of sequence data. *Bioinformatics* **2012**, *28*, 1647–1649. [CrossRef] [PubMed]

69. Clement, M.; Posada, D.; Crandall, K.A. TCS, a computer program to estimate gene genealogies. *Mol. Ecol.* **2000**, *9*, 1657–1659. [CrossRef] [PubMed]

70. Leigh, J.W.; Bryant, D. PopART, Full-feature software for haplotype network construction. *Methods Ecol. Evol.* **2015**, *6*, 1110–1116. [CrossRef]

71. Livak, K.J.; Schmittgen, T.D. Analysis of relative gene expression data using real-time quantitative PCR and the $2^{-\Delta\Delta CT}$ method. *Methods* **2001**, *25*, 402–408. [CrossRef] [PubMed]

72. Jain, M.; Nijhawan, A.; Tyagi, A.K.; Khurana, J.P. Validation of housekeeping genes as internal control for studying gene expression in rice by quantitative real-time PCR. *Biochem. Biophys. Res. Commun.* **2006**, *345*, 646–651. [CrossRef] [PubMed]

4

Overexpression of a Novel *ROP* Gene from the Banana (*MaROP5g*) Confers Increased Salt Stress Tolerance

Hongxia Miao [1,†], **Peiguang Sun** [2,†], **Juhua Liu** [1], **Jingyi Wang** [1], **Biyu Xu** [1,*] and **Zhiqiang Jin** [1,2,*]

[1]   Key Laboratory of Biology and Genetic Resources of Tropical Crops, Ministry of Agriculture, Institute of Tropical Bioscience and Biotechnology, Chinese Academy of Tropical Agricultural Sciences, Xueyuan Road 4, Haikou 571101, China; miaohongxia@itbb.org.cn (H.M.); liujuhua@itbb.org.cn (J.L.); wangjingyi@itbb.org.cn (J.W.)

[2]   Key Laboratory of Genetic Improvement of Bananas, Hainan Province, Haikou Experimental Station, Chinese Academy of Tropical Agricultural Sciences, Xueyuan Road 4, Haikou 570102, China; sunpeiguang@catas.cn

*   Correspondence: biyuxu@126.com (B.X.); jinzhiqiang@itbb.org.cn (Z.J.)

†   These authors contributed equally to this work.

Abstract: Rho-like GTPases from plants (ROPs) are plant-specific molecular switches that are crucial for plant survival when subjected to abiotic stress. We identified and characterized 17 novel ROP proteins from *Musa acuminata* (MaROPs) using genomic techniques. The identified MaROPs fell into three of the four previously described ROP groups (Groups II–IV), with MaROPs in each group having similar genetic structures and conserved motifs. Our transcriptomic analysis showed that the two banana genotypes tested, Fen Jiao and BaXi Jiao, had similar responses to abiotic stress: Six genes (*MaROP-3b, -5a, -5c, -5f, -5g*, and *-6*) were highly expressed in response to cold, salt, and drought stress conditions in both genotypes. Of these, *MaROP5g* was most highly expressed in response to salt stress. Co-localization experiments showed that the MaROP5g protein was localized at the plasma membrane. When subjected to salt stress, transgenic *Arabidopsis thaliana* overexpressing *MaROP5g* had longer primary roots and increased survival rates compared to wild-type *A. thaliana*. The increased salt tolerance conferred by *MaROP5g* might be related to reduced membrane injury and the increased cytosolic K^+/Na^+ ratio and Ca^{2+} concentration in the transgenic plants as compared to wild-type. The increased expression of salt overly sensitive (SOS)-pathway genes and calcium-signaling pathway genes in *MaROP5g*-overexpressing *A. thaliana* reflected the enhanced tolerance to salt stress by the transgenic lines in comparison to wild-type. Collectively, our results suggested that abiotic stress tolerance in banana plants might be regulated by multiple *MaROPs*, and that *MaROP5g* might enhance salt tolerance by increasing root length, improving membrane injury and ion distribution.

Keywords: banana (*Musa acuminata* L.); ROP; genome-wide identification; abiotic stress; salt stress; *MaROP5g*

1. Introduction

Small GTPases (GTP)-binding proteins, present in a wide variety of eukaryotes, are the central regulators of numerous signal transduction processes [1,2]. These proteins are structurally classified into at least five families, including Rat sarcoma (RAS), Ras homolog (RHO), Rat brain (RAB), RAS-related nuclear (RAN), and adenosine diphosphate (ADP) ribosylation factor (ARF) [1–3]. In all reported eukaryotes, RAS and RHO families are signaling switches, whereas these proteins in other families are primarily involved in the regulation of vesicle and large molecule movement [1,2]. However, higher plants have a unique RHO subfamily of small GTP-binding proteins known as ROPs

(Rho-like GTPases from plants) [4,5]. These proteins are also known as RAS-related C3 botulinum toxin substrates (RACs), due to their sequence similarity to Rac GTPases [6].

ROPs are plant-specific molecular switches that regulate intracellular signaling pathways by cycling between an active form and an inactive, guanosine diphosphate (GDP)-bound form. Biological activities associated with ROPs are diverse, which include polar growth, development, environmental stress responses, and host-pathogen interactions [7–13]. Since the first *ROP* gene was isolated from peas [14], multiple *ROPs*, with typical RhoGEF domains and molecular masses between 21 and 24 kDa, have been described in numerous plant species: 11 in *Arabidopsis thaliana* [15], 9 in *Zea mays* [16], 11 in *Brassica napus* [17], 7 in *Vitis vinifera* [9], 7 in *Oryza sativa* [18], 7 in *Medicago truncatula* [7], 6 in *Nicotiana tabacum* [7], 5 in *Hevea brasiliensis* [19], and 9 in *Solanum lycopersicum* [20]. *ROPs* can be classified into four groups (I–IV) based on their molecular structure and motif conservation [5,7,16].

ROP expression levels and biological functions are affected by various abiotic stressors. When exposed to cold, the transcriptional expression of apple *ROP* increases, leading to a decrease in the concentrations of ethylene and reactive oxygen species in the fruits [21]. In *A. thaliana*, *ROP11* expression affected seed germination, seedling growth, stomatal closure, abscisic acid (ABA)-mediated responses, and drought stress responses [22]. The overexpression of *ROP1* in tobacco increased salt sensitivity in response to salt stress by increasing H_2O_2 production [23]. The Na^+/K^+ ratio in transgenic *A. thaliana* expressing the *Medicago falcate* small GTPase gene (*MfARL1*) was lower than that in wild-type (WT) *A. thaliana*; the transgenic plants consequently had an increased tolerance for salt stress [24]. Knock out of the *A. thaliana* ROP effector (RIC1) increased the survival rate of plants under salt stress by improving the reassembly of depolymerized microtubules [25]. Taken together, these studies have revealed the many important roles of ROPs in the regulation of plant response to abiotic stresses.

The banana (*Musa acuminata*) is one of the most intensively produced and globally important fruit crops [26]. As a large monocotyledonous herbaceous annual, banana plants are frequently harmed or destroyed by various abiotic stress conditions during growth and development [27]. In particular, saline soil is a major abiotic stressor which limits banana cultivation worldwide [28,29]. Genome-wide identification of genes involved in the resistance of banana plants to cold, drought, and salt stress increases our knowledge of plant mechanisms for environmental stress tolerance, while the functional identification of relevant genes acts as a framework for future genetic studies focused on increasing the resistance of the banana plant to these stressors [30–33]. However, genome-wide investigations of the *ROP* gene family, and thus an integrated assessment of the potential functions of this important molecular switch, are still lacking in banana.

To address this information gap, we identified *ROPs* genome-wide in *M. acuminata*, known as *MaROPs*, and analyzed their phylogenetic relationships, gene structures, protein motifs, and expression changes in response to a number of abiotic stressors, including cold, drought, and salt. More importantly, we noted that the expression of the *MaROP5g* gene was associated with salt stress in banana. The overexpression of *MaROP5g* in *A. thaliana* conferred increased salt tolerance by lengthening roots, improving recovery of membrane injury and ion distributions. This comprehensive study of *MaROPs* in *M. acuminata* enhances our understanding of *ROPs* in response to abiotic stress conditions in banana plants, and provides a foundation for future studies aiming to improve the abiotic stress resistance of crop plants, especially with respect to salt stress.

2. Results

2.1. Identification and Phylogenetic Analysis of Banana MaROP Genes

We used the basic local alignment search tool (BLAST) and the hidden Markov models (HMM) to identify MaROPs with typical RhoGEF domains (PF00621) in the *M. acuminata* genome, using the sequences of AtROPs and OsROPs as queries [34,35]. We identified 17 MaROPs in the *M. acuminata* genome, and designated these MaROP-2a, -2b, -2c, -3a, -3b, -4, -5a, -5b, -5c, -5d, -5e, -5f, -5h, -5g,

-6, -7a, and -7b, following the nomenclature of their respective orthologous proteins in *O. sativa*. The 17 predicted MaROP proteins ranged from 195 amino acid residues (MaROP5d) to 214 amino acid residues (MaROP4), with relative molecular masses between 21.297 kDa (MaROP5d) and 23.784 kDa (MaROP3a), and isoelectric points between 8.61 and 9.43 (Table S1).

To investigate the evolutionary relationships among ROP proteins, we constructed a maximum-likelihood (ML) phylogenetic tree based on our multiple sequence alignment of 17 ROP proteins from *M. acuminata*, 11 from *A. thaliana*, and 7 from *O. sativa*. The MaROP proteins fell into 3 distinct groups (Figure 1): Group II contained 6 MaROPs (MaROP-2a, -2b, -2c, -3a, -3b, and -4), 3 AtROPs (AtROP-9, -10, and -11), and 4 OsROPs (OsAPL-1, -2, -3, and -4); Group III contained MaROP-7a, -7b, AtROP7, and OsROP7; and Group IV contained 9 MaROPs (MaROP-5a, -5b, -5c, -5d, -5e, -5f, -5g, -5h, and -6), 6 AtROPs (AtROP-1, -2, -3, -4, -5, and -6), and 3 OsROPs (OsROP-2, -5, and -6). Group I containing AtROP8 served as an outgroup to the phylogenetic analysis.

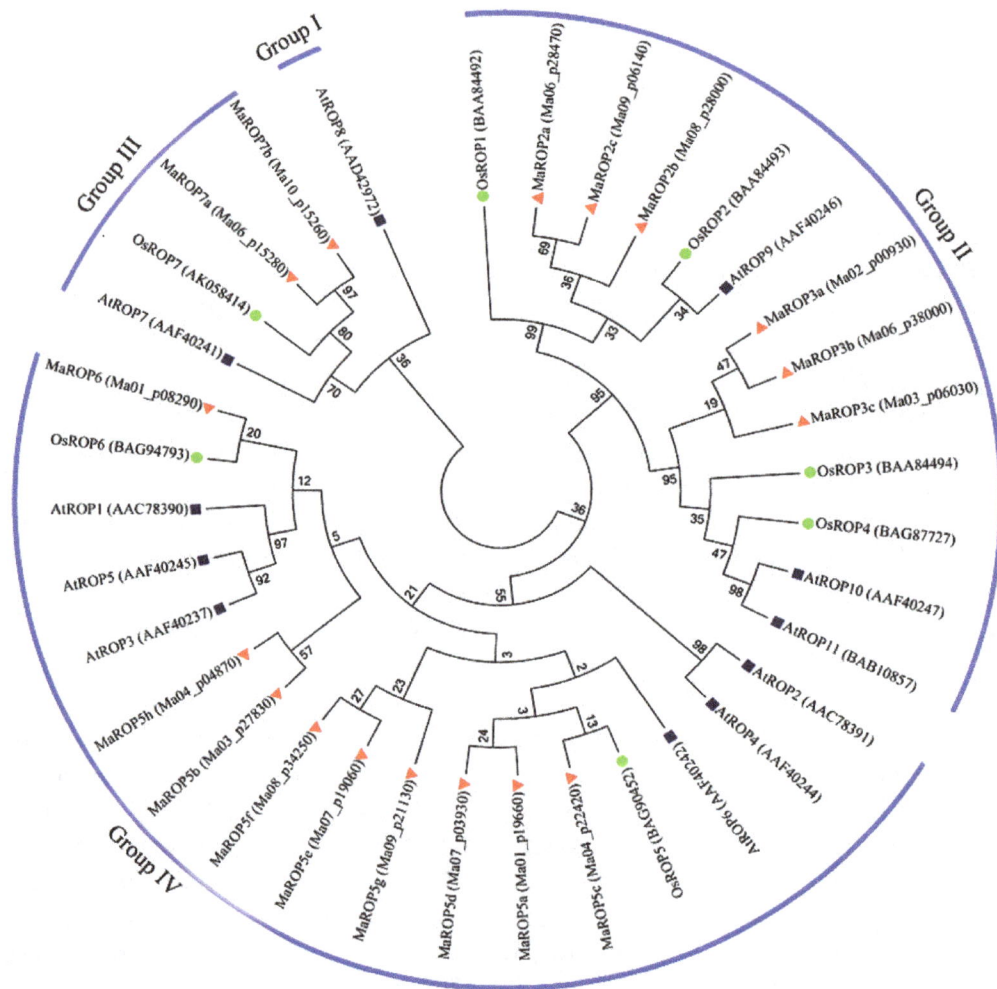

Figure 1. Phylogenetic analysis of Rho-like GTPases from plants (ROPs) from *A. thaliana*, rice, and bananas. The maximum-likelihood phylogenetic tree was drawn with MEGA5.2, using 1000 bootstraps. Four subgroups were identified (Groups I–IV). Circles, squares, and triangles represent ROP proteins from rice, *A. thaliana*, and bananas, respectively.

2.2. Gene Structure and Conserved Motif Analysis of Banana MaROP Genes

Evolutionary analysis supported the classification of the 17 *MaROP* genes into three distinct groups (Groups II–IV), which is consistent with their exon–intron structural divergence within families (Figure 2). Our analysis of the exon–intron structure using the Gene Structure Display Server showed that the *MaROP* genes contained 8 exons in Group II, 7 exons in Group III, and 6–7 exons in Group IV, suggesting the conservation of the exon–intron structures of *MaROPs* within the same group.

To explore MaROPs structural diversity and potential functionality, we analyzed the conserved motifs of the identified MaROPs and predicted their functional annotations. We identified 10 conserved motifs across the 17 MaROP proteins with Multiple Em (Motif Elicitation), which are annotated with InterPro (Figure 2; Table S2). Motifs 1–4 were annotated as a RhoGEF domain (PF00621), the characteristic domain of the ROP protein family. Motifs 1–5 were found across all of the 17 MaROPs, while motif 6 was only present in MaROP-2a, -2b, and -2c; motifs 6–8 were found in MaROP-3a and -3b; and motifs 9 and 10 were found in MaROP4. It is probable that this pattern of conservation and variation across the motifs reflected the evolutionary relatedness and functional divergence of the 17 MaROPs.

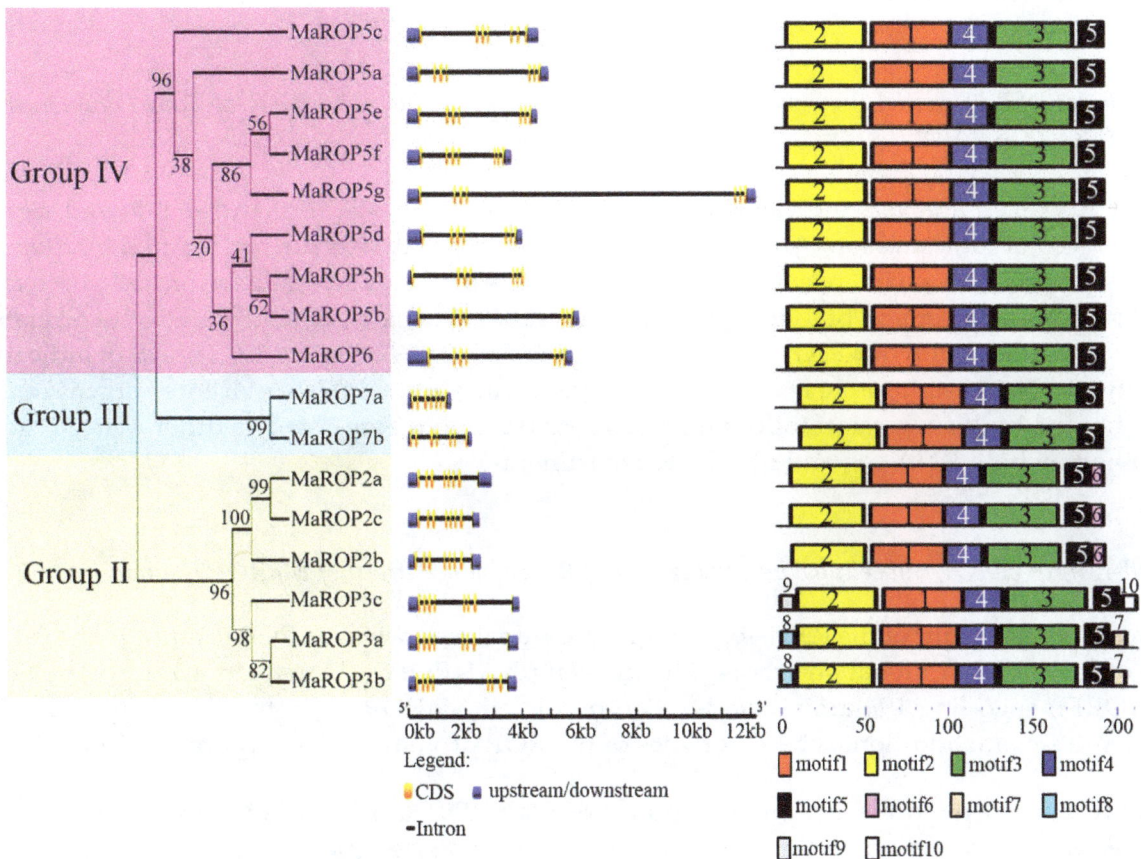

Figure 2. Phylogenetic, gene structure, and motif analyses of banana *Musa acuminata* ROP (MaROP) proteins. MaROPs were classified into Groups II–IV based on their phylogenetic relationships. Exon–intron structure analyses were performed with the Gene Structure Display Server (GSDS). Blue boxes indicate upstream/downstream; yellow boxes indicate exons; black lines indicate introns. All of the proteins were identified using the Multiple EM for Motif Elicitation (MEME) database, using the complete predicted amino acid sequences of each MaROP. Motifs 1–4 were annotated as a RhoGEF domain.

2.3. Expression Analysis of MaROP Genes in Response to Cold, Salt, and Osmotic Stresses

To investigate the response of the *MaROP* genes in response to different abiotic stressors, we analyzed the *MaROP* expression in banana leaves following exposure to cold, salt, and osmotic stress conditions (Figure 3A; Table S3). Compared to the control, significant differences in the expression of

14 *MaROP* genes (82%) were detected following the exposure to abiotic stress treatments (Figure 3A; Table S3). In BaXi Jiao (BX), the expression levels of *MaROP-3b, -5a, -5c, -5f, -5g,* and *-6* were significantly upregulated, as indicated by the fragments per kilobase of exon per million fragments mapped (FPKM) value, which is higher than 2.0 by all three of the abiotic stressors. *MaROP5d* was upregulated by cold treatment only (FPKM > 2.0), and *MaROP2c* was downregulated by osmotic treatment only (FPKM < 0.5). In Fen Jiao (FJ), *MaROP-3b, -5a, -5c, -5f, -5g, -6* were significantly expressed (FPKM > 8.9) in the presence of abiotic stressors. Under osmotic treatment, *MaROP6* was upregulated (FPKM > 2.3) in FJ, but maintained a low level of expression in BX (FPKM < 0.66). In addition, *MaROP-3a* and *-5h* are significantly downregulated in the presence of stress in BX and FJ, compared to control. *MaROP5g* had a higher level of expression (FPKM > 24) in response to salt stress than any of the other *ROPs* across both the BX and FJ genotypes, implying *MaROP5g* might play an important role in the regulation of salt stress tolerance in banana plants.

2.4. Validation of Differential Expression of Six MaROP Genes by Quantitative Real-Time Polymerase Chain Reaction (qRT-PCR) Analysis

Our RNA-seq analysis indicated that *MaROP-3b, -5a, -5c, -5f, -5g,* and *-6* showed significant expressions by abiotic stressors. Such a feature of these six genes was further verified by quantitative real-time polymerase chain reaction (qRT-PCR) analysis. After normalization, all the examined *MaROPs,* with the exception of *MaROP3b* in FJ-salt and *MaROP5c* in FJ-salt, were well-correlated and generally consistent with our RNA-seq analyses ($r = 0.8789–0.9992$; Figure 3B; Table S4), indicating the reliability of our transcriptomic results in both banana varieties. Of particular interest was *MaROP5g,* which showed high expression following salt stress treatment compared to other *MaROP* genes, as determined by both RNA-seq and qRT-PCR experiments.

2.5. Full-Length cDNA, Subcellular Localization, and Expression Pattern of MaROP5g under Salt Stress

Based on the results of the RNA-seq and qRT-PCR analyses, we used PCR to amplify the full-length cDNA of *MaROP5g* from banana roots. The full-length *MaROP5g*cDNA had a 591 bp open reading frame (ORF), encoding 196 amino acids. The predicted MaROP5g protein had a typical RhoGEF domain and several additional characteristics of the ROP protein family (Figure S1; Table S2).

We measured the transcriptional response of *MaROP5g* in BX and FJ plant roots to salt stress. Compared to 0 h (no stress condition), the roots of the BX plants became black following 6 h of salt stress treatment (Figure 4A). However, there was no discernible phenotypic change in the roots of FJ plants under the same treatment (Figure 4C). The expression of *MaROP5g* quickly increased from 0 h, reached the maximum level at 4 h, and then gradually decreased at 6 h (Figure 4B). The expression pattern of *MaROP5g* was similar between FJ and BX (Figure 4D), but *MaROP5g* showed lower expression in FJ compared to BX under salt stress. These results suggested that the regulation of *MaROP5g* expression by salt treatment was genotype-dependent, as the roots of the two tested banana genotypes may have variable sensitivity to salt stress treatments. BX showed more sensitivity than FJ under salt stress treatment.

To localize the MaROP5g protein in the cell, we introduced the *MaROP5g* ORF into a pCAMBIA1304-GFP vector upstream of the *GFP* gene to create a MaROP5g-GFP fusion construct, which was used to transform *A. thaliana*. Co-localization experiments showed that the MaROP5g-GFP (green fluorescent protein) fusion protein was localized to the FM4-64-labeled plasma membrane in *A. thaliana* root tips (Figure 4E).

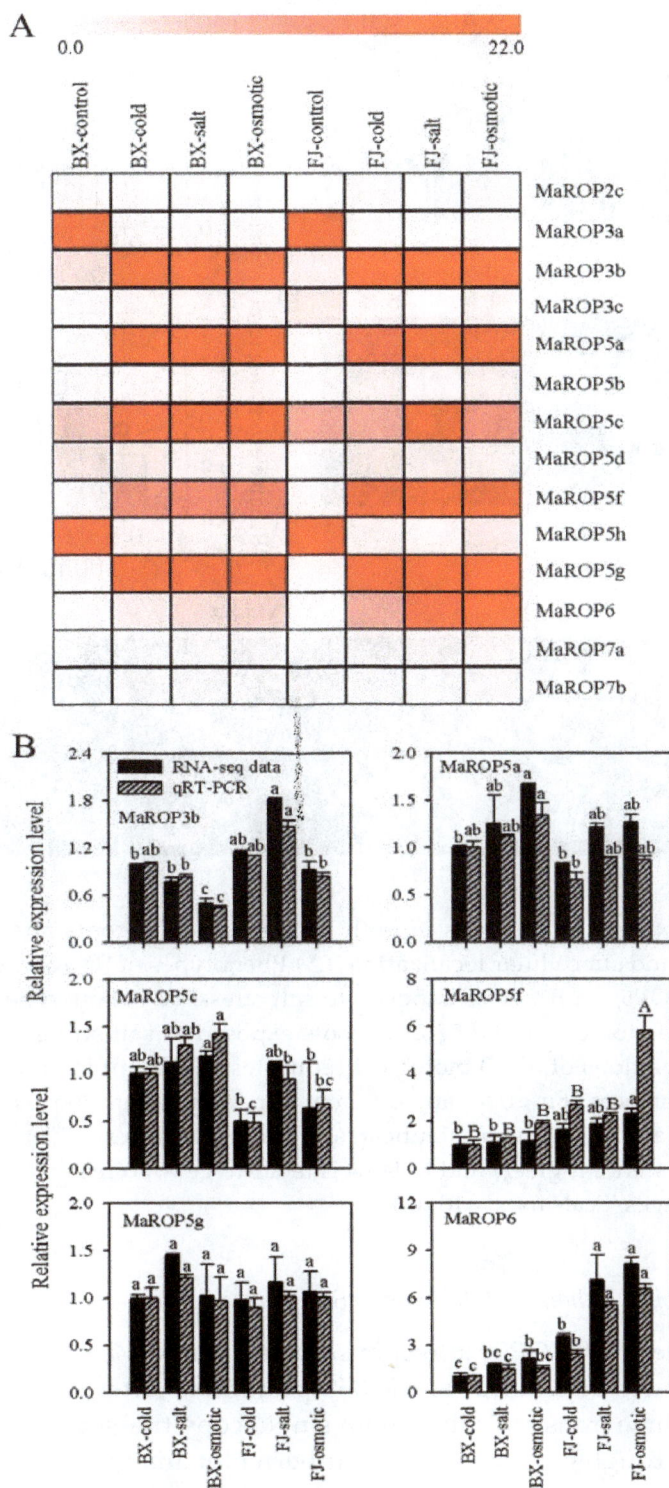

Figure 3. Differential expression of Banana *MaROPs* in response to cold, salt, and osmotic stresses in BaXi Jiao (BX; *M. acuminata* cv. Cavendish; AAA group) and Fen Jiao (FJ; *M. acuminata*; group AAB) banana varieties, as determined with transcriptomic analysis and qRT-PCR. (**A**) Heat map clusters were created based on the fragments per kilobase of exon per million fragments mapped (FPKM) value of the *MaROPs*. Magnitude of differences in gene expression is indicated with a one-color scheme. (**B**) Data are presented as means ± standard deviations of *n* = 3 biological replicates. Different lowercase letters above bars indicate significant differences at $p < 0.05$, and different uppercase letters above bars indicate extremely significant differences at $p < 0.01$, using Duncan's multiple range tests.

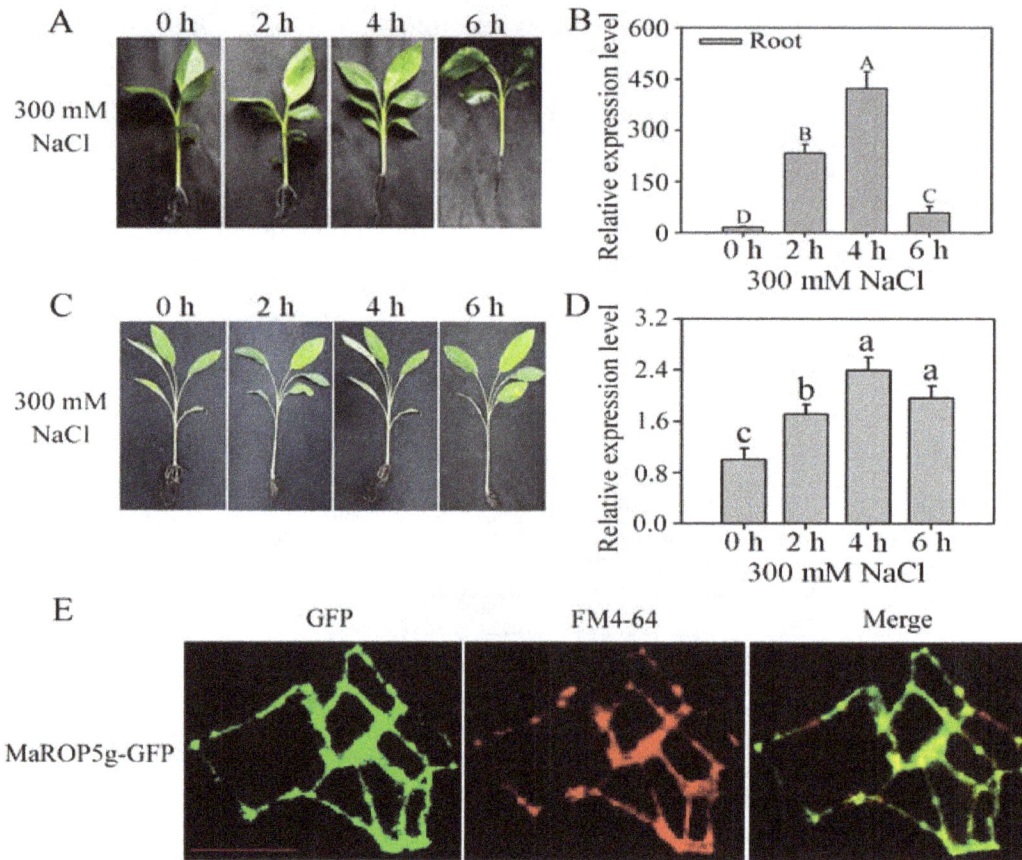

Figure 4. *MaROP5g* expression analyses in both banana varieties roots after different periods of exposure to salt stress and subcellular localization. (**A**) Phenotypes of BX roots exposed to salt stress; (**B**) expression of *MaROP5g* in BX roots exposed to salt stress; (**C**) phenotypes of FJ roots exposed to salt stress; (**D**) expression of *MaROP5g* in FJ roots exposed to salt stress. Data are presented as means ± standard deviations of $n = 3$ biological replicates. Different lowercase letters above bars indicate significant differences at $p < 0.05$, and different uppercase letters above bars indicate extremely significant differences at $p < 0.01$, using Duncan's multiple range tests. (**E**) MaROP5g subcellular localization. GFP fluorescence is green, and FM4-64 is red. Merge was created by merging the GFP and FM4-64 fluorescent images. Scale bars = 10 μm.

2.6. MaROP5g Overexpression Enhances Tolerance to Salt Stress

To investigate the role of *MaROP5g* in response to salt stress, *MaROP5g* was introduced into the pCAMBIA1304 vector under the control of the 35S promoter. After a floral-dip transformation of *A. thaliana*, we analyzed three transgenic lines with single-copy transgene (R3, R8, and R42) from the T3 generation, selected through genomic DNA Southern blot analysis (Figure S2A). The expression level of *MaROP5g* in three transgenic lines was 89–109 folds compared to WT and empty vector (VC), as revealed by qRT-PCR analysis (Figure S2B).

Under no-salt (control) and high-salt conditions, the seed germination rate and root growth were greater in the transgenic seedlings as compared to those in the WT seedlings. After salt treatments ranging from 100 to 200 mM, the transgenic seedlings had grown longer primary roots, as compared to WT seedlings (Figure 5A–C). Furthermore, when adult *A. thaliana* growing in soil were treated with 350 mM salt daily for 15 days, the transgenic lines grew better (Figure 5D,E) and were more likely to survive (Figure 5F), as compared to those of WT. Thus, transgenic *A. thaliana* lines overexpressing *MaROP5g* were more tolerant to salt stress than those of WT.

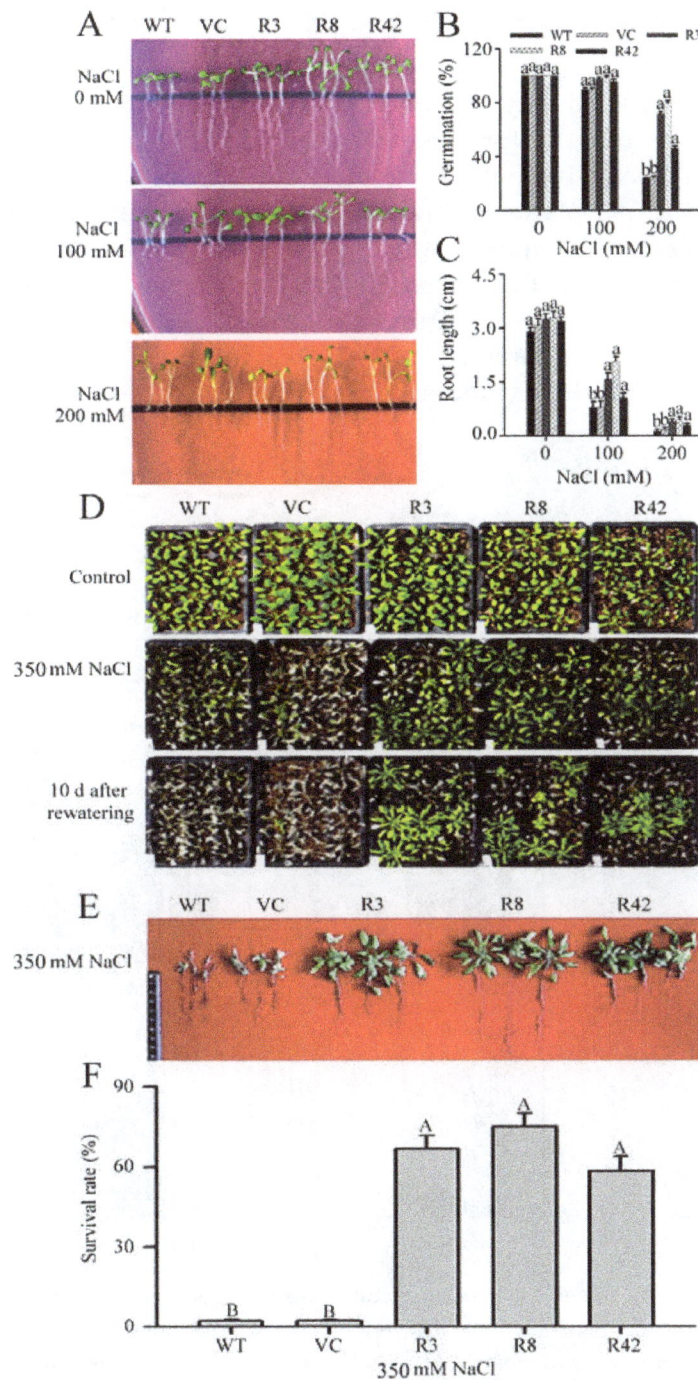

Figure 5. Comparison of wild-type and *A. thaliana* overexpressing *MaROP5g* under standard growing conditions (control) and different salt stress treatments. (**A**) Phenotypes of WT and transgenic lines under control and salt conditions; (**B**) Germination of WT and transgenic lines under control and salt conditions; (**C**) Root length of WT and transgenic lines under control and salt conditions; (**D,E**) Phenotypes of WT and transgenic mature plants under control or salt conditions or after rewatering; (**F**) Survival rates of WT and transgenic mature plants under control or salt conditions. WT: wild-type. VC: vector. R3, R8, R42: *MaROP5g* transgenic plants. ANOVA was used to compare the significance of differences, using Dunnett's tests in the comparison between WT and each overexpression lines. Data are presented as means ± standard deviations of $n = 3$ biological replicates. Different lowercase letters above bars indicate significant differences at $p < 0.05$, different uppercase letters above bars indicate extremely significant differences at $p < 0.01$.

2.7. MaROP5g Overexpression Reduced Malonaldehyde (MDA) Content and Ion Leakage (IL), Increased Ca^{2+} and K^+/Na^+ Ratio under Salt Stress

Malonaldehyde (MDA) is usually employed as an index of oxidative damages in plants [36]. Ion leakage (IL) is an important indicator of membrane injury [36]. To investigate whether *MaROP5g* expression influences MDA and IL content, we measured MDA and IL content in the shoots and roots of transgenic lines and WT plants, following high-salt treatments vs. no-salt control (Figure 6). Following high-salt treatment, MDA content was lower in the shoots and roots of the transgenic plants as compared to those of WT (Figure 6A–G). Similarly, IL value was lower in the shoots and roots of transgenic plants as compared to those of WT under high salt treatment (Figure 6B–H). Taken together, these results suggest that the overexpression of *MaROP5g* in transgenic *A. thaliana* plants may have prevented or reduced membrane injury under salt stress as compared to WT.

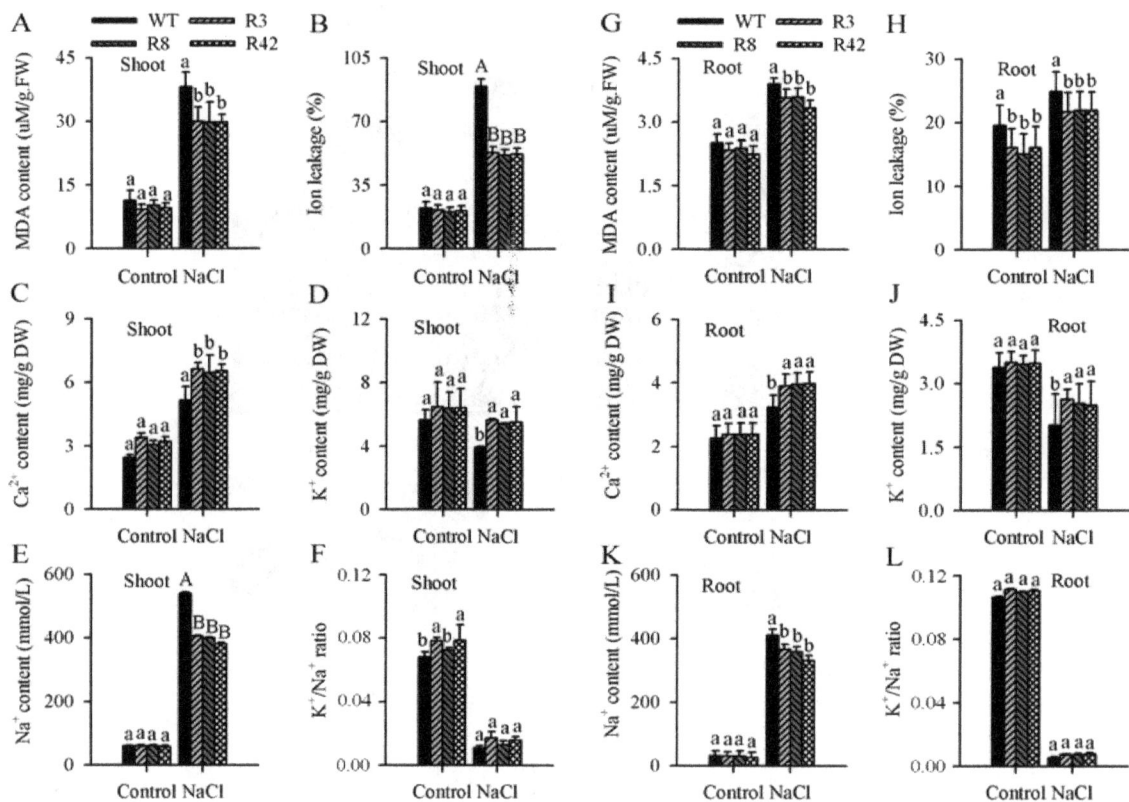

Figure 6. Physiological analyses and ion concentration in roots from wild-type and *A. thaliana* overexpressing *MaROP5g*. (**A–F**) Malonaldehyde content, ion leakage, Ca^{2+} concentration, K^+ concentration, Na^+ ion concentration, and K^+/Na^+ ratio of wild-type and transgenic shoots under normal conditions and salt treatment. (**G–L**) Malonaldehyde content, ion leakage, Ca^{2+} concentration, K^+ concentration, Na^+ ion concentration, and K^+/Na^+ ratio of wild-type and transgenic roots under normal conditions and salt treatment. WT: Wild-type. R3, R8, R42: *MaROP5g* transgenic plants. ANOVA was used to compare the significance of differences, using Student's *t* tests in the comparison between control and NaCl treatment. Data are presented as means ± standard deviations of $n = 3$ biological replicates. Different lowercase letters above bars indicate significant differences at $p < 0.05$, and different uppercase letters above bars indicate extremely significant differences at $p < 0.01$.

Under highly saline conditions, plant cells survive by retaining a high cytosolic Ca^{2+} concentration and a high K^+/Na^+ ratio [37,38]. Under high-salt treatment, the concentrations of Ca^{2+} and K^+ in the shoots or roots of transgenic *A. thaliana* plants were greater (Figure 6C,D), while the Na^+ concentration was lower in the shoots or roots of transgenic *A. thaliana* plants as compared to those of WT (Figure 6E). Therefore, the shoots or roots of the transgenic lines maintained higher K^+/Na^+ ratios than did those of the WT plants during salt treatment (Figure 6F). These results suggested that the overexpression of *MaROP5g* in plants subjected to salt stress increased cellular Ca^{2+} and K^+ accumulation, decreased cellular Na^+ accumulation, and improved the K^+/Na^+ ratio.

2.8. MaROP5g Overexpression Increased the Expression of Salt Overly Sensitive (SOS)-Pathway and Ca²⁺-Sensing Genes

To gain an in-depth understanding of *MaROP5g* function in response to salt stress, we measured the expression of three Salt Overly Sensitive (SOS)-pathway genes (namely *SOS1*, *SOS2*, and *SOS3*) and several genes encoding calcium-signaling pathway proteins, including calcineurin B-like (CBL) proteins, CBL-interacting protein kinases (CIPKs), and calcium-dependent protein kinases (CDPKs) [39], in both WT and *MaROP5g*-overexpressing *A. thaliana* (Figure 7). Under standard growth conditions, we observed no significant differences in the transcription levels of the tested genes in the transgenic lines as compared to those in WT plants. However, under salt stress, the gene expression levels of *SOS1*, *SOS2*, *SOS3*, *CBL*, *CIPK*, and *CDPK* were higher in the transgenic lines as compared to those in WT. This indicated that *MaROP5g* overexpression in response to salt stress led to the up-regulation of both SOS-pathway genes and calcium-signaling pathway genes.

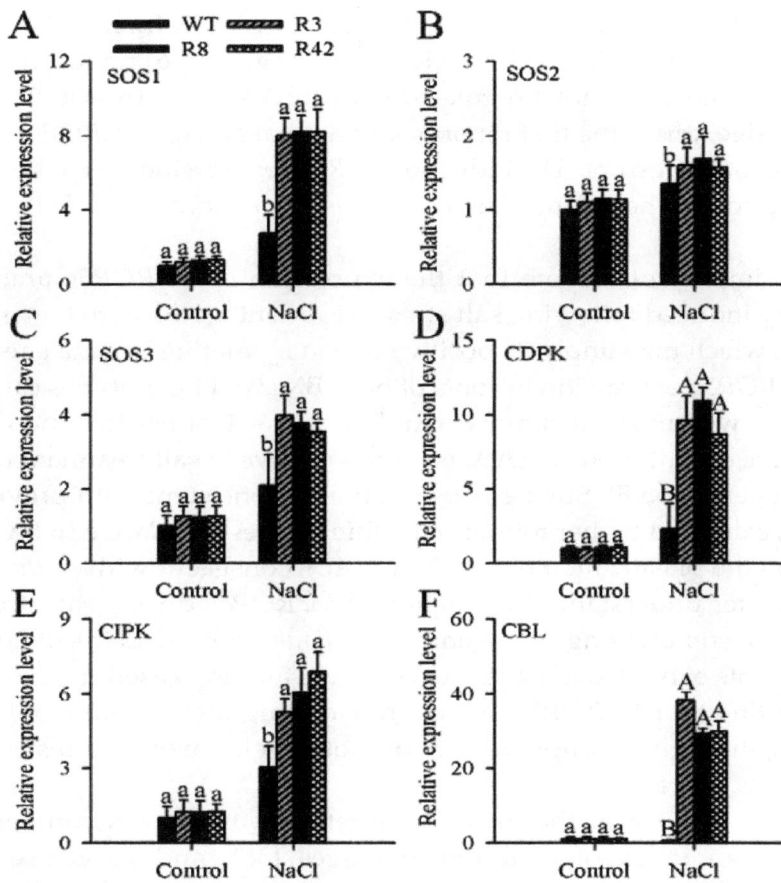

Figure 7. Expression of Salt Overly Sensitive (SOS)- and calcium-signaling genes in wild-type and *A. thaliana* overexpressing *MaROP5g*.WT: Wild-type. R3, R8, R42: *MaROP5g* transgenic plants. (**A–F**) Expression patterns of *SOS1*, *SOS2*, *SOS3*, *CDPK*, *CIPK*, and *CBL* genes in wild-type and transgenic roots under normal conditions and salt treatment. ANOVA was used to compare the significance of differences, using Student's *t* tests in the comparison between control and NaCl treatment. Data are presented as means ± standard deviations of $n = 3$ biological replicates. Different lowercase letters above bars indicate significant differences at $p < 0.05$, and different uppercase letters above bars indicate extremely significant differences at $p < 0.01$.

3. Discussion

Despite its economic and social importance, research on banana plants has generally been slower relative to many other crops, especially with respect to the abiotic stress responses [31,32]. ROP is an important molecular switch involved in plant signal transduction processes, which has been suggested to play crucial roles in the regulation of the environmental stress responses in numerous plant species [8,11,21,40]. We have identified 17 MaROPs by searching the M. acuminata genome, which were classified into three groups (II–IV), following the nomenclature derived for OsROPs [18]. Of the 17 MaROPs, none was categorized into Group I, congruent with ROPs in other higher plants, such as O. sativa [18], Z. mays [16], Medicago truncatula [7], and N. tabacum [7]. The recovered phylogenetic relationships were further supported by our analyses of gene structure and conserved motifs. The MaROP genomic sequences in Groups II–IV were found to contain 6–8 exons and 6–7 introns. Similar structural features have been observed in ROPs of other plant species, including N. tabacum [23], A. thaliana [15], and M. truncatula [7]. Moreover, all of the identified MaROPs had the typical RhoGEF domain (PF00621), and MaROP proteins within each group shared similarly conserved motifs (Figure 2), which is consistent with the observations in M. truncatula [7].

Bananas are extremely sensitive to abiotic stress and can suffer severe losses in yield and quality when exposed to cold, salt, or drought conditions [36]. We found that 82% of the 17 MaROPs showed transcriptional changes following cold, salt, and osmotic stress treatments (Figure 3). Interestingly, except the high expression genes (MaROP-3b, -5a, -5c, -5f, -5g, and -6) under three stress treatments, MaROP-3a and -5h are significantly downregulated in the presence of stress in BX and FJ compared to control. To our knowledge, this is the first report showing that banana MaROPs exhibit extensive and diverse responses to abiotic stressors. The induction of ROP expression by cold, salt, and drought has previously been reported in other plants, such as Malus× domestica Borkh [21], A. thaliana [22], and N. tabacum [23].

It is particularly important to note that the expression of MaROP5g among the 17 MaROP genes was most highly induced following salt stress treatment across both banana genotypes tested (Figures 3 and 4A–D), which may imply its positive role in mediating banana's response to salt stress. Further, although MaROP5g expression in roots of both BX and FJ genotypes can be induced by salt stress treatment, BX showed more sensitivity than FJ under salt stress treatment. This may suggest that BX, with its genome constitution as AAA, is more sensitive to salt treatment in comparison to the B-genome-containing genotype FJ. Such an observation is consistent with previous studies that FJ, with AAB genotypes, exhibited higher tolerance to abiotic stresses relative to BX [27].The MaROP5g protein was located on the plasma membrane (Figure 4E), consistent with A. thaliana ROP2 [41] and rice OsRac5 [18]. To better understand the function of MaROP5g during salt stress, we generated a number of MaROP5g-overexpressing transgenic A. thaliana lines. Under salt stress, the transgenic seedlings and adult plants exhibited a higher survival rate and increased root length as compared to WT (Figure 5), suggesting that MaROP5g overexpression might contribute to the maintenance of a healthy growth status, through the improvement of root development and distributions [24,25], and hence enhance salt stress tolerance.

As cell membranes are one of the primary targets of various environmental stresses, MDA is commonly used as an index of oxidative damages [36], and IL is used as an important indicator of membrane injury in plant research [36]. MDA content and IL were measured to assess the role of MaROP5g overexpression in reducing membrane injury under salt conditions. MaROP5g overexpression resulted in decreased IL and MDA content relative to WT, indicating that MaROP5g-overexpressing plants may experience less membrane injury and maintain a healthy physiological status under salt conditions.

In plants, high K^+ and low Na^+ concentrations are beneficial for the maintenance of physiological processes under salt stress [42]. In recent years, a high cytosolic K^+/Na^+ ratio has become an accepted marker of salinity tolerance [38]. Previous studies have reported that the expression of MfARL1 resulted in a reduced Na^+/K^+ ratio in transgenic A. thaliana as compared to WT, due to a lower accumulation of Na^+ [24]. Under salt stress, *MaROP5g* overexpression decreased the accumulation of cellular Na^+, increased the Ca^{2+} concentration, and improved the K^+/Na^+ ratio in transgenic *A. thaliana* as compared to WT plants (Figure 6). Therefore, the increased salt stress tolerance conferred by *MaROP5g* overexpression may be due not only to the decreased Na^+ accumulation, but also to the increased Ca^{2+} concentration in transgenic lines as compared to WT.

Many different ion transporters and channel proteins, such as SOS, CDPK, CBL, and CIPK, play crucial roles in maintaining ion homeostasis during salt stress [39,42,43]. For example, under salt stress, the SOS1 and SOS2 proteins regulate Na^+/K^+ homeostasis in *A. thaliana*; once the calcium binding protein SOS3 senses an increase in cytosolic calcium concentration, the SOS3–SOS2 protein kinase complex activates the SOS1 ion transporter [42,43]. In addition, calcium (Ca^{2+}), as a second messenger, plays an important role in salt stress processes [39,44]. Ca^{2+} increase can be decoded and recognized by Ca^{2+} sensors, including CBLs, CIPKs, and CDPKs [44]. CBLs recognize the increase in cytosolic Ca^{2+} concentration triggered by Na^+ accumulation [39]. CIPKs and CDPKs may unify and coordinate ionic homeostasis at the cellular and organismal level [44]. We examined the expression of these SOS- and calcium-signaling pathway genes in the *MaROP5g*-overexpressing transgenic *A. thaliana* seedlings in relation to WT seedlings. Following salt treatment, SOS- and calcium-signaling pathway genes were a significant expression in the transgenic seedlings as compared to WT seedlings. This suggested that the *MaROP5g*-overexpressing transgenic plants were more responsive to SOS- and calcium-signaling compared to WT plants, implying that *MaROP5g*-overexpressing plants had improved Na^+ and Ca^{2+} ionic homeostasis under salt stress conditions.

4. Experimental Section

4.1. Plant Materials

BaXi Jiao (BX; *M. acuminata* cv. Cavendish; AAA group) is a triploid banana cultivar that is of high yield and high quality and can be stored for an extended period of time [31,32]. Fen Jiao (FJ; *M. acuminata*; group AAB), another triploid banana cultivar, has good flavor, rapid ripening, and a high tolerance for abiotic stress [31,32]. Both of these banana cultivars were planted and maintained at the banana plantation of the Chinese Academy of Tropical Agricultural Sciences (Danzhou, Hainan, China; 19°11′–19°52′ N, 108°56′–109°46′ E). All of the banana plants were grown in 70% relative humidity at 28 °C, with 200 $\mu mol \cdot m^{-2} \cdot s^{-1}$ light in cycles of 16 h light/8 h dark (Sylvania GRO LUX fluorescent lamps; Utrecht, The Netherlands).

For the salt and drought-simulation experiments, five-leaf stage banana plants of both cultivars were irrigated with 300 $mmol \cdot L^{-1}$ NaCl or 200 $mmol \cdot L^{-1}$ mannitol, respectively, for 7 days, as previously described by Hu et al. [27]. For the cold experiments, five-leaf stage banana plants of both cultivars were subjected to 4 °C for 22 h, as previously described by Hu et al. [27]. For the control experiments, five-leaf stage banana plants of both cultivars were irrigated with equal volume water with the stress groups at 28 °C. After the completion of each treatment, we harvested and immediately froze in liquid nitrogen the leaves and root systems of each plant, which were stored at −80 °C. Twelve five-leaf stage banana plants and three biological replicates were performed for each treatment.

4.2. Identification and Phylogeny of the MaROP Gene Family

The banana ROP protein sequences were downloaded from the DH-Pahang genome database (*M. acuminata*, A-genome, $2n = 22$) (available online: http://banana-genome.cirad.fr) [33]. ROP amino acid sequences from *A. thaliana* (AtROPs) and *O. sativa* (OsROPs) were downloaded from the TAIR (The Arabidopsis Information Resource) (available online: http://www.arabidopsis.org) and RGAP (Rice Genome Annotation Project) (available online: http://rice.plantbiology.msu.edu) databases, respectively. HMMER (available online: http://hmmer.org) was used to predict conserved RhoGEF domains (PF00621; available online: http://pfam.sanger.ac.uk) in the ROP proteins [34]. The basic local alignment search tool (BLAST) (available online: http://www.ncbi.nlm.nih.gov/BLAST/) was used to identify putative MaROPs, based on the sequences of the AtROPs and OsROPs [35]. The conserved domains of the putative MaROPs were identified with the Conserved Domain Database (available online: http://www.ncbi.nlm.nih.gov/cdd) and validated with PFAM (available online: http://pfam.sanger.ac.uk) [45–47]. Identity numbers of all of the putative MaROPs that we have identified was presented in Table S1. All of the MaROP, AtROP, and OsROP sequences were aligned with Multiple Sequence Alignment (MUSCLE), and a bootstrapped ML phylogenetic tree (1000 replicates) was constructed in MEGA 5.2 (available online: http://www.megasoftware.net/) using this alignment [48].

4.3. Protein Properties and Gene Structure

We predicted the molecular masses and isoelectric points of the putative MaROP proteins with the Expert Protein Analysis System database (available online: http://expasy.org/) [49]. We constructed a bootstrapped ML phylogenetic tree (1000 replicates) in MEGA 5.2 software by aligning all MaROP sequences with MUSCLE [47]. MaROP protein motifs were identified with Multiple Em for Motif Elicitation (available online: http://meme-suite.org) and annotated using InterProScan (available online: http://www.ebi.ac.uk/Tools/pfa/iprscan) [50,51]. Structural features of the *MaROP* genes were identified with Gene Structure Display Server (available online: http://gsds.cbi.pku.edu.cn) by comparing the nucleotide sequences to predicted coding regions for all *MaROPs* [52]. *MaROP* promoter sequences were obtained from the banana genome database (available online: http://banana-genome.cirad.fr) [33]. Based on fragments 2000 bp upstream of each *MaROP*, a transcription start site was predicted with the Berkeley *Drosophila* Genome Project database (available online: http://www.fruitfly.org/seq_tools/promoter.html) and the *cis*-acting elements were predicted with PlantCARE (available online: http://bioinformatics.psb.ugent.be/webtools/plantcare/html) [53,54].

4.4. Transcriptomic Analysis

We isolated total RNA from the leaf tissues of the banana seedlings subjected to each of the three treatments (salt, osmosis, and cold) and control (no stress conditions), which were constructed into respective cDNA libraries [31,32]. Deep paired-end sequencing was performed with an Illumina GAII according to manufacturer's instructions. There are two replicates for each sample. The sequencing depth was 5.34X on average. Adaper sequences in the raw reads were removed using FASTX-tookit (Illlumina, San Diego, CA, USA). Using Tophat v.2.0.10, clean reads were mapped to the DH-Pahang genome [33]. The transcriptome assemblies were performed by Cufflinks [27]. Gene expression levels were calculated as fragments per kilobase of exon per million fragments mapped (FPKM). DEGseq was used to identify differently expressed genes [55]. Under three stress conditions, the expression level of each gene was compared with the control. A heat map was created based on the FPKM value of the *MaROPs*, compared to the control by MeV 4.9.0 software (available online: https://sourceforge.net/projects/mev-tm4/).

4.5. QRT-PCR Analysis

The gene expression of *MaROPs* in response to cold, salt, and osmotic (drought) stress was measured with qPCR, using a SYBR Premix ExTaq kit (TaKaRa, Shiga, Japan) on a Stratagene Mx3000P detection system (Stratagene, San Diego, CA, USA). The primer pairs with high specificity and efficiency were selected based on their melting curve and on agarose gel electrophoresis (Table S5). The amplification efficiencies of the primer pairs chosen ranged from 0.9 to 1.1. *MaActin* (EF672732) and *MaUBQ2* (HQ853254) were used as the internal controls. The expression levels of *MaROP* relative to *MaActin* and *MaUBQ2* were calculated with the $2^{-\Delta\Delta CT}$ method [56]. Three biological replicates for each sample were performed.

4.6. Full-Length cDNA of MaROP5g and Gene Expression During Salt Treatment

Based on our RNA-seq results, we selected the *ROP* gene *MaROP5g* for further analysis. The entire coding region of *MaROP5g* was amplified with PCR, using single-stranded cDNA obtained from the roots of banana plants subjected to salt stress as the source template, using a specific primer pair (5′-gcaccatggagatgagcgcgtcgaggt-3′ and 5′-gcgactagtcaatatggagcaacctttc-3′). The resulting *MaROP5g* fragment was verified with DNAMAN (available online: http://www.lynnon.com/) and compared to the genome database of DH-Pahang using BLAST [33,35].

Twelve ex vitro banana plants with uniform growth at the five-leaf stage were selected and divided into four groups for salt treatments, which were irrigated with half-strength Hoagland solution, supplemented with 300 mM NaCl for 0, 2, 4, or 6 h (*n* = 4 per time period), as previously described by Xu et al. [36]. Samples were frozen individually in liquid nitrogen and stored at −70 °C. Compared to the expression of 0 h in BX, the relative expression level of *MaROP5g* at 2, 4, and 6 h under salt stress in BX was calculated. Compared to the expression of 0 h in FJ, the relative expression level of *MaROP5g* at 2, 4, and 6 h under salt stress in FJ banana plants was calculated.

4.7. Subcellular Localization of MaROP5g

The ORF of *MaROP5g* was digested with the restriction enzymes *Nco* I and *Spe* I and inserted into a pCAMBIA1304-GFP expression vector to generate a MaROP5g-GFP fusion protein, under the control of the cauliflower mosaic virus (CaMV) 35S promoter. The recombinant pCAMBIA1304-MaROP5g-GFP plasmid was transferred to *Agrobacterium tumefaciens* strain LBA4404, and used to transform *A. thaliana* through a floral-dip method [57]. Root tips (3–5 mm) of *A. thaliana* seedlings (5-day old) with a stable expression of MaROP5g-GFP were incubated in 1 mL 1/2 Murashige and Skoog (MS) medium containing 10 μg FM4-64 (Invitrogen, Carlsbad, CA, USA) for 5 min at 25 °C, according to the Riqal et al. [58] methods. The GFP (488 nm emission filter) and FM4-64 (543 nm emission filter) signals were visualized using confocal laser scanning microscopy (CLSM; Nikon, A1, Tokyo, Japan). According to the Protein Subcellular Localization Prediction Tool (PSORT) software (available online: www.genscript.com/psort.html) prediction, the XXRR-like motif in the N-terminus of MaROP5g protein was identified as a membrane retention.

4.8. Plant Transformation and Generation of Transgenic Plants

The ORF of *MaROP5g* was digested with the restriction enzymes *Nco* I and *Spe* I, and inserted into a pCAMBIA1304 vector. The recombinant pCAMBIA1304-MaROP5g plasmid was transferred to *A. tumefaciens* strain LBA4404 [57]. Transgenic *A. thaliana* plants were generated using the floral dip-mediated infiltration method [57]. Seeds from T$_0$ transgenic plants were plated in kanamycin selection medium (50 mg·L^{-1}). The homozygous T$_3$ lines were used for further functional investigation of *MaROP5g*.

4.9. Southern Blot Analyses

Genomic DNA isolated from the T$_3$ generation kanamycin-resistant transgenic lines was digested with the *EcoR* I restriction enzyme. A 436 bp region of *MaROP5g* was amplified by PCR, using a pair of specific oligo primers (5′-gtggtggatggtaacacagtta-3′ and 5′-aacctttctgttgctttttttc-3′). Based on this sequence, we prepared a hybridization probe for use with DIG-dUTP (Roche Applied Science, Mannheim, Germany), following the manufacturer's instructions. After hybridization, the HyBond N$^+$ membrane (Amersham) was washed and exposed to X-ray film (Kodak BioMax MS, Kodak Eastman, Rochester, NY, USA), following the method described by Miao et al. [59].

4.10. Salt Stress Treatments in WT and Transgenic Plants

The seeds of both transgenic and WT *A. thaliana* (Columbia ecotype; control) were first vernalized for 2 days at 4 °C in the dark, and surface sterilization in 75% ethanol for 10 min, prior to germination on half-strength MS medium or directly in soil. These *A. thaliana* plants were maintained at 22 °C with 70% humidity and a 16 h light/8 h dark cycle (Sylvania GRO LUX fluorescent lamps; Utrecht, The Netherlands). To analyze *A. thaliana* phenotypes in early seedlings under normal conditions, four day-old seedlings were transferred to 1/2 MS medium for 15 days, photos were taken, and root lengths were measured. To test salt stress tolerance in early seedlings, four day-old seedlings were transferred to either 1/2 MS or 1/2 MS supplemented with 100–200 mM NaCl for 15 days, after which photos were taken and the root length was measured. To test salt stress tolerance in adult plants, 4-week-old *A. thaliana* plants were irrigated with 350 mM NaCl for 15 days, and then photos were taken and survival rates were assessed (leaves fall and roots rot were identified as death). To measure the expression of SOS- and calcium-signal pathway genes in the WT and transgenic lines, 15-day-old seedlings were transferred to 1/2 MS supplemented with 350 mM NaCl for up to 10 h. Whole leaves were used to quantify relative gene expression, using qRT-PCR (see Table S5 for primer sequences).

4.11. Measurement of IL and MDA Content

Four-week-old *A. thaliana* plants were irrigated with 350 mM NaCl for 15 days and leaf samples were collected to examine IL and MDA. IL was detected according to the method described by Xu et al. [36]. Leaf samples were cut into strips and incubated in 10 mL of distilled water at 25 °C for 8 h. The initial conductivity (C1) was determined with a conductivity meter (DDBJ-350). The samples were then boiled for 10 min to yield complete IL. After cooling down, the electrolyte conductivity (C2) was measured. IL was calculated according to the equation: IL (%) = C1/C2:100. MDA content was measured according to the thiobarbituric acid colorimetric method, as described by Xu et al. [36].

4.12. Ca^{2+}, Na^+, and K^+ Concentrations

We irrigated 4-week-old WT and transgenic plants with 350 mM NaCl for 15 days. We then collected the plant roots and washed them with ultrapure water. Plant roots were heated to 105 °C for 10 min and then dried at 80 °C for 48 h. We dissolved 50 mg of each dried sample in 6 mL nitric acid and 2 mL H_2O_2 (30%), and then heated the solution to 180 °C for 15 min. The digested samples were diluted to a total volume of 50 mL with ultrapure water, transferred to clean tubes, and analyzed with atomic absorption spectroscopy (Analyst400, Perkin Elmer, Waltham, MA, USA).

4.13. Statistical Analysis

Three biological replicates for each sample were performed. Statistical analyses were performed using SPSS 19.0 (Chicago, IL, USA). We used analyses of variance (ANOVAs) to compare the significance of differences based on Dunnett's tests or Student's *t* tests. Specifically, Dunnett's tests were used to compare between WT and each overexpression line, while Student's *t* tests were used to compare between control and NaCl treatment. $p < 0.05$ was considered a significant level and $p < 0.01$ an extremely significant level.

5. Conclusions

In this study, for the first time for banana, we identified 17 *MaROP* genes in the *M. acuminata* genome, and classified these genes into three groups (II–IV) based on phylogeny, gene structure, and conserved protein motifs. The expression patterns of the *MaROP* genes in response to abiotic stress as reported here may shed light on the possible involvement of these genes in the regulation of abiotic stress signaling pathways. Of particular interest was *MaROP5g*, the overexpression of which increased the plant's tolerance for salt stress not only by maintaining a healthy growth status, but also by reducing membrane injury and improving ion distribution. Our results lay a foundation for genetic improvements in the banana plant, increasing resistance to various abiotic stressors, particularly salt. It is necessary to point out that these conclusions were drawn from the heterologous expression of banana *MaROP5g* in *A. thaliana* as a model plant system, which may or may not be valid in other plant systems. Further studies are required to characterize the function of *MaROP5g* in banana.

Author Contributions: Conceived and designed the experiments: B.X. and Z.J. Performed the experiments: H.M. and P.S. Analyzed the data: H.M. and P.S. Contributed reagents/materials/analysis tools: H.M., P.S., J.L., J.W., B.X., and Z.J. Wrote the paper: H.M. and B.X.

Acknowledgments: We thank Qing Liu (Commonwealth Scientific and Industrial Research Organization Agriculture and Food, Australia) for technical assistance. This work was supported by the Modern Agro-industry Technology Research System (No. CARS-31), the National Natural Science Foundation of China (NSFC, No. 31401843), the Central Public-interest Scientific Institution Basal Research Fund for Innovative Research Team Program of CATAS (No. 1630052017010), and the Central Public-interest Scientific Institution Basal Research Fund for Chinese Academy of Tropical Agricultural Sciences (No. 1630052016006).

References

1. Takai, Y.; Sasaki, T.; Matozaki, T. Small GTP-binding proteins. *Physiol. Rev.* **2001**, *81*, 153–208. [CrossRef] [PubMed]
2. Ono, E.; Wong, H.L.; Kawasaki, T.; Hasegawa, M.; Kodama, O.; Shimamoto, K. Essential role of the small GTPase Rac in disease resistance of rice. *Proc. Natl. Acad. Sci. USA* **2001**, *98*, 759–764. [CrossRef] [PubMed]
3. Rocha, N.; Payne, F.; Huang-Doran, I.; Sleigh, A.; Fawcett, K.; Adams, C.; Stears, A.; Saudek, V.; O'Rahilly, S.; Barroso, I.; Semple, R.K. The metabolic syndrome-associated small G protein ARL15 plays a role in adipocyte differentiation and adiponectin secretion. *Sci. Rep.* **2017**, *71*, 17593. [CrossRef] [PubMed]
4. Eliáš, M.; Klimeš, V. Rho GTPases: Deciphering the evolutionary history of a complex protein family. *Methods Mol. Biol.* **2012**, *827*, 13–34. [PubMed]
5. Feiquelman, G.; Fu, Y.; Yalovsky, S. RopGTPases structure-function and signaling pathyways. *Plant Physiol.* **2018**, *176*, 57–59. [CrossRef] [PubMed]
6. Winge, P.; Brembu, T.; Bones, A.M. Cloning and characterization of rac-like cDNAs from *Arabidopsis thaliana*. *Plant Mol. Biol.* **1997**, *35*, 483–495. [CrossRef] [PubMed]
7. Liu, W.; Chen, A.M.; Luo, L.; Sun, J.; Cao, L.P.; Yu, G.Q.; Zhu, J.B.; Wang, Y.Z. Characterization and expression analysis of *Medicago truncatula* ROP GTPase family during the early stage of symbiosis. *J. Integr. Plant Biol.* **2010**, *52*, 639–652. [CrossRef] [PubMed]
8. Zheng, Z.L.; Yang, Z. The rop GTPase switch turns on polar growth in pollen. *Trends Plant Sci.* **2000**, *5*, 298–303. [CrossRef]
9. Abbal, P.; Pradal, M.; Sauvage, F.X.; Chatelet, P.; Paillard, S.; Canaguier, A.; Adam-Blondon, A.F.; Tesniere, C. Molecular characterization and expression analysis of the Rop GTPase family in *Vitis vinifera*. *J. Exp. Bot.* **2007**, *58*, 2641–2652. [CrossRef] [PubMed]
10. Zhang, Y.; McCormick, S. The regulation of vesicle trafficking by small GTPases and phospholipids during pollen tube growth. *Sex Plant Reprod.* **2010**, *23*, 87–93. [CrossRef] [PubMed]
11. Huang, J.B.; Liu, H.; Chen, M.; Li, X.; Wang, M.; Yang, Y.; Wang, C.; Huang, J.; Liu, G.; Liu, Y.; et al. Rop3 GTPase contributes to polar auxin transport and auxin responses and is important for embryogenesis and seedling growth in Arabidopsis. *Plant Cell* **2014**, *26*, 3501–3518. [CrossRef] [PubMed]
12. Poraty-Gavra, L.; Zimmermann, P.; Haigis, S.; Bednarek, P.; Hazak, O.; Stelmakh, O.R.; Sadot, E.; Schulze-Lefert, P.; Gruissem, W.; Yalovsky, S. The Arabidopsis Rho of plants GTPase AtROP6 functions in developmental and pathogen response pathways. *Plant Physiol.* **2013**, *161*, 1172–1188. [CrossRef] [PubMed]
13. Ma, Q.H.; Zhu, H.H.; Han, J.Q. Wheat ROP proteins modulate defense response through lignin metabolism. *Plant Sci.* **2017**, *262*, 32–38. [CrossRef] [PubMed]
14. Yang, Z.; Watson, J.C. Molecular cloning and characterization of rho, a ras-related small GTP-binding protein from the garden pea. *Proc. Natl. Acad. Sci. USA* **1993**, *90*, 8732–8736. [CrossRef] [PubMed]
15. Li, H.; Shen, J.J.; Zheng, Z.L.; Lin, Y.K.; Yang, Z.B. The Rop GTPase switch controls multiple developmental processes in Arabidopsis. *Plant Physiol.* **2001**, *126*, 670–684. [CrossRef] [PubMed]
16. Christensen, T.M.; Vejlupkova, Z.; Sharma, Y.K.; Arthur, K.M.; Spatafora, J.W.; Albright, C.A.; Meeley, R.B.; Duvick, J.P.; Quatrano, R.S.; Fowler, J.E. Conserved subgroups and developmental regulation in the monocot *rop* gene family. *Plant Physiol.* **2003**, *133*, 1791–1808. [CrossRef] [PubMed]
17. Chan, J.; Pauls, P.K. Brassica napus Rop GTPases and their expression in microspore cultures. *Planta* **2007**, *225*, 469–484. [CrossRef] [PubMed]
18. Chen, L.; Shiotani, K.; Togashi, T.; Miki, D.; Aoyama, M.; Wong, H.L.; Kawasaki, T.; Shimamoto, K. Analysis of the Rac/Rop small GTPase family in rice: Expression, subcellular localization and role in disease resistance. *Plant Cell Physiol.* **2010**, *51*, 585–595. [CrossRef] [PubMed]

19. Qin, Y.X.; Huang, Y.C.; Fang, Y.J.; Qi, J.Y.; Tang, C.R. Molecular characterization and expression analysis of the small GTPase ROP members expressed in laticifers of the rubber tree (*Hevea brasiliensis*). *Plant Physiol. Biochem.* **2014**, *74*, 193–204. [CrossRef] [PubMed]

20. Liang, Q.X.; Cao, G.Q.; Zhao, S.P.; Huang, Q.C.; Ying, F.Q.; Chen, W. Analysis of ROP signaling in the leaf epidermis of mutant tomato with low-energy ion beam. *Genet. Mol. Res.* **2015**, *14*, 3807–3816. [CrossRef] [PubMed]

21. Zermiani, M.; Zonin, E.; Nonis, A.; Begheldo, M.; Ceccato, L.; Vezzaro, A.; Baldan, B.; Trentin, A.; Masi, A.; Fadanelli, L.; et al. Ethylene negatively regulates transcript abundance of ROP-GAP rheostat-encoding genes and affects apoplastic reactive oxygen species homeostasis in epicarps of cold stored apple fruits. *J. Exp. Bot.* **2015**, *66*, 7255–7270. [CrossRef] [PubMed]

22. Li, Z.; Kang, J.; Sui, N.; Liu, D. ROP11 GTPase is a negative regulator of multiple ABA responses in Arabidopsis. *J. Integr. Plant Biol.* **2012**, *54*, 169–179. [CrossRef] [PubMed]

23. Cao, Y.; Li, Z.; Chen, T.; Zhang, Z.; Zhang, J.; Chen, S. Overexpression of a tobacco small G protein gene NtRop1 causes salt sensitivity and hydrogen peroxide production in transgenic plants. *Sci. China C Life Sci.* **2008**, *51*, 383–390. [CrossRef] [PubMed]

24. Wang, T.Z.; Xia, X.Z.; Zhao, M.G.; Tian, Q.Y.; Zhang, W.H. Expression of a Medicago falcata small GTPase gene, MfARL1 enhanced tolerance to salt stress in *Arabidopsis thaliana*. *Plant Physiol. Biochem.* **2013**, *63*, 227–235. [CrossRef] [PubMed]

25. Li, C.; Lu, H.; Li, W.; Yuan, M.; Fu, Y. A ROP2-RIC1 pathway fine-tunes microtubule reorganization for salt tolerance in Arabidopsis. *Plant Cell Environ.* **2017**, *40*, 1127–1142. [CrossRef] [PubMed]

26. Paul, J.Y.; Khanna, H.; Kleidon, J.; Hoang, P.; Geijskes, J.; Daniells, J.; Zaplin, E.; Rosenberg, Y.; James, A.; Mlalazi, B.; et al. Golden bananas in the field: Elevated fruit pro-vitamin A from the expression of a single banana transgene. *Plant Biotechnol. J.* **2017**, *15*, 520–532. [CrossRef] [PubMed]

27. Hu, W.; Wang, L.; Tie, W.; Yan, Y.; Ding, Z.; Liu, J.; Li, M.; Peng, M.; Xu, B.; Jin, Z. Genome-wide analyses of the bZIP family reveal their involvement in the development, ripening and abiotic stress response in banana. *Sci. Rep.* **2016**, *6*, 30203. [CrossRef] [PubMed]

28. Sreedharan, S.; Shekhawat, U.K.; Ganapathi, T.R. Constitutive and stress-inducible overexpression of a native aquaporin gene (MusaPIP2;6) in transgenic banana plants signals its pivotal role in salt tolerance. *Plant Mol. Biol.* **2015**, *88*, 41–52. [CrossRef] [PubMed]

29. Lee, W.S.; Gudimella, R.; Wong, G.R.; Tammi, M.T.; Khalid, N.; Harikrishna, J.A. Transcripts and MicroRNAs responding to salt stress in *Musa acuminata* Colla (AAA Group) cv. Berangan roots. *PLoS ONE* **2015**, *10*, e0127526. [CrossRef] [PubMed]

30. Hu, W.; Yan, Y.; Shi, H.; Miao, H.; Tie, W.; Ding, Z.; Wu, C.; Liu, Y.; Wang, J.; Xu, B.; Jin, Z. The core regulatory network of the abscisic acid pathway in banana: Genome-wide identification and expression analyses during development, ripening, and abiotic stress. *BMC Plant Biol.* **2017**, *17*, 145. [CrossRef] [PubMed]

31. Miao, H.X.; Sun, P.G.; Liu, Q.; Miao, Y.L.; Liu, J.H.; Zhang, K.X.; Hu, W.; Zhang, J.B.; Wang, J.Y.; Wang, Z.; et al. Genome-wide analyses of SWEET family proteins reveal involvement in fruit development and abiotic/biotic stress responses in banana. *Sci. Rep.* **2017**, *7*, 3536. [CrossRef] [PubMed]

32. Miao, H.X.; Sun, P.G.; Liu, Q.; Liu, J.H.; Xu, B.Y.; Jin, Z.Q. The AGPase family proteins in banana: Genome-wide identification, phylogeny, and expression analyses reveal their involvement in the development, ripening, and abiotic/biotic stress responses. *Int. J. Mol. Sci.* **2017**, *18*, 1581. [CrossRef] [PubMed]

33. D'Hont, A.; Denoeud, F.; Aury, J.M.; Baurens, F.C.; Carreel, F.; Garsmeur, O.; Noel, B.; Bocs, S.; Droc, G.; Rouard, M.; et al. The banana (*Musa acuminata*) genome and the evolution of monocotyledonous plants. *Nature* **2012**, *488*, 213–217. [CrossRef] [PubMed]

34. Eddy, S.R. A new generation of homology search tools based on probabilistic inference. *Genome Inform.* **2009**, *23*, 205–211. [PubMed]

35. Altschul, S.F.; Gish, W.; Miller, W.; Myers, E.W.; Lipman, D.J. Basic local alignment search tool. *J. Mol. Biol.* **1990**, *215*, 403–410. [CrossRef]

36. Xu, Y.; Hu, W.; Liu, J.H.; Zhang, J.B.; Jia, C.H.; Miao, H.X.; Xu, B.Y.; Jin, Z.Q. A banana aquaporin gene, MaPIP1; 1, is involved in tolerance to drought and salt stresses. *BMC Plant Biol.* **2014**, *14*, 59. [CrossRef] [PubMed]

37. Han, S.; Wang, C.W.; Wang, W.L.; Jiang, J. Mitogen-activated protein kinase 6 controls root growth in Arabidopsis by modulating Ca2+-based Na+ flux in root cell under salt stress. *J. Plant Physiol.* **2014**, *171*, 26–34. [CrossRef] [PubMed]

38. Ruiz-Lozano, J.M.; Porcel, R.; Azcón, C.; Aroca, R. Regulation by arbuscular mycorrhizae of the integrated physiological response to salinity in plants: New challenges in physiological and molecular studies. *J. Exp. Bot.* **2012**, *63*, 4033–4044. [CrossRef] [PubMed]

39. Miranda, R.S.; Alvarez-Pizarro, J.C.; Costa, J.H.; Paula, S.O.; Pirsco, J.T.; Gomes-Filho, E. Putative role of glutamine in the activation of CBL/CIPK signaling pathways during salt stress in sorghum. *Plant Signal. Behav.* **2017**, *12*, e1361075. [CrossRef] [PubMed]

40. Hoefle, C.; Huesmann, C.; Schultheiss, H.; Bornke, F.; Hensel, G.; Kumlehn, J.; Huckelhoven, R. A barley ROP GTPase ACTIVATING PROTEIN associates with microtubules and regulates entry of the barley powdery mildew fungus into leaf epidermal cells. *Plant Cell* **2011**, *23*, 2422–2439. [CrossRef] [PubMed]

41. Jeon, B.W.; Hwang, J.U.; Hwang, Y.; Song, W.Y.; Fu, Y.; Gu, Y.; Bao, F.; Cho, D.; Kwak, J.M.; Yang, Z.; et al. The Arabidopsis small G protein ROP2 is activated by light in guard cells and inhibits light-induced stomatal opening. *Plant Cell* **2008**, *20*, 75–87. [CrossRef] [PubMed]

42. Shi, H.; Ishitani, M.; Kim, C.; Zhu, J.K. The *Arabidopsis thaliana* salt tolerance gene *SOS1* encodes a putative Na$^+$/H$^+$ antiporter. *Proc. Natl. Acad. Sci. USA* **2000**, *97*, 6896–6901. [CrossRef] [PubMed]

43. Liu, J.; Ishitani, M.; Halfter, U.; Kim, C.S.; Zhu, J.K. The *Arabidopsis thaliana SOS2* gene encodes a protein kinase that is required for salt tolerance. *Proc. Natl. Acad. Sci. USA* **2000**, *97*, 3730–3734. [CrossRef] [PubMed]

44. Köster, P.; Wallrad, L.; Edel, K.H.; Faisal, M.; Alatar, A.A.; Kudla, J. The battle of two ions: Ca^{2+} signaling against Na$^+$ stress. *Plant Biol.* **2018**, *7*. [CrossRef] [PubMed]

45. Marchler-Bauer, A.; Bo, Y.; Han, L.; He, J.; Lanczycki, C.J.; Lu, S.; Lu, S.; Chitsaz, F.; Derbyshire, M.K.; Geer, R.C.; et al. CDD/SPARCLE: Functional classification of proteins via subfamily domain architectures. *Nucleic Acids Res.* **2017**, *45*, 200–203. [CrossRef] [PubMed]

46. Finn, R.D.; Coggill, P.; Eberhardt, R.Y.; Eddy, S.R.; Mistry, J.; Mitchell, A.L.; Potter, S.C.; Punta, M.; Qureshi, M.; Sangrador-Vegas, A.; et al. The Pfam protein families database: Towards a more sustainable future. *Nucleic Acids Res.* **2016**, *44*, 279–285. [CrossRef] [PubMed]

47. Larkin, M.A.; Blackshields, G.; Brown, N.P.; Chenna, R.; Mcgettigan, P.A.; Mcwilliam, H.; Valentin, F.; Wallace, I.M.; Wilm, A.; Lopez, R.; et al. Clustal W and Clustal X version 2.0. *Bioinformatics* **2007**, *23*, 2947–2948. [CrossRef] [PubMed]

48. Tamura, K.; Peterson, D.; Peterson, N.; Stecher, G.; Nei, M.; Kumar, S. MEGA5: Molecular evolutionary genetic analysis using maximum likelihood, evolutionary distance, and maximum parsimony methods. *Mol. Biol. Evol.* **2011**, *28*, 2731–2739. [CrossRef] [PubMed]

49. Gasteiger, E.; Gattiker, A.; Hoogland, C.; Ivanyi, I.; Appel, R.D.; Bairoch, A. ExPASy: The proteomics server for in-depth protein knowledge and analysis. *Nucleic Acids Res.* **2003**, *31*, 3784–3788. [CrossRef] [PubMed]

50. Bailey, T.L.; Boden, M.; Buske, F.A.; Frith, M.; Grant, C.E.; Clementi, L.; Ren, J.; Li, W.W.; Noble, W.S. MEME SUITE: Tools for motif discovery and searching. *Nucleic Acids Res.* **2009**, *37*, 202–208. [CrossRef] [PubMed]

51. Jones, P.; Binns, D.; Chang, H.Y.; Fraser, M.; Li, W.; McAnulla, C.; McWilliam, H.; Maslen, J.; Mitchell, A.; Nuka, G.; et al. InterProScan 5: Genome-scale protein function classification. *Bioinformatics* **2014**, *30*, 1236–1240. [CrossRef] [PubMed]

52. Hu, B.; Jin, J.; Guo, A.Y.; Zhang, H.; Luo, J.; Gao, G. GSDS 2.0: An upgraded gene feature visualization server. *Bioinformatics* **2015**, *31*, 1296–1297. [CrossRef] [PubMed]

53. Celniker, S.E.; Wheeler, D.A.; Kronmiller, B.; Carlson, J.W.; Haipern, A.; Patel, S.; Adams, M.; Champe, M.; Dugan, S.P.; Frise, E.; et al. Finishing a whole-genome shotgun: Release 3 of the *Drosophila melanogaster* euchromatic genome sequence. *Genome Biol.* **2002**, *3*, RESEARCH0079. [CrossRef] [PubMed]

54. Lescot, M.; Déhais, P.; Thijs, G.; Marchal, K.; Moreau, Y.; Peer, Y.V.; Rouzé, P.; Rombauts, S. PlantCARE, a database of plant *cis*-acting regulatory elements and a portal to tools for in silico analysis of promoter sequences. *Nucleic Acids Res.* **2002**, *30*, 325–327. [CrossRef] [PubMed]

55. Wang, L.; Feng, Z.; Wang, X.; Wang, X.; Zhang, X. DEGseq: An R package for identifying differentially expressed genes from RNA-seq data. *Bioinformatics* **2010**, *26*, 136–138. [CrossRef] [PubMed]

56. Livak, K.J.; Schmittgen, T.D. Analysis of relative gene expression data using real-time quantitative PCR and the 2$^{-\Delta\Delta CT}$ Method. *Methods* **2001**, *25*, 402–408. [CrossRef] [PubMed]

57. Clough, S.J.; Bent, A.F. Floral dip: A simplified method for *Agrobacterium*-mediated transformation of *Arabidopsis thaliana*. *Plant J.* **1998**, *16*, 735–743. [CrossRef] [PubMed]

58. Riqal, A.; Doyle, S.M.; Robert, S. Live cell imaging of FM4-64, a tool for tracing the endocytic pathways in Arabidopsis root cells. *Methods Mol. Biol.* **2015**, *1242*, 93–103.

59. Miao, H.X.; Qin, Y.H.; Teixeira da Silva, J.A.; Ye, Z.X.; Hu, G.B. Cloning and expression analysis of *S-RNase* homologous gene in *Citrus reticulata* Blanco cv. Wuzishatangju. *Plant Sci.* **2011**, *180*, 358–367. [CrossRef] [PubMed]

5

Nitric Oxide is Required for Melatonin-Enhanced Tolerance against Salinity Stress in Rapeseed (*Brassica napus* L.) Seedlings

Gan Zhao [1], **Yingying Zhao** [1], **Xiuli Yu** [1], **Felix Kiprotich** [1], **Han Han** [1], **Rongzhan Guan** [2], **Ren Wang** [3] and **Wenbiao Shen** [1,*]

[1] College of Life Sciences, Laboratory Center of Life Sciences, Nanjing Agricultural University, Nanjing 210095, China; 2015116111@njau.edu.cn (G.Z.); 2017116114@njau.edu.cn (Y.Z.); 2016116109@njau.edu.cn (X.Y.); kiprotichfelix@yahoo.com (F.K.); martinhan956@gmail.com (H.H.)

[2] National Key Laboratory of Crop Genetics and Germplasm Enhancement, Jiangsu Collaborative Innovation Center for Modern Crop Production, Nanjing Agricultural University, Nanjing 210095, China; guanrzh@njau.edu.cn

[3] Institute of Botany, Jiangsu Province and Chinese Academy of Sciences, Nanjing 210014, China; wangren@126.com

* Correspondence: wbshenh@njau.edu.cn

Abstract: Although melatonin (*N*-acetyl-5-methoxytryptamine) could alleviate salinity stress in plants, the downstream signaling pathway is still not fully characterized. Here, we report that endogenous melatonin and thereafter nitric oxide (NO) accumulation was successively increased in NaCl-stressed rapeseed (*Brassica napus* L.) seedling roots. Application of melatonin and NO-releasing compound not only counteracted NaCl-induced seedling growth inhibition, but also reestablished redox and ion homeostasis, the latter of which are confirmed by the alleviation of reactive oxygen species overproduction, the decreases in thiobarbituric acid reactive substances production, and Na^+/K^+ ratio. Consistently, the related antioxidant defense genes, *sodium hydrogen exchanger* (*NHX1*), and *salt overly sensitive 2* (*SOS2*) transcripts are modulated. The involvement *S*-nitrosylation, a redox-based posttranslational modification triggered by NO, is suggested. Further results show that in response to NaCl stress, the increased NO levels are strengthened by the addition of melatonin in seedling roots. Above responses are abolished by the removal of NO by NO scavenger. We further discover that the removal of NO does not alter endogenous melatonin content in roots supplemented with NaCl alone or together with melatonin, thus excluding the possibility of NO-triggered melatonin production. Genetic evidence reveals that, compared with wild-type Arabidopsis, the hypersensitivity to NaCl in *nia1/2* and *noa1* mutants (exhibiting null nitrate reductase activity and indirectly reduced endogenous NO level, respectively) cannot be rescued by melatonin supplementation. The reestablishment of redox homeostasis and induction of *SOS* signaling are not observed. In summary, above pharmacological, molecular, and genetic data conclude that NO operates downstream of melatonin promoting salinity tolerance.

Keywords: Arabidopsis; *Brassica napus*; ion homeostasis; melatonin; NaCl stress; nitric oxide; redox homeostasis

---

1. Introduction

Soil salinity is a major factor that significantly influences global agricultural production [1]. High salinity (mainly NaCl) provokes two primary effects on plants, including ionic and oxidative effects [1–4]. In general, high NaCl stress disturbs the ionic environment of plant cells, notably forming a higher Na^+/K^+ ratio [5]. Plants usually remove excessive Na^+ by Na^+/H^+ antiporters, and genetic evidence revealed that overexpressing these antiporter genes can improve salt tolerance [6,7]. For example, *SOS* signaling is a well-known pathway responsible for initiating transport of Na^+ out of the cells, or activating an unknown transporter, thus leading to the sequestration of Na^+ in the vacuole [8–10]. Another type of Na^+/H^+ antiporter belongs to the Na^+/H^+ exchanger (NHX) family, and constitutive overexpression of *NHX* can increase Na accumulation in vacuoles, and thus, enhance salt tolerance [10]. Meanwhile, a large number of reactive oxygen species (ROS), such as superoxide anion, hydrogen peroxide, and hydroxy1 radicals, are induced under salinity conditions [11]. To combat salt-induced oxidative stress, the enzymatic antioxidant system provides a highly efficient and specific ROS scavenging approach for plants. For example, superoxide dismutase (SOD), catalase (CAT), and guaiacol peroxidase (POD), are very important parts of this enzymatic system, and normally, plants decrease ROS by upregulating activities of these enzymes [11–13].

Rapeseed (*Brassica napus* L.) is one of the most widely cultivated oil crops in the world because of the healthy fatty acid composition of its oil and high protein content of its meal. It is classified as a moderate salinity-tolerant crop [14]. During the growth period, rapeseed plants are challenged by salt stress, and ionic and redox imbalance are two major effects associated with salinity toxicity [11,14–16]. As more land becomes salinized, the studying of related mechanisms (the reestablishment of ionic and redox balance) and the application of effective methods (including reclamation of saline soils by use of chemicals or plant growth-promoting bacteria, or by growing salt tolerant cultivars in the saline soils, etc.) for improving salt tolerance in rapeseed plants [14–18], are becoming increasingly significant [19–25].

Melatonin (*N*-acetyl-5-methoxytryptamine) was discovered, and isolated from the bovine pineal gland in 1958 [26]. With a large set of functions in animals (circadian rhythms, seasonal rhythms, and alleviating oxidative stress; [27–30]), this compound was also detected in plants, and used as both a plant growth regulator and a biostimulator to alleviate abiotic and biotic stresses, including salinity, cold, drought, chemical pollutants, and defense against bacterial pathogen infection [31–33]. Previous results revealed that exogenous application of melatonin not only increased endogenous melatonin levels, but also improved the salt tolerance in Arabidopsis, soybean, Chinese crab apple, rice, cucumber, and bermudagrass [19,20,34–37]. Above beneficial roles of melatonin are normally associated with enhanced activities of antioxidant enzymes, as well as upregulating transcripts of ion channel genes, or sugar and glycolysis metabolism-related genes [36,37]. However, the corresponding detailed mechanism, especially the crosstalk with other signaling components and related transduction cascade, is still not fully characterized.

It is well known that nitric oxide (NO), one of the important gasotransmitters controlling a diverse range of physiological functions in plants [38–40], can also enhance salinity tolerance [21,41–44]. Previous reports revealed that there are at least two major enzymatic sources of NO: a nitrate/nitrite-dependent pathway and an L-Arg-dependent pathway [38,39]. Further experiments with Arabidopsis single and triple mutants, exhibiting null nitrate reductase (NR) activity (*nia1/2*) and indirectly reduced endogenous NO level (*nitric oxide associated1*; *noa1*), revealed that NO production is associated with salinity tolerance in Arabidopsis [42,44,45]. Importantly, protein post-translational modification by *S*-nitrosylation was preliminarily used to explain the physiological functions of NO in both animals and plants [46], including adaptation against biotic and abiotic stresses in plants [41,47,48]. However, it is not clear whether NO-dependent *S*-nitrosylation is also associated with melatonin responses in plants.

Although previous pharmacological data showed the interplay between melatonin and NO leading to plant tolerance against NaCl stress in sunflower seedlings [49,50], no genetic study has yet provided definitive proof of a role of endogenous NO in melatonin signaling governing salinity tolerance. In this study, we firstly evaluated the role of NO in melatonin-triggered salinity tolerance in rapeseed seedlings by using pharmacological and biochemical approaches. The reestablishment of redox and ion homeostasis was confirmed. The involvement of *S*-nitrosylation is also discovered, suggesting that NO is involved in melatonin signaling as a downstream messenger. Afterwards, both *nia1/2* and *noa1* Arabidopsis mutants were utilized to investigate the relationship between NO and melatonin in salinity tolerance. We thus concluded that NO acts downstream of melatonin signaling to enhance tolerance against salinity.

2. Results

2.1. Salt Stress Stimulates Melatonin and NO Production

To assess the sensitivity of rapeseed seedling growth to NaCl stress, the effects of varying concentrations (100, 150, 200, and 250 mM) of NaCl on root growth were investigated. As shown in Figure 1, the exposure of seedlings to NaCl resulted in dose-dependent decreases in the root elongation and root fresh weight. Since approximate 50% inhibition in above parameters was observed in 200 mM NaCl-treated seedlings, this concentration of NaCl was applied in the following experiments.

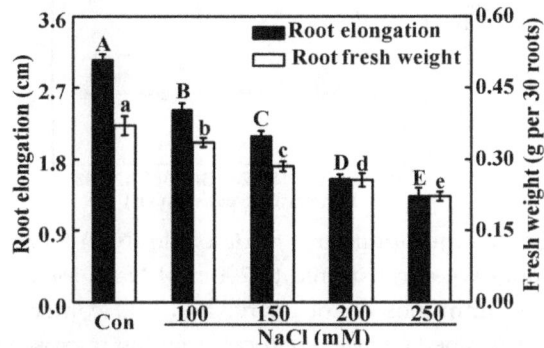

Figure 1. Growth inhibition of seedling roots upon NaCl stress. Three-day-old rapeseed seedlings were transferred to 100, 150, 200, and 250 mM NaCl for 2 days. Afterwards, the root elongation (left) and root fresh weight (right) were measured. The sample without chemicals was the control (Con). Values are means ± standard error (SE) of three independent experiments with at least three replicates for each. Bars with different letters are significant different at $p < 0.05$ according to Duncan's multiple range test.

Griess reagent (visible spectrophotography) and laser scanning confocal microscopy (LSCM) with the specific probe (4-amino-5-methylamino-2′,7′-difluorofluorescein diacetate; DAF-FM DA), are the most frequently used methods for the determination of NO production in plants. Subsequently, the time course experiments for 48 h revealed the rapid burst of endogenous melatonin (Figure 2A; detected by enzyme-linked immunosorbent assay) and NO (Figure 2B,C; respectively determined by visible spectrophotography and LSCM) accumulation in NaCl-treated root tissues, peaking at 6 h and 12 h of stress, compared to the control sample (Con). We also noticed that the increases of melatonin and NO were still evident until 48 h, although both of them were decreased after the peaking points.

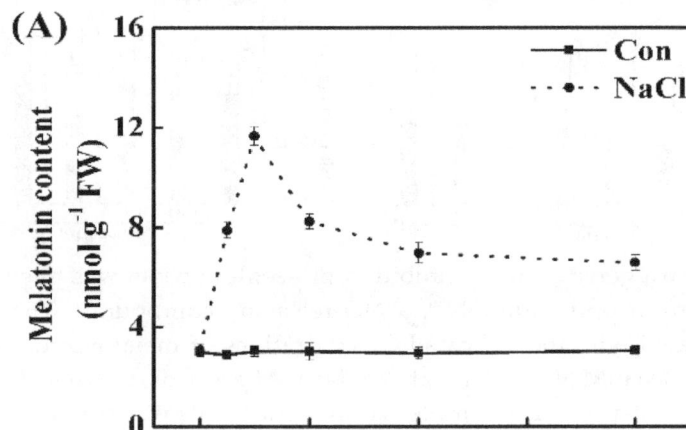

Figure 2. *Cont.*

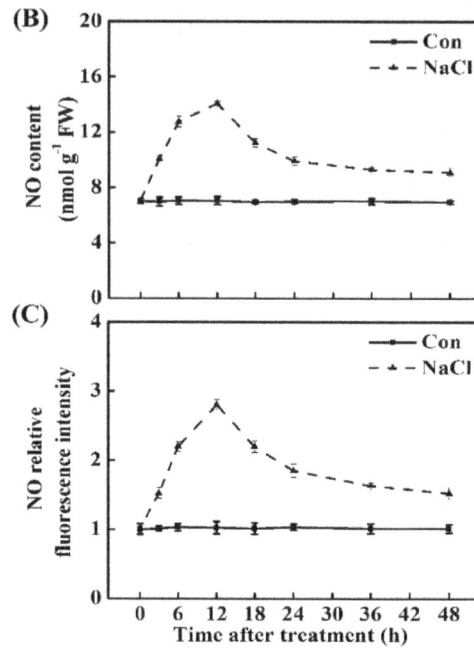

Figure 2. Changes in endogenous melatonin and nitric oxide (NO) levels in response to NaCl stress. Three-day-old rapeseed seedlings were transferred to 200 mM NaCl for 2 days. Meanwhile, melatonin (**A**) detected by enzyme-linked immunosorbent assay; and NO contents (**B**) determined by visible spectrophotography, and (**C**) determined by laser confocal scanning microscopy, and expressed as relative fluorescence intensity) in seedling roots were analyzed. The sample without chemicals was the control (Con). Values are means $\pm$ SE of three independent experiments with at least three replicates for each.

2.2. Melatonin and NO Alleviate NaCl-Induced Seedling Growth Inhibition

It was well known that to discern the role of melatonin in the alleviation of salt stress, a dose–response study of exogenous melatonin in vitro was firstly established. As shown in Figure 3, we observed that the addition of melatonin (0.1, 1, and 10 μM) not only promoted seedling root growth under the normal growth condition, but also differentially alleviated the growth inhibition in roots triggered by NaCl stress, while no significant rescuing effects were observed in 0.01 and 100 μM melatonin-pretreated seedlings. Among these pretreatments, the responses of 1 μM melatonin was maximal, and this concentration was further applied in the following test.

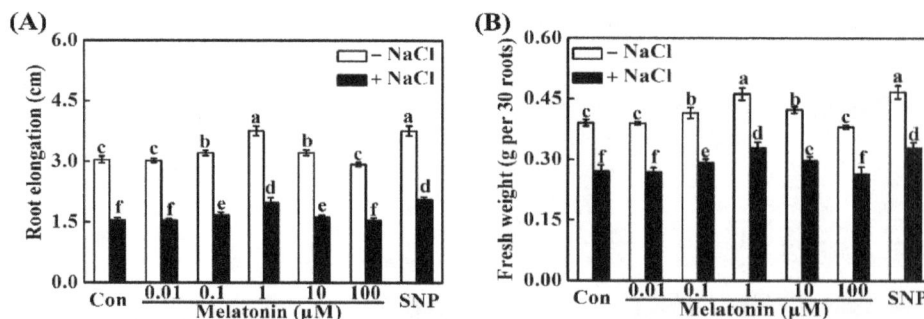

Figure 3. NaCl stress-triggered growth inhibition of seedling roots was alleviated by exogenous melatonin and sodium nitroprusside (SNP; a NO-releasing compound). Three-day-old rapeseed seedlings were pretreated with the indicated concentrations of melatonin or 10 μM SNP for 12 h, and then transferred to 200 mM NaCl for another 2 days. Afterwards, the root elongation (**A**) and root fresh weight (**B**) were measured. The sample without chemicals was the control (Con). Values are means $\pm$ SE of three independent experiments, with at least three replicates for each. Bars with different letters are significant different at $p < 0.05$ according to Duncan's multiple range test.

Meanwhile, the treatment with three types of NO-releasing compounds, namely sodium nitroprusside (SNP), diethylamine NONOate (NONOate), and S-nitrosoglutathione (GSNO), produced similar positive responses in the stressed condition (Figure 3 and Supplementary Materials Figure S1). While, old SNP (a negative control of SNP) failed to influence root growth inhibition. Above results thus suggested the beneficial role of exogenous NO in the plant tolerance against salinity stress. Considering the cost of chemicals, SNP was used as a NO-releasing compound in the following experiment.

2.3. PTIO-Dependent Removal of NO Production Impairs the Response of Melatonin

To assess the possible link between melatonin and NO in the alleviation of NaCl stress, the effects of the NO scavenger 2-phenyl-4,4,5,5-tetramethylimidazoline-1-oxyl-3-oxide (PTIO), on the abovementioned melatonin and SNP responses, were investigated and compared. The results shown in Figure 4 revealed that both melatonin- and SNP-alleviated root growth inhibition was greatly reduced in the presence of PTIO, which was similar to the phenotypes in NaCl-stressed alone conditions. In comparison with NaCl stress, the addition of PTIO aggravated root growth inhibition.

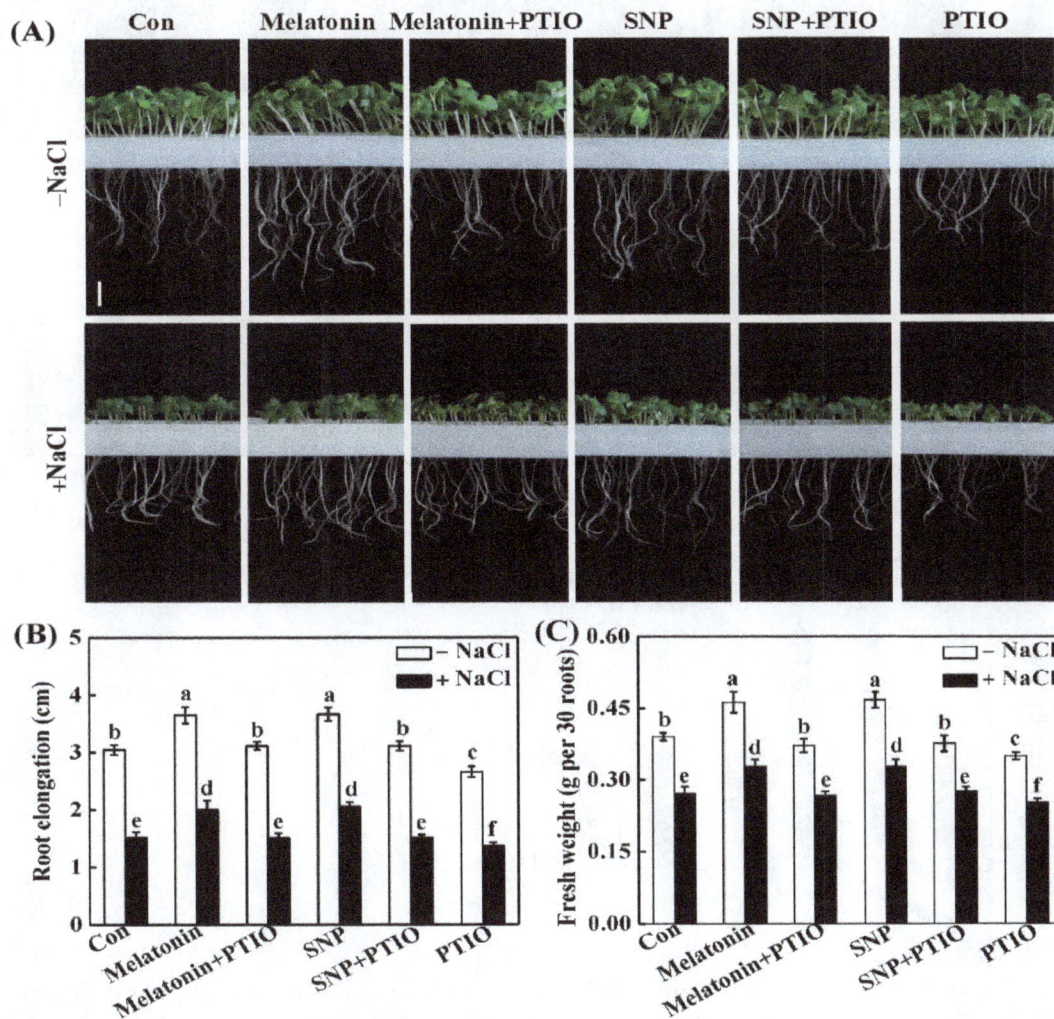

Figure 4. Exogenous melatonin-alleviated root growth inhibition caused by NaCl stress was sensitive to the removal of NO. Three-day-old rapeseed seedlings were pretreated with 1 μM melatonin, 10 μM SNP, 200 μM PTIO, alone or their combinations for 12 h, and then transferred to 200 mM NaCl for 2 days. Afterwards, corresponding photographs were taken ((A); bar: 1 cm). The root elongation (B) and root fresh weight (C) were measured. The sample without chemicals was the control (Con). Values are means ± SE of three independent experiments with at least three replicates for each. Bars with different letters are significant different at $p < 0.05$ according to Duncan's multiple range test.

The role of NO in melatonin-enhanced salinity tolerance was further examined by monitoring NO synthesis in response to applied melatonin and SNP in the presence or absence of PTIO. Similar to the response of SNP, a significant increase in NO-induced fluorescence was observed in stressed seedling roots compared with the control tissue, demonstrating melatonin-mediated NO production (Figure 5A,B). Importantly, melatonin-induced NO synthesis was abolished by co-incubation with PTIO, correlating these data with those from phenotypic analysis (Figure 4). The above results were further confirmed by Griess reagent method (Figure 5C). Together, the pharmacological evidence revealed that PTIO-dependent removal of NO production impairs the response of melatonin.

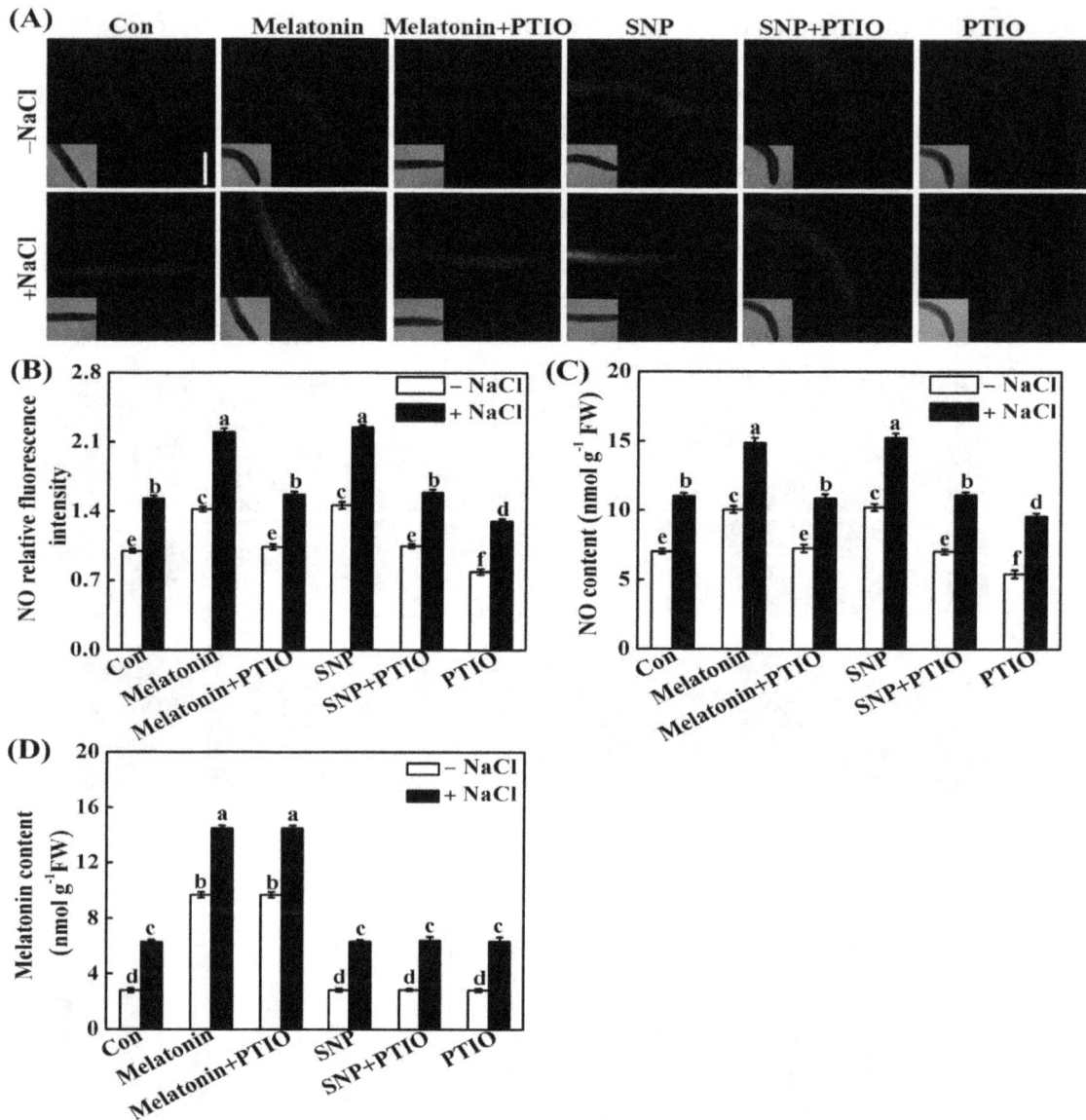

Figure 5. The removal of NO did not alter endogenous melatonin level, but melatonin triggered NO production. Three-day-old rapeseed seedlings were pretreated with 1 μM melatonin, 10 μM SNP, 200 μM PTIO, alone or their combinations for 12 h, and then transferred to 200 mM NaCl for another 2 days. Afterwards, NO ((A); determined by laser confocal scanning microscopy; (C); determined by visible spectrophotography) and melatonin contents ((D); detected by enzyme-linked immunosorbent assay) in root tissues were detected. Scale bar = 1 mm. DAF-FM DA fluorescence densities according to (A) were also given (B). The sample without chemicals was the control (Con). Values are means ± SE of three independent experiments with at least three replicates for each. Bars with different letters are significant different at $p < 0.05$ according to Duncan's multiple range test.

2.4. NO Does Not Alter Melatonin Synthesis

To further confirm above hypothesis, the effects of SNP and PTIO on endogenous melatonin levels were analyzed. Unlike the inducible responses of exogenous melatonin, the treatment with SNP had no effect on either basal or NaCl-induced melatonin production (Figure 5D). Interestingly, the co-incubation with PTIO did not influence melatonin levels in response to either melatonin or SNP when applied exogenously, no matter if seedlings were with or without the treatment of NaCl.

2.5. Redox Balance Is Reestablished by Melatonin via NO

To unravel the molecular mechanism underlying melatonin-triggered salinity tolerance, subsequent histochemical detection of hydrogen peroxide (H_2O_2; DAB staining) and superoxide anion (O_2^-; NBT staining) was applied. Similar to the positive responses of SNP, NaCl-induced H_2O_2 and O_2^- overproduction in roots, confirmed by the dark brown (Figure 6A) and purple-blue (Figure 6B) color precipitates, was differentially abolished by melatonin. Contrasting results were observed when PTIO was added together. These were in accordance with the results of TBARS contents (Figure 6C).

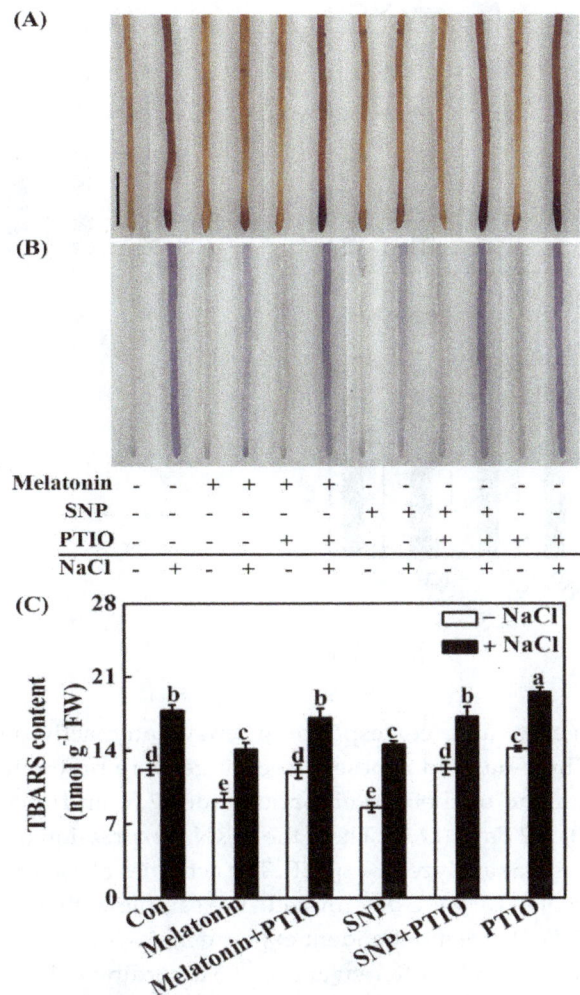

Figure 6. Redox balance was reestablished by melatonin via NO. Three-day-old rapeseed seedlings were pretreated with 1 μM melatonin, 10 μM SNP, 200 μM PTIO, alone or their combinations for 12 h, and then transferred to 200 mM NaCl for another 2 days. Afterwards, seedling roots were stained with DAB (**A**) and NBT (**B**) to detect H_2O_2 and O_2^-. Scale bar = 1 cm. TBARS content (**C**) were also determined. The sample without chemicals was the control (Con). Values are means ± SE of three independent experiments with at least three replicates for each. Bars with different letters are significant different at $p < 0.05$ according to Duncan's multiple range test.

Molecular and biochemical experiments revealed that treatment with PTIO almost completely blocked the increases in the expression of the antioxidant genes *APX*, *MnSOD*, *Cu/ZnSOD* (Figure 7A–C), and the activities of APX and SOD (Figure 7D,E) in NaCl-stressed root tissues. Combined with the results in histochemical detection and TBARS content analysis (Figure 6), these clearly suggested the requirement of NO in melatonin-reestablished redox balance.

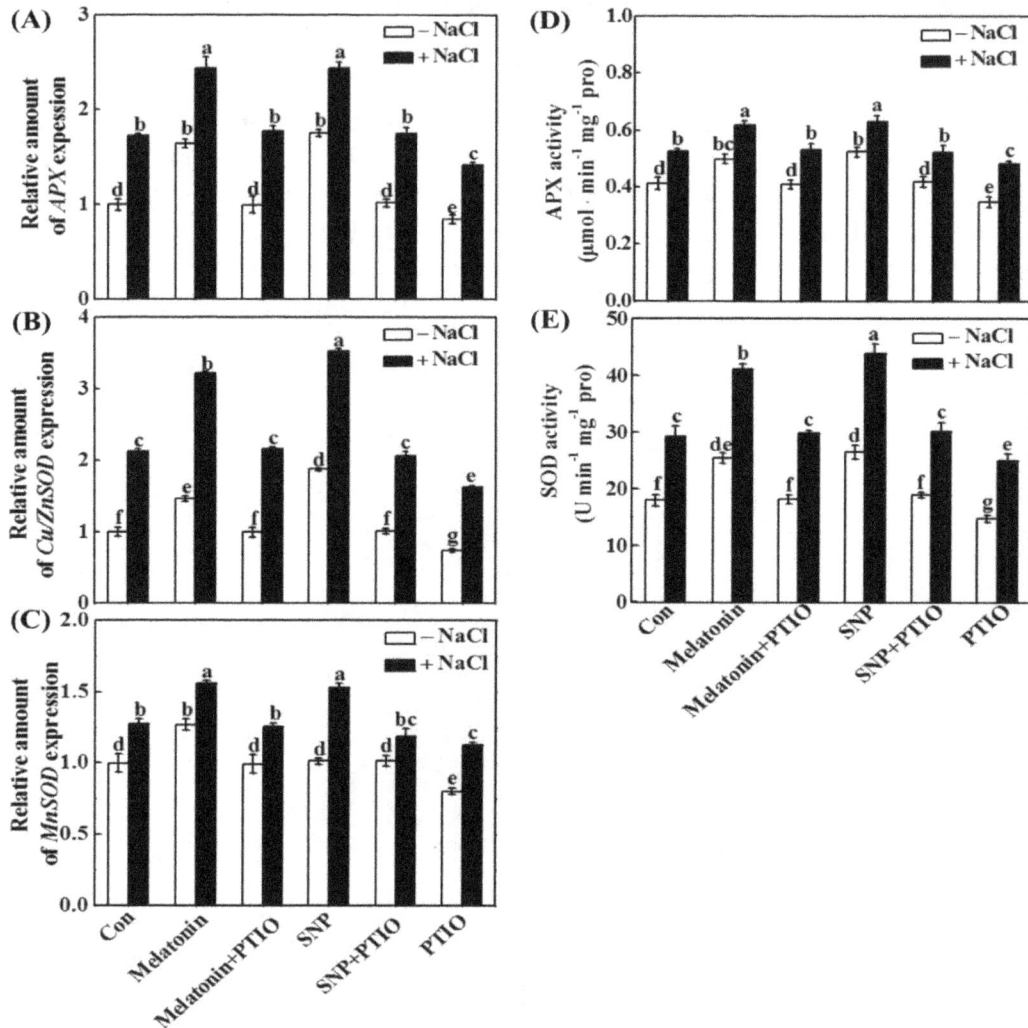

Figure 7. Antioxidant genes and corresponding enzymatic activities were modulated by melatonin-mediated NO. Three-day-old rapeseed seedlings were pretreated with 1 µM melatonin, 10 µM SNP, 200 µM PTIO, alone or their combinations for 12 h, and then transferred to 200 mM NaCl for another 12 h (A–C) or 2 days (D,E). Then, the mRNA expression of *APX* (A), *Cu/ZnSOD* (B), and *MnSOD* (C) in root tissues was analyzed by qPCR. The activities of ascorbate peroxidase (APX; (D)) and superoxide dismutase (SOD; (E)) were determined. The sample without chemicals was the control (Con). Values are means ± SE of three independent experiments with at least three replicates for each. Bars with different letters are significant different at $p < 0.05$ according to Duncan's multiple range test.

2.6. Melatonin Modulates Ion Homeostasis via NO

The keeping of ion homeostasis is essential for plants to resist salt stress. Therefore, the effects of melatonin and NO, as well as their interplay on ion homeostasis, were investigated. Compared with the control, NaCl stress significantly increased the Na^+ accumulation and decreased the K^+ accumulation, thus leading to a higher Na^+/K^+ ratio (Figure 8A) in seedling roots. By contrast, the addition of melatonin and SNP helped seedling roots to reduce the accumulation of Na^+ and improve K$^+$ assimilation, resulting in lower Na$^+$/K$^+$ ratio, compared with NaCl stressed alone, both of which could be reversed by PTIO.

To confirm the cause of this phenomenon, transcripts of Na$^+$ transporter *NHX1* and *SOS2* were analyzed. As expected, changes in *NHX1* and *SOS2* were consistent with the results in Na$^+$/K$^+$ ratio. For example, the removal of endogenous NO by PTIO completely impaired the effects of melatonin and SNP on activating *NHX1* and *SOS2* mRNA (Figure 8B,C).

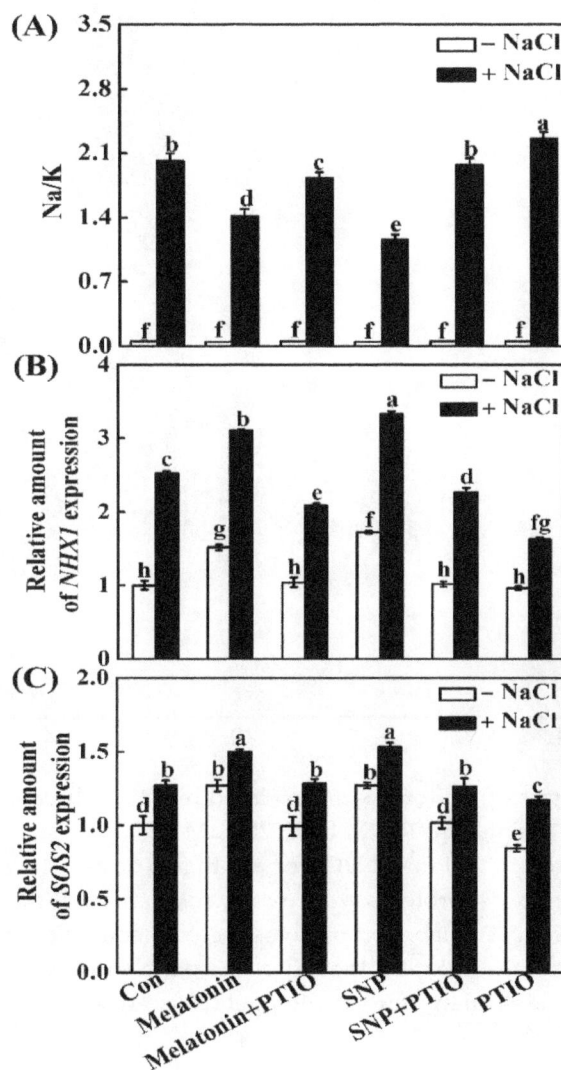

Figure 8. Melatonin modulated ion homeostasis via NO. Three-day-old rapeseed seedlings were pretreated with 1 μM melatonin, 10 μM SNP, 200 μM PTIO, alone or their combinations for 12 h, and then transferred to 200 mM NaCl for another 2 days (**A**) or 12 h (**B,C**). Afterwards, Na$^+$ to K$^+$ ratio (**A**) in seedling roots were detected by ICP-OES. The mRNA expression of *NHX1* (**B**) and *SOS2* (**C**) were analyzed by qPCR. The sample without chemicals was the control (Con). Values are means ± SE of three independent experiments with at least three replicates for each. Bars with different letters are significant different at $p < 0.05$ according to Duncan's multiple range test.

2.7. The Possible Involvement of NO-Dependent S-Nitrosylation

An important bioactivity of NO is implemented by regulating the activity of targeted proteins through S-nitrosylation, a redox-based posttranslational modification. For further confirming the involvement of NO in above melatonin responses, the profiles in S-nitrosylation were analyzed by using the modified biotin switch technique. Figure 9 showed that both melatonin and SNP increased

S-nitrosylation under NaCl stress, which could be partially blocked by the PTIO-dependent removal of NO production. Alone, PTIO supplementation slightly decreased *S*-nitrosylation, compared to the control samples.

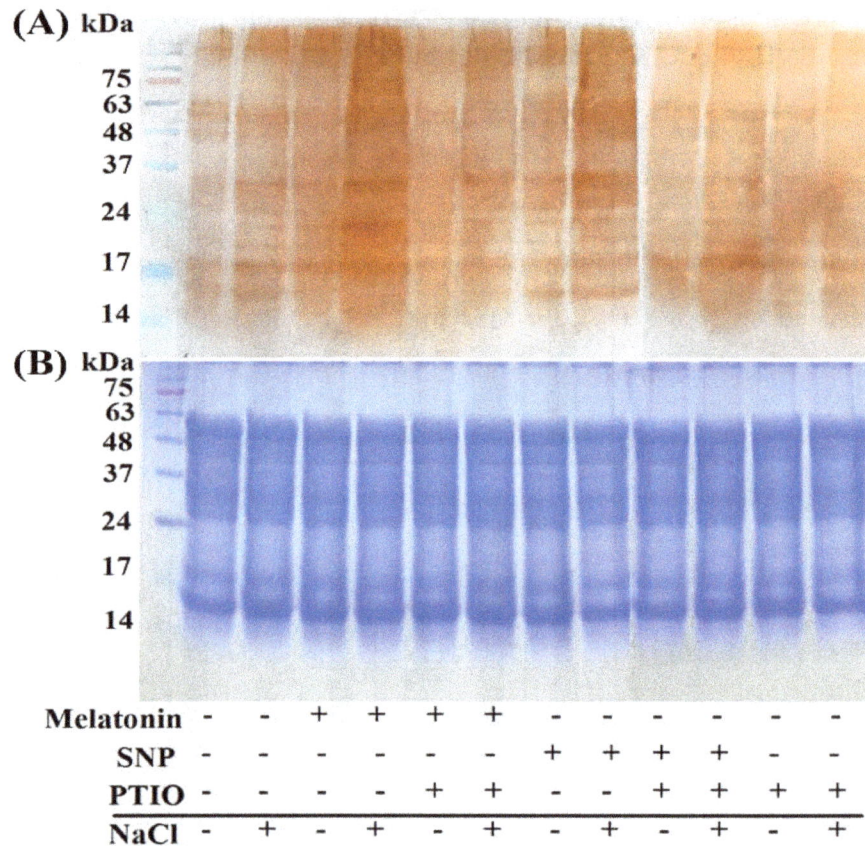

Figure 9. Immunoblot analysis of the total *S*-nitrosylated protein. Three-day-old rapeseed seedlings were pretreated with 1 μM melatonin, 10 μM SNP, 200 μM PTIO alone, or in various combinations, for 12 h, and then transferred to 200 mM NaCl for another 2 days. The sample without chemicals was the control (Con). Afterwards, proteins were extracted from seedling roots, and subjected to the modified biotin switch method. The labelled proteins were detected using protein blot analysis with antibodies against biotin (**A**). Numbers on the left of the panels indicate the position of the protein markers in kDa. A Coomassie Brilliant Blue-stained gel (**B**) is present to show that equal amounts of proteins were loaded.

2.8. Genetic Evidence Reveals That NO Is Required for Melatonin-Induced Salinity Tolerance

To complement above results, we evaluated the role of NO in melatonin-triggered salinity tolerance by using Arabidopsis wild-type (WT) and mutants exhibiting null nitrate reductase (NR) activity (*nia1/2*) and indirectly reduced endogenous NO level (*nitric oxide associated1*; *noa1*). Figure 10A shows WT and *nia1/2* and *noa1* mutants challenged by NaCl stress in the presence or absence of melatonin. As expected, the *nia1/2* and *noa1* mutants exhibited more sensitivity to salinity stress than WT, as measured by the responses in primary root elongation (Figure 10B) and chlorophyll content (Figure 10C) in seeding leaves. It was subsequently observed that the NaCl-induced toxicity in WT plants was obviously rescued by the pretreatment with melatonin. By contrast, no significant rescuing responses were observed in the stressed *nia1/2* and *noa1* mutants with melatonin. Since there are at least two distinct pathways responsible for NO synthesis in plants, the NR- and L-Arg-dependent pathways, our genetic evidence further confirmed the central role of NO in salt tolerance triggered by melatonin.

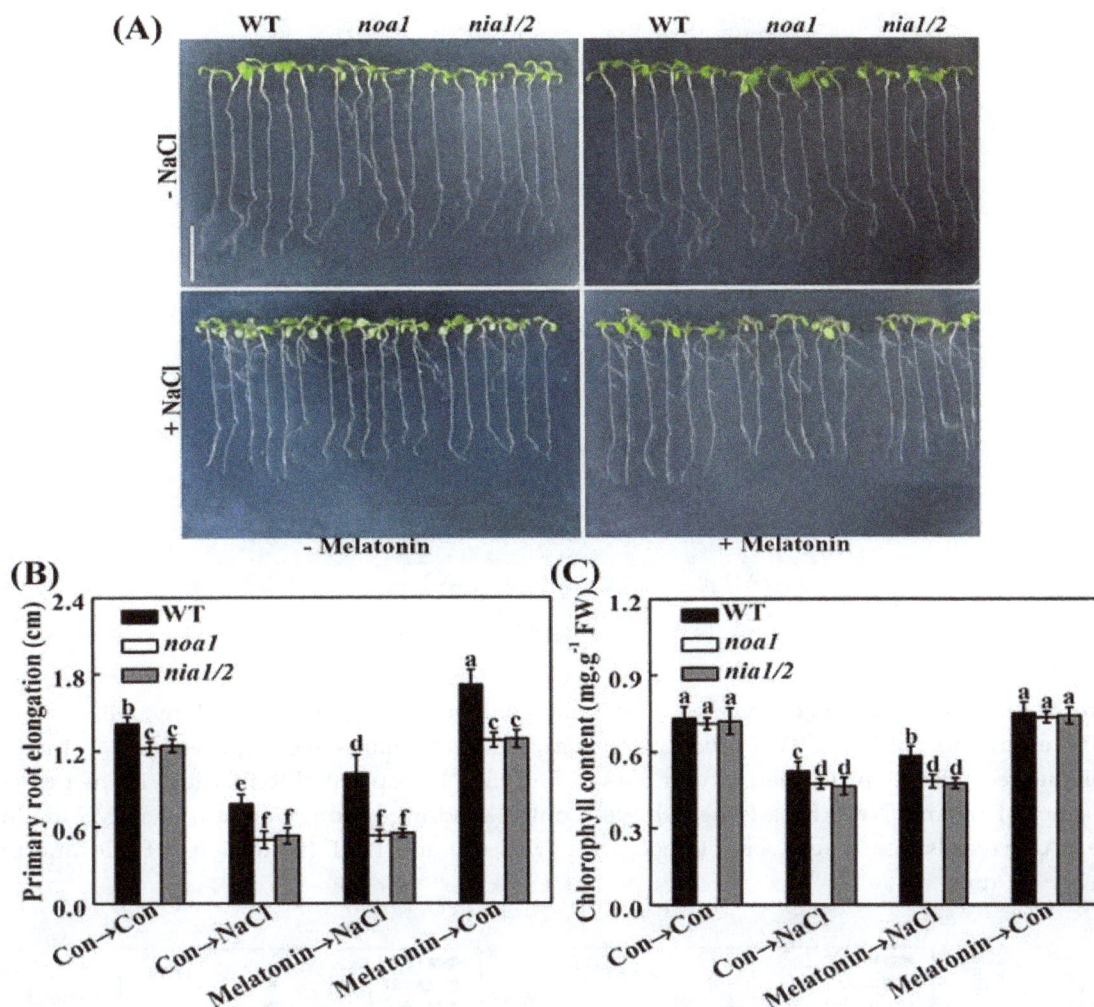

Figure 10. Genetic evidence supported the requirement of NO in melatonin-alleviated NaCl stress. Five-day-old wild-type (WT), and *noa1* and *nia1/2* mutant plants were grown on MS medium supplemented with 1.0 μM melatonin for 5 days, and then transplanted to medium in the presence or absence of 125 mM NaCl for another 5 days. Primary root elongation (**B**) and total chlorophyll content in leaves (**C**) were then determined to assess changes in salt tolerance described in (**A**). Control seedlings (Con) were grown in MS medium alone. Scale bar = 1 cm. Data are means ± SE of three independent experiments with at least three replicates for each. Bars with different letters are significant different at $p < 0.05$ according to Duncan's multiple range test.

In order to further assess whether melatonin-reestablished redox and ion homeostasis is associated with NO signaling, histochemical detection and molecular approach were adopted. As shown in Figure 11, unlike the WT plants, the NaCl-triggered H_2O_2 and O_2^- overproduction in *nia1/2* and *noa1* mutants was largely insensitive to melatonin supplementation. Consistently, NaCl-induced *APX1*, *APX2*, *CAT1*, and *FSD1* transcripts were strengthened by melatonin in wild-type (Figure 12). In comparison, no such induction conferred by melatonin appeared in the mutant plants upon NaCl stress. Molecular evidence further showed that *SOS1*, *SOS2*, and *SOS3* transcripts were upregulated in wild-type and mutant seedlings upon NaCl stress (Figure 13). In the presence of melatonin, above induction was more pronounced in wild-type, but not in *nia1/2* and *noa1* mutants. Combined with the changes in phenotypes (Figure 10), the above genetic evidence thus suggested that endogenous NO mainly produced by NR and NOA1 is required for melatonin-triggered salinity tolerance in Arabidopsis, and the reestablished redox and ion homeostasis were suggested.

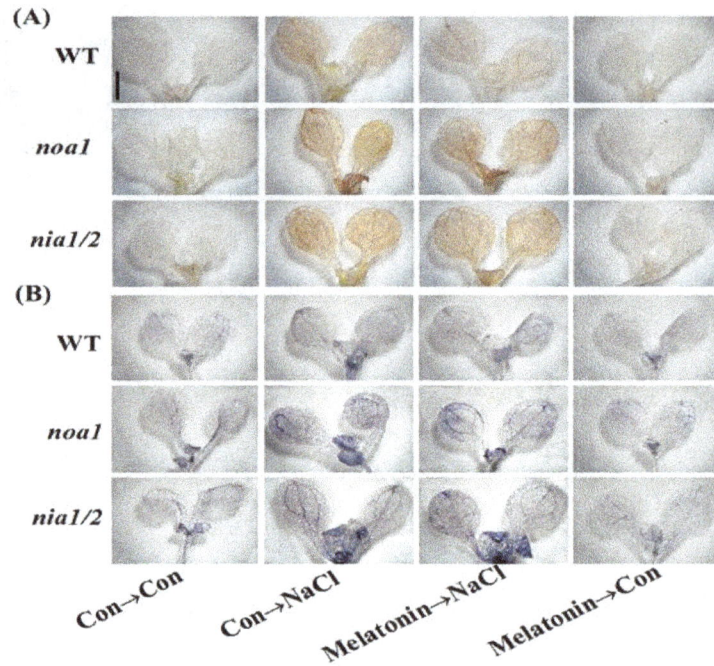

Figure 11. Genetic evidence revealed that redox balance was reestablished by melatonin via NO. Five-day-old wild-type (WT), *noa1*, and *nia1/2* mutant plants were grown on MS medium supplemented with 1.0 µM melatonin for 5 days, and then transplanted to medium in the presence or absence of 125 mM NaCl for another 5 days. Control seedlings (Con) were grown in MS medium alone. Afterwards, seedlings were stained with DAB (**A**) and NBT (**B**) to detect H_2O_2 and O_2^-. Scale bar = 1 mm.

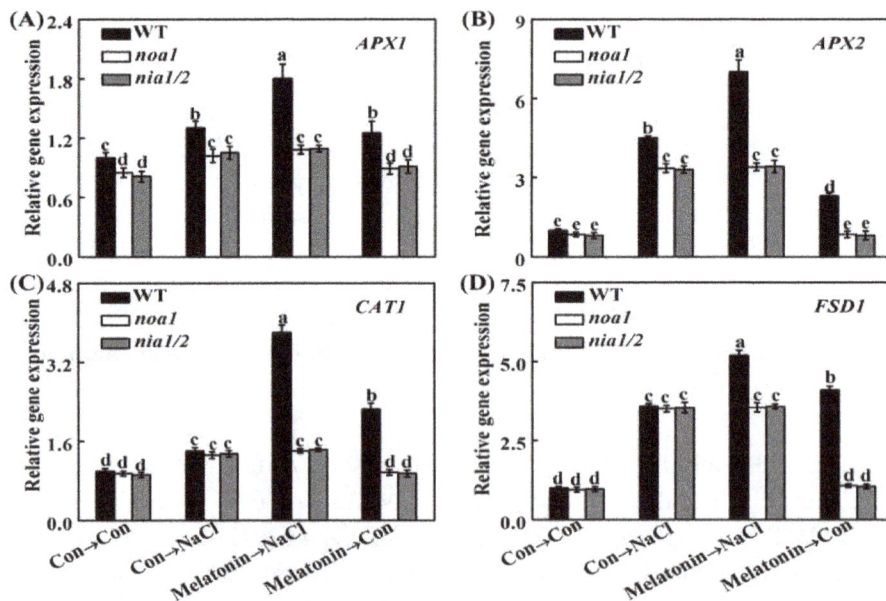

Figure 12. Changes in antioxidant gene expression. Five-day-old wild-type (WT), *noa1*, and *nia1/2* mutant plants were grown on MS medium supplemented with 1.0 µM melatonin for 5 days, and then transplanted to medium in the presence or absence of 125 mM NaCl for another 24 h. The mRNA expression of *APX1* (**A**), *APX2* (**B**), *CAT1* (**C**), and *FSD1* (**D**) in root tissues were analyzed by qPCR. Control seedlings (Con) were grown in MS medium alone. Data are means ± SE of three independent experiments with at least three replicates for each. Bars with different letters are significant different at $p < 0.05$ according to Duncan's multiple range test.

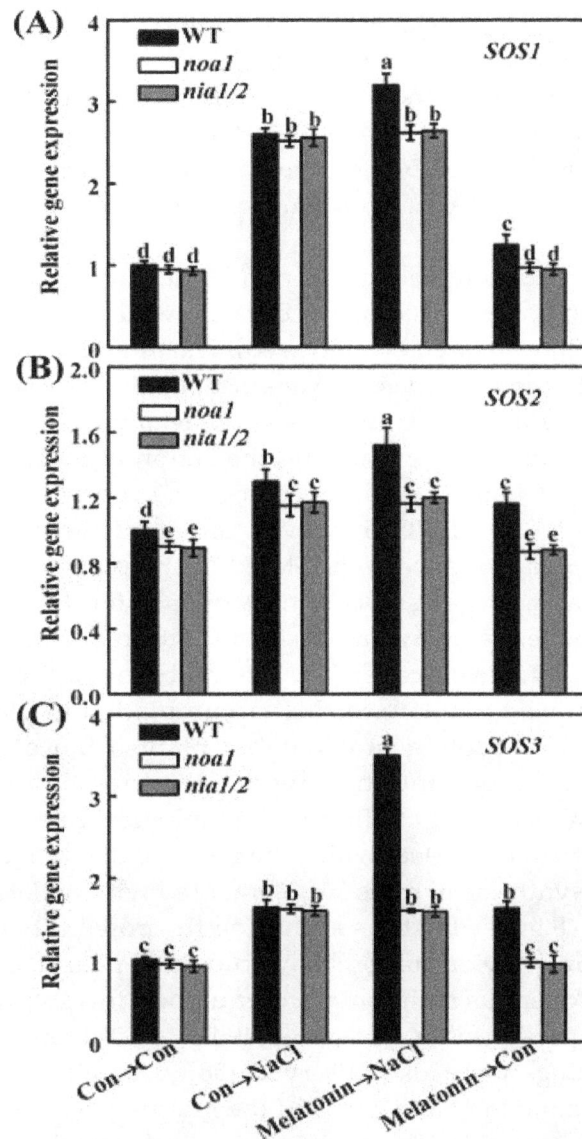

Figure 13. Gene evidence indicated that *SOS* signaling pathway was regulated by melatonin via NO. Five-day-old wild-type (WT), *noa1*, and *nia1/2* mutant plants were grown on MS medium supplemented with 1.0 μM melatonin for 5 days, and then transplanted in medium in the presence or absence of 125 mM NaCl for another 24 h. The mRNA expression of *SOS1* (**A**), *SOS2* (**B**), and *SOS3* (**C**) in root tissues were analyzed by qPCR. Control seedlings (Con) were grown in MS medium alone. Data are means ± SE of three independent experiments with at least three replicates for each. Bars with different letters are significant different at $p < 0.05$ according to Duncan's multiple range test.

3. Discussion

In this work, the molecular basis of melatonin-mediated plant salinity tolerance was investigated. The pharmacological and genetic data presented here show that rapeseed seedlings respond to salinity stress by increasing the concentration of melatonin and NO successively, and that NO is required for the melatonin-mediated plant tolerance against salinity stress.

First, the result shows that an increase in the concentration of melatonin followed by the induction of NO production, is one of the earliest responses involved in the signaling transduction elicited by NaCl stress in rapeseed seedlings (Figure 2). Although the major source(s) of NO have not been investigated in rapeseed seedlings, a cause-effect relationship between melatonin and NO production in salinity tolerance was further established.

Compared to the beneficial role of melatonin in the terms of the alleviation of rapeseed root growth inhibition caused by NaCl stress (Figure 3 and Figure S1), the application of SNP, NONOate, and GSNO, three well-known NO-releasing compounds, could result in the similar responses. The above effect was not found in seedlings supplemented with old SNP, a negative control of SNP, which contains no NO, but nitrate, nitrite, and ferrocyanide [51,52]. These results confirmed the previous conclusion that NO acts as an important gaseous molecule with multiple biological functions in plants, especially in salt tolerance [34–38].

Previously, the interplay between melatonin and NO in plant responses against stress has been controversial [53]. For example, pharmacological data showed that exogenous melatonin induces the production of NO in alkaline stressed tomato seedlings, and NO might be a downstream signal involved in the enhancement of tomato tolerance against alkaline stress triggered by melatonin [54]. A similar conclusion was obtained in Arabidopsis innate immunity against bacterial pathogen (Pst DC3000) [33]. In the present study, mimicking the action of SNP, the increased NO synthesis in response to melatonin in NaCl-stressed rapeseed seedling roots is demonstrated (Figure 5A–C). The removal of endogenous NO by PTIO, however, impaired the effects of melatonin and SNP. Most importantly, the above processes were correlated to the phenotypic responses, showing that the enhancement of plant tolerance against NaCl stress by melatonin is associated with endogenous NO levels (Figure 4). The above results were consistent with reported by Wen et al. [55], in which they found that the increased NO levels caused by the inhibition of S-nitrosoglutathione reductase (GSNOR), which could negatively regulate the NO accumulation in plants [56], were involved in melatonin-triggered adventitious rooting in cucumber plants. Since abiotic stress could trigger adventitious root formation, an important phenotype of the stress-induced morphogenic response (SIMR) in plants [57], our results further confirmed the biological roles of NO in both development and plant responses against stress [38]. Meanwhile, the removal of endogenous NO by PTIO does not alter endogenous melatonin synthesis in roots supplemented with melatonin or NaCl, neither alone or in different combinations (Figure 5D), thus excluding the possibility of NO-triggered melatonin production. By contrast, several previous reports have provided pharmacological evidence indicating that NO could stimulate endogenous melatonin accumulation in sunflower seedling cotyledons as a long-distance signaling response under salt stress [49,50]. NO-dependent melatonin synthesis in cadmium-stressed rice seedlings were also observed [58]. The above differences may reflect the complexity of melatonin signaling in plants [52], and the interplay between melatonin and NO might be dependent on the doses of stresses and exposure times, and even different plant species.

Lozano-Juste and Leŏn [45] reported that nia1/2 and noa1 mutants were more sensitive to NaCl-inhibited germination than wild-type seeds. Thus, we obtained these mutants to further study the role of endogenous NO in Arabidopsis salinity tolerance achieved by melatonin. Figure 10 shows that nia1/2 and noa1 seedlings seemed to be more sensitive than wild-type to NaCl stress, in terms of primary root elongation and chlorophyll degradation (in particular). Most importantly, exogenous melatonin-enhanced salt tolerance was markedly impaired in above Arabidopsis mutants, which were impaired in NIA/NR- and AtNOA1-dependent NO biosynthesis [42,44,45,51,59]. Thus, this genetic evidence supported the requirement of NO in melatonin-alleviated NaCl stress, at least in our experimental conditions, although the possibility of the direct scavenging or the inhibition of NO synthesis by melatonin could not be easily ruled out [60–62].

Upon NaCl stress, redox imbalance caused by ROS overproduction normally occurred in plants, thus leading to growth stunt and even cell death [1–4]. Further genetic evidence strongly revealed that plant tolerance against NaCl stress is closely associated with a more efficient antioxidant defense [13,20,35,63]. Previously, the antioxidant properties of melatonin against the overproduction of ROS have been confirmed in plant responses against salinity stress [34–37]. Similarly to the above discoveries, our results revealed that melatonin counteracted NaCl stress-induced oxidative damage in rapeseed seedlings, which could be confirmed by the decreased ROS production and lipid peroxidation (Figure 6), as well as the induction of representative antioxidant gene expression, including *APX*, *Cu/ZnSOD*, and *MnSOD* (Figure 7A–C). By using biochemical determination, it was further suggested that melatonin was able to increase APX and SOD activities (Figure 7D,E), not only at the transcriptional levels. By contrast, the above changes were prevented by PTIO, an NO scavenger [41,42], indicating that NO is involved in the reestablishment redox balance triggered by melatonin in NaCl stressed rapeseed seedlings. Consistently, we further showed that melatonin was able to reestablish redox balance in Arabidopsis wild-type seedlings upon NaCl stress, but was ineffective in *nia1/2* and *noa1* plants, two NO-deficient mutants (Figures 11 and 12), both of which were used to dissect physiological function of NO in plants [42,44,45]. Together, the above genetic evidence further confirmed that the counteracting effect of melatonin on oxidative damage induced by NaCl stress is NO-dependent.

Maintenance of ion homeostasis, in particular, the Na^+ to K^+ ratio, is another important approach for plants to resist NaCl stress [5,63]. Previous results showed that SOS1, a Na^+/H^+ antiporter, mainly mediated Na efflux, and the activation of SOS2/SOS3 complex can regulate *SOS1* gene expression [63,64]. The reestablishment of ion homeostasis by NO was previously confirmed in NaCl-stressed reed plants, showing that the NaCl-enhanced Na^+ to K^+ ratio is decreased by NO [41]. The application with melatonin exhibited a similar action [34,35], which was confirmed in rapeseed seedlings upon NaCl stress (Figure 8A). Interestingly, changes in rapeseed *NHX1* and *SOS2* transcripts were consistent with the results in Na^+/K^+ ratio (Figure 8B,C). However, the above changes were impaired by the removal of NO, suggesting the NHX-mediated Na^+-sequestration [10] and SOS-mediated Na^+ efflux [64] triggered by NO, might be two important strategies for plant tolerance against NaCl stress when supplemented with melatonin. The above conclusion was partially supported by the changes of *SOS1*, *SOS2*, and *SOS3* transcripts in wild-type, and *nia1/2* and *noa1* Arabidopsis mutants (Figure 13).

NO-mediated *S*-nitrosylation, a redox-related modification of cysteine thiol, is regarded as one of the important post-translational modifications to regulate enzyme activity and interactions among proteins in animals [46]. In higher plants, the formation of *S*-nitrosylation is associated with a wide range of physiological responses, including stress tolerance and root organogenesis [40,47,48,62]. Although a previous paper points to the possible involvement of NO-dependent *S*-nitrosylation in melatonin signaling [33], no study has yet provided definitive proof. In the subsequent work, we discovered that NaCl-induced *S*-nitrosylation was intensified by melatonin and SNP, but impaired by the removal of NO, respectively (Figure 9). Although we have not investigated the detailed target proteins of *S*-nitrosylation, our results strengthened the role of NO in melatonin responses.

Overall, we here presented the first genetic and pharmacological evidence showing the involvement of NO in melatonin signaling in salinity tolerance, and the model was shown in Figure 14. This model proposes that, upon NaCl stress, the increased melatonin triggers a signaling cascade that leads to an induction in NR- and NOA1-dependent NO concentration, thus resulting in the enhancement of salt stress tolerance. Meanwhile, the reestablishment of redox and *SOS*-mediated ion homeostasis are regarded as the important mechanism. NO-dependent *S*-nitrosylation was also illustrated in melatonin responses, and this is a new finding. Thus, the possibility that *miRNA398* modulated antioxidant gene expression [65] and *S*-nitrosylation targeted APX [48], in the above melatonin-mediated NO action, could not be easily ruled out. Further investigation may reveal melatonin-mediated NO-targeted proteins for genetic modification following biotechnological approaches, ultimately aiming to enhance plant tolerance against NaCl stress. Additionally, related studies using potted or field grown rapeseed plants incubated in soil/potting media with a prolonged experimental time would further help in understanding the melatonin and NO application in improving abiotic stress tolerance, thus providing potential benefits in agriculture.

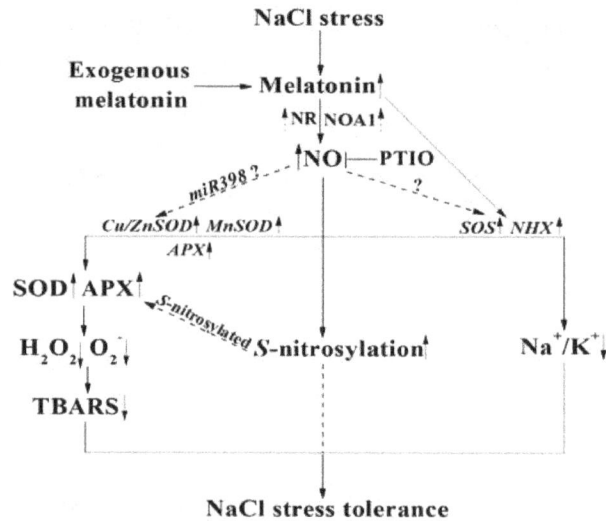

Figure 14. A model depicting the requirement of NR- and NOA1-dependent NO in melatonin-enhanced tolerance against salinity. The reestablishment of redox and ion homeostasis was involved. Dashed lines denote indirect or still undescribed pathways, including *miRNA398*-modulated gene expression [65] and *S*-nitrosylated antioxidant enzymes (APX, [48]).

4. Materials and Methods

4.1. Chemicals

All chemicals were purchased from Sigma-Aldrich (St. Louis, MO, USA) unless stated otherwise. The chemicals used for treatment were melatonin, sodium nitroprusside (SNP, a NO-releasing compound), 2-phenyl-4,4,5,5,-tetramethylimidazoline-1-oxyl-3-oxide (PTIO, a scavenger of NO), diethylamine NONOate (NONOate, a NO donor), and *S*-nitrosoglutathione (GSNO, a NO donor). An SNP solution was used as a negative control of SNP, by maintaining a separated solution of SNP for at least 10 days in light, in a specific open tube, to eliminate all NO [66]. The concentrations of chemicals used in this study were determined in pilot experiments from which maximal induced responses were obtained.

4.2. Plant Materials, Growth Condition, and Experimental Design

Rapeseed (*Brassica napus* L. zhongshuang 11) were kindly supplied by Chinese Academy of Agricultural Sciences. Rapeseed were surface-sterilized with 5% NaClO for 10 min, and rinsed comprehensively in distilled water, then germinated for 2 days at 25 °C in the darkness. Subsequently, the uniform seedlings were cultured with half-strength Hoagland solution in an illuminating incubator (16 h light with a light intensity of 200 $\mu mol \cdot m^{-2} \cdot s^{-1}$, 25 ± 1 °C, and 8 h dark, 23 ± 1 °C). Three-day-old rapeseed seedlings were treated with the indicated chemicals as shown in legends. Three independent experiments with at least three replicates for each were performed, and 30 samples were included in each replicate. Afterwards, rapeseed plants were photographed, and seedling roots were measured immediately after various treatments, or sampled for other analysis.

Arabidopsis thaliana noa1 (CS6511, Col-0) and *nia1/2* (CS2356, Col-0) mutants were obtained from the Arabidopsis Biological Resource Center (http://www.arabidopsis.org/, 5 June 2018). Seeds were surface-sterilized by sodium hypochlorite and rinsed three times with sterile water, then cultured on the solid Murashige and Skoog (MS, pH 5.8) medium containing 1% (*w/v*) agar and 1% (*w/v*) sucrose. For culturing NR-related mutant, the nitrogen in the MS medium included 1 mM NH_4^+ and 1.94 mM NO_3^- [44]. Plates containing seeds were kept at 4 °C for 2 days, and then transferred into the growth chamber with a 16/8 h (23/21 °C) day/light regimes at 150 $\mu mol \cdot m^{-2} \cdot s^{-1}$ irradiation. Five-day-old wild-type (WT), and *noa1*, and *nia1/2* mutant plants grown on MS medium were treated with the indicated chemicals, as shown in legends. Three independent experiments with at least three replicates for each were performed, and about 80 samples were included in each replicate. Afterwards, Arabidopsis plants were photographed, and seedling leaf and root parts were measured immediately, or sampled for other analysis.

4.3. Determination of Melatonin by Enzyme-Linked Immunosorbent Assay (ELISA)

Melatonin was extracted from root tissues by using an acetone-methanol method, and determined by enzyme-linked immunosorbent assay (ELISA) [33,35]. After centrifugation at $12,000 \times g$ for 15 min at 4 °C, the extract was used for quantification of melatonin using the Melatonin ELISA Kit (Jiangsu Baolai Biotechnology, Nanjing, China).

4.4. Determination of NO by Griess Reagent

NO production in root tissues was determined by using Griess reagent [66,67]. Identical filtrate pretreated with 2-(4-carboxyphenyl)-4,4,5,5-tetramethylimidazoline-1-oxyl-3-oxide potassium salt (cPTIO), the specific scavenger of NO, for 15 min, was used as blanks. Absorbance was assayed at 540 nm, and NO content was calculated by comparison to a standard curve of $NaNO_2$.

4.5. Laser Confocal Determination of Endogenous NO Production

Using a fairly specific NO fluorescent probe 4-amino-5-methylamino-2′,7′-difluorofluorescein diacetate (DAF-FM DA), the endogenous NO level of root tissues was detected by a TCS-SP2 confocal laser scanning microscopy (Leica Lasertechnik GmbH, Heidelberg, Germany; excitation at 488 nm, emission at 500–530 nm) [67,68]. Results are from five replicates per experiment. Fluorescence was expressed as relative fluorescence units using the Leica Confocal Software 2.5 (Leica Lasertechnik GmbH, Heidelberg, Germany).

4.6. ROS Detection

H_2O_2 and O_2^- were histochemically detected by 3,3′-diaminobenzidine (DAB) and nitroblue tetrazolium (NBT) staining [35]. Finally, all samples were observed using a light microscope (model Stemi 2000-C; Carl Zeiss, Oberkochen, Germany).

4.7. Assay of Thiobarbituric Acid Reactive Substances (TBARS) Content

Oxidative damage was estimated by measuring the concentration of TBARS as previously described [69].

4.8. Determination of Antioxidant Enzyme Activities

Root tissues were crushed into fine powder in a mortar and pestle under liquid N_2. Soluble proteins were extracted and homogenized in 50 mM PBS (pH 7.0) containing 1 mM EDTA and 1% polyvinylpyrrolidone (PVP), or with the addition of 1 mM ascorbic acid (ASA) in the case of ascorbate peroxidase (APX) activity determination. APX and SOD activities were measured as described previously [35,70]. Protein content was determined according to the method described by Bradford [71].

4.9. Real-Time Quantitative RT-PCR (qPCR) Analysis

According to our pervious method [35], total RNA was extracted from seedling roots using a Tranzol up kit (TransGen Biotech, Beijing, China). RNA concentration and quality were determined using the NanoDrop 2000 (Thermo Fisher Scientific, Wilmington, DE, USA), and then incubated with RNase-free DNase (TaKaRa Bio Inc., Dalian, China) to eliminate traces of DNA. cDNAs were then synthesized using an oligo(dT) primer and a SuperScript First-Strand Synthesis System (Invitrogen, Carlsbad, CA, USA).

By using the gene-specific primers (Supplementary Materials Tables S1 and S2), qPCR was conducted using a Mastercycler ep® *realplex* real-time PCR system (Eppendorf, Hamburg, Germany) with *TransStart®* Green qPCR SuperMix (TransGen Biotech, Beijing, China) [35]. In rapeseed plants, the expression levels of genes were normalized to two internal control gene *Actin* and *GADPH* transcript levels, and presented as values relative to corresponding control samples in NaCl-free conditions. In Arabidopsis, the expression levels of genes were normalized to two internal control gene *Actin* 2 and *GADPH* transcript levels, and presented as values relative to corresponding control of wild-type samples in NaCl-free conditions.

4.10. Determination of Ion Contents

Fresh seedling roots were harvested and washed four times by deionized water after treatments. According to the previous method [21,35], Na and K element contents were measured with an Inductively Coupled Plasma Optical Emission Spectrometer (ICP-OES, Perkin Elmer Optima 2100 DV; PerkinElmer, Shelton, CT, USA).

4.11. Quantification of Chlorophyll Content

Total chlorophyll was extracted using 95% (v/v) ethanol for 24 h in darkness, and then calculated by examining the absorbance at 649 nm and 665 nm [35].

4.12. Modified Biotin Switch Method

Modified biotin switch method was carried out according to the previous method [40,62]. After treatment, total protein was extracted from fresh seedling roots, and *S*-nitrosylated and biotin-labelled proteins were separated by using non-reducing sodium dodecyl sulfate polyacrylamide gel electrophoresis (SDS-PAGE; 12%). Western blotting was then performed to detect the proteins. Anti-biotin antibody horseradish peroxidase (Abcam antibodies, Cambridge, UK) was diluted 1:6000. Coomassie Brilliant Blue-stained gels were used to show that equal amounts of proteins were loaded.

4.13. Statistical Analysis

Values are means ± SE of three independent experiments with at least three replicates for each. Statistical analysis was performed using SPSS 16.0 software (IBM Corporation, Armonk, NY, USA). Differences among treatments were analyzed by one-way analysis of variance (ANOVA), taking $p < 0.05$ as significant according to Duncan's multiple range test.

Author Contributions: G.Z., Y.Z., X.Y., and W.S. designed and refined the research; G.Z., Y.Z., X.Y., H.H., R.G., and R.W. performed research; G.Z. and Y.Z. prepared the mutant materials; G.Z., Y.Z., F.K., and W.S. analyzed data; G.Z., Y.Z., R.G., and W.S. wrote the article. All authors discussed the results and comments on the manuscript.

Acknowledgments: This work was partly supported by the National Key Research and Development Plan (2016YFD0101306), the Fundamental Research Funds for the Central Universities (KYTZ201402), and the Priority Academic Program Development of Jiangsu Higher Education Institutions (PAPD).

Abbreviations

ASA	Ascorbic acid
CAT	Catalase
cPTIO	2-(4-carboxyphenyl)-4,4,5,5-tetramethylimidazoline-1-oxyl-3-oxide potassium salt
DAB	3,3'-diaminobenzidine
GSNO	*S*-nitrosoglutathione
NBT	Nitroblue tetrazolium
NHX1	Sodium hydrogen exchanger
noa1	Nitric oxide associated1
NO	Nitric oxide
NONOate	Diethylamine
NR	Nitrate reductase
POD	Guaiacol peroxidase
PTIO	2-phenyl-4,4,5,5,-tetramethylimidazoline-1-oxyl-3-oxide
ROS	Reactive oxygen species
SNP	Sodium nitroprusside
SOD	Superoxide dismutase
SOS	Salt overly sensitive

References

1. Zhu, J.K. Plant salt tolerance. *Trends Plant Sci.* **2001**, *6*, 66–71. [CrossRef]
2. Zhu, J.K. Salt and drought stress signal transduction in plants. *Annu. Rev. Plant Biol.* **2002**, *53*, 247–273. [CrossRef] [PubMed]
3. Parida, A.K.; Das, A.B. Salt tolerance and salinity effects on plants: A review. *Ecotoxicol. Environ. Saf.* **2005**, *60*, 324–349. [CrossRef] [PubMed]
4. Hasegawa, P.M.; Bressan, R.A.; Zhu, J.K.; Bohnert, H.J. Plant cellular and molecular responses to high salinity. *Annu. Rev. Plant Biol.* **2000**, *51*, 463–499. [CrossRef] [PubMed]
5. Zhu, J.K. Regulation of ion homeostasis under salt stress. *Curr. Opin. Plant Biol.* **2003**, *6*, 441–445. [CrossRef]
6. Ohta, M.; Hayashi, Y.; Nakashima, A.; Hamada, A.; Tanaka, A.; Nakanura, T.; Hayakawa, T. Introduction of a Na$^+$/H$^+$ antiporter gene from *Atriplex gmelini* confers salt tolerance to rice. *FEBS Lett.* **2002**, *532*, 279–282. [CrossRef]
7. Shi, H.; Lee, B.; Wu, S.J.; Zhu, J.K. Overexpression of a plasma membrane Na$^+$/H$^+$ antiporter gene improves salt tolerance in *Arabidopsis thaliana*. *Nat. Biotechnol.* **2003**, *21*, 81–85. [CrossRef] [PubMed]
8. Shi, H.; Ishitani, M.; Kim, C.; Zhu, J.K. The *Arabidopsis thaliana* salt tolerance gene *SOS1* encodes a putative Na$^+$/H$^+$ antiporter. *Proc. Natl. Acad. Sci. USA* **2000**, *97*, 6896–6901. [CrossRef] [PubMed]
9. Shi, H.; Quintero, F.J.; Pardo, J.M.; Zhu, J.K. The putative plasma membrane Na$^+$/H$^+$ antiporter SOS1 controls long-distance Na$^+$ transport in plants. *Plant Cell* **2002**, *14*, 465–477. [CrossRef] [PubMed]
10. Munns, R.; Tester, M. Mechanisms of salinity tolerance. *Annu. Rev. Plant Biol.* **2008**, *59*, 651–681. [CrossRef] [PubMed]
11. Ashraf, M.; Ali, Q. Relative membrane permeability and activities of some antioxidant enzymes as the key determinants of salt tolerance in canola (*Brassica napus* L.). *Environ. Exp. Bot.* **2008**, *63*, 266–273. [CrossRef]
12. Miller, G.; Suzuki, N.; Ciftci-Yilmaz, S.; Mittler, R. Reactive oxygen species homeostasis and signalling during drought and salinity stresses. *Plant Cell Environ.* **2010**, *33*, 453–467. [CrossRef] [PubMed]
13. Mittler, R.; Vanderauwera, S.; Suzuki, N.; Miller, G.; Tognetti, V.B.; Vandepoele, K.; Gollery, M.; Shulaev, V.; Breusegem, F.V. ROS signaling: The new wave? *Trends Plant Sci.* **2011**, *16*, 300–309. [CrossRef] [PubMed]
14. Naeem, M.S.; Jin, Z.L.; Wan, G.L.; Liu, D.; Liu, H.B.; Yoneyama, K.; Zhou, W.J. 5-Aminolevulinic acid improves photosynthetic gas exchange capacity and ion uptake under salinity stress in oilseed rape (*Brassica napus* L.). *Plant Soil* **2010**, *332*, 405–415. [CrossRef]
15. Ruiz, J.M.; Blumwald, E. Salinity-induced glutathione synthesis in *Brassica napus*. *Planta* **2002**, *214*, 965–969. [CrossRef] [PubMed]

16. Dai, Q.; Chen, C.; Feng, B.; Liu, T.; Tian, X.; Gong, Y.; Sun, Y.; Wang, J.; Du, S. Effects of different NaCl concentration on the antioxidant enzymes in oilseed rape (*Brassica napus* L.) seedlings. *Plant Growth Regul.* **2009**, *59*, 273–278. [CrossRef]

17. Kagale, S.; Divi, U.K.; Krochko, J.E.; Keller, W.A.; Krishna, P. Brassinosteroid confers tolerance in *Arabidopsis thaliana* and *Brassica napus* to a range of abiotic stresses. *Planta* **2007**, *225*, 353–364. [CrossRef] [PubMed]

18. Jalili, F.; Khavazi, K.; Pazira, E.; Nejati, A.; Rahmani, H.A.; Sadaghiani, H.R.; Miransarf, M. Isolation and characterization of ACC deaminase-producing fluorescent pseudomonads, to alleviate salinity stress on canola (*Brassica napus* L.) growth. *J. Plant Physiol.* **2009**, *166*, 667–674. [CrossRef] [PubMed]

19. Wei, W.; Li, Q.T.; Chu, Y.N.; Reiter, R.J.; Yu, X.M.; Zhu, D.H.; Zhang, W.K.; Ma, B.; Lin, Q.; Zhang, J.S.; et al. Melatonin enhances plant growth and abiotic stress tolerance in soybean plants. *J. Exp. Bot.* **2015**, *66*, 695–707. [CrossRef] [PubMed]

20. Liang, C.; Zheng, G.; Li, W.; Wang, Y.; Hu, B.; Wang, H.; Wu, H.; Qian, Y.; Zhu, X.G.; Tan, D.X.; et al. Melatonin delays leaf senescence and enhances salt stress tolerance in rice. *J. Pineal Res.* **2015**, *59*, 91–101. [CrossRef] [PubMed]

21. Xie, Y.; Ling, T.; Liu, K.; Zheng, Q.; Huang, L.; Yuan, X.; He, Z.; Hu, B.; Fang, L.; Shen, Z.; et al. Carbon monoxide enhances salt tolerance by nitric oxide-mediated maintenance of ion homeostasis and up-regulation of antioxidant defence in wheat seeding roots. *Plant Cell Environ.* **2008**, *31*, 1864–1881. [CrossRef] [PubMed]

22. Kumar, M.; Choi, J.; An, G.; Kim, S.R. Ectopic expression of *OsSta2* enhances salt stress tolerance in rice. *Front. Plant Sci.* **2017**, *8*, 316. [CrossRef] [PubMed]

23. Li, H.; Chang, J.; Chen, H.; Wang, Z.; Gu, X.; Wei, C.; Zhang, Y.; Ma, J.; Yang, J.; Zhang, X. Exogenous melatonin confers salt stress tolerance to watermelon by improving photosynthesis and redox homeostasis. *Front. Plant Sci.* **2017**, *8*, 295. [CrossRef] [PubMed]

24. Zeng, L.; Cai, J.; Li, J.; Lu, G.; Li, C.; Fu, G.; Zhang, X.; Ma, H.; Liu, Q.; Zou, X.; et al. Exogenous application of a low concentration of melatonin enhances salt tolerance in rapeseed (*Brassica napus* L.) seedlings. *J. Integr. Agric.* **2018**, *17*, 328–335. [CrossRef]

25. Kumar, M.; Choi, J.Y.; Kumari, N.; Pareek, A.; Kim, S.R. Molecular breeding in *Brassica* for salt tolerance: Importance of microsatellite (SSR) markers for molecular breeding in *Brassica*. *Front. Plant Sci.* **2015**, *6*, 688. [CrossRef] [PubMed]

26. Lerner, A.B.; Case, J.D.; Takahashi, Y.; Lee, T.H.; Mori, W. Isolation of melatonin, the pineal gland factor that lightens melanocytes. *J. Am. Chem. Soc.* **1958**, *80*, 2587. [CrossRef]

27. Reiter, R.J.; Tan, D.X.; Galano, A. Melatonin: Exceeding expectations. *Physiology* **2014**, *29*, 325–333. [CrossRef] [PubMed]

28. Jung-Hynes, B.; Reiter, R.J.; Ahmad, N. Melatonin and circadian rhythms: Building a bridge between aging and cancer. *J. Pineal Res.* **2010**, *48*, 9–19. [CrossRef] [PubMed]

29. Cozzi, B.; Morei, G.; Ravault, J.P.; Chesneau, D.; Reiter, R.J. Circadian and seasonal rhythms of melatonin production in mules (*Equus asinus* × *Equus caballus*). *J. Pineal Res.* **1991**, *10*, 130–135. [CrossRef] [PubMed]

30. Santofimia-Castaño, P.; Ruy, D.C.; Garcia-Sanchez, L.; Jimenez-Blasco, D.; Fernandez-Bermejo, M.; Bolaños, J.P.; Salido, G.M.; Gonzalez, A. Melatonin induces the expression of Nrf2-regulated antioxidant enzymes via PKC and Ca^{2+} influx activation in mouse pancreatic acinar cells. *Free Radic. Biol. Med.* **2015**, *87*, 226–236. [CrossRef] [PubMed]

31. Reiter, R.J.; Tan, D.X.; Zhou, Z.; Cruz, M.H.C.; Fuentes-Broto, L.; Galano, A. Phytomelatonin: Assisting plants to survive and thrive. *Molecules* **2015**, *20*, 7396–7437. [CrossRef] [PubMed]

32. Bajwa, V.S.; Shukla, M.R.; Sherif, S.M.; Murch, S.J.; Saxena, P.K. Role of melatonin in alleviating cold stress in *Arabidopsis thaliana*. *J. Pineal Res.* **2014**, *56*, 238–245. [CrossRef] [PubMed]

33. Shi, H.; Chen, Y.; Tan, D.X.; Reiter, R.J.; Chan, Z.; He, Z. Melatonin induces nitric oxide and the potential mechanisms relate to innate immunity against bacterial pathogen infection in *Arabidopsis*. *J. Pineal Res.* **2015**, *59*, 102–108. [CrossRef] [PubMed]

34. Li, C.; Wang, P.; Wei, Z.; Liang, D.; Liu, C.; Yin, L.; Jia, D.; Fu, M.; Ma, F. The mitigation effects of exogenous melatonin on salinity-induced stress in *Malus hupehensis*. *J. Pineal Res.* **2012**, *53*, 298–306. [CrossRef] [PubMed]

35. Chen, Z.; Xie, Y.; Gu, Q.; Zhao, G.; Zhang, Y.; Cui, W.; Xu, S.; Wang, R.; Shen, W. The *AtrbohF*-dependent regulation of ROS signaling is required for melatonin-induced salinity tolerance in *Arabidopsis*. *Free Radic. Biol. Med.* **2017**, *108*, 465–477. [CrossRef] [PubMed]

36. Zhang, H.J.; Zhang, N.; Yang, R.C.; Wang, L.; Sun, Q.Q.; Li, D.B.; Cao, Y.Y.; Wedda, S.; Zhao, B.; Ren, S.; et al. Melatonin promotes seed germination under high salinity by regulating antioxidant systems, ABA and GA$_4$ interaction in cucumber (*Cucumis sativus* L.). *J. Pineal Res.* **2014**, *57*, 269–279. [CrossRef] [PubMed]

37. Shi, H.; Jiang, C.; Ye, T.; Tan, D.X.; Reiter, R.J.; Zhang, H.; Liu, R.; Chan, R. Comparative physiological, metabolomic, and transcriptomic analyses reveal mechanisms of improved abiotic stress resistance in bermudagrass [*Cynodon dactylon* (L.). Pers.] by exogenous melatonin. *J. Exp. Bot.* **2015**, *66*, 681–694. [CrossRef] [PubMed]

38. Besson-Bard, A.; Pugin, A.; Wendehenne, D. New insights into nitric oxide signaling in plants. *Annu. Rev. Plant Biol.* **2008**, *59*, 21–39. [CrossRef] [PubMed]

39. Gupta, K.J.; Fernie, A.R.; Kaiser, W.M.; Dengon, J. On the origins of nitric oxide. *Trends Plant Sci.* **2011**, *16*, 160–168. [CrossRef] [PubMed]

40. Su, J.; Zhang, Y.; Nie, Y.; Chen, D.; Wang, R.; Hu, H.; Chen, J.; Zhang, J.; Du, Y.; Shen, W. Hydrogen-induced osmotic tolerance is associated with nitric oxide-mediated proline accumulation and reestablishment of redox balance in alfalfa seedlings. *Environ. Exp. Bot.* **2018**, *147*, 249–260. [CrossRef]

41. Zhao, L.; Zhang, F.; Guo, J.; Yang, Y.; Li, B.; Zhang, L. Nitric oxide functions as a signal in salt resistance in the calluses from two ecotypes of reed. *Plant Physiol.* **2004**, *134*, 849–857. [CrossRef] [PubMed]

42. Zhao, M.G.; Tian, Q.Y.; Zhang, W.H. Nitric oxide synthase-dependent nitric oxide production is associated with salt tolerance in *Arabidopsis*. *Plant Physiol.* **2007**, *144*, 206–217. [CrossRef] [PubMed]

43. Wang, Y.; Li, L.; Cui, W.; Xu, S.; Shen, W.; Wang, R. Hydrogen sulfide enhances alfalfa (*Medicago sativa*) tolerance against salinity during seed germination by nitric oxide pathway. *Plant Soil* **2012**, *351*, 107–119. [CrossRef]

44. Xie, Y.; Mao, Y.; Lai, D.; Zhang, W.; Zheng, T.; Shen, W. Roles of NIA/NR/NOA1-dependent nitric oxide production and HY1 expression in the modulation of *Arabidopsis* salt tolerance. *J. Exp. Bot.* **2013**, *64*, 3045–3060. [CrossRef] [PubMed]

45. Lozano-Juste, J.; León, J. Enhanced abscisis acid-mediated responses in *nia1nia2noa1-2* triple mutant impaired in NIA/NR- and AtNOA1-dependent nitric oxide biosynthesis in Arabidopsis. *Plant Physiol.* **2010**, *152*, 891–903. [CrossRef] [PubMed]

46. Jaffrey, S.R.; Erdjument-Bromage, H.; Ferris, C.D.; Tempst, P.; Snyder, S.H. Protein *S*-nitrosylation: A physiological signal for neuronal nitric oxide. *Nat. Cell Biol.* **2001**, *3*, 193–197. [CrossRef] [PubMed]

47. Yun, B.W.; Feechan, A.; Yin, M.; Saidi, N.B.B.; Bihan, T.L.; Yu, M.; Moore, J.W.; Kang, J.G.; Kwon, E.; Spoel, S.H.; et al. *S*-nitrosylation of NADPH oxidase regulates cell death in plant immunity. *Nature* **2011**, *478*, 264–268. [CrossRef] [PubMed]

48. Lindermayr, C.; Saalbach, G.; Durner, J. Proteomic identification of *S*-nitrosylated proteins in Arabidopsis. *Plant Physiol.* **2005**, *137*, 921–930. [CrossRef] [PubMed]

49. Kaur, H.; Bhatla, S.C. Melatonin and nitric oxide modulate glutathione content and glutathione reductase activity in sunflower seedling cotyledons accompanying salt stress. *Nitric Oxide* **2016**, *59*, 42–53. [CrossRef] [PubMed]

50. Arora, D.; Bhatla, S.C. Melatonin and nitric oxide regulate sunflower seedling growth under salt stress accompanying differential expression of Cu/Zn SOD and Mn SOD. *Free Radic. Biol. Med.* **2017**, *106*, 315–328. [CrossRef] [PubMed]

51. Han, B.; Yang, Z.; Xie, Y.; Nie, L.; Cui, J.; Shen, W. *Arabidopsis* HY1 confers cadmium tolerance by decreasing nitric oxide production and improving iron homeostasis. *Mol. Plant* **2014**, *7*, 388–403. [CrossRef] [PubMed]

52. Chen, M.; Cui, W.; Zhu, K. Hydrogen-rich water alleviates aluminum-induced inhibition of root elongation in alfalfa via decreasing nitric oxide production. *J. Hazard. Mater.* **2014**, *267*, 40–47. [CrossRef] [PubMed]

53. Arnao, M.B.; Hernández-Ruiz, J. Melatonin and its relationship to plant hormones. *Ann. Bot. Lond.* **2018**, *121*, 195–207. [CrossRef] [PubMed]

54. Liu, N.; Gong, B.; Jin, Z.; Wang, X.; Wei, M.; Yang, F.; Li, Y.; Shi, Q. Sodic alkaline stress mitigation by exogenous melatonin in tomato needs nitric oxide as a downstream signal. *J. Plant Physiol.* **2015**, *186*, 68–77. [CrossRef] [PubMed]

55. Wen, D.; Gong, B.; Sun, S.; Liu, S.; Wang, X.; Wei, M.; Yang, F.; Li, Y.; Shi, H. Promoting roles of melatonin in adventitious root development of *Solanum lycopersicum* L. by regulating auxin and nitric oxide signaling. *Front. Plant Sci.* **2016**, *7*, 718. [CrossRef] [PubMed]

56. Corpas, F.J.; Alché, J.D.; Barroso, J.B. Current overview of *S*-nitrosoglutathione (GSNO) in higher plants. *Front. Plant Sci.* **2013**, *4*, 126. [CrossRef] [PubMed]

57. Potters, G.; Pasternak, T.P.; Guisez, Y.; Palme, J.K.; Jansen, M.A.K. Stress-induced morphogenic responses: Growing out of trouble? *Trends Plant Sci.* **2007**, *12*, 98–105. [CrossRef] [PubMed]

58. Lee, K.; Choi, G.H.; Back, K. Cadmium-induced melatonin synthesis in rice requires light, hydrogen peroxide, and nitric oxide: Key regulatory roles for tryptophan decarboxylase and caffeic acid *O*-methyltransferase. *J. Pineal Res.* **2017**, *63*, e12441. [CrossRef] [PubMed]

59. Xie, Y.; Mao, Y.; Zhang, W.; Lai, D.; Wang, Q.; Shen, W. Reactive oxygen species-dependent nitric oxide production contributes to hydrogen-promoted stomatal closure in Arabidopsis. *Plant Physiol.* **2014**, *165*, 759–773. [CrossRef] [PubMed]

60. Noda, Y.; Mori, A.; Liburdy, R.; Packer, L. Melatonin and its precursors scavenge nitric oxide. *J. Pineal Res.* **1999**, *27*, 159–163. [CrossRef] [PubMed]

61. Kang, Y.S.; Kang, Y.G.; Park, H.J.; Wee, H.J.; Jang, H.O.; Bae, M.K.; Bae, S.K. Melatonin inhibits visfatin-induced inducible nitric oxide synthase expression and nitric oxide production in macrophages. *J. Pineal Res.* **2013**, *55*, 294–303. [CrossRef] [PubMed]

62. Guerrero, J.M.; Reiter, R.J.; Ortiz, G.G.; Pablos, M.I.; Sewerynek, E.; Chuang, J.I. Melatonin prevents increases in neural nitric oxide and cyclic GMP production after transient brain ischemia and reperfusion in the Mongolian gerbil (*Meriones unguiculatus*). *J. Pineal Res.* **1997**, *23*, 24–31. [CrossRef] [PubMed]

63. Deinlein, U.; Stephan, A.B.; Horie, T.; Luo, W.; Xu, G.; Schroeder, J.I. Plant salt-tolerance mechanisms. *Trends Plant Sci.* **2014**, *19*, 371–379. [CrossRef] [PubMed]

64. Qiu, Q.S.; Guo, Y.; Dietrich, M.A.; Schumaker, K.S.; Zhu, J.K. Regulation of SOS1, a plasma membrane Na^+/H^+ exchanger in *Arabidopsis thaliana*, by SOS2 and SOS3. *Proc. Natl. Acad. Sci. USA* **2002**, *99*, 8436–8441. [CrossRef] [PubMed]

65. Sunkar, R.; Kapoor, A.; Zhu, J.K. Posttranscriptional induction of two Cu/Zn superoxide dismutase genes in *Arabidopsis* is mediated by downregulation of miR398 and important for oxidative stress tolerance. *Plant Cell* **2006**, *18*, 2051–2065. [CrossRef] [PubMed]

66. Zhou, B.; Guo, Z.; Xing, J.; Huang, B. Nitric oxide is involved in abscisic acid-induced antioxidant activities in *Stylosanthes guianensis*. *J. Exp. Bot.* **2005**, *56*, 3223–3228. [CrossRef] [PubMed]

67. Balcerczyk, A.; Soszynski, M.; Bartosz, G. On the specificity of 4-amino-5-methylamino-2′,7′-difluorofluorescein as a probe for nitric oxide. *Free Radic. Biol. Med.* **2005**, *39*, 327–335. [CrossRef] [PubMed]

68. Qi, F.; Xiang, Z.; Kou, N.; Cui, W.; Wang, R.; Zhu, D.; Shen, W. Nitric oxide is involved in methane-induced adventitious root formation in cucumber. *Physiol. Plant.* **2017**, *159*, 366–377. [CrossRef] [PubMed]

69. Han, Y.; Zhang, J.; Chen, X.; Gao, Z.; Xuan, W.; Xu, S.; Ding, X.; Shen, W. Carbon monoxide alleviates cadmium-induced oxidative damage by modulating glutathione metabolism in the roots of *Medicago sativa*. *New Phytol.* **2008**, *177*, 155–166. [CrossRef] [PubMed]

70. Nakano, Y.; Asada, K. Hydrogen peroxide is scavenged by ascorbate-specific peroxidase in spinach chloroplasts. *Plant Cell Physiol.* **1981**, *22*, 867–880.

71. Bradford, M.M. A rapid and sensitive method for the quantitation of microgram quantities of protein utilizing the principle of protein-dye binding. *Anal. Biochem.* **1976**, *72*, 248–254. [CrossRef]

6

CaDHN5, a Dehydrin Gene from Pepper, Plays an Important Role in Salt and Osmotic Stress Responses

Dan Luo, Xiaoming Hou, Yumeng Zhang, Yuancheng Meng, Huafeng Zhang, Suya Liu, Xinke Wang and Rugang Chen *

College of Horticulture, Northwest A&F University, Yangling 712100, China; danluonwafu@163.com (D.L.); 15230286139@163.com (X.H.); Kexuanzhangyumeng@163.com (Y.Z.); YuanchengMeng07@126.com (Y.M.); 18848966687@163.com (H.Z.); YaSuLiu@126.com (S.L.); W1942399775@126.com (X.W.)
* Correspondence: rugangchen@nwsuaf.edu.cn

Abstract: Dehydrins (*DHNs*), as a sub-family of group two late embryogenesis-abundant (LEA) proteins, have attracted considerable interest owing to their functions in enhancing abiotic stress tolerance in plants. Our previous study showed that the expression of *CaDHN5* (a dehydrin gene from pepper) is strongly induced by salt and osmotic stresses, but its function was not clear. To understand the function of *CaDHN5* in the abiotic stress responses, we produced pepper (*Capsicum annuum* L.) plants, in which *CaDHN5* expression was down-regulated using VIGS (Virus-induced Gene Silencing), and transgenic *Arabidopsis* plants overexpressing *CaDHN5*. We found that knock-down of *CaDHN5* suppressed the expression of manganese superoxide dismutase (*MnSOD*) and peroxidase (*POD*) genes. These changes caused more reactive oxygen species accumulation in the VIGS lines than control pepper plants under stress conditions. *CaDHN5*-overexpressing plants exhibited enhanced tolerance to salt and osmotic stresses as compared to the wild type and also showed increased expression of salt and osmotic stress-related genes. Interestingly, our results showed that many salt-related genes were upregulated in our transgenic *Arabidopsis* lines under salt or osmotic stress. Taken together, our results suggest that *CaDHN5* functions as a positive regulator in the salt and osmotic stress signaling pathways.

Keywords: *Capsicum annuum* L.; *CaDHN5*; salt stress; osmotic stress; dehydrin

1. Introduction

Plants live in an open environment and cannot move from one place to another. As a result, plants are exposed to various biotic and abiotic stresses. These stresses individually, or in combination, result in huge losses in terms of growth, development, and yield, and sometimes threaten the survival of the plant. Amongst the abiotic factors, water stress is the most important [1,2]. Some earlier studies treated drought and salinity as similar stresses because plants respond in a similar manner to salt and drought stresses, and signaling mechanisms overlap [3].

Dehydrin, a highly hydrophilic plant protein, belongs to the second sub-family of the late embryogenesis developmental protein family (LEA II) [4]. The protein sequence has conserved K-segments (consisting of EKKGIMDKIKEKLPG located near the C-terminus, rich in lysine), Y-segments (consisting of T/VDEYGNP located close to the N-terminus), and S-segments (rich in serine) [5]. Dehydrins are classified into five categories: Y_nSK_n, K_n, SK_n, Y_nK_n, and K_nS [6]. Every type of dehydrin has a different function. For example, SK_n dehydrins can not only bind phospholipids, protect enzyme stability, and prevent heat-induced degeneration, but they are also crucial for plant growth, development, and resistance to low temperature stress responses [7]. Various abiotic stresses and hormones can strongly influence expression of dehydrin [8]. Studies have shown that Y_nSK_n is an alkaline or neutral protein that is highly upregulated by cold stress [9], and has a unique RRKK motif (a nuclear localization signal), which is key for the localization of Y_nSK_n-type dehydrins in the nucleus. Many studies have shown that there is a positive interaction between the expression of dehydrins and resistance to abiotic stresses [10]. *Cicer pinnatifidum* Y_2K-type dehydrin *CpDHN1* and white spruce S_8K_4-type dehydrin *PgDHN1* were induced by methyl jasmonate (MeJA) and salicylic acid (SA) [11,12]. *CaDHN5* belongs to the YSK_2 category of dehydrins. It is a neutral or basic protein that can be induced by osmotic or salt stresses, and exogenous abscisic acid (ABA), as shown in our previous study [13].

In recent years, many studies have explored the functional importance of dehydrin in plant stress resilience [14–20]. In addition, since the expression of various dehydrin genes can be induced by exogenous ABA treatment, dehydrins are also considered as ABA-responsive proteins (ABR) [14]. In tomato plants overexpressing dehydrin gene, drought resistance was enhanced without influencing tomato growth traits [15]. In barley, cold acclimation was due to faster *DHN5* accumulation rates in the winter lines compared to that of spring lines [16]. In *Physcomitrella*, under salt and mannitol stresses, the expression of *PpDHNA* and *PpDHNB* were strongly up-regulated [17]. Studies in transgenic *Arabidopsis thaliana* plants showed that overexpression of *AmDHN* (*Ammopiptanthus mongolicus* dehydrin) improved osmotic stress tolerance and drought resistance [18]. *MusaDHN-1*, an SK_3-type dehydrin gene in banana, contributes positively towards drought and salt stress tolerance, and responses to abscisic acid, ethylene, and methyl jasmonate [19]. Similarly, in *Boea crassifolia*, overexpression of $YNSK_2$-type dehydrin, *BcDh2* enhanced tolerance to mechanical stress, mediated by salicylic acid and jasmonic acid [20].

In a previous study, it was found that specific *DHNs* in pepper are differentially induced in response to different stresses [21]. Among the seven *Capsicum annuum* dehydrin genes, *CaDHN5* was significantly up-regulated under salt and osmotic stress treatments [13]. Therefore, in this study, we further explored the relationship between *CaDHN5* and salt and osmotic stresses via overexpressing and gene silencing techniques. The results showed that *CaDHN5*-silenced pepper plants were less tolerant to salt and osmotic stress, while *CaDHN5*-overexpressing *Arabidopsis* plants showed significantly increased tolerance to these stresses. These results suggest that *CaDHN5* functions as a positive regulator in salt and osmotic stress signaling pathways.

2. Results

2.1. Analysis of Silencing Efficiency of CaDHN5 in Pepper

Virus-Induced Gene Silencing (VIGS) technique was used to investigate the function of *CaDHN5* under salt and osmotic stresses [22,23]. About 310 bp specific sequences from *CaDHN5* were used to construct the vector pTRV2:*CaDHN5*. Phytoene desaturase (*PDS*) was used as a marker of gene silencing, due to its ability to induce a bleached phenotype after successfully silencing plants [22]. Approximately four weeks after the induction of TRV-mediated gene silencing, pepper plants induced with TRV2:*CaPDS* began to show an albino phenotype, while plants with the TRV2 empty vector and TRV2:*CaDHN5* showed no difference in phenotype (Figure 1a). We detected the expression level of the other six genes within the dehydrin family, and found that only *DHN5* expression level was down-regulated (Figure 1b). From the result of *CaDHN5* expression, it can be seen that the expression decreased by about 80% in the fourth week after induction in *CaDHN5*-silenced plants.

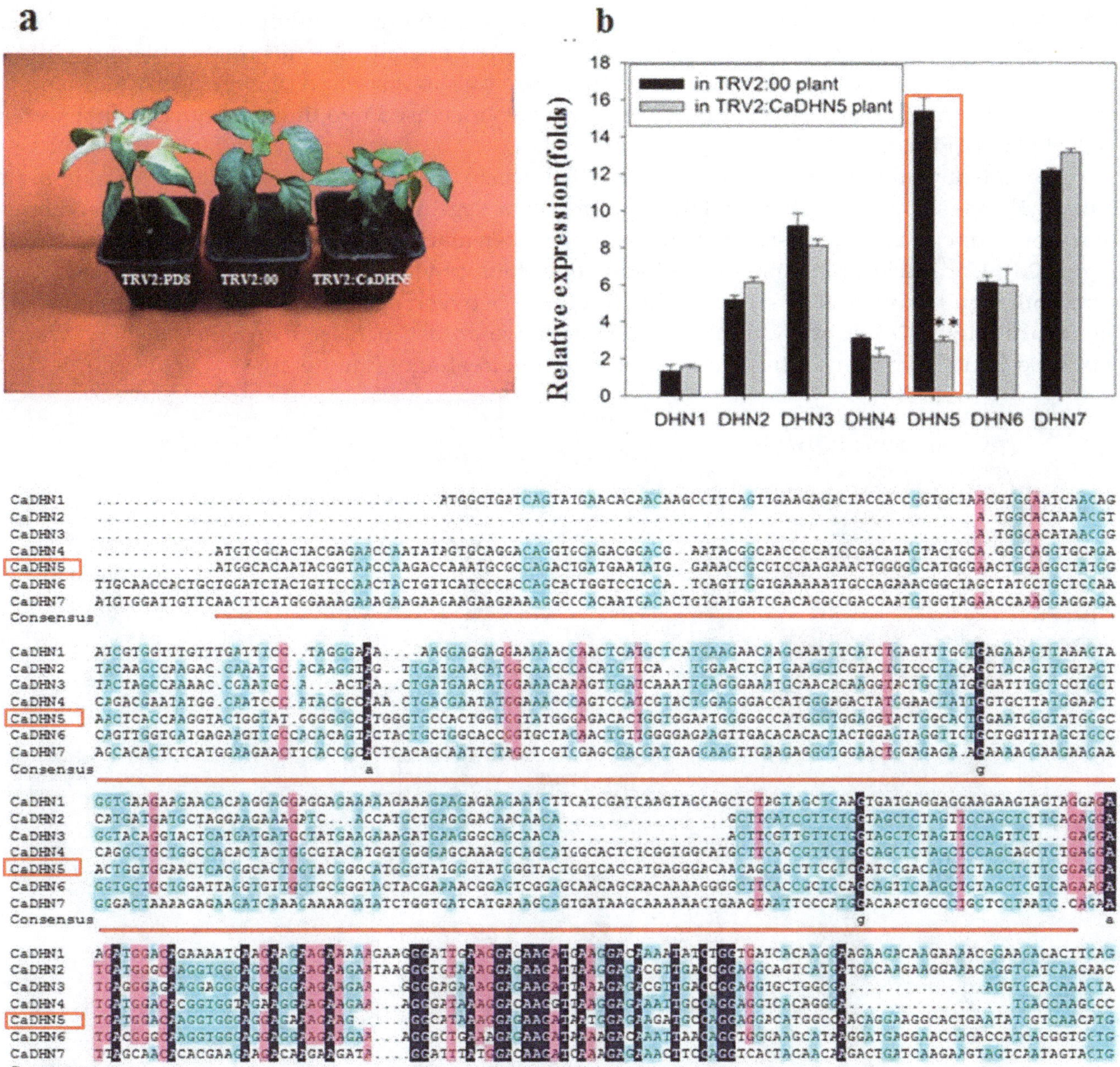

Figure 1. Phenotypes of *CaDHN5*-silenced plants and detection of gene silencing efficiency. (**a**) The phenotypes of silenced pepper plants (about four weeks after injection); (**b**) The relative expression of *CaDHN5* in silenced pepper plants and silencing sequence analysis of *CaDHN5*. The red frame is expression of *CaDHN5* in silenced and control plants. The black areas represent homology level 100%. The pink areas represent a level of homology greater than or equal to 75%. The blue areas represent a level of homology greater than or equal to 50%. The underlined part is a sequence of silencing. The results are the means ± standard deviation (S.D.), replicated three times. The means were compared using Student's test. Note: * indicates significant differences compared with the control at $p < 0.01$.

2.2. Influence of Silencing CaDHN5 on Tolerance of Salt and Osmotic Stresses in Pepper

To investigate the effect of *CaDHN5* silencing on the osmotic tolerance of pepper plants, control and *CaDHN5*-silenced plants were treated with 250 mM mannitol under continuous lighting conditions for three days. *CaDHN5*-silenced plants wilted considerably more than control plants, and some leaves became yellow after three days in mannitol-treated plants (Figure 2a).

In order to compare the differences between *CaDHN5*-silenced and control plants under mannitol treatment, we measured relative electrolyte leakage, rate of water loss, malondialdehyde (MDA), and chlorophyll content of these plants (Figure 2b–f). Following mannitol treatment, MDA levels in the control plants increased about four-fold, compared to control water treated plants, while in *CaDHN5*-silenced pepper plants, MDA levels increased by about six-fold (Figure 2b). After mannitol treatment, the proline content of silenced pepper plants increased two times as much as that seen in control plants (Figure 2c).

The rate of water loss and relative electrolyte leakage are indicators of the degree of membrane injury [24]. As can be seen from the relative electrolyte leakage measurements, the degree of membrane injury of pepper plants was significantly higher under mannitol treatment compared to controls (Figure 2d). Under 250 mM mannitol treatment, total chlorophyll content in control and silenced plants were both significantly decreased, and the difference between control and silenced plants was not significant (Figure 2e). In silenced plants, the rate of water loss increased, and the rate of water loss was three-fold lower than control plants after mannitol treatment (Figure 2f).

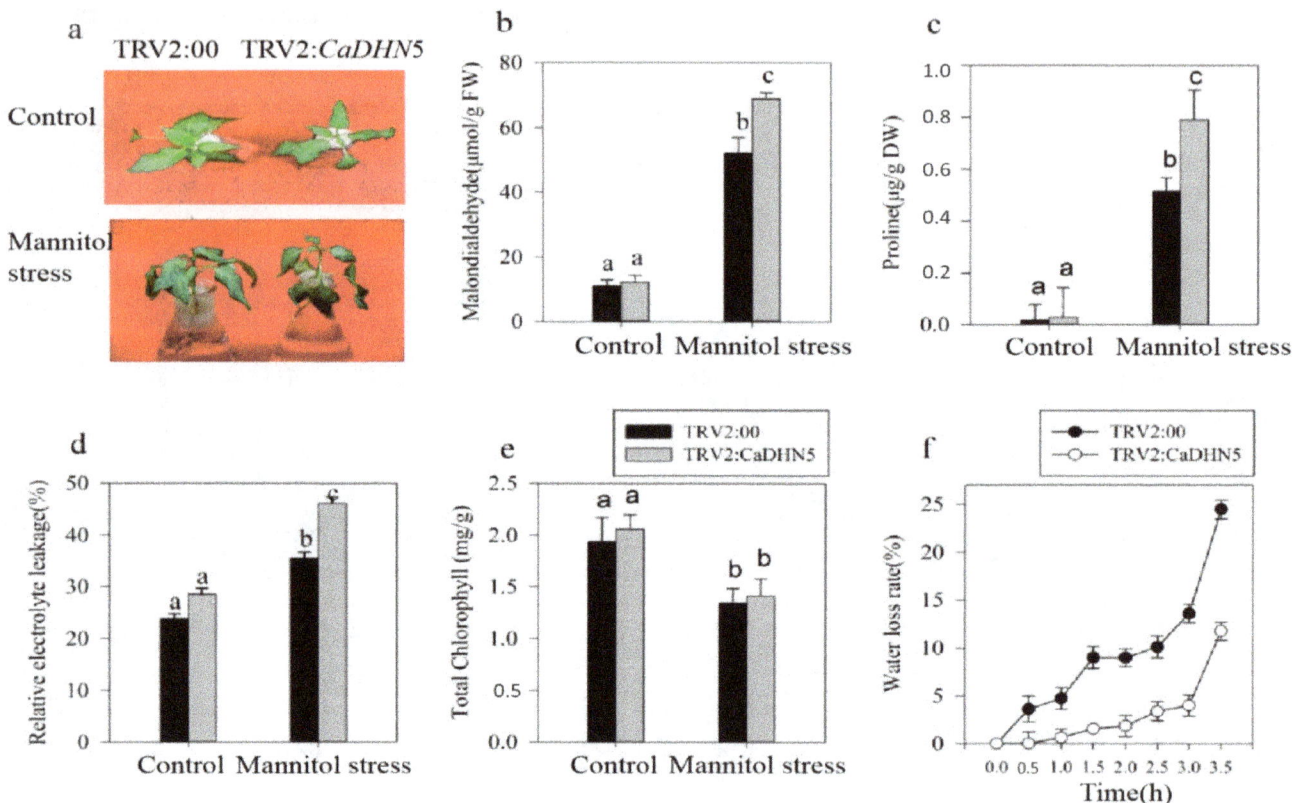

Figure 2. Effects of osmotic stress on plant phenotypes. (**a**) The phenotype of *CaDHN5*-silenced pepper plants under osmotic stress; (**b**) MDA content; (**c**) proline levels; (**d**) relative electrolytic leakage; (**e**) total Chlorophyll content; (**f**) water loss. Data that are significantly different are indicated with letters above the error bars (±S.D.). The different letters with the bars indicate significant differences as determined using Tukey HSD's multiple range tests ($p < 0.05$).

In normal condition the activities of superoxide dismutase (SOD) and peroxidase (POD) were not significantly different between control and silenced plants. However, after mannitol treatment, the activities of SOD and POD increased to scavenge superoxide anions and H_2O_2 produced in the plant. Therefore, the enzyme activities could reflect the ability of the plant to scavenge superoxide anions and H_2O_2. From Figure 3b,c, it can be clearly seen that in the *CaDHN5* silenced plant, the increase in enzyme activity was significantly lower than that of the control plant. The staining results of NBT also showed that after gene silencing, more superoxide anions accumulated in the leaves (Figure 3a).

We also analyzed the expression of the stress and antioxidant system-related genes (*MnSOD*, *POD*, and *ERD15* [23]) in control and silenced lines. There was no significant difference in the expression of

POD between control and silenced lines before treatment. After mannitol treatment, the expression of *POD* in control and silenced lines both increased, but in silenced plants it only increased two-fold, while in control plants this increase was four-fold (Figure 3d). A similar result was found for the expression of *MnSOD*. After *CaDHN5* silencing, the increased expression of *MnSOD* in silenced plants was only half compared to control plants (Figure 3e). In addition, the expression of *ERD15* was significantly higher in control plants treated with mannitol, while in silenced plants, increased expression of *ERD15* was significantly higher than control but less than mannitol-treated silenced pepper plants (Figure 3f).

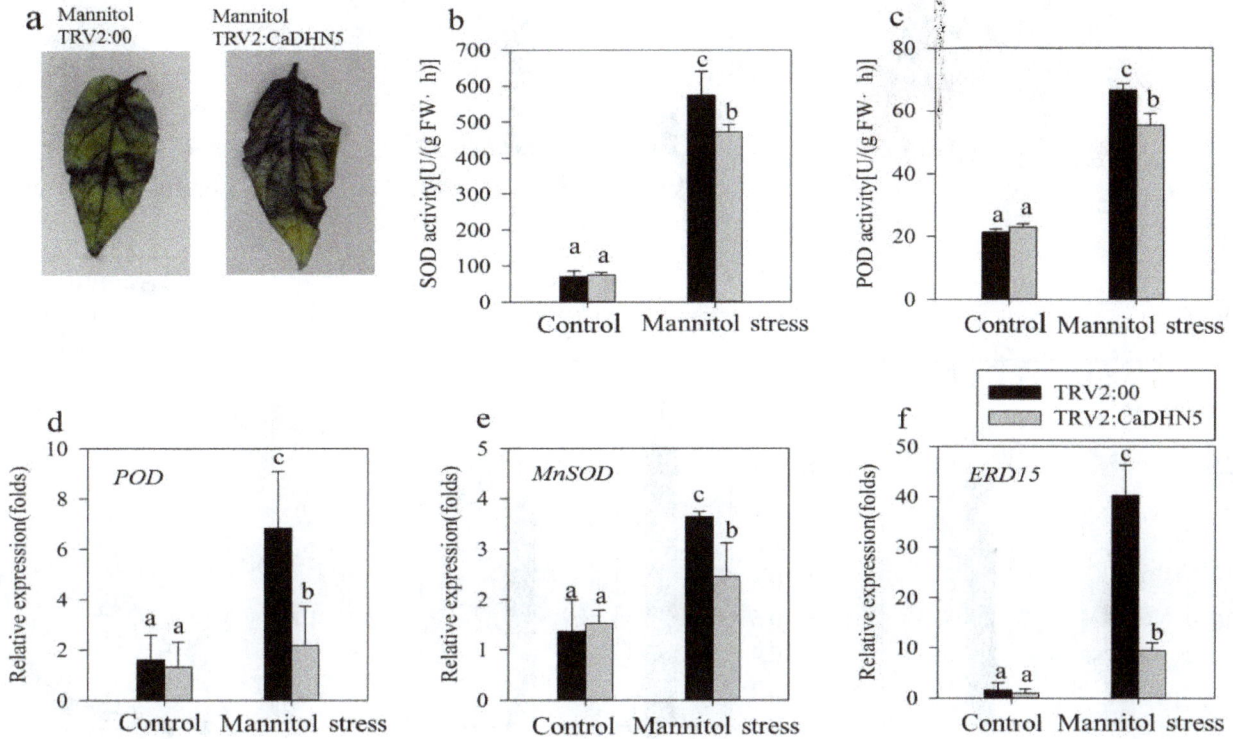

Figure 3. Determination of oxidative stress resistance and stress-related gene expression in mannitol treatment. (**a**) Results of pepper plants stained with NBT under mannitol treatment; (**b**) SOD activity under osmotic stress; (**c**) POD activity under mannitol treatment; (**d**) relative expression of *POD* under mannitol treatment; (**e**) *MnSOD* relative expression under mannitol treatment; (**f**) relative expression of *ERD15* under mannitol treatment. Data that are significantly different are indicated with letters above the error bars (±S.D.). The different letters with the bars indicate significant differences as determined using Tukey HSD's multiple range tests ($p < 0.05$).

In order to investigate the effect of silencing *CaDHN5* on salt stress in pepper, we measured the same physiological indices as those measured for mannitol stress, and performed the same analysis. Silenced and control plants were treated with 250 mM NaCl solution. Regarding the phenotype, the wilting conditions of silenced pepper plants were more evident under NaCl treatment (Figure 4a). Under normal conditions, there was no significant difference in the MDA content between control and silenced plants. However, after NaCl treatment, MDA levels in both control and silenced plants increased significantly; this increase in silenced plants was one and a half times more than that in control plants. Therefore, it appears that *CaDHN5*-silenced pepper plants experienced more serious membrane lipid peroxidation than control plants (Figure 4b). After NaCl treatment, the proline content of silenced pepper plants increased five-fold compared to control plants (Figure 4c). The chlorophyll content of *CaDHN5*-silenced pepper plants decreased more rapidly (Figure 4e). In silenced plants under NaCl treatment, the rate of water loss was four-fold faster than control plants (Figure 4f).

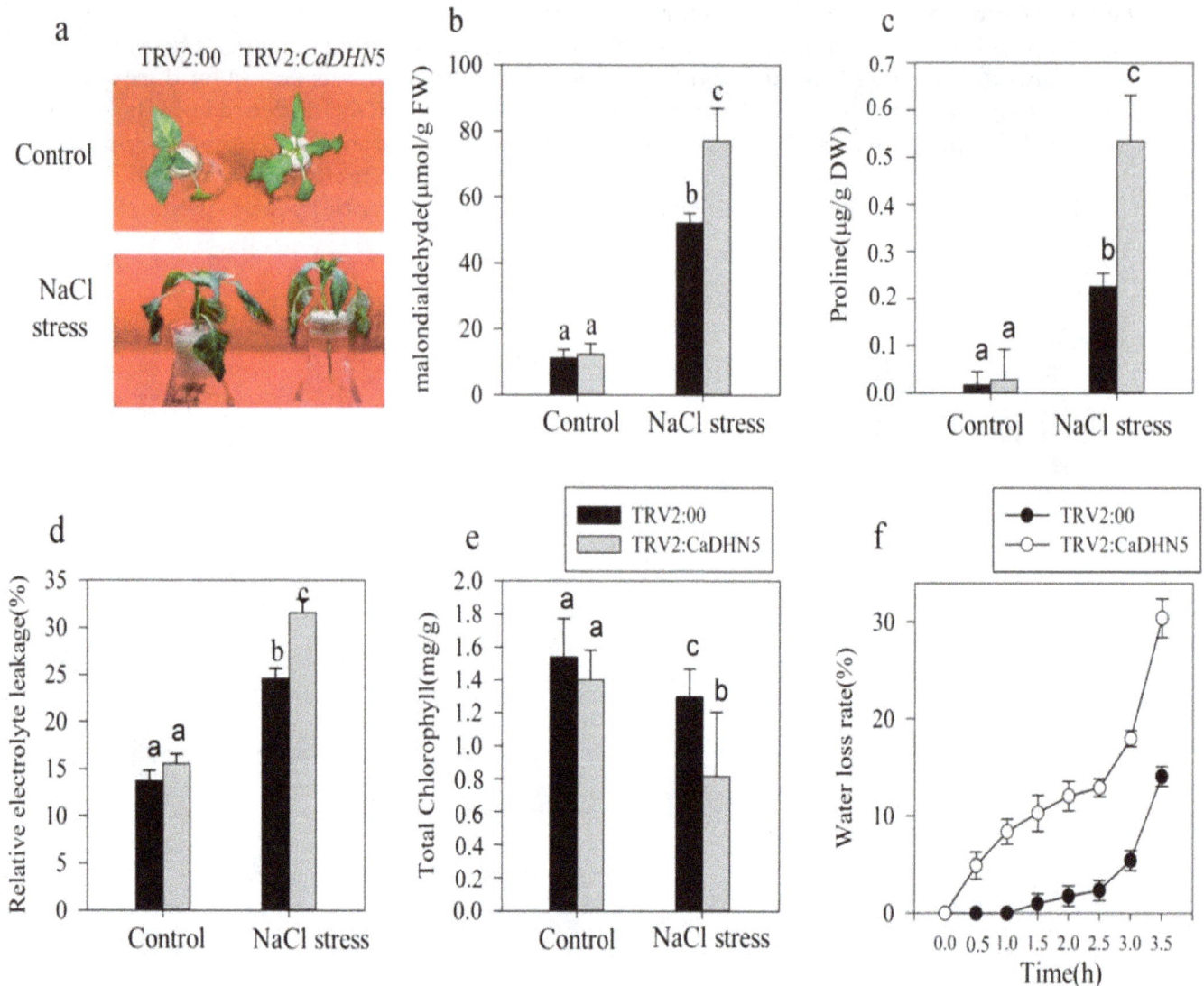

Figure 4. Effects of salt stress on plant phenotypes. (**a**) The phenotype of *CaDHN5*-silenced pepper plants under NaCl treatment; (**b**) MDA content; (**c**) proline levels; (**d**) relative electrolytic leakage; (**e**) total chlorophyll content; (**f**) water loss. Data that are significantly different are indicated with letters above the error bars ($\pm$S.D.). The different letters with the bars indicate significant differences as determined using Tukey HSD's multiple range tests ($p < 0.05$).

The activities of SOD and POD were measured under NaCl treatment. Results show that the activities increased in both silenced and control plants following NaCl treatment, but were slightly lowered in silenced plants compared to control pepper plants. These results suggested that the ability of plants to remove superoxide anions and H_2O_2 decreased slightly in silenced plants (Figure 5b,c). The NBT staining data reflected these observations (Figure 5a). In silenced plants treated with NaCl, NBT-stained leaves were more than that in the control lines. These data suggested that more superoxide anion accumulation occurred in silenced plants. We monitored the expression of POD and ERD15 (Figure 5d,f), which showed a significant increase in control plants treated with NaCl. MnSOD also exhibited a significant increase in expression, although this increase was less than that of POD and ERD15.

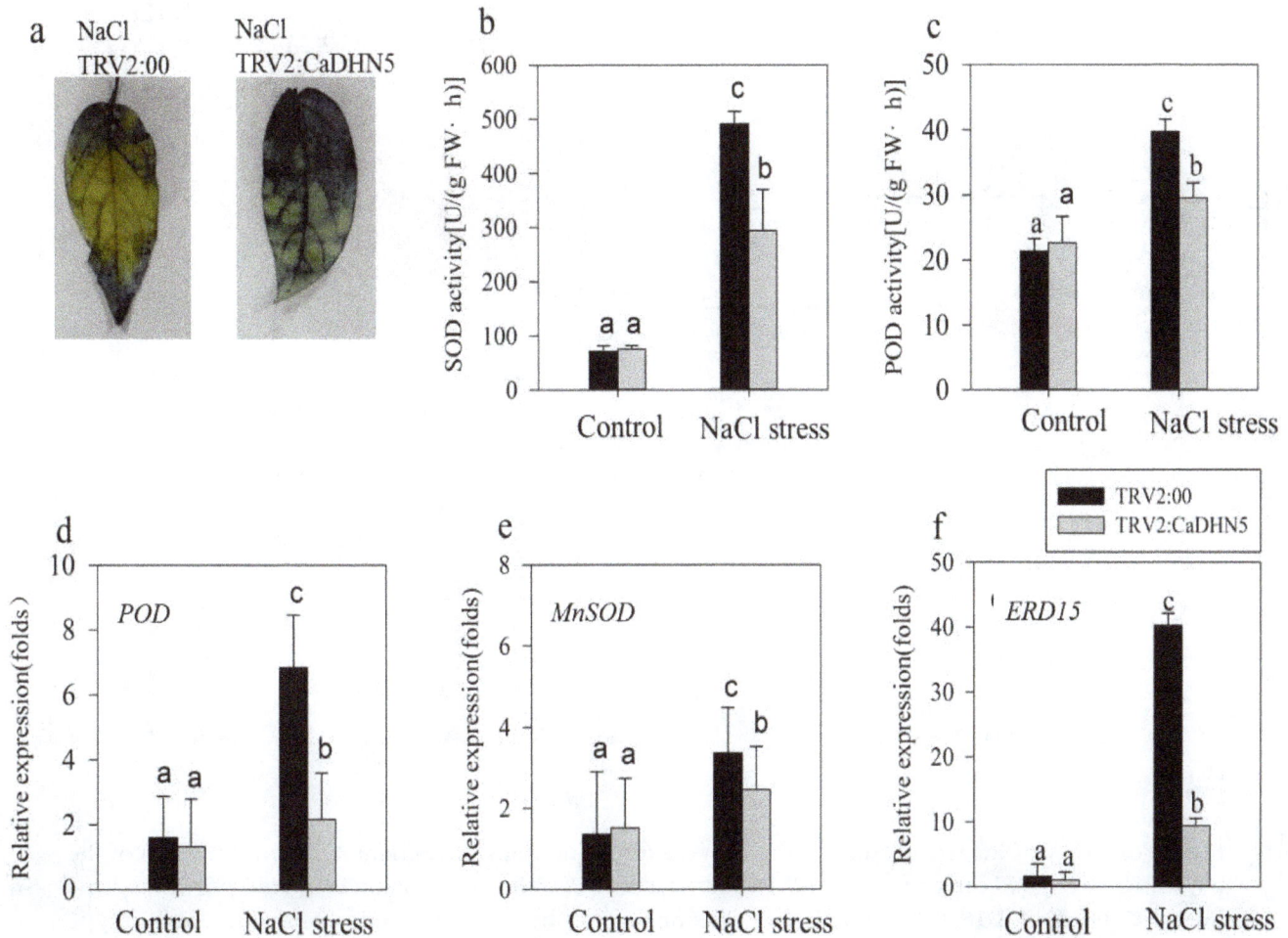

Figure 5. Determination of oxidative stress resistance and stress-related gene expression in NaCl treatment. (**a**) Results of NBT-stained pepper plants under NaCl treatment; (**b**) SOD activity under NaCl stress; (**c**) POD activity under NaCl treatment; (**d**) *POD* relative expression under NaCl treatment; € *MnSOD* relative expression under NaCl treatment; (**f**) *ERD15* relative expression under NaCl treatment. Data significantly different are indicated with letters above the error bars (±S.D.). The different letters with the bars indicate significant differences as determined using Tukey HSD's multiple range tests (*p* < 0.05).

2.3. Analysis of CaDHN5-Overexpression Arabidopsis

We constructed the overexpression vector pVBG2307:*CaDHN5*. The schematic diagram of the vector is shown in Figure 6a. After *Agrobacterium*-mediated transformation, the expression level of *CaDHN5* was estimated by qRT-PCR (Figure 6b). Expression of *CaDHN5* was higher in the lines D6 and D16 than other lines, so their homozygous T3 generation plants were chosen for further physiological analyses.

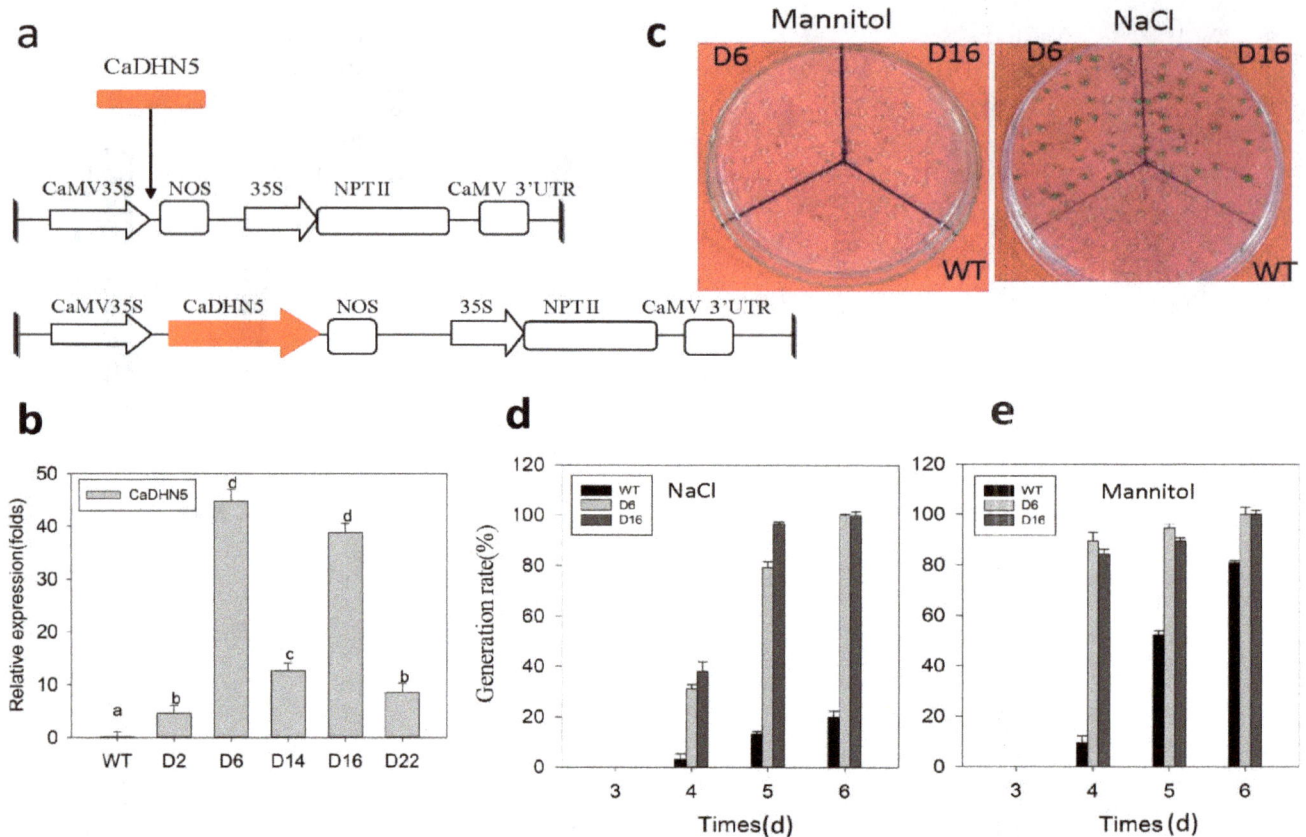

Figure 6. Assay of the transgenic *CaDHN5*-overexpressing lines. (**a**) Schematic representation of the pVBG2307:*CaDHN5* construct; (**b**) qRT-PCR analysis of *CaDHN5* expression in *Arabidopsis* transgenic lines (D2, D6, D14, D16, D22), with WT as control; (**c**) the phenotype of seed germination in wild type and two transgenic *Arabidopsis* plants (D6 and D16) subjected to salt (NaCl) and osmotic (mannitol) stress for five days; (**d,e**) seed germination rates of different lines subjected to salt (NaCl) and osmotic (mannitol) stress. Data significantly that are different are indicated with letters above the error bars (±S.D.). The different letters with the bars indicate significant differences as determined using Tukey HSD's multiple range tests (*p* < 0.05).

2.4. Seed Germination under Osmotic and Salt Stress Conditions

Transgenic *Arabidopsis* seeds were germinated on MS/2 agar medium containing 200 mM NaCl or mannitol solutions, and the germination rate was calculated (Figure 6c). Under NaCl treatment, at five days, almost all transgenic-seeds were germinated. However, only about 13% of WT seeds germinated (Figure 6d). Meanwhile, after six days, almost all transgenic seeds were germinated, while only 20% of the WT seeds were germinated. A similar trend was followed in the presence of mannitol, where the transgenic lines showed better germination compared to the WT seeds. At five days, almost all transgenic-seeds were germinated, while only 13% of WT seeds were germinated. In the following days, the germination rate of WT gradually increased, eventually reaching 80%, and transgenic seeds reached 100% (Figure 6e). These data show that the transgenic D6 and D16 lines displayed a better rate of seed germination than WT under salinity or osmotic stress.

2.5. Increased Tolerance of CaDHN5-Overexpressing Transgenic Arabidopsis Plants towards Salt and Osmotic Stresses

After three days treatment with 250 mM NaCl or mannitol, we observed the phenotype of *CaDHN5*-overexpressing transgenic *Arabidopsis* plants and measured physiological parameters (Figure 7).With mannitol treatment, the phenotypes of all *Arabidopsis thaliana* plants showed varying degrees of water loss, which occurred in all *Arabidopsis thaliana* plants, and the whole plants became brittle. After three days of mannitol treatment, WT leaves had suffered severe water loss and became brittle, while the leaves of transgenic plants remained moist (Figure 7a).

As can be seen from the content of MDA and chlorophyll (Figure 8b,c), the injury of wild-type plants was more serious under 250 mM mannitol treatment. Due to the influence of mannitol and salt, the chlorophyll content decreased to a similar extent in both WT and transgenic lines compared to controls (Figure 7b). The MDA content in WT following mannitol treatment increased by 15-fold compared to the control conditions, while in two transgenic lines D6 and D16, the increase was recorded to be about five-fold compared to controls (Figure 7c). After 250 mM NaCl treatment, the MDA content in WT increased by 12-fold, whereas in the two transgenic lines this increase was only about four-fold.

Based on previous research, we also selected several stress-related genes, *AtDREB2A*, *AtDREB2B* [25], *AtERD7*, and *AtMYC2* [26], to assess responses to osmotic stress, and *AtATR1/MYB34* [26], *AtSOS1* [25], *AtRITF1* [27], and *AtRSA1* for responses to salt stress in *Arabidopsis*. The relative expression levels of the above-mentioned genes were measured in WT and *CaDHN5*-overexpressing plants under stress and mannitol stresses (Figure 8). Only genes with significant changes are shown in Figure 8. *AtATR1/MYB34*, *AtSOS1*, and *AtRSA1* were up-regulated in both osmotic and salt stresses.

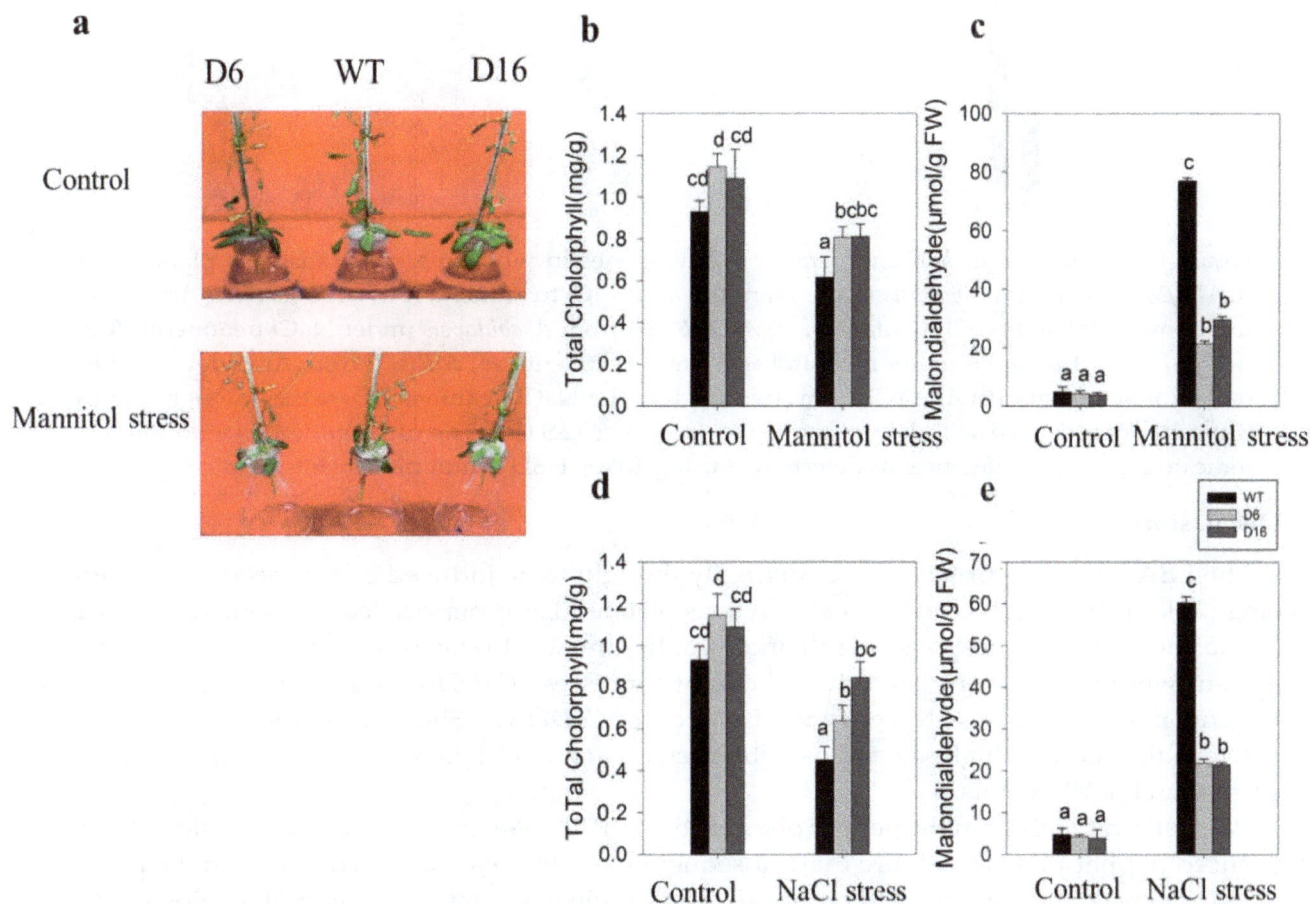

Figure 7. Related-physiological indices of *Arabidopsis thaliana* under salt and osmotic stress treatments. (**a**) The phenotypes of wild type (WT) and *CaDHN5*-overexpressing transgenic plants (D6 and D16) under mannitol treatment; (**b**) effects of mannitol treatment on total chlorophyll in transgenic *Arabidopsis* plants; (**c**) effects of mannitol treatment chlorophyll content in transgenic *Arabidopsis* plants; (**d**) effects of NaCl treatment on MDA content in transgenic *Arabidopsis* plants; (**e**) effects of NaCl treatment on total chlorophyll content in transgenic *Arabidopsis* plants. Data that are significantly different are indicated with letters above the error bars (±S.D.). The different letters with the bars indicate significant differences as determined using Tukey HSD's multiple range tests ($p < 0.05$).

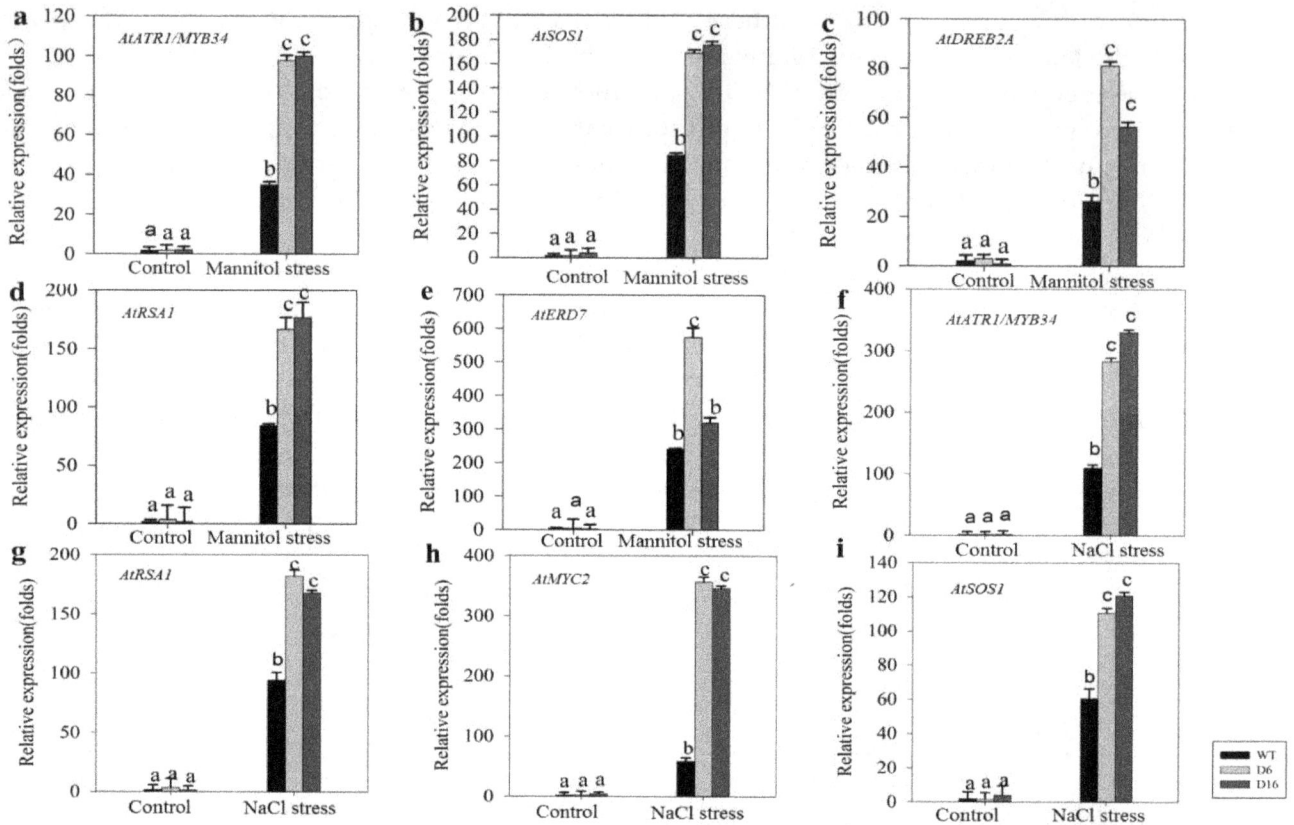

Figure 8. Expression of salt and osmotic related-genes in wild-type and transgenic plants. (**a,f**) *AtATR/MYB34* from *Arabidopsis* under mannitol and NaCl treanment; (**b,i**) *AtSOS1* from *Arabidopsis* under mannitol and NaCl treatments; (**c**) *AtDREB2A* from *Arabidopsis* under NaCl treatments; (**d,g**) *AtRSA1* from *Arabidopsis* under mannitol and NaCl treatments; (**e**) *AtERD7* from Arabidopsis under mannitol treatments; (**h**) *AtMYC2* from *Arabidopsis* under NaCl treanment. Data that are significantly different are indicated with letters above the error bars (±S.D.). The different letters with the bars indicate significant differences as determined using Tukey HSD's multiple range tests ($p < 0.05$).

3. Discussion

The LEA family of proteins were originally thought to be induced during seed maturation and drying [28]. In this study, *CaDHN5* cDNA was isolated from pepper leaves. Our results indicate *CaDHN5* shows a strong response to salt and osmotic stresses. In the experiments, NaCl and mannitol treatment were used to simulate salt and osmotic stresses. *CaDHN5* silenced and overexpressing transgenic plants were used to verify the function of *CaDHN5*. Silenced pepper plants were more sensitive to the effects of high salt and osmotic stresses, and *CaDHN5*-over-expressed plants were more tolerant than the WT plants.

We silenced *CaDHN5* in the pepper plant cultivar "P70". We first examined expression of *CaDHN5* in silenced pepper plants to ensure that subsequent experiments were carried out on the premise of successful gene silencing. Plant tolerance to stress is closely related to some physiological indices. It is well known that plants with strong stress tolerance usually have higher chlorophyll content and lower content of electrolyte leakage, proline, and MDA under stress situations. Under salt and osmotic stress conditions, the different trends in the decrease or increase of chlorophyll content, MDA, and conductivity suggest that *CaDHN5* may be involved in salt and osmotic stress responses. Meanwhile, these results indicated that the membrane damage and leaf senescence of the silenced *CaDHN5* pepper plants were higher under salt and osmotic stresses. Other studies in different plants have shown similar results [29,30]. Under salt stress and osmotic stresses, many *DHNs* were up-regulated in transgenic *Arabidopsis* plants, which showed high tolerance to these stresses [29]. It has also been found that the barley dehydrin *DHN3* responds to various stresses [30]. *POD* and *MnSOD* are important genes that function in the process of scavenging ROS. We found that when *CaDHN5* was silenced in pepper, the expression levels of these two genes were significantly lower than those of the control plants under salt and osmotic stress. This indicates that *CaDHN5* positively regulates the expression of these genes. In addition, results of enzyme activity and staining with NBT indicated that gene-silenced plants had higher levels of superoxide anion. *CaDHN5*-silenced pepper plants had lower tolerance to salt and osmotic stresses.

Further, we generated *CaDHN5* over-expressing transgenic *Arabidopsis*. Under high salinity and osmotic stress, it was found that when *CaDHN5* was overexpressed in *Arabidopsis*, it resulted in increased tolerance to salt and osmotic stress. Previous reports have described the increased anti-stress ability of different LEA genes in various plants, such as rice, wheat, and *Arabidopsis* [31,32]. The *MusaDHN1* gene of banana is not only induced by drought, salt, cold, oxidation, and heavy metal stress, but can also be induced by abscisic acid, ethylene, and methyl jasmonate [19]. In this study, *CaDHN5* transgenic lines are more tolerant under high concentrations of NaCl and mannitol. Transgenic seeds germinate rapidly under 200 mM mannitol compared to WT. Studies have shown that the dehydrin in Chinese cabbage has a similar function [33]. When we studied the influence of *CaDHN5* on salt stress tolerance, we found significant differences in MDA content between WT and transgenic plants. Meanwhile, overexpression of *CaDHN5* in *Arabidopsis* resulted in decreasing Chlorophyll degradation under NaCl treatment, but had no significant effect under osmotic stress. This could have resulted from the low accumulation of MDA in transgenic lines. In salt and osmotic stresses, the germination rates of *CaDHN5*-overexpressing *Arabidopsis* plants in the presence of NaCl and mannitol were significantly higher than those of WT plants. Monitoring the expression of other salts and osmotic stress-related genes showed that when *CaDHN5* was overexpressed, the expression levels of these stress-related genes also increased to varying degrees. Other studies have also shown similar results. When transgenic *Arabidopsis thaliana* transformed with wheat *TaDHN1* and *TaDHN3* genes were treated with salt and mannitol, the transgenic plants grew better and the root lengths were longer than wild type [34]. *HbDHN1*, *HbDHN2* were also transformed into *Arabidopsis thaliana*. *Arabidopsis thaliana* was transformed with *HbDHN1*, and *HbDHN2* reduced electrolyte leakage of cells and accumulation of ROS by increasing the SOD and POD activity, thereby resisting salt and osmotic stress [35]. In fact, among the expression of the eight related genes that we recorded, five genes (*AtATR1/MYB34*, *AtSOS1*, *AtDERB2A*, *AtRSA1*, and *AtERD7*) were up-regulated under osmotic stress, and four (*AtATR1/MYB34*, *AtRSA1*, *AtMYC2*, and *AtSOS1*) were up-regulated under salt stress. As mentioned earlier, signaling pathways in response to salt and osmotic stresses overlap [3]. The expression of *AtSOS1*, which encoded a plasma membrane Na^+/H^+ antiporter essential for salt tolerance. [27], was significantly increased in transgenic lines, both under salt and osmotic stresses. The expression of the transcription factor *AtDREB2A* in the ABA signaling pathway was also significantly increased. *AtRSA1* and *AtRITF1* are interacting genes that not only participate in the regulation of the transcription of several genes in the ROS scavenging system, but also regulate the expression of *AtSOS1*. It is worth noting that although the expression of *AtRSA1* gene was increased under salt and osmotic stresses, the interacting partners of *AtRSA1* and *AtRITF1* were only up-regulated under salt treatment.

4. Materials and Methods

4.1. Plant Materials, Growth Conditions

Seeds (wild-type: Columbia ecotype) and pepper (*Capsicum annuum* L.) cultivar "P70" were used in the current work, which were provided by Vegetable Plant Biotechnology and Germplasm Innovation laboratory, Northwest A&F University-China. The *Arabidopsis thaliana* seeds were treated as per Brini's method [32]. The pepper seedlings were cultured in a growth chamber by maintaining them in 16 h/8 h light/dark at 25 °C/20 °C [23]. The control plants were grown in the same environment and treated with corresponding solvents.

4.2. Isolation CaDHN5

According to the full-length CaDHN5 ORF sequence (GenBank accession No.: XM016705201), forward and reverse primers were designed as 5'-AGGAGATGGCACAATACGGT-3'AND5'-ATCCTTTGTTTTCATTTTCAGC-3', respectively. PCR products were cloned into the pMD19-T vector (TaKaRa, Dalian, China) and sequenced (Xi'an AuGCT Biotechnologies Co. Xi'an, China).

4.3. Silencing Efficiency Analysis of CaDHN5 in Pepper

The pTRV2: *CaDHN5* construct was engineered to include a 310 bp sequence in *CaDHN5* cloned from a pepper cDNA template, using the forward primers 5'-ATGGCACAATACGGTAACC-3'and the reverse primers 5'-CCGAAGAGCTAGAGCTGTC-3'. The recombinant plasmid pTRV2: *CaDHN5* was constructed by combining CaDHN5 and pTRV2. *Agrobacterium tumefaciens* GV3101 containing pTRV2:*CaDHN5* was injected into pepper plants after combining GV3101 with pTRV1, and plants were grown as described previously. Fifty plants were used for the silencing assay [21].

4.4. Generation of Transgenic Arabidopsis Plants

The vector pVBG2307:*CaDHN5* contains the kanamycin resistance gene as a selectable marker between the 35S promoter and terminator (Figure 8a). Agrobacterium-mediated transformation was performed via the floral dipping technique of *Arabidopsis thaliana* (ecotype Columbia) [36]. Over-expressing transgenic plants were selected by growing seeds on MS/2 agar medium containing 50 mg/L kanamycin, which were grown up to the T3 generation to identify plants homozygous for the transgene.

4.5. Isolation of RNA, qRT-PCR

Total RNA was extracted from 200 mg of young leaves from *Arabidopsis transgenic* lines or silenced pepper plants using the RNeasy total RNA isolation kit (TianGen, Beijing, China). The cDNA was made by using PrimScript RT Kit (TaKaRa, Dalian, China). Primers are presented in Supplementary Table S1. The qRT-PCR was carried out as described previously [23]. The CaUbi3 gene (GenBank Accession No. AY486137.1) encoding the ubiquitin-conjugating protein was amplified from pepper plants as a reference gene for normalization of the *CaDHN5* cDNA samples [37], and the Atactin gene (GenBank Accession No. AY572427.1) was used as an internal control in *Arabidopsis* [38]. The relative fold difference in mRNA levels was determined using the $2^{-\Delta\Delta CT}$ method.

4.6. Measurement of Correlative Physiological Indices

4.6.1. Determination of MDA Content

Approximately 0.5 g of pepper leaves were weighed and rapidly ground with pre-chilled 10% trichloroacetic acid solution. Finally, to the mixed solution, 10% trichloroacetic acid was added to reconstitute the solution to 10 mL, and centrifuged at 4000 rpm for 10 min at 4 °C. A volume of 2 mL supernatant was taken and mixed with 2 mL 0.6% thiobarbituric acid solution. The mixed solution was heated in boiling water for 15 min and rapidly cooled. Following centrifugation at 4000 rpm for 10 min at 4 °C, the absorbance of the supernatant was measured at 532 nm, 450 nm, and 600 nm, according to the method described previously [39].

4.6.2. Total Chlorophyll Content

Pepper leaves (0.1 g) were immersed in 95% ethanol. After the leaves were completely decolored, the absorbance of the supernatant was measured at 470 nm, 649 nm, and 665 nm, as described previously [40].

4.6.3. Relative Electrolyte Leakage

Electrolyte leakage was measured according to the method described previously [41]. Leaves from treated and control plants were selected; 10 leaf discs were made by using a perforator, and the leaf discs were placed in a 50 mL centrifuge tube containing 10 mL of distilled water. After being kept at room temperature for 2 h, electrolyte leakage (EC1) was measured. The centrifuge tubes were heated in boiling water for 30 min after cooling, and the conductivity measurement value (EC2) was measured. Relative electrolyte leakage was calculated as (EC1/EC2) × 100.

4.6.4. Enzyme Activity

The SOD and POD activities were measured according to a previously described method [42]. Fresh leaves (0.5 g) were mixed with 8 mL PBS pH 7.8 and the mixture was centrifuged at 10,000 rpm for 15 min. The supernatant was considered as the crude enzyme extract. In the presence of hydrogen peroxide, POD can oxidize guaiacol to produce colored substances; the product concentration was calculated and POD activity was measured. SOD activity was determined by a similar principle, with NBT as the reaction substance.

4.6.5. NBT Staining

The NBT staining method was as used as described previously [43]. The plant leaves were immersed in a 0.1 mg/L NBT solution in Tris-HCl, pH 7.8, and vacuum infiltrated for about 1 min. After being incubated for 1 h in the dark, the leaves were placed in 80% ethanol, which was changed twice. After complete removal of chlorophyll, the degree of leaf staining was observed.

4.6.6. Water Loss Rate

Isolated plant leaves were placed on the laboratory bench (20–22 °C, humidity 45–60%) and their weight was measured every 30 min, as described previously [44]. The initial fresh weight of the leaves was recorded as W0, and thereafter weighed every 30 min. The leaf weight after 4 h was recorded as Wt. The water loss rate per 30 min was calculated as: (W0−Wt)/W0 × 100.

4.6.7. Proline Content

Approximately 0.5 g leaves were mixed with 5 mL of 3% sulfosalicylic acid; the mixture was placed in a 100 °C water bath for 10 min. After cooling, the mixture was centrifuged at 3000 rpm for 10 min. The supernatant (extraction solution, 2 mL) was mixed with a color rendering agent, indene (2 mL), and glacial acetic acid (2 mL). The mixed solution was heated in boiling water for about 40 min. A volume of 5 mL toluene was added into the mixing solution after cooling and the absorbance value was measured at 520 nm, according to a previously described method [45].

4.7. Statistical Analysis

The qRT-PCR data analysis was carried out using SPSS (Chicago, IL, USA). The relative expression levels of CaDHN5 under salt and osmotic stress are shown as mean ± SD of three biological replicate samples. Each replicate sample was a composite of leaves from three individual seedlings. Statistical analyses were performed using the SPSS (Chicago, IL, USA), and the means were compared using Tukey's HSD multiple range test, taking $p < 0.05$ as a significant difference.

5. Conclusions

In conclusion, although the physiological function of *CaDHN5* at a molecular level has not yet been identified, here we show that *Arabidopsis* plants overexpressing *CaDHN5* have higher survival rates in salt and osmotic stress conditions. These results suggest a functional role for *CaDHN5* in response to salt and osmotic stress. *Arabidopsis* plants overexpressing *CaDHN5* were significantly superior to WT in various physiological indices measured under salt and osmotic stresses. After gene silenced pepper plants, the tolerance of pepper plants to salt and mannitol were significantly decreased, and the above two factors jointly proved the effect of *CaDHN5* on plant tolerance to salt and osmotic stress.

Author Contributions: D.L., X.H., Y.Z., and R.C. conceived and designed the experiments. D.L., X.J., Y.M., and H.Z. performed the experiments. D.L., S.L., and X.W. analyzed the data. R.C. contributed reagents, materials, analysis tools. D.L. wrote the paper. All authors read and approved the final manuscript.

References

1.   Jaspers, P.; Kangasjarvi, J. Reactive oxygen species in abiotic stress signaling. *Physiol. Plant.* **2010**, *138*, 405–413. [CrossRef]

2.   Ismail, A.M.; Hall, A.E.; Close, T.J. Purification and partial characterization of a dehydrin involved in chilling tolerance during seedling emergence of cowpea. *Plant Physiol.* **1999**, *120*, 237–244. [CrossRef] [PubMed]

3.   Zhu, J.K. Salt and drought stress signal transduction in plants. *Annu. Rev. Plant. Biol.* **2002**, *53*, 247–273. [CrossRef] [PubMed]

4.   Dure, L.; Crouch, M.; Harada, J. Common amino acid se-quence domains among the Lea proteins of higher plants. *Plant Mol. Biol.* **1989**, *12*, 475–486. [CrossRef]

5.   Close, T.J. Dehydrins: A commonalty in the response of plants to dehydration and low temperature. *Physiol. Plant.* **1997**, *100*, 291–296. [CrossRef]

6.   Kosova, K.; Vitamvas, P.; Prasil, I.T. Wheat and barley dehydrins under cold, drought, and salinity—what can LEA-II proteins tell us about plant stress response? *Front Plant Sci.* **2014**, *5*, 343. [CrossRef]

7.   Kovacs, D.; Kalmar, E.; Torok, Z.; Tompa, P. Chaperone activity of ERD10 and ERD14, two disordered stress-related plant proteins. *Plant Physiol.* **2008**, *147*, 381–390. [CrossRef] [PubMed]

8.   Close, T.J. Dehydrins: Emergence of a biochemical role of a family of plant dehydration proteins. *Physiol. Plant.* **1996**, *97*, 795–803. [CrossRef]

9.   Zolotarov, Y.; Strmvik, M. De novo regulatory motif discovery identifies significant motifs in promoters of five classes of plandehydrin genes. *PLoS ONE* **2015**, *10*, 1522–1529. [CrossRef] [PubMed]

10.   Riera, M.; Figueras, M.; López, C.; Goday, A.; Pagès, M. Protein kinase CK2 modulates developmental functions of the abscisic acidresponsive protein Rab17 from maiz. *Proc. Natl. Acad. Sci. USA* **2004**, *101*, 9879–9884. [CrossRef]

11.   Richard, S.; Morency, M.J.; Drevet, C.; Jouanin, L.; Séguin, A. Isolation and characterization of a dehydrin gene from white spruce induced upon wounding, drought and cold stresses. *Plant Mol. Biol.* **2000**, *43*, 1–10. [CrossRef] [PubMed]

12.   Bhattarai, T.; Fettig, S. Isolation and characterization of a dehydrin gene from Cicer pinnatifidum, a drought resistant wild relative of chickpea. *Physiol. Plant.* **2005**, *123*, 452–458. [CrossRef]

13.   Jing, H.; Li, C.; Ma, F.; Ma, J.H.; Khan, A.; Wang, X. Genome-Wide Identification, Expression Diversication of Dehydrin Gene Family and Characterization of CaDHN3 in Pepper (*Capsicum annuum* L.). *PLoS ONE* **2016**, *11*, e0161073. [CrossRef]

14.   Eriksson, S.K.; Kutzer, M.; Procek, J.; Gröbnercand, G.; Harryson, P. Tunable membrane binding of the intrinsically disordered dehydrin lti30, a cold-induced plant stress protein. *Plant Cell* **2011**, *23*, 2391–2404. [CrossRef]

15.   Gerszberg, K.; HnatuszkoKonka, K. Tomato tolerance to abiotic stress: A review of most often engineered target sequences. *Plant Growth Regul.* **2017**, *83*, 175–198. [CrossRef]

16.   Kosova, K.; Tom Prasil, I.; Prasilova, P.; Vitamvas, P.; Chrpova, J. The development of frost tolerance and DHN5 protein accumulation in barley (*Hordeum vulgare*) doubled haploid lines derived from Atlas 68 x Igri cross during cold acclimation. *J. Plant Physiol* **2010**, *67*, 343–350. [CrossRef]

17.   Ruibal, C.; Salamó, I.P.; Carballo, V.; Castro, A.; Bentancor, M.; Borsani, O.; Szabados, L.; Vidal, S. Differential contribution of individual dehydrin genes from Physcomitrella patens to salt and osmotic stress tolerance. *Plant Sci.* **2012**, *190*, 89–102. [CrossRef]

18.   Sun, J.; Nie, L.Z.; Sun, G.Q. Cloning and characterization of dehydrin gene from ammopiptanthus mongolicus. *Mol. Biol. Rep.* **2013**, *40*, 2281–2291. [CrossRef]

19.   Shekhawat, U.K.; Srinivas, L.; Ganapathi, T.R. MusaDHN-1, a novel multiple stress-inducible SK3-type dehydrin gene, contributes affirmatively to drought and salt stress tolerance in banana. *Planta* **2011**, *234*, 915–932. [CrossRef]

20. Shen, Y.; Tang, M.J.; Hu, Y.L.; Lin, Z.P. Isolation and characterization of a dehydrin-like gene from drought tolerant Boea crassifolia. *Plant Sci.* **2004**, *166*, 1167–1175. [CrossRef]

21. Guo, W.L.; Chen, R.G.; Gong, Z.H.; Yin, Y.X.; Li, D.W. Suppression subtractive hybridization analysis of genes regulated by application of exogenous abscisic acid in pepper plant (*Capsicum annuum* L.) leaves under chilling stress. *PLoS ONE* **2013**, *8*, e66667. [CrossRef]

22. Wang, J.E.; Liu, K.K.; Li, D.W.; Zhang, Y.L.; Zhao, Q.; He, Y.M.; Gong, Z.H. A novel peroxidase CanPOD gene of pepper is involved in defense responses to Phytophthora capsici infection as well as abiotic stress tolerance. *Int. J. Mol. Sci.* **2013**, *14*, 3158–3177. [CrossRef]

23. Chen, R.G.; Jing, H.; Guo, W.L.; Wang, S.B.; Ma, F.; Pan, B.G. Silencing of dehydrin CaDHN1 diminishes tolerance to multiple abiotic stresses in *Capsicum annuum* L. *Plant Cell Rep.* **2015**, *34*, 2189–2200. [CrossRef]

24. Griffith, M.; McIntyre, H.C.H. The interrelationship of growth and frost tolerance in winter rye. *Physiol. Plant.* **1993**, *87*, 335–344. [CrossRef]

25. Zhou, G.A.; Chang, R.Z.; Qiu, L.J. Overexpression of soybean ubiquitin-conjugating enzyme gene *GmUBC2* confers enhanced drought and salt tolerance through modulating abiotic stress-responsive gene expression in *Arabidopsis*. *Plant Mol. Biol.* **2010**, *72*, 357–367. [CrossRef]

26. Yuan, Y.; Fang, L.; Karungo, S.K. Overexpression of *VaPAT1*, a GRAS transcription factor from Vitis amurensis, confers abiotic stress tolerance in *Arabidopsis*. *Plant Cell Rep.* **2015**, *35*, 655. [CrossRef]

27. Guan, Q.; Wu, J.; Yue, X. A Nuclear Calcium-Sensing Pathway Is Critical for Gene Regulation and Salt Stress Tolerance in *Arabidopsis*. *PLoS Genetics* **2013**, *9*, e1003755. [CrossRef]

28. Bray, E.A.; BaileySerres, J.; Weretilnyk, E. Responses to abiotic stresses. In *Biochemistry and Molecular Biology of Plants*; Buchanan, B., Gruissem, W., Jones, R., Eds.; American Society of Plant Physiologists: Rockville, MD, USA, 2000; pp. 1158–1176.

29. Santos, A.B.; Mazzafera, P. Dehydrins are highly expressed in water-stressed plants of two coffee species. *Tropical Plant Biol.* **2012**, *5*, 218–232. [CrossRef]

30. Choi, D.W.; Zhu, B.; Close, T.J. The barley (Horderum vulgare L.) dehydrin multigene family: Sequences, allele types, chromosome assignments, and expression characteristics of 11 Dhn genes of cv Dicktoo. *Theor. Appl. Genet.* **1999**, *98*, 1234–1247. [CrossRef]

31. Sivamani, E.; Bahieldin, A.; Wraith, J.M.; AlNiemi, T.; Dyer, W.E.; Ho, T.H.D.; Wu, R. Improved biomass productivity and water use efficiency under water-deficit conditions in transgenic wheat constitutively expressing the barley HVA1 gene. *Plant Sci.* **2000**, *155*, 1–9. [CrossRef]

32. Brini, F.; Hanin, M.; Lumbreras, V.; Amara, I.; Khoudi, H.; Hassairi, A.; Pages, M.; Masmoudi, K. Overexpression of wheat dehydrin DHN-5 enhances tolerance to salt and osmotic stress in Arabidopsis thaliana. *Plant Cell Rep.* **2007**, *26*, 2017–2026. [CrossRef] [PubMed]

33. Park, B.J.; Liu, Z.; Kanno, A.; Kameya, T. Genetic improvement of Chinese cabbage for salt and drought tolerance by constitutive expression of a B. napus LEA gene. *Plant Sci.* **2005**, *169*, 553–558. [CrossRef]

34. Qin, Y.X.; Qin, F. Dehydrins from wheat x Thinopyrum ponticum amphiploid increase salinity and drought tolerance under their own inducible promoters without growth retardation. *Plant Physiol. Bioch.* **2016**, *99*, 142–149. [CrossRef] [PubMed]

35. Cao, Y.; Zhai, J.; Wang, Q.; Yuan, H.; Huang, X. Function of Hevea brasiliensis NAC1 in dehydration-induced laticifer differentiation and latex biosynthesis. *Planta* **2017**, *245*, 31–44. [CrossRef] [PubMed]

36. Clough, S.J.; Bent, A.F. Floral dip: A simplified method for Agrobacterium-mediated transformation of Arabidopsis thaliana. *Plant J.* **1998**, *16*, 735–743. [CrossRef]

37. Wan, H.J.; Yuan, W.; Ruan, M.; Ye, Q.; Wang, R.; Li, Z.; Zhou, G.; Yao, Z.; Zhao, J.; Liu, S.; et al. Identification of reference genes for reverse transcription quantitative real-time PCR normalization in pepper (*Capsicum annuum* L.). *Biochem. Biophy. Res. Commun.* **2011**, *416*, 24–30. [CrossRef] [PubMed]

38. Gutierrez, L.; Mauriat, M.; Gue'nin, S.; Pelloux, J.; Lefebvre, J.F.; Louvet, R.; Rusterucci, C.; Moritz, T.; Guerineau, F.; Bellini, C.; et al. The lack of asystematic validation of reference genes: A serious pitfall undervalued in reverse transcription-polymerase chain reaction (RT-PCR) analysis in plants. *Plant Biotechnol. J.* **2008**, *6*, 609–618. [CrossRef] [PubMed]

39. Dhindsa, R.S.; Plumb-Dhindsa, P.; Thorpe, T.A. Leaf senescence: Correlated with increased levels of membrane permeability and lipid peroxidation, and decreased levels of superoxide dismutase and catalase. *J. Exp. Bot.* **1981**, *32*, 93–101. [CrossRef]

40. Arkus, K.A.J.; Cahoon, E.B.; Jez, J.M. Mechanistic analysis of wheat chlorophyllase. *Arch. Biochem. Biophys.* **2005**, *438*, 146–155. [CrossRef]

41. Danyluk, J.; Perron, A.; Houde, M.; Limin, A.; Fowler, B.; Benhamou, N.; Sarhan, F. Accumulation of an acidic dehydrin in the vicinity of the plasma membrane during cold acclimation of wheat. *Plant Cell* **1998**, *10*, 623–638. [CrossRef]

42. Liang, J.G.; Tao, R.X.; Hao, Z.N.; Wang, L.P.; Zhang, X. Induction of resistance in cucumber against seedling damping-off by plant growth-promoting rhizobacteria (PGPR) Bacillus megaterium strain L8. *Afr. J. Biotechnol.* **2011**, *10*, 6920–6927.

43. Jabs, T.; Dietrich, R.A.; Dangl, J.L. Initiation of runaway cell death in an Arabidopsis mutant by extracellular superoxide. *Science* **1996**, *273*, 1853–1856. [CrossRef]

44. Zhang, L.N.; Zhang, L.C.; Xia, C.; Zhao, G.Y.; Liu, J.; Jia, J.Z.; Kong, X.Y. A novel wheat bZIP transcription factor, Tab ZIP60, confers multiple abiotic stress tolerances in transgenic Arabidopsis. *Physiol. Plant.* **2014**, *153*, 538–554. [CrossRef]

45. Bates, L.S.; Waldren, R.P.; Teeare, I.D. Rapid determination of free Pro for water-stress studies. *Plant Soil* **1973**, *39*, 205–207. [CrossRef]

7

Root Abscisic Acid Contributes to Defending Photoinibition in Jerusalem Artichoke (*Helianthus tuberosus* L.) under Salt Stress

Kun Yan [1],*, Tiantian Bian [1,2], Wenjun He [1], Guangxuan Han [1],*, Mengxue Lv [1], Mingzhu Guo [3] and Ming Lu [3]

[1] Key Laboratory of Coastal Environmental Processes and Ecological Remediation, Yantai Institute of Coastal Zone Research, Chinese Academy of Sciences, Yantai 264003, China; czlbtt@163.com (T.B.); wjhe@yic.ac.cn (W.H.); xuhualing1981@163.com (M.L.)

[2] School of Life Sciences, Ludong University, Yantai 264025, China

[3] College of Life Sciences, Yantai University, Yantai 264005, China; m17616156626@163.com (M.G.); 17865561399@163.com (M.L.)

* Correspondence: kyan@yic.ac.cn (K.Y.); gxhan@yic.ac.cn (G.H.)

Abstract: The aim of the study was to examine the role of root abscisic acid (ABA) in protecting photosystems and photosynthesis in Jerusalem artichoke against salt stress. Potted plants were pretreated by a specific ABA synthesis inhibitor sodium tungstate and then subjected to salt stress (150 mM NaCl). Tungstate did not directly affect root ABA content and photosynthetic parameters, whereas it inhibited root ABA accumulation and induced a greater decrease in photosynthetic rate under salt stress. The maximal photochemical efficiency of PSII (Fv/Fm) significantly declined in tungstate-pretreated plants under salt stress, suggesting photosystem II (PSII) photoinhibition appeared. PSII photoinhibition did not prevent PSI photoinhibition by restricting electron donation, as the maximal photochemical efficiency of PSI ($\Delta MR/MR_0$) was lowered. In line with photoinhibition, elevated H_2O_2 concentration and lipid peroxidation corroborated salt-induced oxidative stress in tungstate-pretreated plants. Less decrease in $\Delta MR/MR_0$ and Fv/Fm indicated that PSII and PSI in non-pretreated plants could maintain better performance than tungstate-pretreated plants under salt stress. Consistently, greater reduction in PSII and PSI reaction center protein abundance confirmed the elevated vulnerability of photosystems to salt stress in tungstate-pretreated plants. Overall, the root ABA signal participated in defending the photosystem's photoinhibition and protecting photosynthesis in Jerusalem artichoke under salt stress.

Keywords: chlorophyll fluorescence; lipid peroxidation; Na^+; photosynthesis; photosystem

1. Introduction

Soil salinity is a serious problem for agricultural cultivation because of the detrimental effects on crop growth and yield. Under salt stress, plants have to tolerate osmotic stress, ionic toxicity, and secondary oxidative stress, and the metabolisms may be disrupted with damaged biological macromolecules [1–3]. Correspondingly, plants have evolved some physiological adaption measures such as Na^+ exclusion, osmolyte synthesis, and antioxidant induction, however, signal molecules which sensitively perceive external stresses are required to activate these protective mechanisms [4,5].

Abscisic acid (ABA) is defined as a stress hormone, because ABA can mediate extrinsic stress signals to improve expression of resistance genes [6–8]. As well documented, ABA plays an important role in regulating stomatal closure to limit water loss from transpiration, which assists in plant acclimatization to osmotic tolerance [8–10]. The positive role of ABA in plant salt tolerance also has been reviewed, and besides stomatal closure, osmolyte synthesis and antioxidant induction usually associate with ABA signal under salt stress despite some inconsistent reports due to species difference [7,11–14]. Na^+ is the primary toxic component for plants upon salt stress [2]. To date, it is still ambiguous whether ABA signal contributes to controlling Na^+ long-distance transportation and exclusion [4,15]. Particularly, Cabot et al. [16] reported that leaf ABA accumulation resulted in higher leaf Na^+ concentration in *Phaseolus vulgaris* under salt stress due to lowered leaf Na^+ exclusion and increased Na^+ translocation from root to shoot. Therefore, ABA function in defending salt-induced ionic toxicity seems not definite in contrast to its role in osmotic tolerance. Moreover, it remains unknown whether root ABA or leaf ABA has a greater effect on plant salt tolerance.

As one of the most important metabolisms for plant growth, photosynthesis is sensitive to salt stress. Photosynthetic analysis seems to be an effective and convenient way for diagnosing plant salt tolerance, because photosynthetic capacity in susceptible cultivars is more liable to be inhibited than tolerant ones [17–22]. Salt-induced stomatal closure initially depressed photosynthesis by lowering CO_2 availability [23,24], and subsequently, the negative effect on Rubisco can further restrict CO_2 fixation [25,26]. Eventually, the declined CO_2 assimilation will elevate excitation pressure in chloroplast through feedback inhibition on photosynthetic electron transport and then bring about photosystems photoinhibition or even irreversible damage with excess ROS production [27,28]. At present, photosystems photoinhibition and interaction under salt stress have been reported. In addition to PSII, PSI is also a crucial photoinhibition site and PSI photoinhibition poses a great threat to the entire photosynthetic apparatus by inducing PSII photoinhibition [20,29]. However, the relationship between the ABA signal and photosystem photoinhibition remains to be disclosed. ABA-induced stomatal limitation may trigger photosystem photoinhibition, but ABA-induced antioxidant activity can prevent from photoinhibition by scavenging reactive oxygen species (ROS). Particularly, the ambiguous function of ABA for regulating Na^+ transportation make it more complex.

Jerusalem artichoke (*Helianthus tuberosus* L.) is a valuable energy crop with high fructose and inulin concentrations in the tuber [30]. Jerusalem artichoke has certain salt tolerance and serves as a promising crop for utilizing coastal marginal land in China [30,31]. According to previous studies, salt stress could induce photosynthetic stomatal limitation, oxidative injury, chlorophyll loss, and ABA accumulation in Jerusalem artichoke [32–34]. However, the importance of ABA for salt tolerance in Jerusalem artichoke has not been tested. At present, gas exchange combined with modulated chlorophyll fluorescence has become a traditional method to examine plant stress tolerance. Recently, a simultaneous measurement of chlorophyll fluorescence transients and modulated 820 nm reflection has been applied to investigate PSII and PSI performance and their coordination, which enriches the traditional photosynthetic analysis [20,35–38]. In this study, we aimed to verify ABA function for salt adaptability in Jerusalem artichoke by photosynthetic analysis after applying a specific ABA synthesis inhibitor to the roots. Simultaneous measurement of chlorophyll fluorescence transients and modulated 820 nm reflection was carried out to complement traditional gas exchange analysis for revealing photosystems performance and coordination. Particularly, the abundance of PSII and PSI reaction center proteins was detected by immunoblot analysis to confirm salt-induced damage on photosystems. We hypothesized that root ABA accumulation helped prevent photosystems photoinhibition and protect photosynthesis by alleviating water loss and ionic toxicity. Our study can deepen the knowledge of salt tolerance in Jerusalem artichoke and may provide a reference for the cultivation in coastal saline land.

2. Results

2.1. Leaf Na$^+$, Relative Water, Malondialdehyde (MDA) and H$_2$O$_2$ Content, and Root Na$^+$ Flux

After four days of salt stress, leaf Na^+ and H_2O_2 content were significantly increased, whereas leaf relative water content was significantly decreased (Table 1). Leaf Na^+, MDA and H_2O_2 content, and root Na^+ flux were not directly affected by tungstate. Upon four days of salt stress, tungstate had no effect on the decreased amplitude of leaf relative water content but amplified the increase in leaf Na^+ and H_2O_2 content (Table 1). Salt-induced significant increase in leaf MDA content was found in tungstate-pretreated plants rather than non-pretreated plants (Table 1). Root Na^+ efflux was significantly elevated by salt stress, but salt-induced increase in Na^+ efflux was greatly reduced in tungstate-pretreated plants (Table 1).

Table 1. H_2O_2, malondialdehyde (MDA), Na^+ and relative water contents in the leaf and average root Na^+ flux in Jerusalem artichoke after four days of salt stress. Data in the table indicate the mean of five replicates ($\pm$SD). Within each column, means followed by the same letters are not significantly different at $p < 0.05$. FW indicates fresh weight. CP indicates control plants without pretreatment and NaCl stress; T1 indicates tungstate-pretreated plants without NaCl stress; T2 indicates non-pretreated plants under 150 mM NaCl stress; T3 indicates tungstate-pretreated plants under 150 mM NaCl stress.

Treatments	H_2O_2 Content ($\mu mol \cdot g^{-1}$ FW)	MDA Content ($nmol \cdot g^{-1}$ FW)	Na^+ Content ($mg \cdot g^{-1}$ FW)	Root Na^+ Efflux ($pmol \cdot cm^{-2} s^{-1}$)	Relative Water Content (%)
CP	$0.11 \pm 0.01c$	$53.00 \pm 5.86b$	$1.08 \pm 0.20c$	$1.80 \pm 0.70c$	$91.41 \pm 3.55a$
T1	$0.10 \pm 0.02c$	$54.13 \pm 5.64b$	$1.23 \pm 0.44c$	$3.84 \pm 1.09c$	$90.30 \pm 1.92a$
T2	$0.18 \pm 0.04b$	$52.83 \pm 4.11b$	$3.58 \pm 0.25b$	$130.13 \pm 23.59a$	$62.93 \pm 5.78b$
T3	$0.27 \pm 0.03a$	$69.01 \pm 6.50a$	$7.22 \pm 0.59a$	$33.79 \pm 6.59c$	$62.49 \pm 4.01b$

2.2. ABA Content in Leaf and Root

Single tungstate pretreatment did not affect root ABA content (Figure 1). After two days of salt stress, root ABA content was significantly increased by 47.8%, and the increase was dampened by tungstate pretreatment (Figure 1). After four days of salt stress, root ABA content was still remarkably lower in tungstate-pretreated plants than non-pretreated plants under salt stress (Figure 1). In all treatment groups, root ABA content after four days was lower than that after two days (Figure 1), which might originate from root development or ABA translocation from root to leaf.

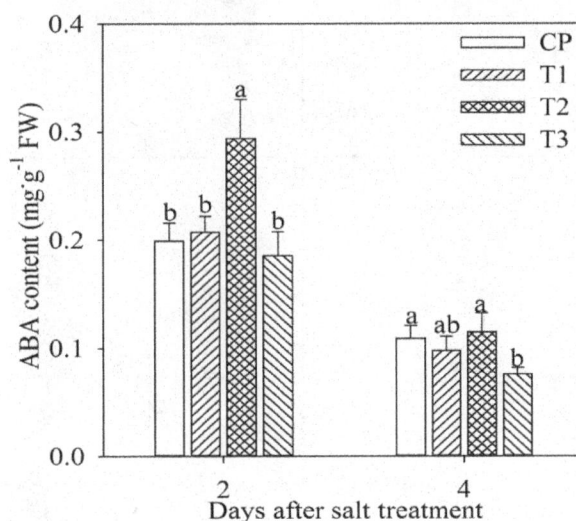

Figure 1. Changes in root abscisic acid (ABA) content in Jerusalem artichoke after salt stress for two days (a) and four days (b). Data in the figure indicate mean of five replicates ($\pm$SD), and different letters on error bars indicate significant difference at $p < 0.05$. CP indicates control plants without pretreatment and NaCl stress; T1 indicates tungstate-pretreated plants without NaCl stress; T2 indicates non-pretreated plants under 150 mM NaCl stress; T3 indicates tungstate-pretreated plants under 150 mM NaCl stress. The symbols, CP, T1, T2, and T3 are also used in the following figures.

2.3. Gas Exchange and Modulated Chlorophyll Fluorescence Parameters

Tungstate did not obviously influenced photosynthetic rate (Pn), stomatal conductance (g_s) and transpiration rate (Tr) (Figure 2a–c). Pn, g_s, and Tr significantly decreased in non-pretreated plants after one day of salt stress, and the decrease was up to 56.54%, 74.31% and 76.86% after four days of salt stress. In contrast, the decrease in Pn, g_s and Tr was remarkably higher in tungstate-pretreated plants upon salt stress (Figure 2a–c).

Under salt stress, decreased actual photochemical efficiency of PSII (ΦPSII) was noted with increased non-photochemical quenching (NPQ) in non-pretreated plants, whereas PSII excitation pressure (1-qP) did not show obvious change (Figure 2d–f). Tungstate did not significantly influenced ΦPSII, 1-qP and NPQ, and salt-induced decrease in ΦPSII was greater in tungstate-pretreated plants than non-pretreated plants. After two and three days of salt stress, 1-qP was significantly increased in tungstate-pretreated plants, but the increase became slight after four days of salt stress (Figure 2e). NPQ was significantly increased in tungstate-pretreated plants after one day of salt stress, but the increase disappeared after 3 days of salt stress (Figure 2f).

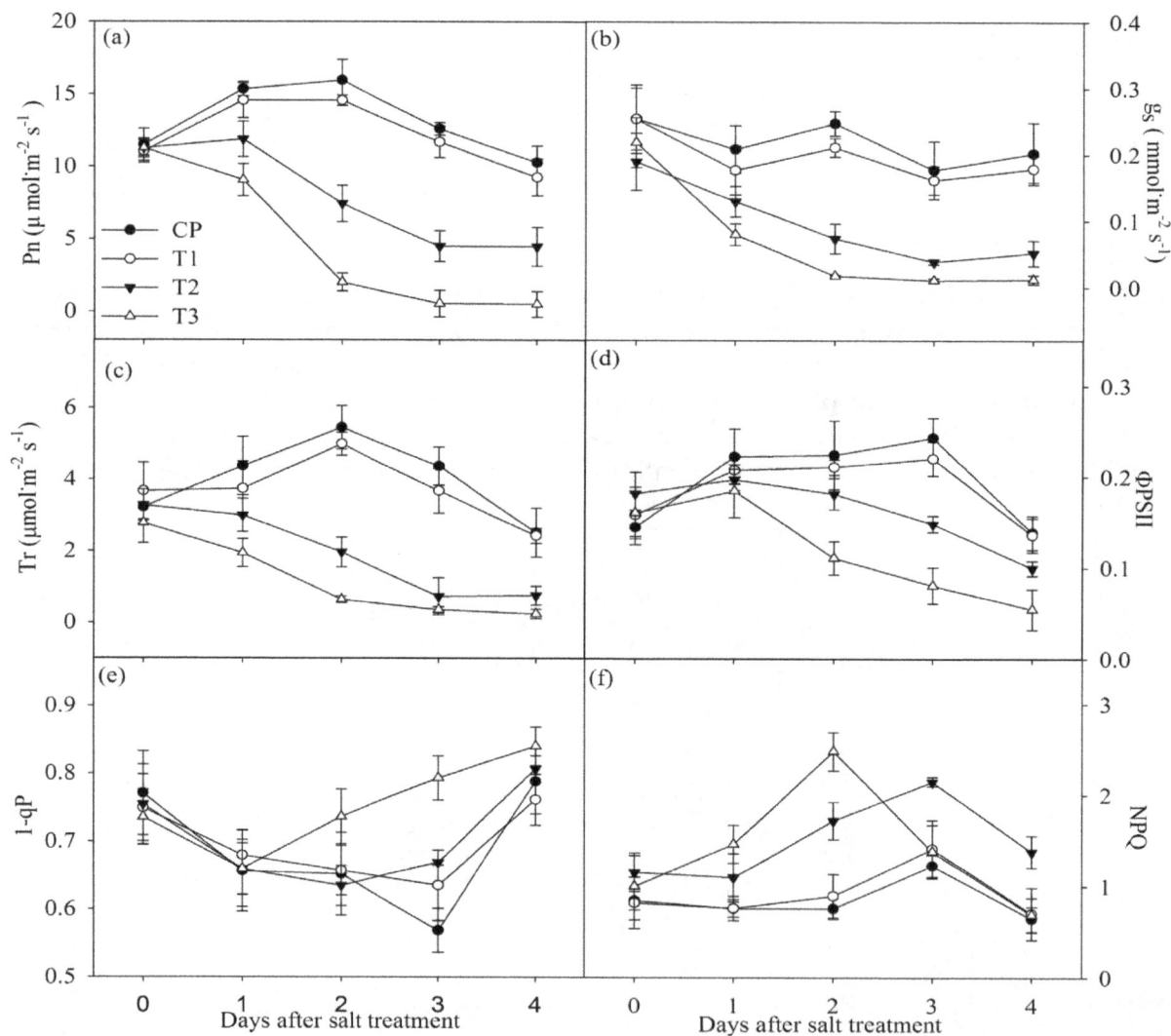

Figure 2. Changes in photosynthetic rate (Pn, (**a**)), stomatal conductance (g_s, (**b**)), transpiration (Tr, (**c**)), actual photochemical efficiency of PSII (ΦPSII, (**d**)), PSII excitation pressure (1-qP, (**e**)) and non-photochemical quenching (NPQ, (**f**)) in Jerusalem artichoke under salt stress. Data in the figure indicate the mean of five replicates ($\pm$SD).

2.4. Chlorophyll Fluorescence and Modulated 820 nm Reflection Transients

After two days of salt stress, chlorophyll fluorescence and modulated 820 nm reflection transients did not exhibit obvious change. The initial decrease in 820 nm reflection signal indicated PSI oxidation process, and the subsequent increase suggested that PSI was gradually re-reduced. After two days of salt stress, chlorophyll fluorescence transient descended in tungstate-pretreated plants (Figure 3a), suggesting PSII capacity was negatively affected. After salt stress for two days, the 820 nm reflection transient also changed in tungstate-pretreated plants, indicated by prolonged PSI oxidation process and lowered PSI re-reduction level (Figure 3b).

After four days of salt stress, chlorophyll fluorescence transient declined, while the PSI oxidation process was shortened (Figure 3c,d). Tungstate pretreatment never induced any change in chlorophyll fluorescence and 820 nm reflection transients, but their variations under salt stress were amplified by tungstate pretreatment (Figure 3c,d).

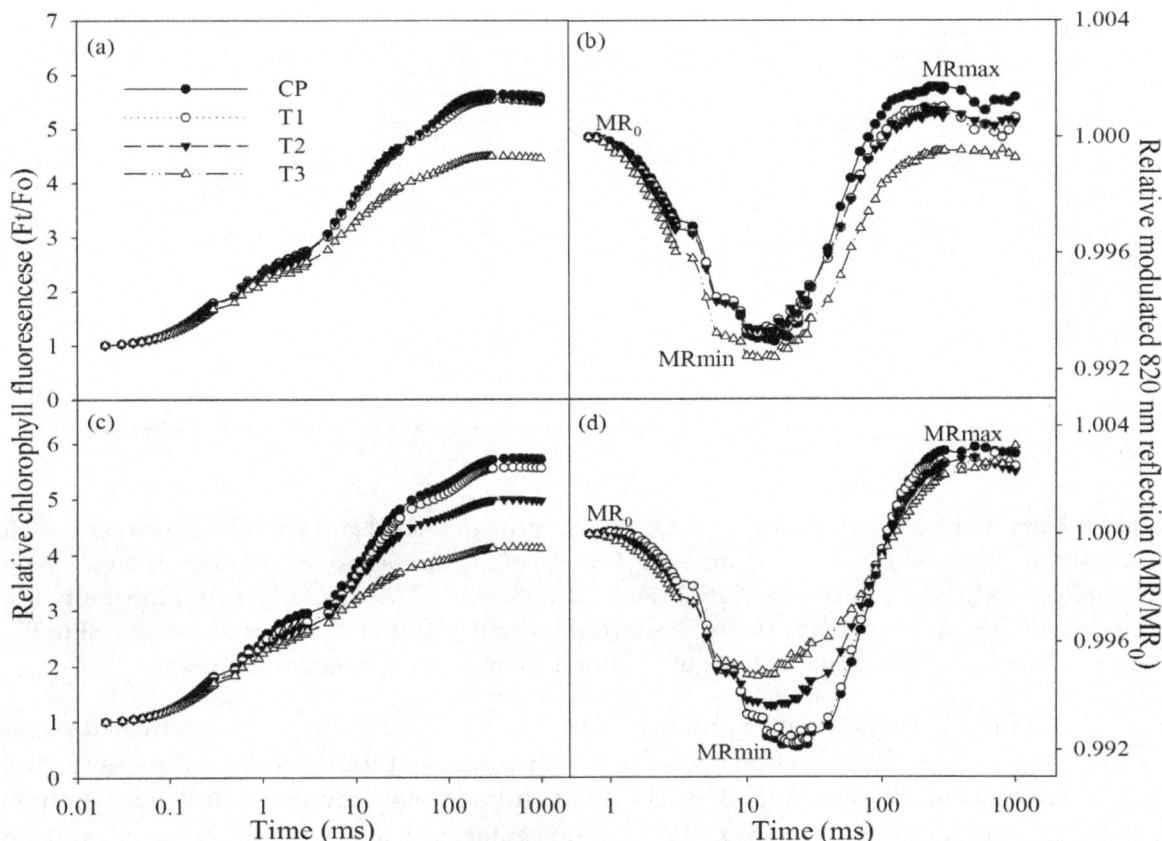

Figure 3. Chlorophyll fluorescence transients and 820 reflection transients during the first 1 s red illumination in Jerusalem artichoke under salt stress for two days (**a,b**) and four days (**c,d**). Ft is chlorophyll fluorescence intensity during the 1 s of red illumination, and Fo is fluorescence intensity at 20 μs, when all reaction centers of PSII are open. MR is the reflection signal during the 1 s of red illumination, and MR_0 is the value of modulated 820 nm reflection at the onset of red light illumination (0.7 ms, the first reliable MR measurement). MRmin and MRmax indicate the maximal point during PSI oxidation and the maximal point during PSI re-reduction, respectively. Data in the figure indicate the mean of five replicates.

2.5. PSII Performance, the Maximal Photochemical Capacity of PSI, and Immunoblot Analysis

Tungstate had no direct effect on the maximal photochemical capacity of PSI ($\Delta MR/MR_0$), the maximal quantum yield of PSII (Fv/Fm), probability that an electron moves further than primary acceptor of PSII (ETo/TRo) and quantum yield for electron transport (ETo/ABS) (Figure 4c–f). After two days of salt stress, Fv/Fm, $\Delta MR/MR_0$, ETo/TRo and ETo/ABS did not obviously change in non-pretreated plants, but significant decrease in Fv/Fm was observed in tungstate-pretreated plants (Figure 4c–f). After four days of salt stress, significant decrease in $\Delta MR/MR_0$ appeared with slightly lowered Fv/Fm, ETo/TRo and ETo/ABS in non-pretreated plants, but the decrease was greater in tungstate-pretreated plants (Figure 4c–f).

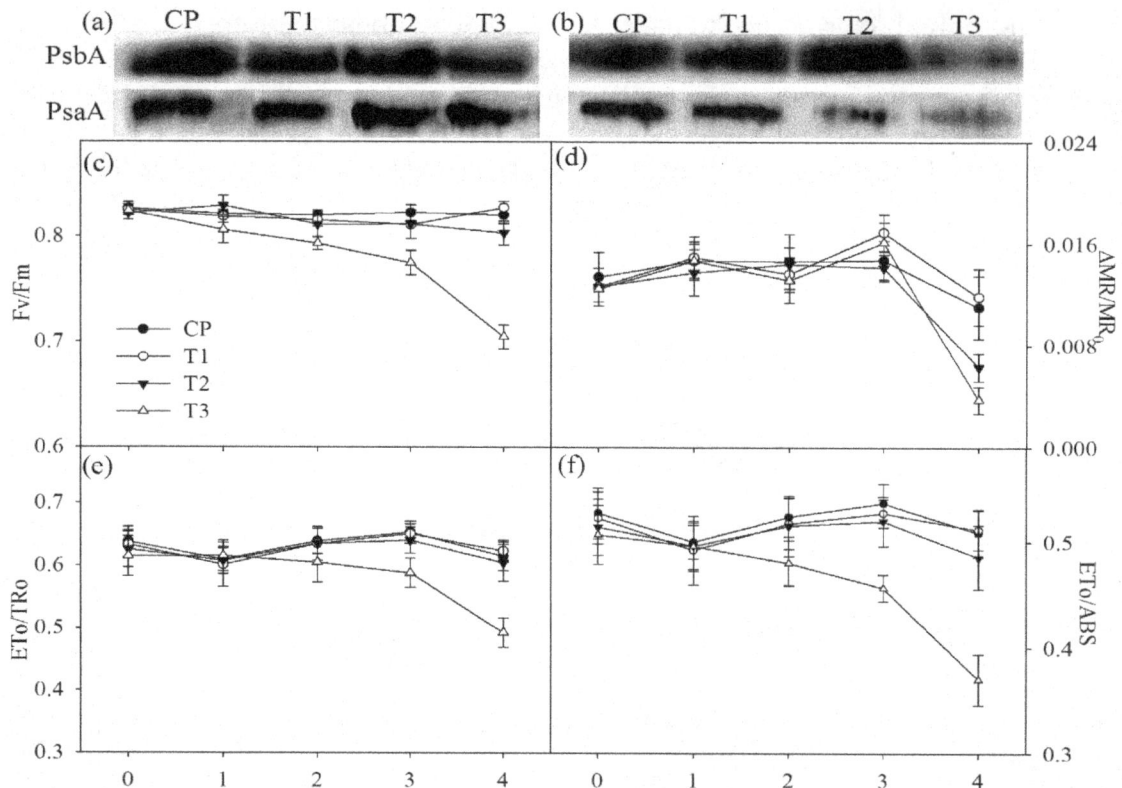

Figure 4. Immunoblot analysis of PSII reaction center protein (PsbA) and PSI reaction center protein (PsaA) abundance after two days (**a**) and four days (**b**) of salt stress and salt-induced changes in the maximal photochemical efficiency of PSII (Fv/Fm, (**c**)) and PSI ($\Delta MR/MR_0$, (**d**)), probability that an electron moves further than Q_A (ETo/TRo, (**e**)) and quantum yield for electron transport (REo/ETo, (**f**)) in Jerusalem artichoke. Data in the figure indicate the mean of five replicates ($\pm$SD).

The amount of PSII reaction center protein (PsbA) was reduced in tungstate-pretreated plants after two days of salt stress, and the reduction became more obvious after four days of salt stress (Figure 4a,b). In contrast, PsbA abundance was not affected by salt stress in plants without tungstate pretreatment (Figure 4a,b). After four days of salt stress, PSI reaction center protein (PsaA) abundance was decreased, and the decrease was greater in tungstate-pretreated than non-pretreated plants (Figure 4b).

3. Discussion

As with common knowledge, salt stress elevated root ABA concentration in Jerusalem artichoke, and tungstate pretreatment prevented salt-induced root ABA accumulation (Figure 1). The salt-induced greater decrease in Pn and ΦPSII in tungstate-pretreated plants than non-pretreated plants suggested that root ABA aided in protecting photosynthetic process in Jerusalem artichoke against salt stress (Figure 2a,d). Under salt stress, leaf stomatal closure reduced water loss from transpiration in Jerusalem artichoke (Figure 2b,c), but could inevitably induce stomatal limitation on photosynthesis. Tungstate-pretreated plants should encounter stronger photosynthetic stomatal limitation under salt stress due to the greater decrease in g_s compared with non-pretreated plants (Figure 2b). Lowered CO_2 assimilation can elevate PSII excitation pressure by feedback inhibition on photosynthetic electron transport and cause oxidative injury with excessive ROS production [28,39]. Under salt stress, PSII excitation pressure did not obviously change in spite of lowered CO_2 assimilation in non-pretreated plants, as the excessive excitation energy was effectively dissipated as heat (Figure 2e,f). In contrast, elevated PSII excitation pressure due to greater lowered CO_2 assimilation and insufficient heat dissipation could bring about photosystems photoinhibition in tungstate-pretreated plants upon salt stress. Notably, elevated PSII excitation pressure disappeared in tungstate-pretreated plants after four days of salt stress (Figure 2e), implying tremendous decrease in trapped energy in reaction center due to severe PSII photoinhibition.

In line with the above deduction, PSII photoinhibition actually occurred in tungstate-pretreated plants upon salt stress, indicated by declined Fv/Fm and chlorophyll fluorescence transient (Figure 3a,c and Figure 4c). Thus, considering slight change in Fv/Fm and chlorophyll fluorescence transient in non-pretreated plants (Figure 3a,c and Figure 4c), root ABA should participate in protecting PSII against photoinhibition in Jerusalem artichoke under salt stress. This positive role of root ABA was corroborated by immunoblot analysis, as lowered and unchanged PsbA abundance appeared, respectively, in tungstate-pretreated and non-pretreated plants upon salt stress (Figure 4a,b). Similar to PSII, PSI photoinhibition also derives from oxidative injury on reaction center proteins [27,28,36]. Along with elevated ROS production and lipid peroxidation (Table 1), PSI photoinhibition appeared after four days of salt stress, indicated by the significant decrease in $\Delta MR/MR_0$ (Figure 3d). In agreement with our recent study on waterlogging [40], PSI was also more vulnerable than PSII in Jerusalem artichoke under salt stress according to less decrease in Fv/Fm than $\Delta MR/MR_0$ (Figure 3c,d). Nonetheless, inhibition on root ABA synthesis led to higher PSII susceptibility to salt stress compared with PSI, as earlier significant decrease was observed in Fv/Fm rather than $\Delta MR/MR_0$ in tungstate-pretreated plants (Figure 3c,d). Prolonged PSI oxidation and lowered PSI re-reduction level in 820 nm reflection transients after two days of salt stress verified greater PSII vulnerability (Figure 2b). Thus, contrary to recent studies [20,29], PSII photoinhibition was not induced by PSI photoinhibition in tungstate-pretreated plants under salt stress. We inferred that photoprotective mechanisms were not adequately induced by salt stress in tungstate-pretreated plants and, as a result, lower heat dissipation appeared with greater excitation pressure on PSII (Figure 2e,f). As a traditional viewpoint, PSII photoinhibition can protect PSI against photoinhibition by restricting photosynthetic electron transport to PSI. In this study, PSII photoinhibition declined electron flow to PSI in tungstate-pretreated plants under salt stress, but PSI photoinhibition was still exacerbated according to greater decrease in $\Delta MR/MR_0$ compared with non-pretreated plants (Figure 4c–f). After four days of salt stress, greater shortened PSI oxidation also implied more severe PSI photoinhibition in tungstate-pretreated plants (Figure 3d), and this result was confirmed by salt-induced greater reduction in PsaA abundance in tungstate-pretreated plants (Figure 4b). Overall, root ABA signal helped defend salt-induced PSII and PSI photoinhibition in Jerusalem artichoke, and the protective way for PSI did not depend on PSII inactivation.

Although osmotic pressure can rapidly depress photosynthesis through stomatal limitation, Na^+ toxicity is more hazardous under salt stress. Na^+ can irreversibly inactivate PSII and PSI by inducing secondary oxidative injury or through direct damage on photosynthetic proteins [41–44]. Particularly, severe PSII photoinhibition without elevated excitation pressure in tungstate-pretreated plants after four days of salt stress may result from the direct effect of Na^+ in large part. In this study, inhibition on root ABA synthesis did not influence leaf water status in Jerusalem artichoke under salt stress, as similar relative leaf water content existed in tungstate-pretreated and non-pretreated plants (Table 1). In contrast, inhibited root ABA accumulation declined Na^+ exclusion from roots and led to prominent increase in leaf Na^+ concentration (Table 1). Therefore, Na^+ toxicity should be responsible for more severe PSII and PSI photoinhibition in tungstate-pretreated plants. However, the signal pathway for regulating Na^+ transport and uptake needs to be revealed in future study.

In agreement with the hypothesis, root ABA signal contributed to defending photosystems photoinhibition and protecting photosynthesis in Jerusalem artichoke under salt stress, but this positive role of root ABA was actualized mainly by reducing Na^+ toxicity.

4. Materials and Methods

4.1. Plant Material and Treatment

Tubers of Jerusalem artichoke were collected in Laizhou Bay, China. The tubers were planted in plastic pots filled with vermiculite (one tuber in each pot) and placed in an artificial climatic room

(Qiushi, China). The vermiculite was kept wet by watering. In the room, day/night temperature and humidity were controlled at 25/18 °C and 70%, and photon flux density was 400 $\mu mol \cdot m^{-2} s^{-1}$ for 12 h per day from 07:00 to 19:00. After one month, the tubers germinated and were daily watered with Hoagland nutrient solution (pH 5.7). One month later, health and uniform plants were selected and separated to four groups. In the first group, plants without tungstate sodium pretreatment were not subjected to NaCl stress. In the second group, plants were pretreated with tungstate sodium but not subjected to NaCl stress. In the third group, plants were exposed to 150 mM NaCl for four days without tungstate sodium pretreatment. In the fourth group, plants were pretreated with tungstate sodium and then subjected to 150 mM NaCl for four days. NaCl was added to nutrient solution incrementally by 50 mM step every day to reach the final concentration. The solution was refreshed every two days, and before refreshing solution, the culture substrate was thoroughly leached using nutrient solution for avoiding ion accumulation. One day before salt treatment, tungstate sodium (1 mM), a specific inhibitor of ABA synthesis, was added to nutrient solution for pretreatment.

4.2. Measurements of Na+, Relative Water Content, and Root Na+ Flux

The extraction of Na^+ was performed according to Song et al. [45]. Deionized H_2O (25 mL) was added to 0.1 g dried leaf powder and boiled for 2 h. The supernatant was diluted 50 times with deionized H_2O for measuring Na^+ content by using an atomic absorption spectrophotometer (TAS-990, Beijing, China). Net Na^+ flux was measured using NMT (Younger, Amherst, MA, USA) and the principle and protocol for measuring root Na^+ flux have been elucidated in detail in our recent study [20]. In this experiment, newly developed root segments were sampled and a vigorous Na^+ flux was identified at 500 μm from the root apex. The measured root position can be visualized under microscope, and tungstate pretreatment dampened salt-induced increase in root Na^+ efflux (Supplemental Figure S1). The average value of Na^+ flux is presented in Table 1.

Fresh leaves were harvested and weighed (fresh weight, FW), and then were immersed in distilled water for 4 h at room temperature to determine saturated fresh weight (SW). Subsequently, the leaves were dried completely in an oven at 70 °C and weighed (dry weight, DW). Relative water content (RWC) was calculated as: RWC = (FW − DW)/(SW − DW) × 100%.

4.3. Measurements of MDA, H2O2, and ABA Content

Leaf tissues (0.5 g) were ground under liquid nitrogen and homogenized in 5 mL 0.1% TCA. The homogenate was centrifuged at 12,000× g and 4 °C for 10 min to collect the supernatant for measurements of MDA and H_2O_2 content. The supernatant (0.5 mL) was mixed with 10 mM potassium phosphate buffer (0.5 mL, pH 7.0) and 1 M KI (1 mL), and the absorbance at 390 nm was recorded for calculating H_2O_2 content [46]. MDA content was determined by thiobarbituric acid reaction to reflect the extent of lipid peroxidation [47].

ABA content was analyzed according to Lopez-Carbonell and Jauregui [48] with some modification. Root and leaf tissues (0.5 g) were ground under liquid nitrogen and homogenized in 3 mL 80% methanol containing 0.1% acetic acid. After agitation for 30 min at 4 °C, the homogenate was centrifuged at 12,000× g and 4 °C for 10 min. The supernatant was filtered through a 0.45 μm polytetrafluoroethylene membrane, and the filtrate (10 μL) was injected into a high performance liquid chromatography instrument equipped with mass spectrometer (Thermo, Waltham, MA, USA). A hypersil C18 column (4.6 mm × 150 mm; particle size, 5.0 μm) was used in the liquid chromatography system, and the mobile phase consisted of water with 0.1% HCO_2H (A) and MeOH with 0.1% HCO_2H (B). A gradient elution program was applied, and the initial gradient of methanol was kept at 30% for 2 min and increased linearly to 100% at 20 min. All the analyses of mass spectrum (MS) were performed using ionspray source in negative ion mode, and MS/MS product ions were produced by collision-activated dissociation of selected precursor ions. Since many compounds could present the same nominal molecular mass, MS/MS method was required to selectively monitor ABA in crude plant extracts by identifying parent mass and unique fragment ion. In this study, MS/MS method was used for the quantitation of ABA by monitoring 263/153 transition, and ABA concentration was determined by using a standard curve plotted with known concentrations of the standards.

4.4. Measurements of Gas Exchange and Modulated Chlorophyll Fluorescence

Gas exchange and modulated chlorophyll fluorescence were simultaneously measured by using an open photosynthetic system (LI-6400XTR, Li-Cor, Lincoln, NE, USA) equipped with a fluorescence leaf chamber (6400-40 LCF, Li-Cor). Temperature, CO_2 concentration and actinic light intensity were, respectively, set at 25 °C, 400 μmol·mol^{-1} and 1000 μmol·m^{-2} s^{-1} in the leaf cuvette. Pn, g$_s$ and Tr were simultaneously noted. After steady-state fluorescence yield was recorded, a saturating actinic light pulse of 8000 μmol·m^{-2} s^{-1} for 0.7 s was used to produce maximum fluorescence yield by temporarily inhibiting PSII photochemistry for measuring ΦPSII. Photochemical quenching coefficient was also recorded for calculating 1-qP. Thereafter, the leaves were dark-adapted for 30 min, and a saturating actinic light pulse of 8000 μmol·m^{-2} s^{-1} for 0.7 s was applied to measure the maximal fluorescence for calculating NPQ [49].

4.5. Simultaneous Measurements of Chlorophyll Fluorescence and Modulated 820 nm Reflection Transients

A multifunctional plant efficiency analyzer (MPEA, Hansatech, Norfolk, UK) was used for the measurements, and its operating mechanism has been described in detail [37]. The leaves were dark-adapted for 30 min, and the leaves were orderly illuminated with 1 s red light (627 nm, 5000 μmol photons·m^{-2} s^{-1}), 10 s far red light (735 nm, 200 μmol photons·m^{-2} s^{-1}) and 2 s red light (627 nm, 5000 μmol photons·m^{-2} s^{-1}). Chlorophyll fluorescence and modulated 820 nm reflection were simultaneously detected during the illumination. Chlorophyll fluorescence and modulated 820 nm reflection transients were simultaneously recorded during the illumination. Fv/Fm, ETo/TRo, and ETo/ABS were calculated according to chlorophyll fluorescence transients [20], and ΔMR/MR$_0$ was determined from modulated 820 nm reflection signal [50–52].

4.6. Isolation of Thylakoid Membranes and Western Blot

Five grams of leaf discs were ground under liquid nitrogen and homogenized in a solution containing 400 mM sucrose, 50 mM HEPES-KOH (pH 7.8), 10 mM NaCl, and 2 mM MgCl$_2$ [53]. The homogenate was filtered through two layers of cheesecloth and then centrifuged at 5000× g and 4 °C for 10 min to collectthylakoid pellets. The pellets were resuspended in the homogenization buffer, and chlorophyll content was measured.

Thylakoid membranes with 10 μg chlorophyll were separated by a 12% (w/w) SDS-PAGE gel. Proteins from the gel were transferred onto polyvinylidene fluoride membrane by semi dry method. After blocking with 5% skimmed milk for 1 h, the membranes were incubated for 2 h with the primary anti-PsbA or anti-PsaA antibodies (PhytoAB, San Francisco, CA, USA) and then incubated with horseradish peroxidase-conjugated anti-rabbit IgG antibody (PhytoAB, USA) for 2 h. BeyoECL Plus substrate (Beyotime Biotechnology, Shanghai, China) was used to test immunoreaction, and the chemiluminescence was detected by a Tanon-5500 cooled CCD camera (Tanon, Shanghai, China).

4.7. Statistical Analysis

One-way ANOVA was carried out by using SPSS 16.0 (SPSS Inc., Chicago, IL, USA) for all sets of data. The values presented are the means of measurements with five replicate plants, and comparisons of means were determined through LSD test. The difference was considered significant at $p < 0.05$.

Author Contributions: K.Y. designed the experiment, performed data analysis and wrote the manuscript. T.B., W.H., M.L. (Mengxue Lv), M.G., and M.L. (Ming Lu) participated in the experiment. G.H. reviewed the manuscript and proposed some critical suggestions.

Abbreviations

ETo/ABS	quantum yield for electron transport
ETo/TRo	probability that an electron moves further than primary acceptor of PSII
Fv/Fm	the maximal quantum yield of PSII
g_s	stomatal conductance
MDA	malondialdehyde
NPQ	non-photochemical quenching
Pn	photosynthetic rate
PSI	Photosystem I
PSII	Photosystem II
Q_A	primary quinone
ROS	reactive oxygen species
Tr	transpiration rate
1-qP	excitation pressure of PSII
$\Delta MR/MR_0$	the maximal photochemical capacity of PSI
ΦPSII	actual photochemical efficiency of PSII

References

1. Hossain, M.S.; Dietz, K.J. Tuning of redox regulatory mechanisms, reactive oxygen species and redox homeostasis under salinity stress. *Front. Plant Sci.* **2016**, *7*, 548. [CrossRef] [PubMed]

2. Munns, R.; Tester, M. Mechanisms of salinity tolerance. *Annu. Rev. Plant Biol.* **2008**, *59*, 651–681. [CrossRef] [PubMed]

3. Zhu, J.K. Regulation of ion homeostasis under salt stress. *Curr. Opin. Plant Biol.* **2003**, *6*, 441–445. [CrossRef]

4. Zhu, J.K. Abiotic Stress Signaling and Responses in Plants. *Cell* **2016**, *167*, 313–324. [CrossRef] [PubMed]

5. Julkowska, M.M.; Testerink, C. Tuning plant signaling and growth to survive salt. *Trends Plant Sci.* **2015**, *20*, 586–594. [CrossRef] [PubMed]

6. Hong, J.H.; Seah, S.W.; Xu, J. The root of ABA action in environmental stress response. *Plant Cell Rep.* **2013**, *32*, 971–983. [CrossRef] [PubMed]

7. Sah, S.K.; Reddy, K.R.; Li, J. Abscisic acid and abiotic stress tolerance in crop plants. *Front. Plant Sci.* **2016**, *7*, 571. [CrossRef] [PubMed]

8. Zhang, J.; Jia, W.; Yang, J.; Ismail, A.M. Role of ABA in integrating plant responses to drought and salt stresses. *Field Crop Res.* **2006**, *97*, 111–119. [CrossRef]

9. Zhang, F.P.; Sussmilch, F.; Nichols, D.S.; Cardoso, A.A.; Brodribb, T.J.; McAdam, S.A.M. Leaves, not roots or floral tissue, are the main site of rapid, external pressure-induced ABA biosynthesis in angiosperms. *J. Exp. Bot.* **2018**, *69*, 1261–1267. [CrossRef]

10. Zhu, J.K. Salt and drought stress signal transduction in plants. *Annu. Rev. Plant Biol.* **2002**, *53*, 247–273. [CrossRef]

11. Ryu, H.; Cho, Y.G. Plant hormones in salt stress tolerance. *J. Plant Biol.* **2015**, *58*, 147–155. [CrossRef]

12. Hong, C.Y.; Chao, Y.Y.; Yang, M.Y.; Cheng, S.Y.; Cho, S.C.; Kao, C.H. NaCl-induced expression of glutathione reductase in roots of rice (*Oryza sativa* L.) seedlings is mediated through hydrogen peroxide but not abscisic acid. *Plant Soil* **2009**, *320*, 103–115. [CrossRef]

13. Kalinina, E.B.; Keith, B.K.; Kern, A.J.; Dyer, W.E. Salt- and osmotic stress-induced choline monooxygenase expression in Kochia scoparia is ABA-independent. *Biol. Plant.* **2012**, *56*, 699–704. [CrossRef]

14. Per, T.S.; Khan, N.A.; Reddy, P.S.; Masood, A.; Hasanuzzaman, M.; Khan, M.I.R.; Anjum, N.A. Approaches in modulating proline metabolism in plants for salt and drought stress tolerance: Phytohormones, mineral nutrients and transgenics. *Plant Physiol. Biochem.* **2017**, *115*, 126–140. [CrossRef]

15. Osakabe, Y.; Yamaguchi-Shinozaki, K.; Shinozaki, K.; Tran, L.S. ABA control of plant macroelement membrane transport systems in response to water deficit and high salinity. *New Phytol.* **2014**, *202*, 35–49. [CrossRef] [PubMed]

16. Cabot, C.; Sibole, J.V.; Barceló, J.; Poschenrieder, C. Abscisic acid decreases leaf Na^+ exclusion in salt-treated *Phaseolus vulgaris* L. *J. Plant Growth Regul.* **2009**, *28*, 187–192. [CrossRef]

17. Chen, P.; Yan, K.; Shao, H.; Zhao, S. Physiological mechanisms for high salt tolerance in wild soybean (*Glycine soja*) from Yellow River Delta, China: Photosynthesis, osmotic regulation, ion flux and antioxidant capacity. *PLoS ONE* **2013**, *8*, e83227. [CrossRef]

18. Kalaji, H.M.; Govindjee Bosa, K.; Koscielniak, J.; Zuk-Golaszewska, K. Effects of salt stress on photosystem II efficiency and CO_2 assimilation of two Syrian barley landraces. *Environ. Exp. Bot.* **2011**, *73*, 64–72. [CrossRef]

19. Stepien, P.; Johnson, G.N. Contrasting responses of photosynthesis to salt stress in the glycophyte *Arabidopsis* and the halophyte *Thellungiella*: Role of the plastid terminal oxidase as an alternative electron sink. *Plant Physiol.* **2009**, *149*, 1154–1165. [CrossRef]

20. Yan, K.; Wu, C.; Zhang, L.; Chen, X. Contrasting photosynthesis and photoinhibition in tetraploid and its autodiploid honeysuckle (*Lonicera japonica* Thunb.) under salt stress. *Front. Plant Sci.* **2015**, *6*, 227. [CrossRef]

21. Yan, K.; Xu, H.; Cao, W.; Chen, X. Salt priming improved salt tolerance in sweet sorghum by enhancing osmotic resistance and reducing root Na^+ uptake. *Acta Physiol. Plant.* **2015**, *37*, 203. [CrossRef]

22. Aparicio, C.; Urrestarazu, M.; Cordovilla, M.D. Comparative physiological analysis of salinity effects in six olive genotypes. *Hortscience* **2014**, *49*, 901–904.

23. Chaves, M.M.; Flexas, J.; Pinheiro, C. Photosynthesis under drought and salt stress: Regulation mechanisms from whole plant to cell. *Ann. Bot.* **2009**, *103*, 551–560. [CrossRef] [PubMed]

24. Loreto, F.; Centritto, M.; Chartzoulakis, K. Photosynthetic limitations in olive cultivars with different sensitivity to salt stress. *Plant Cell. Environ.* **2003**, *26*, 595–601. [CrossRef]

25. Feng, L.L.; Han, Y.J.; Liu, G.; An, B.G.; Yang, J.; Yang, G.H.; Li, Y.S.; Zhu, Y.G. Overexpression of sedoheptulose-1,7-bisphosphatase enhances photosynthesis and growth under salt stress in transgenic rice plants. *Funct. Plant Biol.* **2007**, *34*, 822–834. [CrossRef]

26. Lu, K.X.; Cao, B.H.; Feng, X.P.; He, Y.; Jiang, D.A. Photosynthetic response of salt-tolerant and sensitive soybean varieties. *Photosynthetica* **2009**, *47*, 381–387. [CrossRef]

27. Sonoike, K. Photoinhibition of photosystem I. *Physiol. Plant.* **2011**, *142*, 56–64. [CrossRef] [PubMed]

28. Takahashi, S.; Murata, N. How do environmental stresses accelerate photoinhibition? *Trends Plant Sci.* **2008**, *13*, 178–182. [CrossRef] [PubMed]

29. Yan, K.; Zhao, S.; Liu, Z.; Chen, X. Salt pretreatment alleviated salt-induced photoinhibition in sweet sorghum. *Theor. Exp. Plant Physiol.* **2015**, *27*, 119–129. [CrossRef]

30. Long, X.H.; Liu, L.P.; Shao, T.Y.; Shao, H.B.; Liu, Z.P. Developing and sustainably utilize the coastal mudflat areas in China. *Sci. Total Environ.* **2016**, *569–570*, 1077–1086. [CrossRef]

31. Dias, N.S.; Ferreira, J.F.S.; Liu, X.; Suarez, D.L. Jerusalem artichoke (*Helianthus tuberosus*, L.) maintains high inulin, tuber yield, and antioxidant capacity under moderately-saline irrigation waters. *Ind. Crops Prod.* **2016**, *94*, 1009–1024. [CrossRef]

32. Huang, Z.; Long, X.; Wang, L.; Kang, J.; Zhang, Z.; Zed, R.; Liu, Z. Growth, photosynthesis and H+-ATPase activity in two Jerusalem artichoke varieties under NaCl-induced stress. *Process Biochem.* **2012**, *47*, 591–596. [CrossRef]

33. Li, L.; Shao, T.; Yang, H.; Chen, M.; Gao, X.; Long, X.; Shao, H.; Liu, Z.; Rengel, Z. The endogenous plant hormones and ratios regulate sugar and dry matter accumulation in Jerusalem artichoke in salt-soil. *Sci. Total Environ.* **2017**, *578*, 40–46. [CrossRef]

34. Long, X.; Huang, Z.; Zhang, Z.; Li, Q.; Zed, R.; Liu, Z. Seawater stress differentially affects germination, growth, photosynthesis, and ion concentration in genotypes of Jerusalem Artichoke (*Helianthus tuberosus* L.). *J. Plant Growth Regul.* **2009**, *29*, 223–231. [CrossRef]

35. Li, P.M.; Ma, F.W. Different effects of light irradiation on the photosynthetic electron transport chain during apple tree leaf dehydration. *Plant Physiol. Biochem.* **2012**, *55*, 16–22. [CrossRef] [PubMed]

36. Oukarroum, A.; Bussotti, F.; Goltsev, V.; Kalaji, H.M. Correlation between reactive oxygen species production and photochemistry of photosystems I and II in *Lemna gibba* L. plants under salt stress. *Environ. Exp. Bot.* **2015**, *109*, 80–88. [CrossRef]

37. Strasser, R.J.; Tsimilli-Michael, M.; Qiang, S.; Goltsev, V. Simultaneous in vivo recording of prompt and delayed fluorescence and 820 nm reflection changes during drying and after rehydration of the resurrection plant *Haberlea rhodopensis Biochim. Biophys. Acta* **2010** *1797*, 122. [CrossRef]

38. Zivcak, M.; Brestic, M.; Kunderlikova, K.; Olsovska, K.; Allakhverdiev, S.I. Effect of photosystem I inactivation on chlorophyll a fluorescence induction in wheat leaves: Does activity of photosystem I play any role in OJIP rise? *J. Photochem. Photobiol. B Biol.* **2015**, *152*, 318–324. [CrossRef] [PubMed]

39. Gill, S.S.; Tuteja, N. Reactive oxygen species and antioxidant machinery in abiotic stress tolerance in crop plants. *Plant Physiol. Biochem.* **2010**, *48*, 909–930. [CrossRef] [PubMed]

40. Yan, K.; Zhao, S.; Cui, M.; Han, G.; Wen, P. Vulnerability of photosynthesis and photosystem I in Jerusalem artichoke (*Helianthus tuberosus* L.) exposed to waterlogging. *Plant Physiol. Biochem.* **2018**, *125*, 239–246. [CrossRef] [PubMed]

41. Allakhverdiev, S.I.; Murata, N. Salt stress inhibits photosystems II and I in cyanobacteria. *Photosynth Res.* **2008**, *98*, 529–539. [CrossRef] [PubMed]

42. Munns, R. Comparative physiology of salt and water stress. *Plant Cell Environ.* **2002**, *25*, 239–250. [CrossRef] [PubMed]

43. Murata, N.; Takahashi, S.; Nishiyama, Y.; Allakhverdiev, S.I. Photoinhibition of photosystem II under environmental stress. *Biochim. Biophys. Acta* **2007**, *1767*, 414–421. [CrossRef]

44. Yang, C.; Zhang, Z.S.; Gao, H.Y.; Fan, X.L.; Liu, M.J.; Li, X.D. The mechanism by which NaCl treatment alleviates PSI photoinhibition under chilling-light treatment. *J. Photochem. Photobiol. B Biol.* **2014**, *140*, 286–291. [CrossRef] [PubMed]

45. Song, J.; Shi, G.W.; Gao, B.; Fan, H.; Wang, B.S. Waterlogging and salinity effects on two *Suaeda salsa* populations. *Physiol. Plant.* **2011**, *141*, 343–351. [CrossRef] [PubMed]

46. Velikova, V.; Yordanov, I.; Edreva, A. Oxidative stress and some antioxidant systems in acid rain-treated bean plants—Protective role of exogenous polyamines. *Plant Sci.* **2000**, *151*, 59–66. [CrossRef]

47. Yan, K.; Cui, M.; Zhao, S.; Chen, X.; Tang, X. Salinity stress is beneficial to the accumulation of chlorogenic acids in honeysuckle (*Lonicera japonica* Thunb.). *Front. Plant Sci.* **2016**, *7*, 1563. [CrossRef]

48. Lopez-Carbonell, M.; Jauregui, O. A rapid method for analysis of abscisic acid (ABA) in crude extracts of water stressed Arabidopsis thaliana plants by liquid chromatography—Mass spectrometry in tandem mode. *Plant Physiol. Biochem.* **2005**, *43*, 407–411. [CrossRef]

49. Maxwell, K.; Johnson, G.N. Chlorophyll fluorescence—A practical guide. *J. Exp. Bot.* **2000**, *51*, 659–668. [CrossRef]

50. Schansker, G.; Srivastava, A.; Strasser, R.J. Characterization of the 820-nm transmission signal paralleling the chlorophyll a fluorescence rise (OJIP) in pea leaves. *Funct. Plant Biol.* **2003**, *30*, 785–796. [CrossRef]

51. Yan, K.; Chen, P.; Shao, H.B.; Zhao, S.J. Characterization of photosynthetic electron transport chain in bioenergy crop Jerusalem artichoke (*Helianthus tuberosus* L.) under heat stress for sustainable cultivation. *Ind. Crop. Prod.* **2013**, *50*, 809–815. [CrossRef]

52. Yan, K.; Han, G.; Ren, C.; Zhao, S.; Wu, X.; Bian, T. *Fusarium solani* infection depressed photosystem performance by inducing foliage wilting in apple seedlings. *Front. Plant Sci.* **2018**, *9*, 479. [CrossRef] [PubMed]

53. Zhang, Z.S.; Jin, L.Q.; Li, Y.T.; Tikkanen, M.; Li, Q.M.; Ai, X.Z.; Gao, H.Y. Ultraviolet-B Radiation (UV-B) Relieves Chilling-light-induced PSI photoinhibition and accelerates the recovery Of CO_2 assimilation in cucumber (*Cucumis sativus* L.) leaves. *Sci. Rep.* **2016**, *6*, 34455. [CrossRef] [PubMed]

8

Identification of Salt Stress Responding Genes using Transcriptome Analysis in Green Alga *Chlamydomonas reinhardtii*

Ning Wang [†], Zhixin Qian [†], Manwei Luo, Shoujin Fan, Xuejie Zhang and Luoyan Zhang [*]

Key Lab of Plant Stress Research, College of Life Science, Shandong Normal University, No. 88 Wenhuadong Road, Jinan 250014, China; wangning_sdnu@163.com (N.W.); qianzhixin_sdnu@163.com (Z.Q.); luomanwei_sdnu@163.com (M.L.); fansj@sdnu.edu.cn (S.F.); zxjpublic@sohu.com (X.Z.)
* Correspondence: zhangluoyan@sdnu.edu.cn
† These authors contributed equally to this work.

Abstract: Salinity is one of the most important abiotic stresses threatening plant growth and agricultural productivity worldwide. In green alga *Chlamydomonas reinhardtii*, physiological evidence indicates that saline stress increases intracellular peroxide levels and inhibits photosynthetic-electron flow. However, understanding the genetic underpinnings of salt-responding traits in plantae remains a daunting challenge. In this study, the transcriptome analysis of short-term acclimation to salt stress (200 mM NaCl for 24 h) was performed in *C. reinhardtii*. A total of 10,635 unigenes were identified as being differently expressed by RNA-seq, including 5920 up- and 4715 down-regulated unigenes. A series of molecular cues were screened for salt stress response, including maintaining the lipid homeostasis by regulating phosphatidic acid, acetate being used as an alternative source of energy for solving impairment of photosynthesis, and enhancement of glycolysis metabolism to decrease the carbohydrate accumulation in cells. Our results may help understand the molecular and genetic underpinnings of salt stress responses in green alga *C. reinhardtii*.

Keywords: *Chlamydomonas reinhardtii*; salt stress; transcriptome analysis; impairment of photosynthesis; underpinnings of salt stress responses

1. Introduction

Salinity is one of the most important abiotic stresses threatening agricultural productivity worldwide. Although plants have gradually evolved a series of adaptive molecular, physiology and biochemistry processes to respond to salinity stress, it could threaten 30% of cultivable soils by 2050 [1,2]. Understanding the molecular machineries of salt stress response in model plants of basal taxa, such as green algae, may contribute to finding the evolutionary cues of abiotic stress response in plants and developing salt-resistant crops with additional salt-responding traits [2–9].

Salt stress causes diverse impacts on plant growth by disturbing the osmotic/ionic balance and eliciting Na^+ toxicity [9,10]. Under aquatic saline stress, a series of physical and biochemical processes are recruited by algae to respond to the damage caused by osmotic and ionic stresses, such as photosynthesis inhibition, macromolecular compound synthesis and homeostasis adjustment [6,10–14]. It has been reported that salt stress leads to decreased photosynthetic efficiency [15,16] which influences chlorophyll content in plant leaves [17,18]. In green algae, salt stress remarkably influences the structure and functions of the photosynthetic apparatus in *Scenedesmus obliquus* [19] and reduces the maximum quantum yield of photosystem II (PSII) in *Dunaliella salina* [20]. In alga *Botryococcus braunii*, metabolism of lutein was significantly enhanced under stress conditions [12].

Chlamydomonas reinhardtii is a free-living freshwater alga with unicellular vegetative cell. Previous studies exposed the *C. reinhardtii* strain 21 gr and CC-503 to salt stress and demonstrated the physiological and metabolic processes impacted by ionic toxicity and osmotic stress caused by salt damage [6–8,16,21]. Vega [22] demonstrated that 200 mM NaCl in the culture medium was highly toxic for *C. reinhardtii* productivity. The addition of NaCl immediately blocked the photosynthetic activity of the alga which partially recovered, after 1 h of treatment, remaining high during the following 24 h. However, after 24 h treatment with NaCl 200 mM, the intracellular catalase activity of the alga reached a 20-fold higher level than in the control cells. The physiological data indicate that saline stress induces in *C. reinhardtii* an increase of intracellular peroxide, which parallels a significant inhibition of the photosynthetic-electron flow. However, the related machineries of up-stream regulating and the triggering of appropriate cellular and physiological responses to cope with stress circumstances are still largely unknown.

Transcriptome sequencing is an effective strategy for detecting potential participants of stress response on a genome-wide scale. Hundreds of studies about salt stress responses in model plant *Arabidopsis thaliana* [23–26], crops *Oryza sativa* [23,27] and *Glycine max* [28], and in some halophytes (plants able to complete their life cycles under saline environments) have been widely conducted using sequencing technologies [7,29–41]. The integrations of genes' spatio-temporal expression patterns and responding traits have helped to identify a large number of salt stress-related differentially expressed genes (DEGs) and mechanisms.

Keeping this in mind, the work presented here was carried out to explore the saline stress-responding mechanisms of *C. reinhardtii* by transcriptome sequencing of strains GY-D55 wild type. The aim of this study was to identify dys-regulated genes in *C. reinhardtii* cells under salt stress by RNA-seq, screen physiological and biochemical cues by gene ontology (GO) terms and MapMan functional enrichment analyses, and investigate the physiological adaptions and cellular regulatory networks for salt stress responding.

2. Results

2.1. Transcriptome Profiling of C. reinhardtii

After sequencing with the Illumina HiSeq X platform, a total of 56,438,218, 72,853,712, 47,551,786, 56,962,722, 52,926,804 and 55,998,748 high-quality pair-end reads were obtained from three control and three salt stress treated samples of *C. reinhardtii* (Table 1), respectively. *De novo* transcriptome assembly generated 91,242 unigenes, with an average length of 2691 nt and N50 of 4554. On average, 90.66% of the reads from six samples were mapped to the reference genome (Table 1). The assembled transcriptome information of *C. reinhardtii* is shown in Supplementary Figure S1.

Table 1. Summary of mapping transcriptome reads to reference sequence.

Sample Name	Sample Description	Total Reads	Total Mapped	Ratio of Mapped Reads
C_0_1	Control replication 1	56,438,218	51,454,456	91.17%
C_0_2	Control replication 2	72,853,712	66,008,290	90.60%
C_0_3	Control replication 3	47,551,786	43,268,544	90.99%
S_200_1	Salt stress replication 1	56,962,722	51,633,614	90.64%
S_200_2	Salt stress replication 2	52,926,804	47,815,814	90.34%
S_200_3	Salt stress replication 3	55,998,748	50,507,824	90.19%

2.2. Functional Annotations of Unigenes

Similarity searches were performed to annotate unigenes against different databases using BLASTX. For *C. reinhardtii*, 65,679 (71.98%) unigenes were annotated in at least one database (Figure 1C and Supplementary Figure S1). A total of 52,884 (57.96%) and 58,062 (63.63%) unigenes showed similarity to sequences in NR and PFAM databases with an E-value threshold of 1×10^{-5}.

About 58,651 (64.28%) unigenes were annotated in the GO database by Blast2GO v2.5 with an E-value cutoff of 1×10^{-6} (Figure 1C and Supplementary Figure S1). Unigenes of the *C. reinhardtii* were assigned to *C. reinhardtii* and *A. thaliana* gene IDs for GO annotation mapping and TFs/PKs perdition. By sequence alignment, a total of 48,158 unigenes were aligned to *C. reinhardtii* PLAZA genome genes. A total of 54,509 unigenes were assigned to TAIR10 locus IDs by BLASTP with an E-value cutoff of 1×10^{-5} and classified into GO categories for GO analysis (Supplementary Table S1).

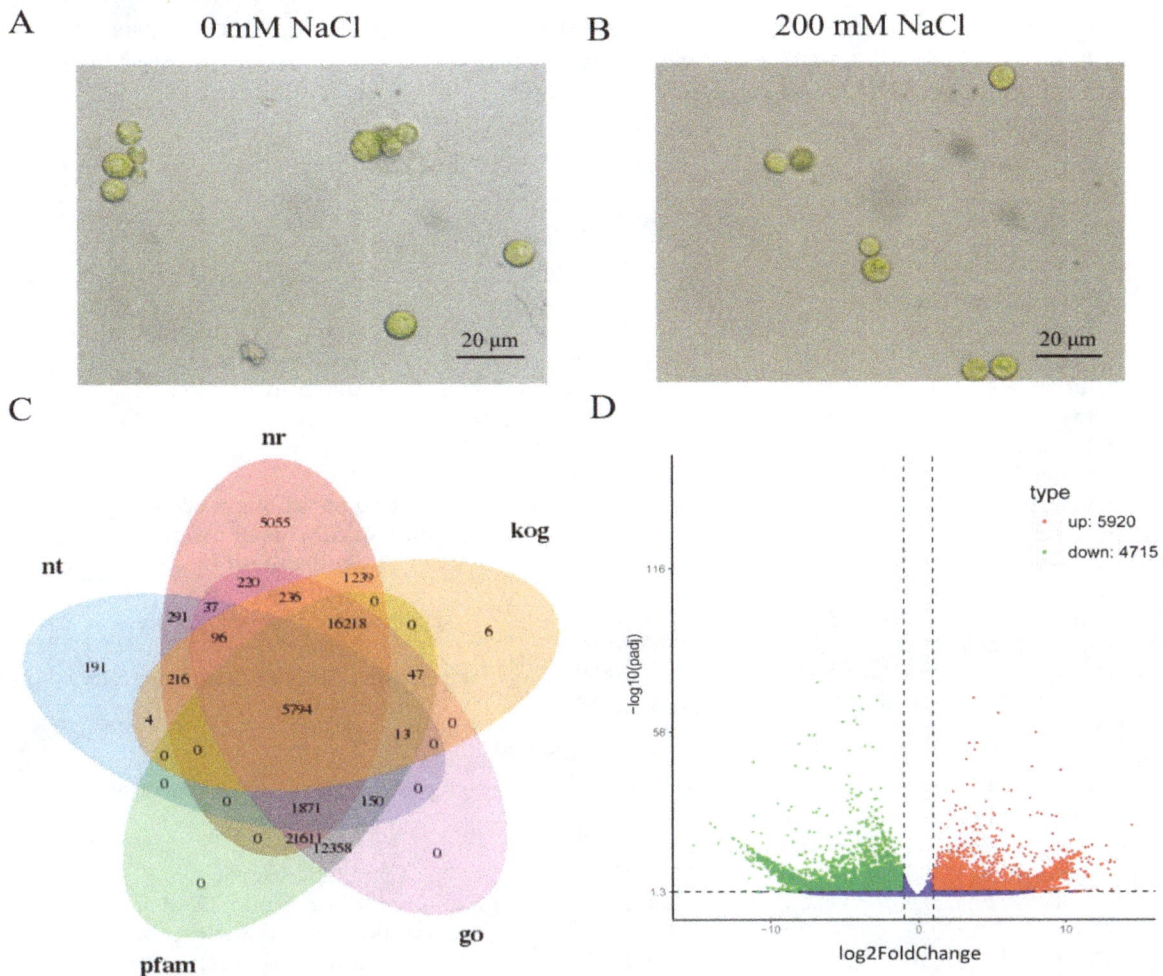

Figure 1. (**A**) The morphology of *C. reinhardtii* cells without addition of NaCl. (**B**) The morphology of *C. reinhardtii* cells under 200 mM NaCl treatment. (**C**) Venn diagram of functional annotations of unigenes in nt (NCBI non-redundant protein sequences), nr (NCBI non-redundant protein sequences), kog (Clusters of Orthologous Groups of proteins), go (Gene Ontology) and pfam (Protein family) databases. (**D**) Expression patterns of differentially expressed genes (DEGs) identified between 200 mM NaCl treated and control. S_200 indicated cells under 200 mM NaCl stressed condition for 24 h; C_0 indicated cells cultured under control condition. Red and green dots represent DEGs, blue dots indicate genes that were not differentially expressed. In total, 10,635 unigenes were identified as DEGs (padj < 0.05) between S_200 and C_0, including 5920 upregulated genes and 4715 downregulated genes.

2.3. Differently Expressed Genes (DEGs) Calculation

To evaluate the relative level of gene expression in *C. reinhardtii* under control or salt stress treatment, the FPKM values were calculated based on the uniquely mapped reads. The FPKM distributions of unigenes in six samples are shown in Supplementary Figure S2. The FPKM value for genes detected in six samples ranged from 0 to 40,486.05, with mean value of 7.08. By comparative analysis, a part of the genes was observed to be differently expressed in 200 Mm NaCl treated samples: 5920 unigenes were calculated as up-regulated in salt treated samples and 4715 filtered as down-regulated genes with the cutoff of padj < 0.05 and | log2(foldchange) | > 1 (Supplementary Table S2).

The most significantly dysregulated 30 genes are recorded in Table 2. The most significantly upregulated unigenes included RNA recognition motif containing gene Cluster-2749.47186 (log2FoldChange [L_2fc] = 3.894), "transcription, DNA-templated" participating gene Cluster-2749.64181 (L_2fc = 5.573) and "potassium ion transport" gene Cluster-2749.61362 (L_2fc = 8.112) (Table 2 and Supplementary Table S2). Downregulated unigenes, included "chlorophyll metabolic process" related gene Cluster-2749.44503 (L_2fc = −8.623) with the lowest p-value, "proteolysis" related gene Cluster-2749.61923 (L_2fc = −6.748) and "regulation of transcription, DNA-templated" participating gene Cluster-2749.45379 (L_2fc = −3.663).

Table 2. Top30 dysregulated genes in *C. reinhardtii* under 200 mM NaCl treated and control conditions.

Gene_ID	L_2fc	$p$val	BP Description
		Up-regulated	
Cluster-2749.47186	3.894	3.77×10^{-75}	
Cluster-2749.64181	5.573	1.55×10^{-69}	transcription, DNA-templated
Cluster-2749.61362	8.112	1.95×10^{-62}	potassium ion transport
Cluster-2749.33332	4.129	1.19×10^{-58}	signal transduction
Cluster-2749.48242	3.610	1.64×10^{-58}	
Cluster-2749.21356	3.975	4.34×10^{-56}	
Cluster-2749.37168	3.413	1.00×10^{-52}	
Cluster-2749.23874	7.849	5.01×10^{-50}	lipid metabolic process
Cluster-2749.57700	9.756	9.76×10^{-49}	iron-sulfur cluster assembly
Cluster-2749.59287	3.459	1.42×10^{-43}	cell adhesion
Cluster-2749.53252	3.877	2.36×10^{-43}	pathogenesis
Cluster-2749.49912	5.957	1.07×10^{-41}	lipoprotein metabolic process
Cluster-2749.84953	6.468	2.29×10^{-41}	
Cluster-2749.82821	2.504	5.20×10^{-41}	regulation of protein kinase activity
Cluster-2749.3203	7.706	1.83×10^{-38}	
		Down-regulated	
Cluster-2749.44503	−8.623	4.01×10^{-178}	chlorophyll metabolic process
Cluster-2749.61923	−6.748	6.07×10^{-81}	proteolysis
Cluster-2749.38883	−3.906	7.54×10^{-76}	
Cluster-2749.44595	−2.699	3.50×10^{-74}	metabolic process
Cluster-2749.45379	−3.663	6.53×10^{-71}	regulation of transcription, DNA-templated
Cluster-2749.49076	−4.268	2.29×10^{-70}	chlorophyll biosynthetic process
Cluster-2749.44117	−4.239	1.30×10^{-66}	oxidation-reduction process
Cluster-2749.42573	−5.023	3.04×10^{-66}	protein glycosylation
Cluster-2749.32226	−4.043	2.67×10^{-65}	proteolysis
Cluster-2749.45636	−7.283	1.98×10^{-61}	
Cluster-2749.44732	−6.934	2.08×10^{-61}	
Cluster-2749.49721	−7.951	3.18×10^{-58}	
Cluster-2749.65261	−3.524	1.91×10^{-57}	
Cluster-2749.36258	−2.996	5.32×10^{-57}	
Cluster-2749.43872	−4.589	1.10×10^{-55}	cell adhesion

Note: Top30 dysregulated genes with the lowest p-value ($p$val) are represented; L_2fc indicates the log2FoldChange of genes differently expressed in 200 mM NaCl treated samples and control samples; BP Description means descriptions of genes' potential participating biological process predicted by sequence similarity search.

2.4. GO Enrichment of DEGs

For uncovering the differences of molecular mechanisms of *C. reinhardtii* under salt stress, the DEGs were then characterized with GO databases. A total of 353 biological processes (BP) terms were enriched by the 5920 up-regulated unigenes, like "oxidation-reduction process" (GO:0055114), "response to cadmium ion" (GO:0046686) and "response to salt stress" (GO:0009651) (Table 3; Supplementary Table S3). The 4715 down-regulated genes were calculated enriched in 313 BP terms, as "photosynthesis, light harvesting in photosystem I" (GO:0009768), "chlorophyll biosynthetic process" (GO:0015995) and "isoleucine biosynthetic process" (GO:0009097) (Table 3; Supplementary Table S3).

Table 3. Top30 biological processes enriched by the up- and down-regulated genes.

GO ID	GO Term	Annotated Gene Number	Enriched Gene Number	p-Value
	Up-Regulated			
GO:0008150	biological process	33682	2820	1.00×10^{-30}
GO:0055114	oxidation-reduction process	3653	385	2.90×10^{-27}
GO:0046686	response to cadmium ion	1317	159	3.40×10^{-18}
GO:0042542	response to hydrogen peroxide	189	41	1.10×10^{-15}
GO:0009408	response to heat	717	122	1.40×10^{-15}
GO:0051259	protein oligomerization	109	25	7.50×10^{-12}
GO:0010090	trichome morphogenesis	131	26	4.60×10^{-10}
GO:0009414	response to water deprivation	668	79	6.70×10^{-10}
GO:0009651	response to salt stress	1488	143	3.90×10^{-09}
GO:0043335	protein unfolding	39	14	1.80×10^{-08}
GO:0016036	cellular response to phosphate starvation	262	40	2.40×10^{-08}
GO:0010030	positive regulation of seed germination	85	20	6.50×10^{-08}
GO:0030866	cortical actin cytoskeleton organization	31	12	7.20×10^{-08}
GO:0016477	cell migration	31	12	7.20×10^{-08}
GO:0045010	actin nucleation	31	12	7.20×10^{-08}
	Down-Regulated			
GO:0008150	biological process	33682	2018	1.00×10^{-30}
GO:0009768	photosynthesis, light harvesting in photosystem I	87	46	1.00×10^{-30}
GO:0009645	response to low light intensity stimulus	72	37	1.00×10^{-30}
GO:0015995	chlorophyll biosynthetic process	242	54	4.40×10^{-29}
GO:0009644	response to high light intensity	393	71	6.70×10^{-22}
GO:0006412	translation	1779	179	3.80×10^{-16}
GO:0009409	response to cold	978	103	5.30×10^{-16}
GO:0009269	response to desiccation	41	18	1.00×10^{-14}
GO:0009769	photosynthesis, light harvesting in photosystem II	36	17	1.30×10^{-14}
GO:0010218	response to far red light	101	25	8.70×10^{-14}
GO:0006364	rRNA processing	742	89	2.10×10^{-12}
GO:0010114	response to red light	159	28	5.90×10^{-11}
GO:0015979	photosynthesis	853	137	2.40×10^{-10}
GO:0009097	isoleucine biosynthetic process	53	16	2.60×10^{-10}
GO:0009099	valine biosynthetic process	43	14	1.10×10^{-09}

2.5. MapMan Enrichment of DEGs

A more specific comparison of metabolic and regulatory pathways was conducted using MapMan. A total of 5920 up- and 4715 down-regulated genes were assigned to 1334 and 1050 homologs in *Arabidopsis thaliana*, respectively. Consequently, these uniquely expressed genes were mapped to 797 pathways by MapMan, of which, 22 pathways were filtered enriched by the dysregulated genes with the cutoff p-value < 0.05 (Figure 3A; Supplementary Table S4). The expression of genes implicated in "TCA/org. transformation.TCA", "Tetrapyrrole synthesis", "Starch" and "Sucrose" were over-expressed in *C. reinhardtii*, while those genes involved in "PS.lightreaction", "PS.lightreaction.photosystem I" and "PS.lightreaction.photosystem I.LHC-I" were down-regulated in *C. reinhardtii* during salt stress responding (Figure 3A).

2.6. KEGG Enrichment of DEGs

To gain a deeper insight into the regulation of photosynthesis underlying salt stress response, down-regulated unigenes involved in "photosynthesis" KEGG pathways (ko00195) were mapped and shown in Figure 2B. Orthologs of 44 genes annotated in this pathway were filtered as down-regulated in the NaCl treated samples in the green alga, such as, photosystem II oxygen-evolving enhancer protein PSBO Cluster-2749.35825 ($L_2fc = -2.5458$) and Cluster-2749.43661 ($L_2fc = -2.1558$), cytochrome b6-f complex iron-sulfur subunit PETC, Cluster-2749.42943 ($L_2fc = -2.7088$), and F-type H+-transporting ATPase subunit ATPF0A, Cluster-2966.0 ($L_2fc = -3.1245$) (Figure 2B; Supplementary Table S2).

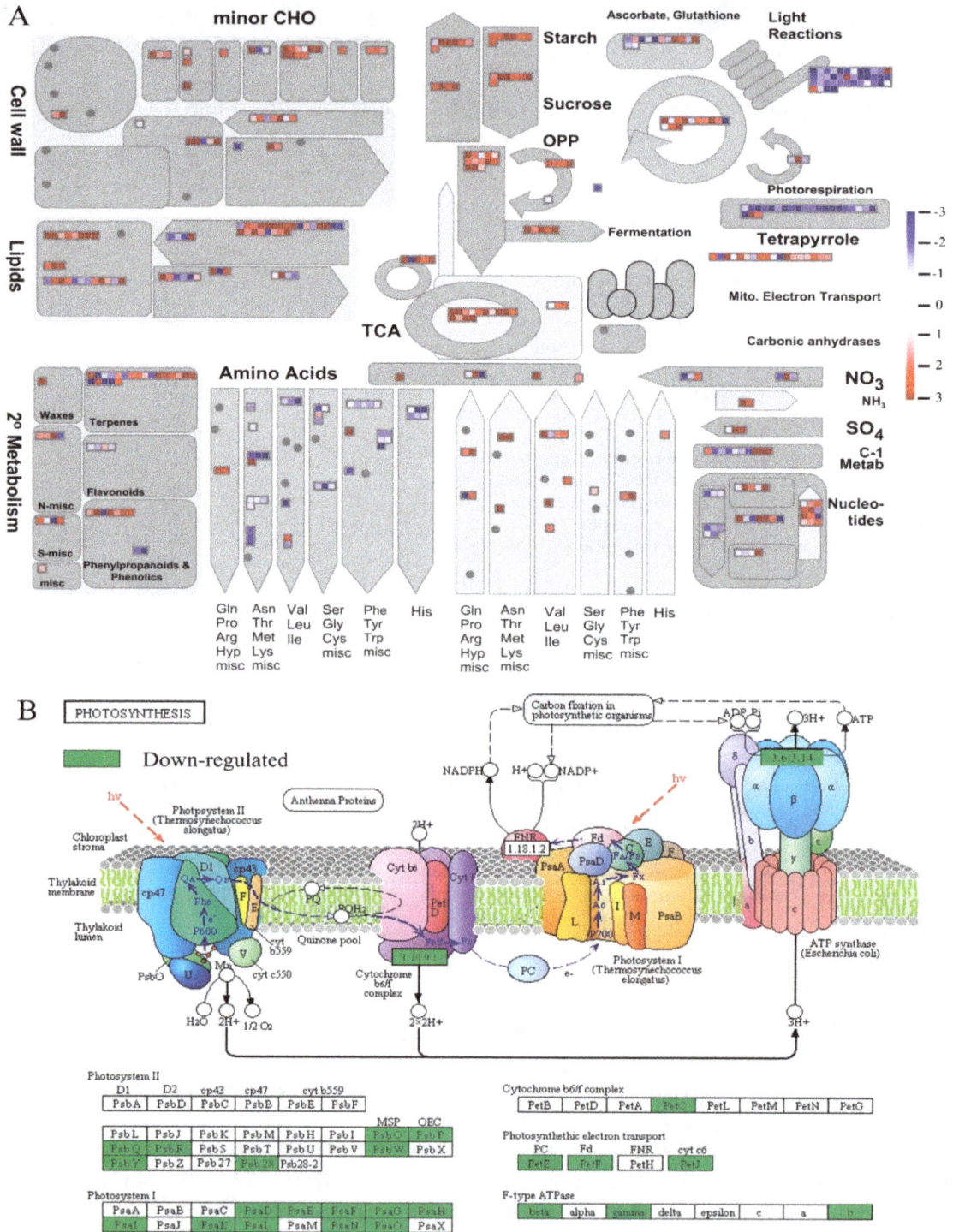

Figure 2. (**A**) Global view of differently expressed genes (DEGs) involved in diverse metabolic pathways. DEGs genes were selected for the metabolic pathways analysis using the MapMan software (3.5.1 R2). The colored boxes indicate the Log$_2$ of expression ratio of DEGs genes. The dys-regulated unigenes were assigned to 1334 and 1050 homologs in Arabidopsis, respectively. These genes were mapped to 797 pathways by MapMan, of which, 22 pathways were filtered enriched by the dys-regulated genes with the cutoff p-value < 0.05. (**B**) The KEGG pathways (ko00195) "photosynthesis" mapped with 44 down-regulated unigenes. The down-regulated genes are marked by a green frame. The black solid line with a black arrow means molecular interaction or relation; the black dash line with a black arrow means indirect link or unknown reaction; the red dash line with a red arrow stands for the light quanta.

2.7. The Differentially Expressed TFs and PKs

Among the expressed unigenes, 2050 and 1624 sequences were assigned to 45 TF families and 78 PK families, respectively (Supplementary Table S5). Of the TF families, MYB family had the largest number of upregulated genes (16 unigenes), including MYB109 ortholog unigenes Cluster-2749.35807 (L_2fc = 2.5798) and Cluster-2749.70085 (L_2fc = 1.3722). In contrast, SET family had the largest number of downregulated genes (16 unigenes). Of the PKs families, TKL-Cr-3 family was uncovered to contain the largest number of upregulated genes. By comparison, CAMK_CDPK and Group-Cr-2 family contained the largest number of downregulated unigenes (Supplementary Table S5).

2.8. Real-Time Quantitative PCR Validation

To verify the RNA-seq results, an alternative strategy was selected for the upregulated unigenes. In total, five over-expressed unigenes were randomly selected for validation by qRT-PCR using the same RNA samples that were used for RNA-seq. Primers were designed to span exon-exon junctions (see Supplementary Table S6 and Figure S3). In most cases, the gene expression trends were similar between these two methods; the result is shown in Figure 3. The ortholog of cytosolic small heat shock protein encoding genes HSP17.6A, Cluster-2749.57700, which was detected by RNA-Seq as up-regulating genes in the salt treated samples (L_2fc = 9.76), was also detected to be significantly over-expressed by qRT-PCR method (Figure 3).

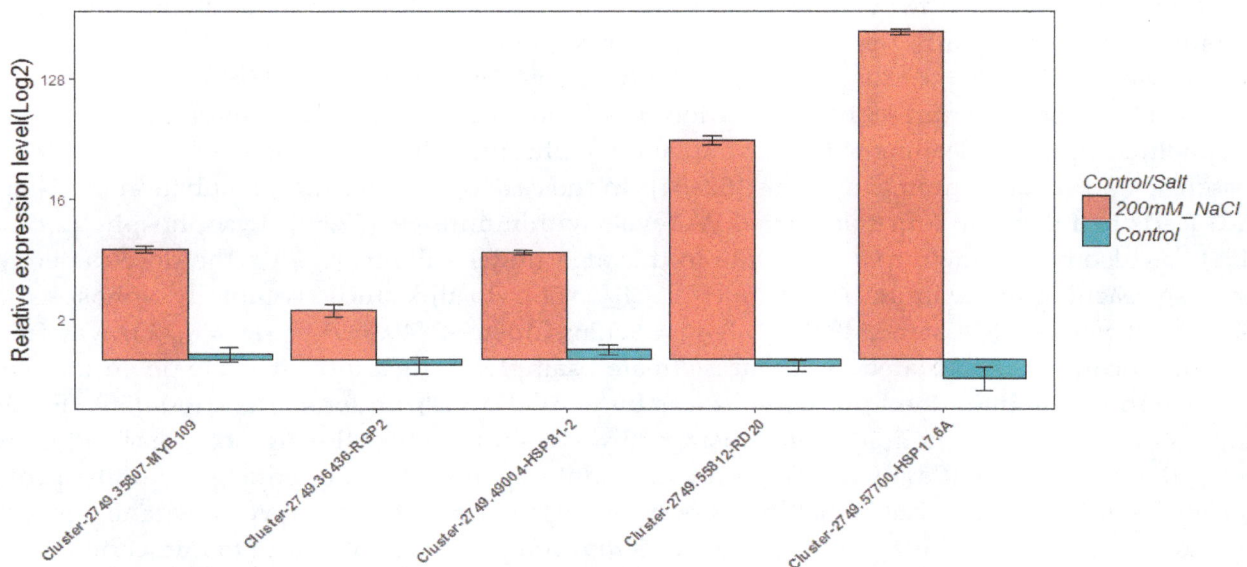

Figure 3. Real-time PCR verification of five up-regulated genes in *C. reinhardtii*. The red bars represent the qPCR results of samples under salt stress condition, while the corresponding blue bars represent the results of control samples. The individual black bars, representing the qPCR data, are the means ± SD of nine measurements (three technical replicates each for three biological samples).

3. Discussion

Salinity is one of the major environmental factors threatening crop productivity and plant growth worldwide [2,9,42]. Due to the complexity of abiotic stress-responding processes, although several hundreds of salt-responding genes have been reported in plants, understanding the genetic underpinnings of salt-responding traits in plantae remains a daunting challenge. The model alga *C. reinhardtii*, which contains one large cup-shaped chloroplast, has the ability to adapt rapidly to changing environmental conditions, such as high salinity, via the generation of novel traits [8,14,43,44]. Given previous results from analysis of salt stress in *C. reinhardtii* and other plants, we analyzed the Illumina RNA-seq data from this alga grown in BG-11 medium with the addition of 200 mM NaCl and analyzed in triplicate after 24 h of incubation [16,22]. In this study, a total of 5920 and

4715 unigenes were identified as up- and down-regulated genes in *C. reinhardtii* under salt stress by RNA-seq. Our study found some molecular cues for reducing the negative effects due to ionic/osmotic toxicity and photosynthesis impairment under saline conditions in *C. reinhardtii*.

Previous studies discovered that the cell density of *C. reinhardtii* cells obviously reduced when stressed by NaCl [8,14,16,21,22]. Neelam et al. demonstrated that at the morphological level, 150 or 200 mM NaCl salt stress led to palmelloid morphology, flagellar resorption, reduction in cell size, and slower growth rate in *C. reinhardtii* [21]. It should be noted that dead and dying cells have dys-regulated mRNA and contribute to transcript levels under saline stress. In our study, programmed cell death (PCD) in the *C. reinhardtii* cell was found with PCD-regulating proteins being significantly up-regulated, e.g., condensin complex subunit (Cluster-2749.11751: L_2fc = 9.368; Cluster-2749.12889: L_2fc = 7.766), sucrose-phosphatase 1 (Cluster-2749.35394: L_2fc = 4.304) and stress tolerance related fibrillin family member (Cluster-2749.70284: L_2fc = 1.920).

Saline stress leads to the overproduction of reactive oxygen species (ROS) in plants which are highly reactive and toxic and cause damage to lipids, carbohydrates, proteins and DNA which ultimately results in oxidative stress [8,9,14,45]. The accumulation of ROS also influences the expression of a number of genes and therefore controls many processes, such as growth, cell cycle, PCD, secondary stress responses and systemic signaling [8,9,14,45]. The excess Na^+ and oxidative stress in the intracellular or extracellular environment activates the acytoplasmic Ca^{2+} signal pathway for regulating an osmotic adjustment or homeostasis regulating of salt stress responses [24,29,39,46–51]. In our study, calcium-related pathway in the *C. reinhardtii* cell was found with several calcium ion binding proteins being significantly upregulated, e.g., peroxygenase 3 (Cluster-2749.55812: L_2fc = 10.431; Cluster-2749.59997: L_2fc = 7.680) and calreticulin (Cluster-2749.35394: L_2fc = 3.082).

The short-term (within 48 h) acclimation to salt stress in *C. reinhardtii* involves activation of phospholipid signaling, leading to the accumulation of phosphatidic acid (PA), which is a lipid second messenger in plant and animal systems [52–54]. In the case of *C. reinhardtii*, incubation in 150 mM NaCl leads to a three- to four-fold rise of PA levels within minutes [52,55]. Lysophosphatidic acid (LPA) has also been shown to accumulate in this alga under salt stress, with the dose-dependent response reaching a maximum at 300 mM NaCl [55,56]. In this study, soluble lysophosphatidic acid acyltransferase (Cluster-2749.8269: L_2fc = 9.126; Cluster-2749.9895: L_2fc = 8.720) was found to be significantly up-regulated in salt stress treated samples, which indicated the potential role of this gene in maintaining the lipid homeostasis by regulating PA under saline stress [55]. Further, analysis of glycerophospholipid metabolism pathways showed that the alga cells had significant up-regulation of FAD (flavin adenine dinucleotide)-dependent oxidoreductase family protein (Cluster-2749.52046; L_2fc = 2.695) that involves storing lipid catabolism and glycerol assimilation, and in glycerol-3-phosphate shuttle, which transports reduced power from cytosol to mitochondrion [8]. This suggests that the intracellular glycerol pool in *C. reinhardtii* cells likely increased as a response to salt stress, similar to what has been shown for the green alga *Dunaliella tertiolecta* [57,58].

Requirement of energy to maintain ion homeostasis is the major metabolic impact of salt stress. The reduction of oxidative stress and osmotic stress, and the up-regulation of heatshock proteins were speculated to aid protein renaturation and recover homeostasis [59–61]. In this study, the stress response is apparent in the *C. reinhardtii* cells with significant up-regulation of genes involved in oxidative/osmotic stress reduction process including glyceraldehyde-3-phosphate dehydrogenase C subunit 1 (Cluster-2749.27769: L_2fc = 1.930) and fumarase 1 (Cluster-2749.35832: L_2fc = 8.306). In bacterium *Escherichia coli*, trehalose is synthesized as a compatible solute and enables cells to exclude toxic cations and to acclimate to high concentrations of salt in the growth medium [62]. For maize, trehalose has helped to reduce the negative effects of saline stress as an osmoprotectant [63]. In our study, enzymes involved in trehalose synthesis significantly up-regulated, e.g., trehalose-6-phosphatase synthase S8 (Cluster-2749.17684: L_2fc = 6.453) and trehalose-6-phosphate synthase (Cluster-2749.61951: L_2fc = 1.123). These results indicated the potential underpinnings for these to maintain homeostasis in *C. reinhardtii* under saline conditions.

In plants, saline stress generally causes ion injury and osmotic stress, which interferes with numerous biochemical and physiological processes, including energy metabolism pathways such as photosynthesis [26,36,64,65] and photorespiration [8]. Previous pigment analyses have demonstrated that photosystem I-light harvesting complexes (LHCs) are damaged by ROS at high salt conditions, and PSII proteins involved in oxygen evolution are impaired [21,45]. In our study, impairment of photosynthesis in the *C. reinhardtii* cell population was found, with several photosystem I-light harvesting complex (LHC) proteins being significantly down-regulated (Figure 2B), e.g., photosystem I light harvesting complex gene LHCA2 (Cluster-2749.32743: $L_2fc = -4.74$; Cluster-2749.52511: $L_2fc = -3.28$), LHCA3 (Cluster-2749.43129: $L_2fc = -6.583$) and LHCA5 (Cluster-2749.40312: $L2fc = -11.375$; Cluster-2749.34085: $L_2fc = -6.553$). Further, we found most of the chloroplast encoded transcripts (e.g., PsaA, B, C, J, M) in photosystem I (PSI) were relatively unchanged in level while the nuclear genes (e.g., PsaD, E, G, F, H) down-regulated under saline conditions (Figure 2B). Existing studies have demonstrated the usage of acetate in the medium as alternative source of energy to compensate for the lowered efficiency in photosynthesis [66]. Consistent with this view, we found that acetyl-CoA synthetase (Cluster-2749.60516: $L_2fc = 5.144$; Cluster-2749.25511: $L_2fc = 2.495$), which combines acetate and CoA to form acetyl-CoA, was significantly up-regulated in the alga cells under saline conditions. In this study, a significant down-regulation was found in a key enzyme of the glyoxylate cycle—isocitrate lyase (ICL, [Cluster-2749.51492; $L_2fc = -3.119$]) [8,45,66,67]—which catalyzes the cleavage of isocitrate to succinate and glyoxylate. Together with malate synthase, ICL bypasses the two decarboxylation steps of the tricarboxylic acid cycle (TCA cycle) [8]. The spatio-temporal expression patterns of genes suggest that in alga cells acetyl-CoA is introduced into energy generation pathways for salt stress responses.

Glycolysis is considered to play an important role in plant development and adaptation to multiple abiotic stresses, such as cold, salt, and drought. It is the key respiratory pathway for generating ATP and carbohydrates metabolites [50,68–73]. In our work, salt stress significantly increased the expression of genes participating in the metabolism of main carbohydrates, such as starch, sucrose, soluble sugar and glucose (Figure 2A). For example, 31 genes of "glycolytic process" (GO:0006096) over-expressed during salt stress responding, including plastidic pyruvate kinase PKP-ALPHA (Cluster-2749.14688: $L_2fc = 8.53$) and PKP-BETA1 (Cluster-2749.26182: $L_2fc = 3.68$). This is consistent with Zhong et al. [68], who reported salt stress significantly increased the main carbohydrate contents of cucumber leaves [53]. Carbohydrates are involved not only in osmotic adjustment, but also can be used as protective agents for homeostasis regulating during salt stress tolerance [24,30,39,48,69,70,74–77]. Given that salt injury caused the destruction of photosynthesis, which might inhibit transport of carbohydrate and accumulate excess starch or sucrose, we speculate *C. reinhardtii* enhanced glycolysis metabolism to decrease carbohydrate accumulation in cells, which would promote the respiratory metabolism and mitochondrial electron transport, thus reducing the effects of ionic toxicity and osmotic stress caused by salt damage.

4. Materials and Methods

4.1. Chlamydomonas Material Preparation, Salt Stress Treatment and RNA Extraction

The *C. reinhardtii* strain GY-D55 wild type from LeadingTec (Shanghai, China) were grown in 150 mL of BG11 media, and placed on a shaking table with 120 rpm and maintained at light (16 h)/dark (8 h) at 23 °C, with an illumination of 100 μmol m^{-2}·s^{-1}. The density of cell cultures was determined by using the blood cell counting plate, with each value being the means of 6 repeats. Under this condition, *C. reinhardtii* cells were grown in BG11 for 14 d.

The methods published by Zhao [16] and Vega [22] were referenced for NaCl treatment in this study. A total of 50 mL medium with 800 mM NaCl was added to the 150 mL culture medium on a shaking table for finishing 200 mM NaCl treatment, the added NaCl was rapidly diluted, and then the pH value was adjusted to 7.0. A parallel set of cells that were unexposed to NaCl stress conditions and cultured in medium served as the experimental control. A total of 50 mL medium without NaCl was added into the control group. Each treatment had 3 repeats. For 24 h, 200 mM NaCl treatment significantly affected the cellular physiology of the alga, such as its photosynthetic and intracellular catalase activity; in this study, the culture time for *C. reinhardtii* under salt stress was 24 h.

After 24 h, 100 mL cell culture medium was extracted from the NaCl treated and control culture bottles, respectively. The collected cells were centrifuged at $3000 \times g$ for 5 min, and the collected cells were resuspended in 25 mL RNAlater (Ambion, Shanghai, China) solution for RNA extraction. The cells of each repeat were mixed and total RNAs were extracted separately using the TRIzol Reagent (Invitrogen, Carlsbad, CA, USA) following the manufacturer's procedures. RNA quality was assessed using the RNA Nano 6000 Assay Kit of the Agilent Bioanalyzer 2100 system (Agilent Technologies, Santa Clara, CA, USA) and the NanoDrop 2000 spectrophotometer (Thermo Scientific, Wilmington, NC, USA).

4.2. Illumina Library Construction and Sequencing

A total amount of 1.5 µg RNA per sample was used as input material for the RNA sample preparations. Sequencing libraries were generated using the NEBNext® Ultra™ RNA Library Prep Kit for Illumina® (NEB, San Diego, CA, USA) by following manufacturer's procedures, and index codes were added to attribute sequences to each sample. Briefly, mRNA was purified from total RNA using poly-T oligo-attached magnetic beads. The random hexamer primer and M-MuLV Reverse Transcriptase (RNase H⁻) were used to synthesize the first strand cDNA and the DNA Polymerase I and RNase H were used for second strand cDNA synthesis. Fragments of 150~200 bp cDNA were purified with the AMPure XP system (Beckman Coulter, Beverly, MA, USA). Then, 3 µL USER Enzyme (NEB, USA) was used with size-selected, adaptor-ligated cDNA at 37 °C for 15 min followed by 5 min at 95 °C. Then, PCR was performed with Phusion High-Fidelity DNA polymerase, Universal PCR primers and Index (X) Primer. Ten cycles were used for PCR enrichment. Finally, PCR products were purified (AMPure XP system) and library quality was assessed on the Agilent Bioanalyzer 2100 system. The clustering of the index-coded samples was performed on a cBot Cluster Generation System using TruSeq PE Cluster Kit v3-cBot-HS (Illumia, San Diego, CA, USA) according to the manufacturer's instructions. After cluster generation, the library preparations were sequenced on an Illumina HiSeq X platform (Illumina, San Diego, CA, USA), according to the manufacturer's procedures. All genetic data have been submitted to the NCBI Sequence Read Archive (SRA) database (https://www.ncbi.nlm.nih.gov/sra), SRA accession: PRJNA490089.

4.3. De Novo Transcriptome Assembling and Unigene Annotation

RNA sequencing and de novo transcriptome assembling were conducted to create reference sequence libraries for *C. reinhardtii*. The RNA sample of each repeat was sequenced separately. cDNA library construction and Illumina pair-end 150 pb sequencing (PE150) were performed at Novogene Co., Ltd. (Shanghai, China), according to instructions provided by Illumina Inc. Reads containing adapter, ploy-N and low-quality reads were removed from raw data for obtaining clean reads. The filtered high-quality reads were used for transcriptome assembling by the Trinity software with default parameters [78]. Clean datasets of 6 samples were pooled for de novo assembling and comprehensive sequence library construction. The Basic Local Alignment Search Tool (BLAST) searches of de novo assembled sequences against public databases (NR, NT, Swiss-Prot, Pfam, KOG/COG, Swiss-Prot, KEGG Ortholog database and Gene Ontology) with an E-value threshold of 10^{-10} were used for unigenes' annotation.

4.4. Calculation and Comparison of Unigene Expression

The independent transcripts libraries of 3 repeats under NaCl treatment conditions and 3 under control conditions were generated for *C. reinhardtii* by a PE150 sequencing analysis. The clean reads were aligned to the de novo assembled transcriptome and estimated by the RSEM [79] method. Gene expression levels were calculated by the fragment per kilobase of exon model per million mapped reads (FPKM) method. DESeq2 [80] was used to compare the expression levels between NaCl treated and control samples with an cutoff of adjusted p-value (padj) < 0.05 and $|\log2(\text{foldchange})| > 1$.

4.5. Gene Ontology (GO), Transcription Factors (TFs) and Protein Kinases (PKs) Prediction

The unigenes were transferred to the *C. reinhardtii* and *A. thaliana* gene IDs by using sequence similarity searching analysis against the genome of *C. reinhardtii* (ftp://ftp.psb.ugent.be/pub/plaza/plaza_public_dicots_04/Fasta/cds.all_transcripts.cre.fasta.gz) and *A. thaliana* (ftp://ftp.psb.ugent.be/pub/plaza/plaza_public_dicots_04/Fasta/cds.all_transcripts.ath.fasta.gz) with an E-value cutoff of 10^{-5}. The classifications of TFs and PKs of *C. reinhardtii* were downloaded from the iTAK database (http://bioinfo.bti.cornell.edu/cgi-bin/itak/index.cgi) [81]. The GO functional annotations file of *A. thaliana* was downloaded from Gene Ontology database (submitted 5 June 2018, http://geneontology.org/gene-associations/gene_association.tair.gz). The TFs and PKs of *C. reinhardtii* genes were transferred to their hit unigenes and the GO functional annotations of *A. thaliana* genes were assigned to their ortholog unigenes in *C. reinhardtii*.

4.6. GO, KEGG and MapMan Annotation and Enrichment

The GO enrichment analysis for DEGs of *C. reinhardtii* was performed by the topGO package of R. KEGG [82] is a database resource for understanding high-level functions and utilities of the biological system, such as the cell, the organism and the ecosystem, from molecular-level information, especially large-scale molecular datasets generated by genome sequencing and other high-throughput experimental technologies (http://www.genome.jp/kegg/). We used KOBAS [83] software to test the statistical enrichment of differential expression genes in KEGG pathways. MapMan (version 3.5.1 R2) [84] was also used to annotate the DEGs onto metabolic pathways. The DEGs of *C. reinhardtii* unigene IDs were transferred to the Arabidopsis Information Resource (TAIR) locus IDs during the MapMan analysis.

4.7. Real-Time Quantitative PCR (qRT-PCR) Verification

Real-time quantitative PCR (qRT-PCR) was performed to verify the expression patterns revealed by the RNA-seq analysis. The purified RNA of samples under salt stress and control conditions were treated with DNaseI and converted to cDNA using the PrimeScript RT Reagent Kit with gDNA Eraser (Takara, Dalian, China) according to the manufacturer's procedures. Five up-regulated unigenes in *C. reinhardtii* were selected for the qRT–PCR assay, including Cluster-2749.49004 (ortholog of HSP81-2), Cluster-2749.57700 (ortholog of HSP17.6A), Cluster-2749.55812 (ortholog of RD20), Cluster-2749.36436 (ortholog of RGP), and Cluster-2749.35807 (ortholog of MYB109). Gene-specific qRT–PCR primers (18–20 bp) (Table S6) were designed using Premier 5.0 software. qPCR was performed using SYBR Green qPCR Master Mix (DBI, Ludwigshafen, Germany) in ABI7500 Real-Time PCR System (ABI, Waltham, MA, USA). Three replicates were performed, and the amplicons were used for melting curve analysis to evaluate the amplification specificity. Relative gene expression was quantified using the $2^{-(\Delta\Delta Ct)}$ method [85]. Ortholog of the *A. thaliana* housekeeping GTP binding Elongation factor Tu family member AT5G60390 in *C. reinhardtii* (Cluster-2749.43263) was used to normalize the amount of template cDNA added in each reaction.

5. Conclusions

We performed a transcriptome analysis of short-term acclimation to salt stress (200 mM NaCl for 24 h) in *C. reinhardtii*. In total, 10,635 unigenes were identified as differentially expressed in *C. reinhardtii* under salt stress by RNA-seq, including 5920 that were up- and 4715 that were down-regulated. A series of molecular cues were screened by GO terms, MapMan and KEGG functional enrichment analyses, which were identified as potential mechanisms for salt stress responses. These mainly include maintaining the lipid homeostasis by regulating phosphatidic acid, acetate being used as an alternative source of energy for solving impairment of photosynthesis and enhancement of glycolysis metabolism to decrease the carbohydrate accumulation in cells. Our results may help understand the molecular and genetic underpinnings of salt responding traits in green alga *C. reinhardtii*.

Supplementary Materials
Figure S1: The assembled transcriptome information of *C. reinhardtii*, Figure S2: The FPKM density distribution of *C. reinhardtii*, Table S1: Unigenes of *C. reinhardtii* annotated in *A. thaliana* genome by BLASTX analysis, Table S2: Information of the 5920 up- and 4715 down-regulated unigenes in *C. reinhardtii*, Table S3: The Gene Ontology (GO) enrichment results of the dys-regulated genes in *C. reinhardtii*, Table S4: The MapMan pathways enrichment results of the dys-regulated genes in *C. reinhardtii*, Table S5: Information of the differently expressed transcription factors (TFs) and protein kinases (PKs), Table S6: Information of the qRT–PCR primers.

Author Contributions: L.Z. and N.W. conceived and designed the study. L.Z., N.W., Z.Q., M.L., X.Z. performed the data collection and analysis. L.Z. and N.W. wrote the paper. L.Z., S.F. and X.Z. reviewed and edited the manuscript. All authors read and approved the manuscript.

Abbreviations

DEGs	differentially expressed genes
TFs	transcriptional factors
PKs	protein kinases
GO	gene ontology
FPKM	fragment per kilobase of exon model per million mapped reads
qRT-PCR	real-time quantitative PCR
BP	biological processes
PSI	photosystem I
PSII	photosystem II
PCD	programmed cell death
ROS	reactive oxygen species
PA	phosphatidic acid

References

1. Munns, R.; Tester, M. Mechanisms of salinity tolerance. *Annu. Rev. Plant Boil.* **2008**, *59*, 651–681. [CrossRef] [PubMed]

2. Song, J.; Wang, B.S. Using euhalophytes to understand salt tolerance and to develop saline agriculture: *Suaeda salsa* as a promising model. *Ann. Bot.* **2015**, *115*, 541–553. [CrossRef] [PubMed]

3. Epstein, E. Salt-tolerant crops: Origins, development, and prospects of the concept. *Plant Soil* **1985**, *89*, 187–198. [CrossRef]

4. Zhang, L.Y.; Zhang, X.J.; Fan, S.J. Meta-analysis of salt-related gene expression profiles identifies common signatures of salt stress responses in Arabidopsis. *Plant Syst. Evol.* **2017**, *303*, 757–774. [CrossRef]

5. Yuan, F.; Leng, B.Y.; Wang, B.S. Progress in Studying Salt Secretion from the Salt Glands in Recretohalophytes: How Do Plants Secrete Salt? *Front. Plant Sci.* **2016**, *7*, 977. [CrossRef] [PubMed]

6. Khona, D.K.; Shirolikar, S.M.; Gawde, K.K.; Hom, E.; Deodhar, M.A.; D'Souza, J.S. Characterization of salt stress-induced palmelloids in the green alga, *Chlamydomonas Reinhardtii*. *Algal Res.* **2016**, *16*, 434–448. [CrossRef]

7. Shen, X.Y.; Wang, Z.L.; Song, X.F.; Xu, J.J.; Jiang, C.Y.; Zhao, Y.X.; Ma, C.L.; Zhang, H. Transcriptomic profiling revealed an important role of cell wall remodeling and ethylene signaling pathway during salt acclimation in Arabidopsis. *Plant Mol. Boil.* **2014**, *86*, 303–317. [CrossRef] [PubMed]

8. Perrineau, M.M.; Zelzion, E.; Gross, J.; Price, D.C.; Boyd, J.; Bhattacharya, D. Evolution of salt tolerance in a laboratory reared population of *Chlamydomonas Reinhardtii*. *Environ. Microbiol.* **2014**, *16*, 1755–1766. [CrossRef] [PubMed]

9. Zhu, J.K. Plant salt tolerance. *Trends Plant Sci.* **2001**, *6*, 66–71. [CrossRef]

10. Shabala, S.; Munns, R.; Shabala, S. Salinity stress: Physiological constraints and adaptive mechanisms. In *Plant Stress Physiology*; CABI: Wallingford, UK, 2012.

11. Young, M.A.; Rancier, D.G.; Roy, J.L.; Lunn, S.R.; Armstrong, S.A.; Headley, J.V. Seeding conditions of the halophyte *Atriplex patula* for optimal growth on a salt impacted site. *Int. J. Phytoremediat.* **2011**, *13*, 674–680. [CrossRef]

12. Rao, A.R.; Sarada, R.; Ravishankar, G.A. Enhancement of carotenoids in green alga *Botyrocccus braunii* in various autotrophic media under stress conditions. *Int. J. Biomed. Pharm. Sci.* **2010**, *4*, 87–92.

13. Stanier, R.Y.; Kunisawa, R.; Mandel, M.; Cohenbazire, G. Purification and properties of unicellular blue-green algae (order Chroococcales). *Bacteriol. Rev.* **1971**, *35*, 171–205. [PubMed]

14. Liu, F.; Jin, Z.; Wang, Y.; Bi, Y.; Melton, R.J. Plastid Genome of Dictyopteris divaricata (Dictyotales, Phaeophyceae): Understanding the Evolution of Plastid Genomes in Brown Algae. *Mar. Biotechnol.* **2017**, *19*, 1–11. [CrossRef] [PubMed]

15. Sayed, O.H. Chlorophyll Fluorescence as a Tool in Cereal Crop Research. *Photosynthetica* **2003**, *41*, 321–330. [CrossRef]

16. Zuo, Z.; Chen, Z.; Zhu, Y.; Bai, Y.; Wang, Y. Effects of NaCl and Na_2CO_3 stresses on photosynthetic ability of *Chlamydomonas Reinhardtii. Biologia* **2014**, *69*, 1314–1322. [CrossRef]

17. Fedina, I.S.; Georgieva, K.; Grigorova, I. Response of Barley Seedlings to UV-B Radiation as Affected by Proline and NaCl. *J. Plant Physiol.* **2003**, *47*, 549–554. [CrossRef]

18. Khan, N.A. NaCl-Inhibited Chlorophyll Synthesis and Associated Changes in Ethylene Evolution and Antioxidative Enzyme Activities in Wheat. *Boil. Plant.* **2003**, *47*, 437–440. [CrossRef]

19. Demetriou, G.; Neonaki, C.; Navakoudis, E.; Kotzabasis, K. Salt stress impact on the molecular structure and function of the photosynthetic apparatus—The protective role of polyamines. *Biochim. Et Biophys. Acta* **2007**, *1767*, 272–280. [CrossRef] [PubMed]

20. Liu, X.D.; Shen, Y.G. Salt shock induces state II transition of the photosynthetic apparatus in dark-adapted *Dunaliella salina* cells. *Environ. Exp. Bot.* **2006**, *57*, 19–24. [CrossRef]

21. Neelam, S.; Subramanyam, R. Alteration of photochemistry and protein degradation of photosystem II from *Chlamydomonas reinhardtii* under high salt grown cells. *J. Photochem. Photobiol. B Boil.* **2013**, *124*, 63–70. [CrossRef] [PubMed]

22. Vega, J.M.; Garbayo, I.; Domínguez, M.J.; Vigara, J. Effect of abiotic stress on photosynthesis and respiration in: Induction of oxidative stress. *Enzym. Microb. Technol.* **2007**, *40*, 163–167. [CrossRef]

23. Atkinson, N.J.; Lilley, C.J.; Urwin, P.E. Identification of genes involved in the response of Arabidopsis to simultaneous biotic and abiotic stresses. *Plant Physiol.* **2013**, *162*, 2028–2041. [CrossRef] [PubMed]

24. Han, N.; Lan, W.J.; He, X.; Shao, Q.; Wang, B.S.; Zhao, X.J. Expression of a *Suaeda salsa* Vacuolar H^+/Ca^{2+} Transporter Gene in Arabidopsis Contributes to Physiological Changes in Salinity. *Plant Mol. Boil. Rep.* **2012**, *30*, 470–477. [CrossRef]

25. Qi, Y.C.; Liu, W.Q.; Qiu, L.Y.; Zhang, S.M.; Ma, L.; Zhang, H. Overexpression of glutathione S-transferase gene increases salt tolerance of arabidopsis. *Russ. J. Plant Physiol.* **2010**, *57*, 233–240. [CrossRef]

26. Zhang, S.R.; Song, J.; Wang, H.; Feng, G. Effect of salinity on seed germination, ion content and photosynthesis of cotyledons in halophytes or xerophyte growing in Central Asia. *J. Plant Ecol.* **2010**, *3*, 259–267. [CrossRef]

27. Lu, T.; Lu, G.; Fan, D.; Zhu, C.; Wei, L.; Qiang, Z.; Qi, F.; Yan, Z.; Guo, Y.; Li, W. Function annotation of the rice transcriptome at single-nucleotide resolution by RNA-seq. *Genome Res.* **2010**, *20*, 1238–1249. [CrossRef] [PubMed]

28. Liu, A.; Xiao, Z.; Li, M.W.; Wong, F.L.; Yung, W.S.; Ku, Y.S.; Wang, Q.; Wang, X.; Xie, M.; Yim, A.K. Transcriptomic reprogramming in soybean seedlings under salt stress. *Plant Cell Environ.* **2018**. [CrossRef] [PubMed]

29. Cui, F.; Sui, N.; Duan, G.; Liu, Y.; Han, Y.; Liu, S.; Wan, S.; Li, G. Identification of Metabolites and Transcripts Involved in Salt Stress and Recovery in Peanut. *Front. Plant Sci.* **2018**, *9*, 217. [CrossRef] [PubMed]

30. Guo, J.; Li, Y.; Han, G.; Song, J.; Wang, B. NaCl markedly improved the reproductive capacity of the euhalophyte *Suaeda Salsa. Funct. Plant Boil.* **2018**, *45*, 350. [CrossRef]

31. Cao, S.; Du, X.H.; Li, L.H.; Liu, Y.D.; Zhang, L.; Pan, X.; Li, Y.; Li, H.; Lu, H. Overexpression of *Populus tomentosa* cytosolic ascorbate peroxidase enhances abiotic stress tolerance in tobacco plants. *Russ. J. Plant Physiol.* **2017**, *64*, 224–234. [CrossRef]

32. Sui, N.; Tian, S.S.; Wang, W.Q.; Wang, M.J.; Fan, H. Overexpression of Glycerol-3-Phosphate Acyltransferase from *Suaeda salsa* Improves Salt Tolerance in Arabidopsis. *Front. Plant Sci.* **2017**, *8*, 1337. [CrossRef] [PubMed]

33. Wang, J.S.; Zhang, Q.; Cui, F.; Hou, L.; Zhao, S.Z.; Xia, H.; Qiu, J.J.; Li, T.T.; Zhang, Y.; Wang, X.J.; et al. Genome-Wide Analysis of Gene Expression Provides New Insights into Cold Responses in *Thellungiella Salsuginea. Front. Plant Sci.* **2017**, *8*, 713. [CrossRef] [PubMed]

34. Yuan, F.; Lyu, M.J.A.; Leng, B.Y.; Zhu, X.G.; Wang, B.S. The transcriptome of NaCl-treated *Limonium bicolor* leaves reveals the genes controlling salt secretion of salt gland. *Plant Mol. Boil.* **2016**, *91*, 241–256. [CrossRef] [PubMed]

35. Yuan, F.; Lyu, M.J.A.; Leng, B.Y.; Zheng, G.Y.; Feng, Z.T.; Li, P.H.; Zhu, X.G.; Wang, B.S. Comparative transcriptome analysis of developmental stages of the *Limonium bicolor* leaf generates insights into salt gland differentiation. *Plant Cell Environ.* **2015**, *38*, 1637–1657. [CrossRef] [PubMed]

36. Feng, Z.T.; Deng, Y.Q.; Fan, H.; Sun, Q.J.; Sui, N.; Wang, B.S. Effects of NaCl stress on the growth and photosynthetic characteristics of *Ulmus pumila* L. seedlings in sand culture. *Photosynthetica* **2014**, *52*, 313–320. [CrossRef]

37. Yuan, F.; Chen, M.; Yang, J.C.; Leng, B.Y.; Wang, B.S. A system for the transformation and regeneration of the recretohalophyte *Limonium bicolor*. *In Vitro Cell. Dev. Boil. Plant* **2014**, *50*, 610–617. [CrossRef]

38. Zhang, Q.; Zhao, C.Z.; Li, M.; Sun, W.; Liu, Y.; Xia, H.; Sun, M.N.; Li, A.Q.; Li, C.S.; Zhao, S.Z.; et al. Genome-wide identification of *Thellungiella salsuginea* microRNAs with putative roles in the salt stress response. *BMC Plant Boil.* **2013**, *13*, 180. [CrossRef] [PubMed]

39. Guo, Y.H.; Jia, W.J.; Song, J.; Wang, D.A.; Chen, M.; Wang, B.S. Thellungilla halophila is more adaptive to salinity than *Arabidopsis thaliana* at stages of seed germination and seedling establishment. *Acta Physiol. Plant.* **2012**, *34*, 1287–1294. [CrossRef]

40. Liu, J.; Zhang, F.; Zhou, J.J.; Chen, F.; Wang, B.S.; Xie, X.Z. Phytochrome B control of total leaf area and stomatal density affects drought tolerance in rice. *Plant Mol. Boil.* **2012**, *78*, 289–300. [CrossRef] [PubMed]

41. Zhou, J.C.; Fu, T.T.; Sui, N.; Guo, J.R.; Feng, G.; Fan, J.L.; Song, J. The role of salinity in seed maturation of the euhalophyte *Suaeda Salsa*. *Plant Biosyst.* **2016**, *150*, 83–90. [CrossRef]

42. Waśkiewicz, A.; Muzolf-Panek, M.; Goliński, P. *Phenolic Content Changes in Plants under Salt Stress*; Springer: New York, NY, USA, 2013; pp. 283–314.

43. Sudhir, P.; Murthy, S.D.S. Effects of salt stress on basic processes of photosynthesis. *Photosynthetica* **2004**, *42*, 481–486. [CrossRef]

44. Maršálek, B.; Zahradníčková, H.; Hronková, M. Extracellular Production of Abscisic Acid by Soil Algae under Salt, Acid or Drought Stress. *Z. Für Naturforschung C* **1992**, *47*, 701–704. [CrossRef]

45. Pineau, B.; Gérard-Hirne, C.; Selve, C. Carotenoid binding to photosystems I and II of Chlamydomonas reinhardtii, cells grown under weak light or exposed to intense light. *Plant Physiol. Bioch.* **2001**, *39*, 73–85. [CrossRef]

46. Ding, F.; Chen, M.; Sui, N.; Wang, B.S. Ca^{2+} significantly enhanced development and salt-secretion rate of salt glands of *Limonium bicolor* under NaCl treatment. *S. Afr. J. Bot.* **2010**, *76*, 95–101. [CrossRef]

47. Feng, Z.T.; Deng, Y.Q.; Zhang, S.C.; Liang, X.; Yuan, F.; Hao, J.L.; Zhang, J.C.; Sun, S.F.; Wang, B.S. K^+ accumulation in the cytoplasm and nucleus of the salt gland cells of *Limonium bicolor* accompanies increased rates of salt secretion under NaCl treatment using NanoSIMS. *Plant Sci.* **2015**, *238*, 286–296. [CrossRef] [PubMed]

48. Han, N.; Shao, Q.; Bao, H.Y.; Wang, B.S. Cloning and Characterization of a Ca^{2+}/H^+ Antiporter from Halophyte *Suaeda salsa* L. *Plant Mol. Boil. Rep.* **2011**, *29*, 449–457. [CrossRef]

49. Yang, S.; Li, L.; Zhang, J.; Geng, Y.; Guo, F.; Wang, J.; Meng, J.; Sui, N.; Wan, S.; Li, X. Transcriptome and Differential Expression Profiling Analysis of the Mechanism of $Ca^{(2+)}$ Regulation in Peanut (*Arachis hypogaea*) Pod Development. *Front. Plant Sci.* **2017**, *8*, 1609. [CrossRef] [PubMed]

50. Zheng, Y.; Liao, C.C.; Zhao, S.S.; Wang, C.W.; Guo, Y. The Glycosyltransferase QUA1 Regulates Chloroplast-Associated Calcium Signaling During Salt and Drought Stress in Arabidopsis. *Plant Cell Physiol.* **2017**, *58*, 329–341. [CrossRef] [PubMed]

51. Zhang, L.Y.; Zhang, Z.; Zhang, X.J.; Yao, Y.; Wang, R.; Duan, B.Y.; Fan, S.J. Comprehensive meta-analysis and co-expression network analysis identify candidate genes for salt stress response in Arabidopsis. *Plant Biosyst.* **2018**. [CrossRef]

52. Arisz, S.A.; Valianpour, F.; van Gennip, A.H.; Munnik, T. Substrate preference of stress-activated phospholipase D in Chlamydomonas and its contribution to PA formation. *Plant J. Cell Mol. Boil.* **2003**, *34*, 595–604. [CrossRef]

53. Zhou, J.J.; Liu, Q.Q.; Zhang, F.; Wang, Y.Y.; Zhang, S.Y.; Cheng, H.M.; Yan, L.H.; Li, L.; Chen, F.; Xie, X.Z. Overexpression of OsPIL15, a phytochromeinteracting factor- like protein gene, represses etiolated seedling growth in rice. *J. Integr. Plant Boil.* **2014**, *56*, 373–387. [CrossRef] [PubMed]

54. Sui, N.; Li, M.; Li, K.; Song, J.; Wang, B.S. Increase in unsaturated fatty acids in membrane lipids of *Suaeda salsa* L. enhances protection of photosystem II under high salinity. *Photosynthetica* **2010**, *48*, 623–629. [CrossRef]

55. Arisz, S.A.; Munnik, T. The salt stress-induced LPA response in Chlamydomonas is produced via PLA2 hydrolysis of DGK-generated phosphatidic acid. *J. Lipid Res.* **2011**, *52*, 2012–2020. [CrossRef] [PubMed]

56. Meijer, H.J.; Arisz, S.A.; Van Himbergen, J.A.; Musgrave, A.; Munnik, T. Hyperosmotic stress rapidly generates lyso-phosphatidic acid in Chlamydomonas. *Plant J.* **2010**, *25*, 541–548. [CrossRef]

57. Takagi, M.; Karseno; Yoshida, T. Effect of Salt Concentration on Intracellular Accumulation of Lipids and Triacylglyceride in Marine Microalgae Dunaliella Cells. *J. Biosci. Bioeng.* **2006**, *101*, 223–226. [CrossRef] [PubMed]

58. Goyal, A. Osmoregulation in Dunaliella, Part II: Photosynthesis and starch contribute carbon for glycerol synthesis during a salt stress in Dunaliella tertiolecta. *Plant Physiol. Biochem.* **2007**, *45*, 705–710. [CrossRef] [PubMed]

59. Yokthongwattana, C.; Mahong, B.; Roytrakul, S.; Phaonaklop, N.; Narangajavana, J.; Yokthongwattana, K. Proteomic analysis of salinity-stressed *Chlamydomonas reinhardtii* revealed differential suppression and induction of a large number of important housekeeping proteins. *Planta* **2012**, *235*, 649–659. [CrossRef] [PubMed]

60. Sun, Z.B.; Qi, X.Y.; Wang, Z.L.; Li, P.H.; Wu, C.X.; Zhang, H.; Zhao, Y.X. Overexpression of TsGOLS2, a galactinol synthase, in *Arabidopsis thaliana* enhances tolerance to high salinity and osmotic stresses. *Plant Physiol. Biochem.* **2013**, *69*, 82–89. [CrossRef] [PubMed]

61. Pang, C.H.; Li, K.; Wang, B.S. Overexpression of SsCHLAPXs confers protection against oxidative stress induced by high light in transgenic *Arabidopsis thaliana*. *Physiol. Plant.* **2011**, *143*, 355–366. [CrossRef] [PubMed]

62. Ferjani, A.; Mustardy, L.; Sulpice, R.; Marin, K.; Suzuki, I.; Hagemann, M.; Murata, N. Glucosylglycerol, a compatible solute, sustains cell division under salt stress. *Plant Physiol.* **2003**, *131*, 1628–1637. [CrossRef] [PubMed]

63. Zeid, I.M. Trehalose as osmoprotectant for maize under salinity-induced stress. *Res. J. Agric. Boil. Sci.* **2009**, *5*, 613–622.

64. Xu, J.J.; Li, Y.Y.; Ma, X.L.; Ding, J.F.; Wang, K.; Wang, S.S.; Tian, Y.; Zhang, H.; Zhu, X.G. Whole transcriptome analysis using next-generation sequencing of model species *Setaria viridis* to support C_4 photosynthesis research. *Plant Mol. Boil.* **2013**, *83*, 77–87. [CrossRef] [PubMed]

65. Sui, N.; Han, G.L. Salt-induced photoinhibition of PSII is alleviated in halophyte Thellungiella halophila by increases of unsaturated fatty acids in membrane lipids. *Acta Physiol. Plant.* **2014**, *36*, 983–992. [CrossRef]

66. Heifetz, P.B.; Boynton, J.E. Effects of Acetate on Facultative Autotrophy in *Chlamydomonas reinhardtii* Assessed by Photosynthetic Measurements and Stable Isotope Analyses. *Plant Physiol.* **2000**, *122*, 1439–1445. [CrossRef] [PubMed]

67. Soussi, M.; Ocaña, A.; Lluch, C. Effects of salt stress on growth, photosynthesis and nitrogen fixation in chick-pea (*Cicer arietinum* L.). *J. Exp. Bot.* **1998**, *49*, 1329–1337. [CrossRef]

68. Zhong, M.; Yuan, Y.; Shu, S.; Sun, J.; Guo, S.; Yuan, R.; Tang, Y. Effects of exogenous putrescine on glycolysis and Krebs cycle metabolism in cucumber leaves subjected to salt stress. *Plant Growth Regul.* **2016**, *79*, 319–330. [CrossRef]

69. Huang, J.; Li, Z.Y.; Biener, G.; Xiong, E.H.; Malik, S.; Eaton, N.; Zhao, C.Z.; Raicu, V.; Kong, H.Z.; Zhao, D.Z. Carbonic Anhydrases Function in Anther Cell Differentiation Downstream of the Receptor-Like Kinase EMS1. *Plant Cell* **2017**, *29*, 1335–1356. [CrossRef] [PubMed]

70. Wang, F.X.; Xu, Y.G.; Wang, S.; Shi, W.W.; Liu, R.R.; Feng, G.; Song, J. Salinity affects production and salt tolerance of dimorphic seeds of *Suaeda Salsa*. *Plant Physiol. Biochem.* **2015**, *95*, 41–48. [CrossRef] [PubMed]

71. Shao, Q.; Han, N.; Ding, T.L.; Zhou, F.; Wang, B.S. SsHKT1;1 is a potassium transporter of the C_3 halophyte *Suaeda salsa* that is involved in salt tolerance. *Funct. Plant Boil.* **2014**, *41*, 790–802. [CrossRef]

72. Li, K.; Pang, C.H.; Ding, F.; Sui, N.; Feng, Z.T.; Wang, B.S. Overexpression of *Suaeda salsa* stroma ascorbate peroxidase in Arabidopsis chloroplasts enhances salt tolerance of plants. *S. Afr. J. Bot.* **2012**, *78*, 235–245. [CrossRef]

73. Liu, S.S.; Wang, W.Q.; Li, M.; Wan, S.B.; Sui, N. Antioxidants and unsaturated fatty acids are involved in salt tolerance in peanut. *Acta Physiol. Plant.* **2017**, *39*, 207. [CrossRef]

74. Chen, M.; Song, J.; Wang, B.S. NaCl increases the activity of the plasma membrane H^+-ATPase in C_3 halophyte *Suaeda salsa* callus. *Acta Physiol. Plant.* **2010**, *32*, 27–36. [CrossRef]

75. Song, J.; Shi, G.W.; Gao, B.; Fan, H.; Wang, B.S. Waterlogging and salinity effects on two *Suaeda salsa* populations. *Physiol. Plant.* **2011**, *141*, 343–351. [CrossRef] [PubMed]

76. Meng, X.; Yang, D.Y.; Li, X.D.; Zhao, S.Y.; Sui, N.; Meng, Q.W. Physiological changes in fruit ripening caused by overexpression of tomato SlAN2, an R2R3-MYB factor. *Plant Physiol. Biochem.* **2015**, *89*, 24–30. [CrossRef] [PubMed]

77. Li, Y.Y.; Ma, X.L.; Zhao, J.L.; Xu, J.J.; Shi, J.F.; Zhu, X.G.; Zhao, Y.X.; Zhang, H. Developmental Genetic Mechanisms of C_4 Syndrome Based on Transcriptome Analysis of C_3 Cotyledons and C_4 Assimilating Shoots in *Haloxylon Ammodendron*. *PLoS ONE* **2015**, *10*, e0117175. [CrossRef] [PubMed]

78. Grabherr, M.G.; Haas, B.J.; Yassour, M.; Levin, J.Z.; Thompson, D.A.; Amit, I.; Adiconis, X.; Fan, L.; Raychowdhury, R.; Zeng, Q.; et al. Full-length transcriptome assembly from RNA-Seq data without a reference genome. *Nat. Biotechnol.* **2011**, *29*, 644–652. [CrossRef] [PubMed]

79. Li, B.; Dewey, C.N. RSEM: Accurate transcript quantification from RNA-Seq data with or without a reference genome. *BMC Bioinform.* **2011**, *12*, 323. [CrossRef] [PubMed]

80. Haas, B.J.; Papanicolaou, A.; Yassour, M.; Grabherr, M.; Blood, P.D.; Bowden, J.; Couger, M.B.; Eccles, D.; Li, B.; Lieber, M. De novo transcript sequence reconstruction from RNA-Seq: Reference generation and analysis with Trinity. *Nat. Protoc.* **2013**, *8*, 1494–1512. [CrossRef] [PubMed]

81. Zheng, Y.; Jiao, C.; Sun, H.; Rosli, H.G.; Pombo, M.A.; Zhang, P.; Banf, M.; Dai, X.; Martin, G.B.; Giovannoni, J.J.; et al. iTAK: A Program for Genome-wide Prediction and Classification of Plant Transcription Factors, Transcriptional Regulators, and Protein Kinases. *Mol. Plant* **2016**, *9*, 1667–1670. [CrossRef] [PubMed]

82. Okuda, S.; Yamada, T.; Hamajima, M.; Itoh, M.; Katayama, T.; Bork, P.; Goto, S.; Kanehisa, M. KEGG Atlas mapping for global analysis of metabolic pathways. *Nucleic Acids Res.* **2008**, *36*, W423–W426. [CrossRef] [PubMed]

83. Wu, J.; Mao, X.; Cai, T.; Luo, J.; Wei, L. KOBAS server: A web-based platform for automated annotation and pathway identification. *Nucleic Acids Res.* **2006**, *34*, W720–W724. [CrossRef] [PubMed]

84. Thimm, O.; Blasing, O.; Gibon, Y.; Nagel, A.; Meyer, S.; Kruger, P.; Selbig, J.; Muller, L.A.; Rhee, S.Y.; Stitt, M. MAPMAN: A user-driven tool to display genomics data sets onto diagrams of metabolic pathways and other biological processes. *Plant J. Cell Mol. Boil.* **2004**, *37*, 914–939. [CrossRef]

85. Livak, K.J.; Schmittgen, T.D. Analysis of relative gene expression data using real-time quantitative PCR and the 2(-Delta Delta C (T)) Method. *Methods* **2001**, *25*, 402–408. [CrossRef] [PubMed]

9

The Salt-Stress Response of the Transgenic Plum Line J8-1 and its Interaction with the Salicylic Acid Biosynthetic Pathway from Mandelonitrile

Agustina Bernal-Vicente [1], Daniel Cantabella [1,2], Cesar Petri [3], José Antonio Hernández [1,]* and Pedro Diaz-Vivancos [1,4]

[1]   Biotechnology of Fruit Trees Group, Department Plant Breeding, CEBAS-CSIC, Campus Universitario de Espinardo, 25, 30100 Murcia, Spain; tina.cartagena@hotmail.com (A.B.-V.); daniel.cantabella@irta.cat (D.C.); pdv1@um.es (P.D.-V.)

[2]   IRTA, XaRTA-Postharvest, Edifici Fruitcentre, Parc Científic i Tecnològic Agroalimentari de Lleida, 25003 Lleida, Catalonia, Spain

[3]   Departamento de Producción Vegetal, Universidad Politécnica de Cartagena, Paseo Alfonso XIII, 48, 30203 Cartagena, Spain; cesar.petri@upct.es

[4]   Department of Plant Biology, Faculty of Biology, University of Murcia, Campus de Espinardo, E-30100 Murcia, Spain

*   Correspondence: jahernan@cebas.csic.es

Abstract: Salinity is considered as one of the most important abiotic challenges that affect crop productivity. Plant hormones, including salicylic acid (SA), are key factors in the defence signalling output triggered during plant responses against environmental stresses. We have previously reported in peach a new SA biosynthetic pathway from mandelonitrile (MD), the molecule at the hub of the cyanogenic glucoside turnover in *Prunus* sp. In this work, we have studied whether this new SA biosynthetic pathway is also present in plum and the possible role this pathway plays in plant plasticity under salinity, focusing on the transgenic plum line J8-1, which displays stress tolerance via an enhanced antioxidant capacity. The SA biosynthesis from MD in non-transgenic and J8-1 micropropagated plum shoots was studied by metabolomics. Then the response of J8-1 to salt stress in presence of MD or Phe (MD precursor) was assayed by measuring: chlorophyll content and fluorescence parameters, stress related hormones, levels of non-enzymatic antioxidants, the expression of two genes coding redox-related proteins, and the content of soluble nutrients. The results from in vitro assays suggest that the SA synthesis from the MD pathway demonstrated in peach is not clearly present in plum, at least under the tested conditions. Nevertheless, in J8-1 NaCl-stressed seedlings, an increase in SA was recorded as a result of the MD treatment, suggesting that MD could be involved in the SA biosynthesis under NaCl stress conditions in plum plants. We have also shown that the plum line J8-1 was tolerant to NaCl under greenhouse conditions, and this response was quite similar in MD-treated plants. Nevertheless, the MD treatment produced an increase in SA, jasmonic acid (JA) and reduced ascorbate (ASC) contents, as well as in the coefficient of non-photochemical quenching (qN) and the gene expression of *Non-Expressor of Pathogenesis-Related 1* (*NPR1*) and *thioredoxin H* (*TrxH*) under salinity conditions. This response suggested a crosstalk between different signalling pathways (NPR1/Trx and SA/JA) leading to salinity tolerance in the transgenic plum line J8-1.

Keywords: chlorophyll fluorescence; J8-1 plum line; mandelonitrile; *Prunus domestica*; redox signalling; salicylic acid; salt-stress; soluble nutrients

1. Introduction

Salinity or salt stress significantly affects crop productivity, and it is considered as one of the most important abiotic challenges that plant scientists must confront today. Due to the use of saline waters for irrigation, the percentage of land affected by salinity is continuously growing worldwide. When plants are submitted to salt stress conditions, physiological, biochemical, and nutritional disorders occur, limiting plant growth and development and, ultimately, productivity. These deleterious effects are due to the accumulation of toxic ions (Na^+ and Cl^-), leading to especially Ca^{2+} and K^+ deficiency among other nutrient imbalances, and the reduced water uptake produced by osmotic stress [1,2]. In addition, salinity also induced an oxidative stress mediated by reactive oxygen species (ROS) at the subcellular level [2].

Furthermore, it is well known that plant hormones are key factors in the defence signalling output triggered during both abiotic and biotic environmental stress conditions. Among these hormones, SA has attracted much attention, although other plant hormones, such as abscisic acid (ABA) and jasmonic acid (JA), have also been suggested as modulators of plant defence responses.

Considering the important roles of SA during plant responses against stress conditions, SA is of potential agro-economic interest as a modulator of plant plasticity. Although the regulation of SA biosynthesis and the SA-mediated stress tolerance mechanism have not been fully characterised [3], researchers have found that the exogenous application of SA or analogues induce tolerance to several stress conditions [4]. In the same line, we have previously reported in peach (*P. persica* L.) plants that mandelonitrile (MD) is also involved in SA biosynthesis and improves plant performance under biotic and abiotic stress conditions [5]. In *Prunus*, MD is at the hub of cyanogenic glycoside (CNglcs) synthesis and turnover [6]. CNglcs are specialized secondary metabolites that have been linked to plant plasticity improvement against environmental stress conditions. However, CNglcs turnover is highly species dependent [6]. As a result, further studies must be performed to elucidate whether this new SA biosynthetic pathway from MD is also present in other *Prunus* species, and to determine its possible role in plant plasticity under stress conditions. Accordingly, other authors have suggested that SA biosynthesis varies depending on different factors, including the plant species and the environmental conditions [7–9].

One common consequence of exposure to stress conditions is the establishment of oxidative signalling that triggers transduction cascades controlling plant development and defence [10]. The major low-molecular-weight antioxidants ascorbate (ASC) and glutathione (GSH) determine the specificity of this oxidative signalling. Thus, ASC and GSH have been shown to be multifunctional metabolites that are important in redox homeostasis and signalling as well as in developmental and defence reactions [11]. The *NON-EXPRESSOR OF PR-PROTEINS1* (*NPR1*) transcription factor, which is activated by SA, is one of the few known redox-regulated signalling proteins in plants, highlighting the crosstalk between the antioxidant metabolism and plant hormones during environmental stress responses. On the other hand, the roles of thioredoxins (Trx) in redox signalling as regulators of scavenging mechanisms and as components of signalling pathways are well established [12]. It has been suggested that SA signalling activates Trx-h5, leading to NPR1 reduction and releasing active monomers that are translocated from the cytosol into the nucleus; this, in turn, activates the expression of defence genes [13,14].

In the present manuscript, we have analysed whether the SA biosynthetic pathway from MD, previously observed in peach [5], is also present in plum (*P. domestica* cv. *Claudia verde*) plants in the presence or absence of NaCl. Moreover, in order to gain deeper knowledge of the SA-mediated defence network in *Prunus*, we have used transgenic plum plants over-expressing four copies of the cytosolic ascorbate peroxidase gene. These transgenic plants with enhanced antioxidant capacity, named line J8-1, have shown higher regeneration efficiency and enhanced vigour as well as tolerance to salt stress under in vitro conditions [15,16]. Moreover, line J8-1 has displayed enhanced tolerance to water stress under greenhouse conditions [17]. Thus, line J8-1 could be an excellent model to study the crosstalk among stress tolerance, oxidative stress and SA in plum plants. Taking into account all the mentioned above, we analysed the effect of MD and Phe (MD precursor) treatments on plant performance (chlorophyll content, chlorophyll fluorescence parameters and leaf and root water contents), on the content of stress-related hormones, on the redox state and the expression of two redox-related genes, and on the soluble leaf and root nutrient content in the transgenic J8-1 line under control and salt stress conditions.

2. Results

The following experiments were designed in order to elucidate whether MD could be a precursor of SA biosynthesis in plum plants, as occurred in peach [5]. Moreover, to further study the crosstalk among stress tolerance, oxidative stress, and SA under salinity conditions, the effect of MD and Phe (MD precursor) has been investigated in the transgenic plum line J8-1 submitted to NaCl.

2.1. Metabolomic Analysis of SA Biosynthesis in Plum Plants

We have previously described in peach that the cyanogenic glycoside (CNglcs) pathway is involved in SA biosynthesis, suggesting the existence of a third SA biosynthetic pathway, being MD the intermediary molecule between both pathways [5]. Taking into account the fact that the CNglcs pathway is highly dependent on the plant species [6], here we have studied whether this SA biosynthetic pathway, from MD, is also functional in plum plants under control and salinity conditions.

When micropropagated non-transgenic plum (cv. *Claudia verde*) shoots were fed with [^{13}C]Phe or with [^{13}C]MD, in the absence of NaCl, increased levels of Phe, MD and benzoic acid were recorded, whereas amygdalin only increased in Phe-treated shoots (Figure 1). However, contrary to that which we observed in peach [5], none of the treatments produced a significant rise in SA content (Figure 1). In the presence of NaCl, benzoic acid (BA) content only increased in MD-treated micropropagated shoots, whereas only the Phe treatment produced an accumulation of MD and SA (Figure 1).

The SA biosynthesis from the CNglcs pathway was also studied in micropropagated shoots from the transgenic plum line J8-1. In the absence of stress, the [^{13}C]MD treatment decreased MD and BA levels, while [^{13}C]Phe-fed micropropagated J8-1 shoots displayed increased amounts of Phe and amygdalin and lesser amounts of BA (Figure 2). Similar to results in cv. *Claudia verde* and contrary to that which occurred in peach plants [5], neither [^{13}C]MD nor [^{13}C]Phe increased the SA content under in vitro conditions. Salt stress induced a significant decrease in Phe, MD, and amygdalin in both control and treated (MD or Phe) J8-1 shoots. However, both treatments ameliorated the decrease in benzoic acid observed in control shoots (Figure 2). Regarding SA levels, no statistically significant differences were observed in NaCl-submitted shoots (Figure 2).

Under our experimental conditions, we were able to detect [^{13}C]-Phe, -MD and -SA, but no [^{13}C]-benzoic acid was observed in either plum plant, cv. *Claudia verde* or the J8-1 line. Regarding the percentage of [^{13}C]-labelled compounds, similar values were recorded in both plum plants, and no significant differences were observed among the different treatments and conditions. We observed less than 10% of [^{13}C]Phe, and [^{13}C]MD but [^{13}C]SA values ranged between 20% and 25% of the total amount detected (Supplemental Figure S1). It is noteworthy to mention that, although differences were not statistically significant, the highest levels of [^{13}C]MD and [^{13}C]SA were observed in [^{13}C]MD-fed micropropagated shoots. Moreover, under salinity conditions, no [^{13}C]Phe was detected, probably because its rapid turnover under stress conditions (Supplemental Figure S1).

These results suggest that the SA synthesis from the MD pathway demonstrated in peach is not clearly present in plum, at least under in vitro conditions. For this reason, further experiments in order to investigate the crosstalk among stress tolerance, oxidative stress and SA under salinity were performed on the transgenic line J8-1, displaying an enhanced antioxidant capacity.

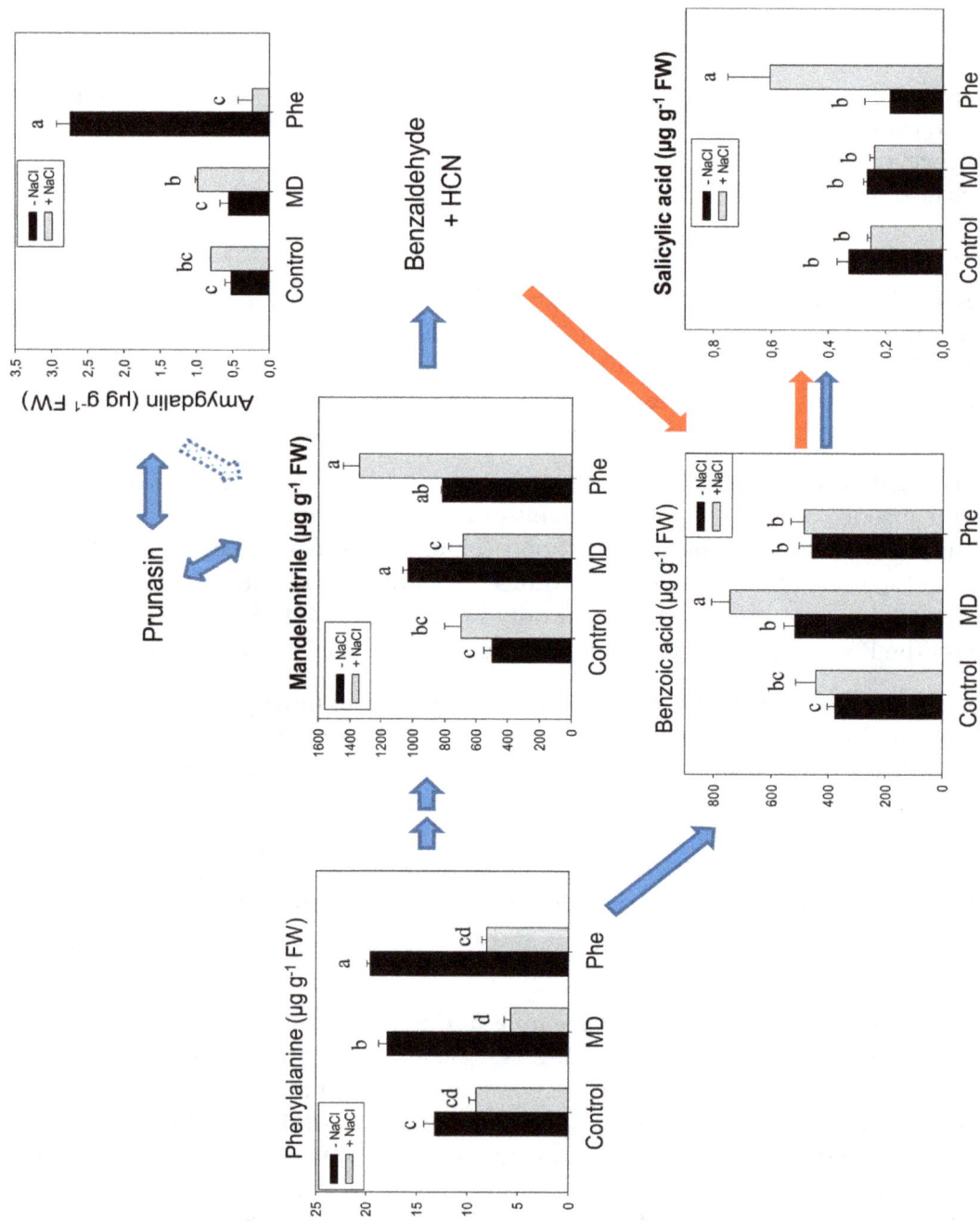

Figure 1. Salicylic acid (SA) biosynthetic and cyanogenic glucoside (CNglcs) pathways in salt-stressed (100 mM NaCl) plum cv. *Claudia verde* shoots micropropagated in the presence or absence of [13C]MD or [13C]Phe. Total levels (μM g^{-1} FW) of amygdalin, benzoic acid, mandelonitrile, phenylalanine, and salicylic acid are shown. Data represent the mean $\pm$ SE of at least 12 repetitions of each treatment. Different letters indicate significant differences in each graph according to Duncan's test ($p \leq 0.05$). Blue arrows indicate the previously described SA biosynthesis in higher plants [3] (dot arrow, putative), whereas red arrows show the recently described pathway [5].

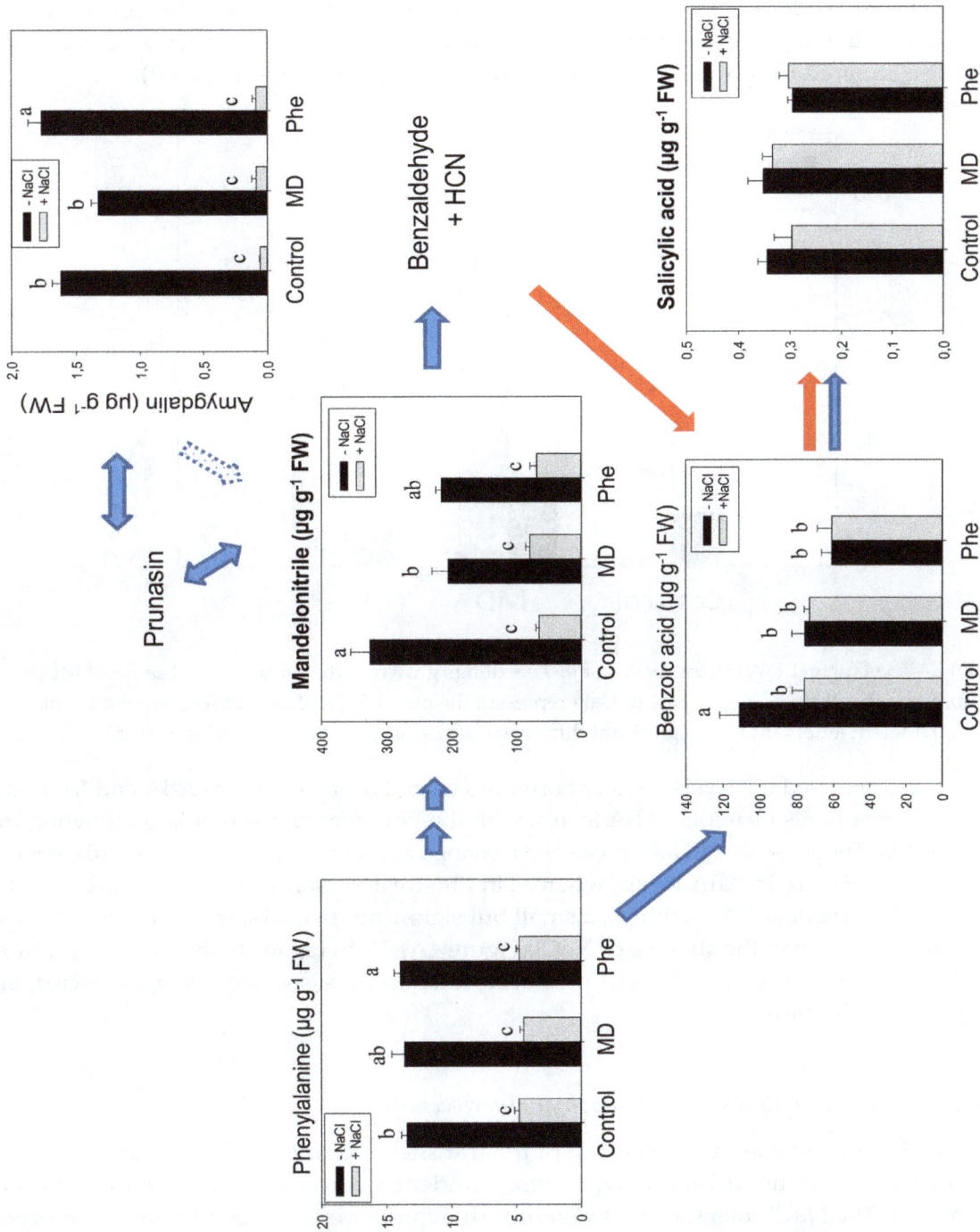

Figure 2. Salicylic acid (SA) biosynthetic and cyanogenic glucoside (CNglcs) pathways in salt-stressed (100 mM NaCl) transgenic J8-1 plum shoots micropropagated in the presence or absence of [13C]MD or [13C]Phe. Total levels (μM g^{-1} FW) of amygdalin, benzoic acid, mandelonitrile, phenylalanine, and salicylic acid are shown. Data represent the mean $\pm$ SE of at least 12 repetitions of each treatment. Different letters indicate significant differences in each graph according to Duncan's test ($p \leq 0.05$). Blue arrows indicate the previously described SA biosynthesis in higher plants [3] (dot arrow, putative), whereas red arrows show the recently described pathway [5].

2.2. Effect on Stress-Related Hormones: SA, ABA and JA

It is known that cross-talk among different hormonal signals is involved in different physiological responses as well as in response to environmental challenges. In that sense, the content of SA and other well-known stress-related hormones like ABA and JA was determined in leaves from J8-1 seedlings.

In the absence of NaCl, similar to that observed in micropropagated shoots, neither MD nor Phe affected SA levels (Figure 3). Under NaCl stress, however, a significant increase in SA concentration was observed, especially in the presence of MD. In fact, MD-treated J8-1 seedlings showed a 2.3- and 1.7-fold SA increase compared to control and Phe-treated plants, respectively (Figure 3).

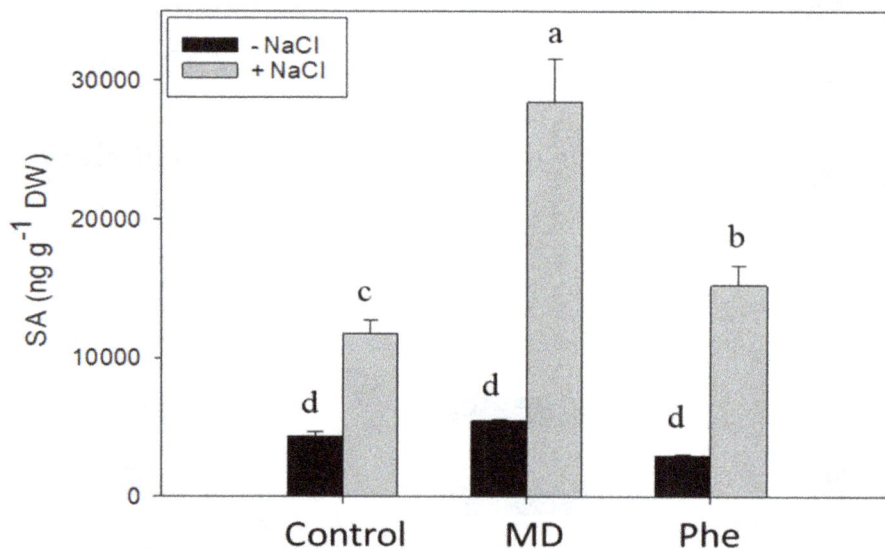

Figure 3. Total SA level (ng g-1 DW) in the leaves of J8-1 seedlings grown in the presence or absence of MD or Phe and submitted to salt stress (6 g/L NaCl). Data represent the mean ± SE of at least four repetitions of each treatment. Different letters indicate significant differences according to Duncan's test ($p \leq 0.05$).

In addition, we also analysed the levels of other hormones related to stress such as ABA and JA. In the absence of NaCl, both treatments increased ABA levels, with the Phe-treated J8-1 seedlings showing the highest levels (Figure 4A). The presence of NaCl produced a change in this response. In that regards, control plants showed a 1.7-fold increase in ABA levels, whereas in Phe-treated plants, ABA levels declined up to 1.8-fold. Regarding MD-treated J8-1 seedlings, a small but significant decrease in ABA was recorded, in relation to the levels observed in the absence of NaCl (Figure 4A). With regard to the JA concentration, only the MD treatment in the presence of NaCl produced statistically significant changes, increasing considerably the JA levels (Figure 4B).

2.3. Plant Growth, Chlorophyll Contents, and Chlorophyll Fluorescence

In previous works, we reported the tolerance of the transgenic plum line J8-1 to salinity (up to 150 mM) under in vitro conditions [16] and to water-stress under ex vitro conditions (up to 15 days of water deprivation) [17]. The NaCl-tolerance was also confirmed under ex vitro conditions as observed by the effect of MD and Phe (MD precursor) treatments on plant performance (chlorophyll content, chlorophyll fluorescence parameters and leaf and root water contents) under salinity stress conditions. Accordingly, NaCl treatment (6 g/L) did not have a significant effect on plant growth (Supplemental Figure S2) or on the leaf water content either in the absence or presence of MD or Phe treatments (Supplemental Figure S3). On the other hand, salinity increased the root water content in control and MD-treated seedlings (Supplemental Figure S3).

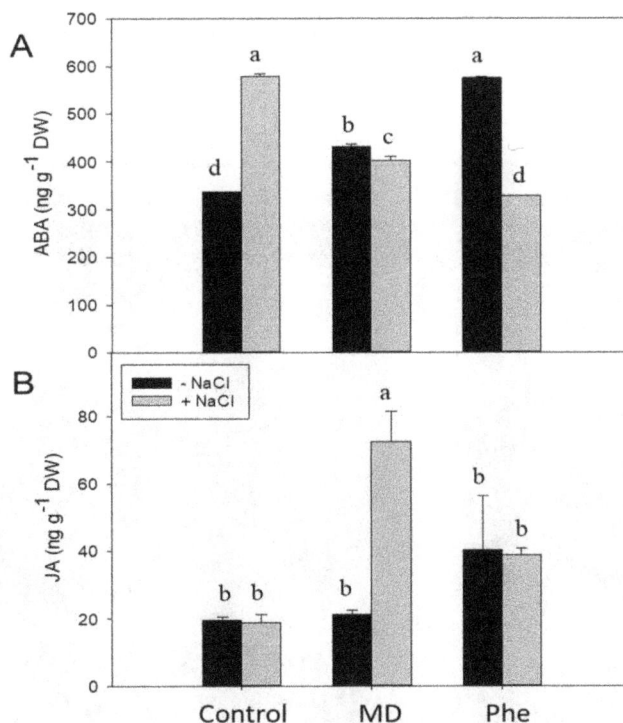

Figure 4. Effect on the stress-related hormones ABA and JA. Total ABA (**A**) and JA (**B**) levels (ng g^{-1} DW) in the leaves of J8-1 seedlings grown in the presence or absence of MD or Phe submitted to salt stress (6 g/L NaCl). Data represent the mean ± SE of at least four repetitions of each treatment. Different letters indicate significant differences according to Duncan's test ($p \leq 0.05$).

We also analysed the effect of NaCl in the presence or absence of MD and Phe treatments on the chlorophyll content in leaves from J8-1 seedlings. In the absence of NaCl, MD treatment increased the Chla content, whereas Phe produced a rise in Chla and Chlb. Under salinity conditions, an increase in Chla and Chlb was observed in non-treated plants as well as in the presence of MD. However, a decrease in Chla was produced in Phe-treated plants (Figure 5).

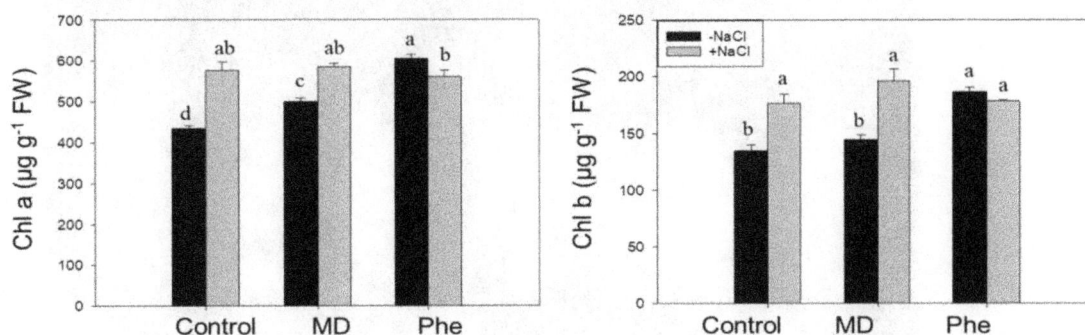

Figure 5. Effect of NaCl (6 g/L) on Chla and Chlb content in control, MD, and Phe treated J8-1 plum seedlings. Data represents the mean ± SE of at least four repetitions. Different letters indicate statistical significance according to Duncan's test ($p < 0.05$).

In addition to chlorophyll content determination, different photochemical [Y(II) and qP] and non-photochemical [(Y(NPQ) and qN] quenching chlorophyll fluorescence related parameters were also analysed. Under control conditions, both treatments increased qP, whereas a decrease in qN occurred in Phe-treated plants. Under NaCl stress, non-treated plants showed an increase in qP and qN (Figure 6). The MD treatment decreased the photochemical quenching parameters, but an increase in the non-photochemical quenching parameters took place. Regarding the Phe treatment, a decrease in Y(II) was observed, but qP did not show statistically significant changes, whereas, similar to the MD treatment, a significant increase in the non-photochemical quenching parameters occurred (Figure 6).

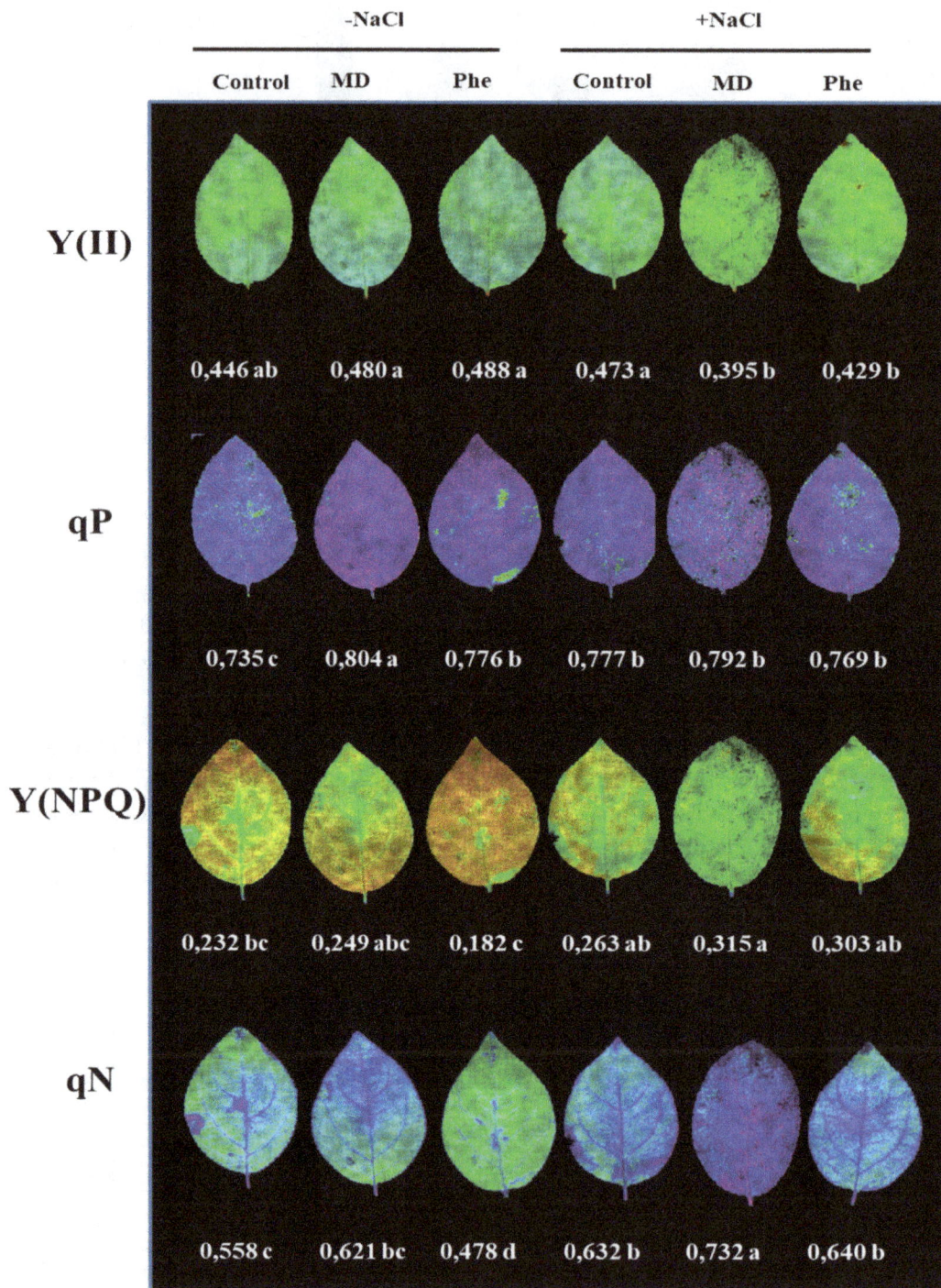

Figure 6. The effect of salt stress (6 g/L NaCl) on the chlorophyll fluorescence parameters in J8-1 seedling leaves. Representative images of the quantum yield of photochemical energy conversion in PS II [Y(II)], the photochemical quenching (qP) and the quantum yield of regulated non-photochemical energy loss in PS II and its coefficient [y(NPQ) and qN] are shown. Zero represents the lowest value and 1 the maximum value for each parameter. The averages of the values of the different parameters analysed are displayed below each image. Data represent the mean $\pm$ SE of at least six repetitions. Different letters indicate statistical significance according to Duncan's test ($p < 0.05$).

2.4. Redox State and the Gene Expression of Redox-Related Genes

It is well known that the stress hormone SA can interact with the antioxidant metabolism modulating cellular redox homeostasis. For this raison, we determined the redox state in micropropagated J8-1 shoots and in leaves from J8-1 seedlings by analysing the ascorbate and glutathione levels in the absence and in the presence of NaCl.

Under control conditions, micropropagated shoots did not show significant changes in ascorbate or glutathione levels (Tables 1 and 2). In the presence of NaCl, MD treatment decreased the total (TASC) and reduced ascorbate (ASC) levels, whereas Phe increased the TASC content. As a result, a decrease in the redox state of ascorbate (ASC/TASC) occurred in Phe-treated plants (Table 1). Regarding glutathione levels, an increase in its reduced form (GSH), as well as in the total glutathione (TGSH) level, was only observed in MD-treated micropropagated shoots under salt stress. However, no changes in the redox state of glutathione (GSH/TGSH) were observed in any treatments (Table 2).

Table 1. Effect of salt stress (100 mM NaCl), in the presence or absence of MD and Phe treatments, on total ascorbate (TASC) and reduced ascorbate (ASC) content in micropropagated J8-1 shoots. Data represent the mean $\pm$ SE of at least four repetitions. Different letters in the same column indicate significant differences according to Duncan's test ($p \leq 0.05$).

	Treatment	TASC (μmol g^{-1} FW)	ASC (μmol g^{-1} FW)	Ascorbate Redox State
−NaCl	Control	1.2 ± 0.16 c	0.9 ± 0.08 ab	0.74 ± 0.03 a
	MD	1.3 ± 0.03 bc	0.9 ± 0.04 ab	0.73 ± 0.01 a
	Phe	1.7 ± 0.03 c	1.2 ± 0.05 a	0.69 ± 0.02 a
+NaCl	Control	1.7 ± 0.09 ab	1.2 ± 0.03 a	0.69 ± 0.05 a
	MD	1.2 ± 0.06	0.8 ± 0.04 b	0.72 ± 0.00 a
	Phe	ab 2.1 ± 0.26 a	1.2 ± 0.16 a	0.57 ± 0.02 b

Table 2. Effect of salt stress (100 mM NaCl), in the presence or absence of MD and Phe treatments, on total glutathione (TGSH) and reduced glutathione (GSH) content in micropropagated J8-1 shoots. Data represent the mean $\pm$ SE of at least four repetitions. Different letters in the same column indicate significant differences according to Duncan's test ($p \leq 0.05$).

	Treatment	TGSH (nmol g^{-1} FW)	GSH (nmol g^{-1} FW)	Glutathione Redox State
−NaCl	Control	91.1 ± 4.82 c	86.7 ± 4.25 c	0.95 ± 0.01 ab
	MD	90.1 ± 1.94 c	86.1 ± 1.47 c	0.96 ± 0.00 a
	Phe	104.9 ± 7.60 c	98.3 ± 6.99 bc	0.93 ± 0.00 ab
+NaCl	Control	104.9 ± 6.66 bc	96.9 ± 6.62 bc	0.92 ± 0.00 b
	MD	140.5 ± 6.40 a	131.7 ± 6.38 a	0.94 ± 0.00 ab
	Phe	120.5 ± 6.55 ab	112.0 ± 5.63 ab	0.93 ± 0.01 ab

The response was rather different in J8-1 seedlings. In this case, under our experimental conditions, we were not able to detect oxidised ascorbate, so only ASC content is shown (Table 3). In absence of NaCl, and similar to that observed under in vitro conditions, no significant differences were apparent in the ASC and GSH levels. When plants were subjected to saline stress, MD treatment increased ASC but decreased GSH. However, MD and Phe treatments induced an accumulation of oxidised glutathione (GSSG), leading to a decrease in the redox state of glutathione in both cases (Table 3).

Table 3. Effect of salt stress (100 mM NaCl), in the presence or absence of MD and Phe treatments, on ascorbate (ASC) and glutathione (GSH, reduced; GSSG, oxidized) contents in the leaves of J8-1 seedlings. Data represent the mean ± SE of at least four repetitions. Different letters in the same column indicate significant differences according to Duncan's test ($p \leq 0.05$).

	Treatment	ASC (μmol g^{-1} FW)	GSH (nmol g^{-1} FW)	GSSG (nmol g^{-1} FW)	Glutathione Redox State
−NaCl	Control	4.6 ± 0.7 b	109.5 ± 1.9 a	11.4 ± 1.1 b	0.91 ± 0.01 a
	MD	3.7 ± 0.2 bc	99.6 ± 5.2 a	37.5 ± 1.4 a	0.72 ± 0.02 c
	Phe	4.5 ± 0.6 b	114.5 ± 5.4 a	30.1 ± 2.8 a	0.79 ± 0.02 b
+NaCl	Control	3.4 ± 0.4 bc	103.5 ± 7.5 a	13.5 ± 1.0 b	0.88 ± 0.01 a
	MD	7.1 ± 1.5 a	79.3 ± 8.0 b	36.3 ± 3.3 a	0.68 ± 0.02 c
	Phe	1.8 ± 0.2 c	114.2 ± 6.9 a	30.2 ± 2.1 a	0.79 ± 0.01 b

We also studied the effect of MD and Phe treatments on the *Non-Expressor of Pathogenesis-Related 1* (*NPR1*) and *thioredoxin H* (*TrxH*) gene expression levels in NaCl-stressed J8-1 micropropagated plum shoots and in leaves from J8-1 plum seedlings. In the absence of NaCl, micropropagated shoots treated with Phe showed reduced *NPR1* expression but induced *TrxH* expression (Figure 7A,B). In the presence of NaCl, both treatments increased the expression of the studied redox-related genes. The induction was especially striking for the effect of Phe on *NPR1* expression (nearly a five-fold increase) and for the increase in *TrxH* expression (13-fold) observed as a result of MD treatment (Figure 7A,B).

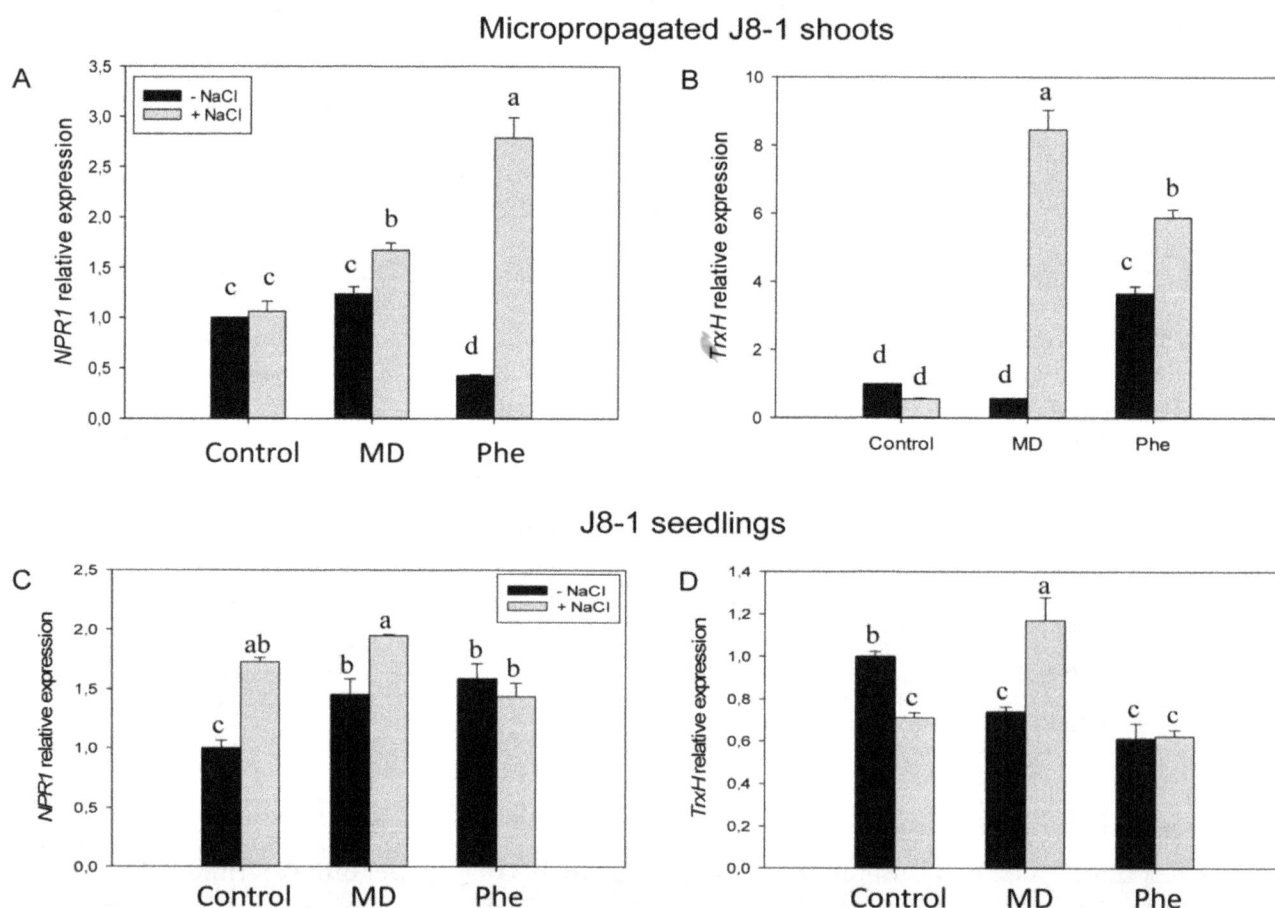

Figure 7. Gene expression of *NPR1* and *TrxH* in micropropagated J8-1 shoots (**A,B**) and in the leaves of J8-1 seedlings (**C,D**) grown in the presence or absence of MD or Phe and submitted to salt stress. Data represent the mean ± SE of at least five repetitions of each treatment. Different letters indicate significant differences in each graph according to Duncan's test ($p \leq 0.05$).

The effect of the treatments on the expression of both redox-related genes in the transgenic plum seedlings was somewhat different. In the absence of NaCl, both MD and Phe treatments induced *NPR1* expression but reduced *TrxH* expression in a similar manner (Figure 7C,D). Under salinity conditions, control and MD-treated seedlings increased *NPR1* expression, whereas *TrxH* expression was repressed in control seedlings but was again induced by MD (Figure 7C,D).

2.5. Effect of MD and Phe on Soluble Leaf and Root Nutrient Content under Salt Stress Conditions

Salt stress produced ion toxicity associated with excess Cl^- and Na^+ uptake, leading to Ca^{2+} and K^+ deficiency and other nutrient imbalances [2]. Therefore, the effect of MD and Phe treatments on soluble K^+, Ca^{2+}, Na^+ and Cl^- levels was analysed in leaves and roots from transgenic plum seedlings grown in the presence and absence of NaCl. Under control conditions, MD and Phe increased leaf K^+ content but decreased leaf Ca^{2+} content. No effects of MD or Phe treatments on Na^+ and Cl^- levels were observed (Figure 8). In NaCl-stressed seedlings, an increase in all the analysed nutrients occurred in the leaves from non-treated plants. Similar results were observed in the MD and Phe treatments, in which the leaves of salt-stressed seedlings also displayed increased Ca^{2+}, Na^+, and Cl^- levels, although the K^+ level slightly decreased due to MD and was not affected by Phe. It is important to note the low leaf Na^+ levels found in the transgenic plum seedlings (Figure 8).

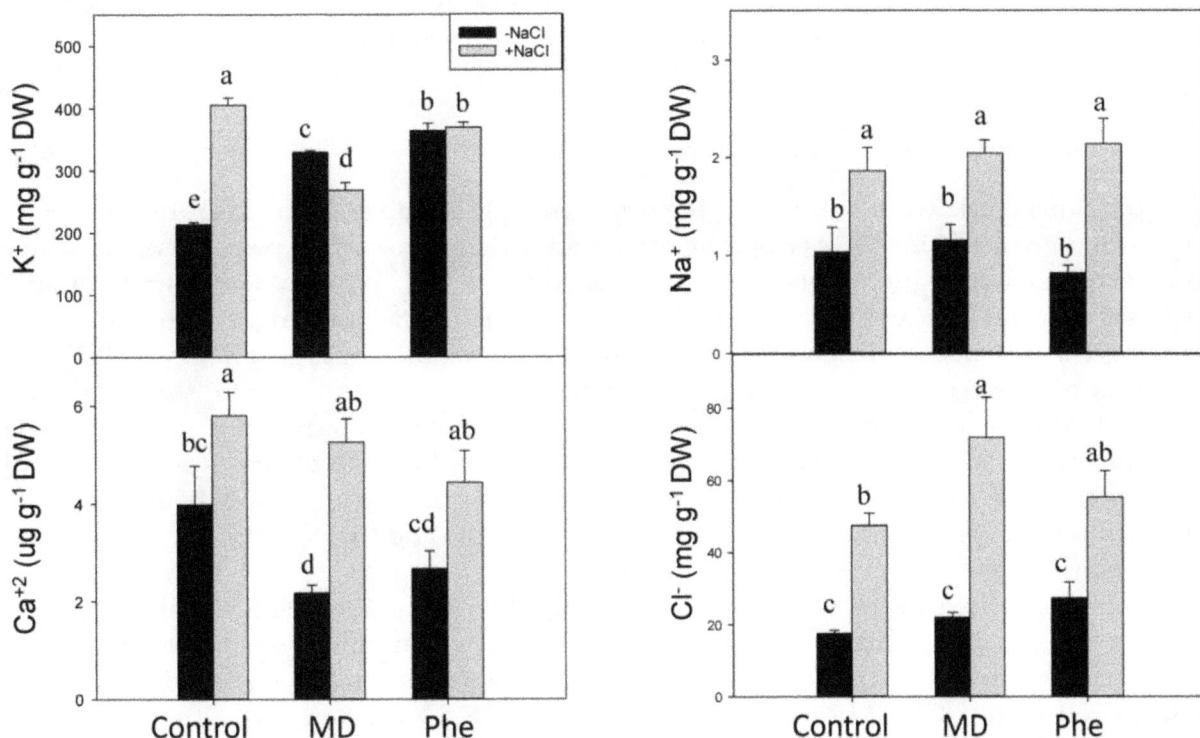

Figure 8. Effect of salt stress (6 g/L NaCl) on soluble K^+, Ca^{2+}, Na^+, and Cl^- contents in leaves from control, MD- and Phe-treated J8-1 seedlings. Data represent the mean $\pm$ SE of at least four repetitions. Different letters indicate statistical significance according to Duncan's test ($p < 0.05$).

In the absence of NaCl stress, the only significant change observed in roots was an increase in soluble K^+ in Phe-treated plants (Figure 9). In the presence of NaCl, an accumulation of the phytotoxic ions Na^+ and Cl^- was observed in control and MD- and Phe-treated seedlings. However, the Na^+ accumulation in roots was lower in MD-treated plants than in the other treatments. The K^+ content increased in non-treated NaCl-stressed roots, whereas no significant differences were observed in Ca^{2+} levels in any case (Figure 9).

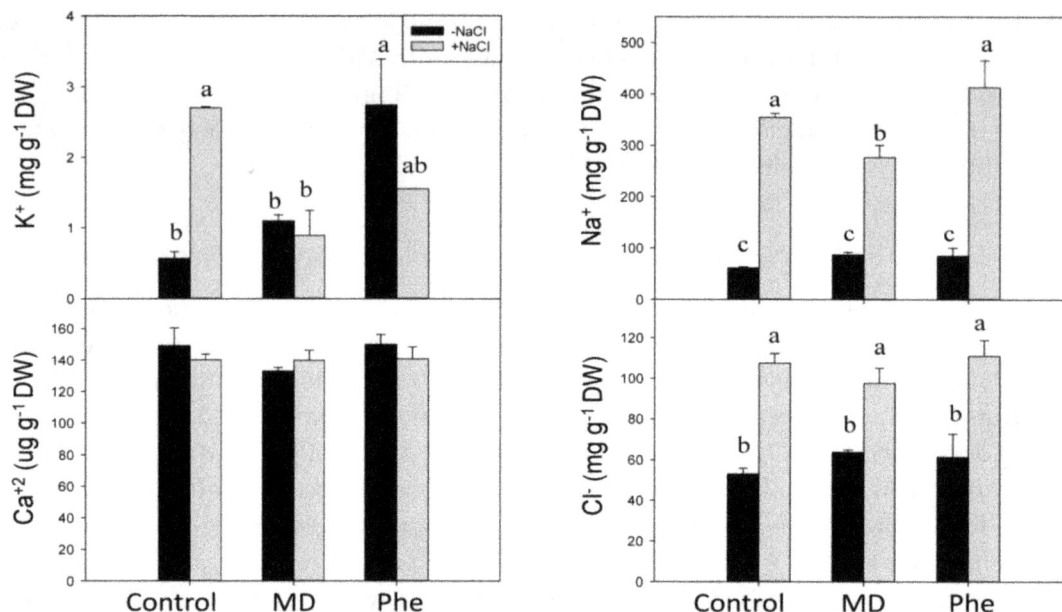

Figure 9. Effect of salt stress (6 g/L NaCl) on soluble K^+, Ca^{2+}, Na^+, and Cl^- contents in roots from control, MD- and Phe-treated J8-1 seedlings. Data represent the mean ± SE of at least four repetitions. Different letters indicate statistical significance according to Duncan's test ($p < 0.05$).

3. Discussion

3.1. Involvement of MD on SA Biosynthesis in Plum

In a previous work, we reported that the transgenic plum line J8-1 was tolerant up to 150 mM NaCl under in vitro conditions. This response correlated with high ascorbate peroxidase (APX) activity and gene expression and glutathione and ascorbate contents [16]. We also demonstrated that APX overexpression in line J8-1 can play a major role in the response of J8-1 seedlings to drought conditions by inducing changes at the physiological, biochemical, proteomic, and genetic levels [17]. In our opinion, it is of interest to characterise the response of this transgenic line to NaCl stress under ex vitro conditions. In addition, we recently reported that MD is the intermediary molecule between a suggested new SA biosynthetic pathway and CNglcs turnover in peach plants [5]. All these findings led us to investigate not only the response of the J8-1 line to salinity, but also whether the new SA pathway described in peach plants [5] is also present in this line and the possible role of this pathway on plant performance.

In micropropagated peach shoots fed with [^{13}C]MD, nearly 20% of the total SA quantified appeared as [^{13}C]SA, demonstrating that MD can be an intermediary molecule in this novel pathway controlling amygdalin and SA biosynthesis [5] However, when micropropagated plum *"Claudia Verde"* shoots were fed with [^{13}C]MD or [^{13}C]Phe, no increases in SA were detected, although significant increases in MD, Phe and benzoic acid (SA-precursor) were observed. We also assayed this possibility using micropropagated J8-1 shoots. However, the results concerning the involvement of MD as a putative intermediary of SA biosynthesis in this plum line were negative. These results led to the hypothesis that the new SA synthesis pathway from MD, previously demonstrated in peach, seemed not to be operative in plum under in vitro conditions. Nevertheless, we observed that NaCl stress affected the CNglcs pathway, mainly at the MD and amygdalin levels. Due to the fact that amygdalin is derived from MD by the addition of two glucose molecules, it is logical to assume that the amygdalin content decreases. The glucose molecules could be used for osmotic adjustment or to obtain energy for different metabolic processes.

As a conclusion, it seems that MD is not an intermediary for SA biosynthesis in micropropagated plums as it was found to be in micropropagated peach shoots, unless the synthesized SA appears as conjugated form. However, an increase in SA was produced in J8-1 plum seedlings when NaCl stress was imposed, especially in MD-treated plants, suggesting that MD could be involved in the SA biosynthesis in plum grown in the presence of NaCl under greenhouse conditions. These findings are different to the results found in peach, where the contribution of this pathway to the SA pool does not seem to be relevant under salt stress or *Plum pox virus*-infection conditions [18].

3.2. Plant Performance of Plum under NaCl Stress

As expected, line J8-1 was also tolerant to high NaCl levels under greenhouse conditions, as observed by the lack of negative effects on plant growth (measured as shoot biomass fresh weight) or in the leaf and root water levels, both in control and MD- and Phe-treated plants. In addition, chlorophyll levels can be seen as a biochemical marker of salt tolerance in plants; the maintenance or the increase in Chl content under NaCl stress can be considered as a protection mechanism for the photosynthesis process. In this regards, both non-treated and MD-treated plants showed increases in Chla and Chlb levels. Moreover, J8-1 plants increased/and or maintained the non-photochemical quenching parameters under NaCl stress, especially in the presence of MD. The maintenance of the non-photochemical quenching parameters under stress situations has been associated with a capacity to dissipate light energy safely, and it can be seen as an adaptive mechanism to protect the chloroplasts under NaCl conditions, avoiding the over-generation of ROS, as described for other plant species [2,19–21].

Thus, according to plant growth, leaf and root water content, chlorophyll contents and chlorophyll fluorescence data, the transgenic plum line J8-1 can be considered as salt-tolerant.

3.3. Stress-Related Hormones and NaCl Response

The effect of MD and Phe treatments on the stress-related hormones ABA and JA in line J8-1 was quite different to that observed in peach plants. In peach, in the absence of NaCl, the treatments had no effect on ABA and reduced the JA levels. Under saline conditions, MD decreased ABA and JA concentrations, whereas Phe produced a decrease in JA [18]. The transgenic line J8-1 showed contrasting results. In this case, the differences could be due to the use of different plant species showing a different NaCl tolerance and the different NaCl levels used: peach plants were subjected to 2 g/L NaCl, whereas plum plants were treated with 6 g/L NaCl.

Some studies have suggested a positive interaction between SA, JA, ethylene and ABA signalling pathways, improving the response of plants to environmental stresses [22]. JA and SA can regulate plant responses to abiotic stresses. Accordingly, the exogenous application of both JA and SA has been found to enhance salt-tolerance in some plant species by increasing their antioxidative capacity [23, 24]. In addition, an increase in the SA/JA ratio has been suggested as a marker of salt stress [25]. In the current study, an increased SA/JA ratio due to salinity was observed in control and MD- and Phe-treated seedlings, the increase being much greater in MD-treated plants. This response was mainly due to the sharp increase in SA levels in MD-treated J8-1 seedlings under salinity conditions. In peach plants, the MD treatment slightly decreased the SA/JA ratio, and no effect of NaCl on plant development was observed [18]. In a study of the salt-tolerant sweet-potato genotype ND98, the JA concentrations in the leaves and roots increased after 12 h of saline treatment (200 mM NaCl), and this response correlated with a regulated stomatal closure [26].

ABA is a well-known regulator of stomatal regulation and, hence, of the abiotic stress response. In the current study, Phe treatment decreased the ABA content, whereas a small decrease occurred in MD-treated plants and an increase was observed in untreated seedlings under salinity conditions. Moreover, SA is also involved in stomatal regulation, and SA treatment has been found to decrease the stomatal aperture in Arabidopsis [27]. Accordingly, the SA increase, and the scanty effect in ABA observed in MD-treated plants in the current study suggests more efficient stomatal regulation under saline conditions.

Considered together, all of this data suggests that MD could have a positive effect on the J8-1 response to salinity through an increase in SA and JA and tight control of the ABA levels.

3.4. NaCl Effects on Redox State and Ion Homeostasis

In micropropagated J8-1 shoots, GSH levels increased under salt stress, especially in the MD treatment. Furthermore, this increase correlated with the induction of *NPR1* and *TrxH*, suggesting a role for GSH in the stress-induced expression of these redox-related genes. It seems that GSH could play a role in *NPR1* induction under in vitro conditions. In micropropagated peach shoots, treatment with the artificial precursor of cysteine, L-2-oxothiazolidine-4-carboxylic acid (OTC), produced an increase in GSH and in the GSH/GSSG ratio as well as in the *NPR1* expression both in healthy and in *Plum pox virus*-infected shoots [28]. On the other hand, in J8-1 seedlings, the increase in *NPR1* and *TrxH* expression due to the MD treatment correlated with a decrease in the GSH redox state. As mentioned, under stress conditions, the MD treatment produced a sharp increase in SA content, and researchers have shown that SA-induced changes in glutathione lead to a more oxidised environment that modulates the plant defence responses [29–31]. Similarly, MD treatment also leads to a more oxidised environment in peach plants via changes in non-enzymatic and enzymatic antioxidant levels that could be responsible for the modification of the function of redox-regulated proteins such as NPR1 [5]. In absence of stress this MD-induced oxidised environment was due to a decrease in ASC content and GSH redox state in both peach [5] and J8-1 seedlings, whereas in J8-1 seedlings submitted to salinity the decrease in the GSH redox state was accompanied by an increase in ASC. This increase in ASC could be also related to the salt stress tolerance displayed by J8-1 seedlings [4,11,16].

Thioredoxins (Trx) are ubiquitous disulfide reductases that regulate the redox status of target proteins and seem to be involved in the protection of plant cells in stress situations that induce oxidative stress [12]. Trx can prevent the oxidative damage of important macromolecules, thus protecting plants against the stress-induced lipid peroxidation of membranes or repairing oxidised proteins [12]. Proteomic tools have made it possible to identify many potential targets of Trx, including many proteins related with important cellular processes [32]. One of the proteins regulated by thioredoxins is NPR1. Cytosolic Trxs catalyze the redox changes in NPR1 from oligomeric to monomeric forms, with SA inducing TRX-5h to catalyse NPR1 monomer release and to prevent re-oligomerization [13]. Different studies have reported the induction of TrxHs by abiotic or biotic stresses [33], suggesting that these proteins can act as antioxidants in vivo [34].

In the current study, the induction of *TrxH* gene expression by MD was more evident under in vitro conditions, where the absence of the root system leads to more severe symptoms under saline conditions. In plum seedlings (ex vitro conditions), the induction of *TrxH* was lower (only a 1.6-fold increase), which was similar to that observed in the induction of *Trxh1* in rice plants treated with 100 mM NaCl [35]. In addition, the effect of salinity on *TrxH* expression was very similar to that observed in peach [18]: in the absence of chemical treatments, salinity reduced the expression of *TrxH*, whereas induction occurred in the presence of MD, and no changes were produced in the presence of Phe.

Elevated SA levels may mediate adaptive responses against salt stress through NPR1-dependent and NPR1-independent pathways. Salt stress (100 mM NaCl) was found to have a strong effect on plant growth in the Arabidopsis *npr1-5* mutant, which lacks the NPR1-dependent SA signalling pathway. However, the effect of NaCl stress on the plant growth of the Arabidopsis *nudt7* mutant, which constitutively expressed NPR1-dependent and NPR1-independent SA signalling, was more attenuated [36]. In addition, the *npr1-5* mutant was unable to control the Na^+ influx and prevent K^+ loss in shoots and roots, in contrast to the results observed in the *nudt7* mutant. These authors concluded that the constitutive expression of NPR1-dependent SA signalling enhanced salt tolerance by controlling Na^+ entry into roots and shoots as well as minimising K^+ loss during NaCl challenges, which is an important component of salt and oxidative stress tolerance in Arabidopsis [36]. This information agrees with our results, since under saline conditions, MD-treated plants showed elevated SA levels as well as *NPR1* and *TrxH* expression and less Na^+ accumulation in roots than the other treatments. This effect of SA on Na^+ levels has also been observed in pea plants [37]. These authors reported that SA treatment reduced Na^+ accumulation in pea roots in the presence on 70 mM NaCl.

In the salt-tolerant sweet potato genotype ND98, the JA content increased in the leaves and roots after 12 h of saline treatment (200 mM NaCl), and this response correlated with regulated stomatal closure, reduced Na^+ accumulation and increased K^+ concentrations. Furthermore, this genotype showed a more balanced ion homeostasis than the salt-sensitive genotype [26]. In the current study, this response was also partially observed in MD-treated plants under saline stress. In this case, the increase in JA produced by MD treatment in the presence of NaCl correlated with a lower Na^+ accumulation in roots compared with the other treatments.

In MD- and Phe-treated peach seedlings, an accumulation of saline ions in roots was recorded, suggesting that both treatments could trigger different mechanisms leading to the development of adaptive responses against salinity [38]. These results contrast to those obtained in MD-treated J8-1 seedlings submitted to salinity conditions, which showed a strong increase in leaf soluble Ca^{2+}, which correlated with increased SA and JA contents. However, the soluble Ca^{2+} accumulation in leaves was also observed in control and Phe-treated plants that showed increased SA but no change in JA levels. It is important to note that the Ca^{2+} levels observed in leaves from MD- and Phe-treated J8-1 seedlings were lower than in control plants, and a similar response was observed in peach seedlings, suggesting that Ca^{2+} ions could be chelated by organic molecules like MD and Phe [38]. In pea plants, treatment with exogenous SA (50–100 µM) induced an increase of Ca^{2+} in shoots but not in roots, although under NaCl stress (70 mM), the presence of SA did not prevent a NaCl-induced decrease in Ca^{2+} levels [37]. It has been suggested that SA-induced Ca^{2+} contents can also lead to stomatal closure [39]. In J8-1 plants, increases in leaf SA levels under salinity conditions seem to be related to increases in Ca^{2+} in leaves that could lead to tight stomatal control, thus providing protection to membranes under stress conditions.

4. Material and Methods

4.1. Plant Material

The assays were performed on micropropagated plum [*Prunus domestica* cv. *Claudia verde* and transgenic line J8-1 [16,17]] shoots and J8-1 seedlings, which were submitted to NaCl stress in the presence or absence of MD and Phe (MD precursor) treatments.

The micropropagated plum shoots were subcultured at four-week intervals for micropropagation and samples were taken at the end of the second subculture in the presence of MD and Phe treatments. In the micropropagated shoots, salt stress was imposed by adding 100 mM NaCl to the micropropagation media in the presence or absence of 200 µM [^{13}C]MD or [^{13}C]Phe (Campro Scientific GmbH, Germany), as described in Diaz-Vivancos et al. (2017) [5]. Seedlings were obtained from rooted and acclimatized to ex vitro conditions J8-1 plantlets. Under greenhouse conditions, J8-1 seedlings were grown in 2 L pots during two months. Then seedlings were submitted to an artificial rest period (eight weeks) in a cold chamber to ensure uniformity and fast growth. After the rest period, seedlings were irrigated once a week with 6 g/L NaCl in the presence or absence of 1 mM MD or Phe for seven weeks. Samples were taken at the end of this period. For all the conditions, 12 seedlings were assayed, and another 12 plants were kept as control.

4.2. Metabolomic Analysis

Micropropagated shoots leaf samples (0.5 g FW) were extracted in 50% methanol (1/3 *w/v*) and then filtered in PTFE 0.45 µm filters (Agilent Technologies, Palo Alto, CA, USA).The levels of Phe, MD, amygdalin, benzoic acid and SA were determined in micropropagated shoots at the Metabolomics Platform at CEBAS-CSIC (Murcia, Spain) using an Agilent 1290 Infinity UPLC system coupled to a 6550 Accurate-Mass quadrupole TOF mass spectrometer (Agilent Technologies, Palo Alto, CA, USA) [5]. The hormone levels (ABA, JA and SA) in the dry leaves of J8-1 seedlings treated with MD or Phe (0.2 g DW) were determined using a UHPLC-mass spectrometer (Q-Exactive, ThermoFisher Scientific, Barcelona, Spain) at the Plant Hormone Quantification Platform at IBMCP (Valencia, Spain).

4.3. Chlorophyll Determination and Chlorophyll Fluorescence

Approximately 0.2 g of the leaves from the J8-1 seedlings submitted to NaCl stress in the presence or absence of MD and Phe were incubated in 50 mL of 80% acetone (v/v) for 72 h under darkness. The chlorophyll a (Chla) and chlorophyll b (Chlb) content was analysed by measuring the absorbance at 663 and 645 nm [40].

Chlorophyll fluorescence parameters were measured in detached leaves from J8-1 seedlings submitted to NaCl stress in the presence or absence of MD and Phe treatments using a chlorophyll fluorimeter (IMAGIM-PAM M-series, Heinz Walz, Effeltrich, Germany). After a dark incubation period (15 min), the leaves' minimum and maximum fluorescence yields were monitored. Kinetic analyses were carried out as previously described [21], and the effective PSII quantum yield [Y(II)], the coefficients of photochemical quenching (qP) and non-photochemical quenching (qN), and the quantum yield of regulated energy dissipation [(Y(NPQ)] were recorded.

4.4. Ascorbate and Glutathione Analysis

Micropropagated J8-1 line shoot and seedling leaf samples were snap-frozen in liquid nitrogen and stored at -80 °C until use. The frozen samples were homogenised (1/3 w/v) with 1 M $HClO_4$ containing 1 mM polyvinylpolypyrrolidone and 1 mM EDTA. Homogenates were centrifuged at 12,000× g for 10 min, and the supernatant was neutralised with 5 M K_2CO_3 to pH 5.5–6. The homogenate was centrifuged at 12,000× g for 1 min to remove $KClO_4$. The supernatant obtained was used for ascorbate and glutathione determination as previously described [41,42].

4.5. Gene Expression

We studied the expression levels of the redox-regulated genes NPR1 (Non-Expressor of Pathogenesis-Related Gene 1) and TrxH (thioredoxin H). Briefly, micropropagated shoots and leaf samples from line J8-1 were snap-frozen in liquid nitrogen and stored at -80 °C until use. RNA was extracted using the Power Plant RNA Isolation kit (Mo Bio), according to the manufacturer's instructions. The primer sequences were as follows: NPR1 (forward 5′-tgcacgagctcctttagtca-′3; reverse 5′-cggcttactgcgatcctaag-′3); TrxH (forward 5′-tggcggagttggctaagaag-′3; 5′-ttcttggcacccacaacctt-′3); β-actin (forward 5′tgcctgccatgtatgttgccatcc′3; reverse 5′aacagcaaggtcagacgaaggat′3).

The expression levels of NPR1, TrxH, and the β-actin gene, used for normalisation, were determined as described in [15] by real-time RT-PCR using the GeneAmp 7500 sequence detection system (Applied Biosystems, Foster City, CA, USA). Relative quantification of gene expression was calculated by the Delta-Delta Ct method.

4.6. Determination of Soluble K^+, Ca^{2+}, Na^+, and Cl^- Content

The effect of NaCl stress in the presence and absence of MD and Phe on soluble K^+, Ca^{2+}, Na^+, and Cl^- content was determined in leaves and roots of J8-1 seedlings grown under greenhouse conditions. First, leaf and root samples (at least five replicates per treatment) were oven-dried at 65°C and ground to a fine powder. Then, approximately 0.1 g was extracted with milliQ water (1/10 w/v) at 50 °C for 3 h and shake-incubated for 24 h at 30 °C.

The concentrations of the soluble nutrients analysed were determined by ion-selective electrodes (IonMeter, Nsensors ©) that were previously calibrated with standard solutions of NaCl (for Na^+ and Cl^-), $CaCl_2$ (for Ca^{+2}), and KCl (for K^+).

4.7. Statistical Analysis

The data were analysed by one-way or two-way ANOVA using SPSS 22 software (Chicago, IL, USA). Means were separated with the Duncan's Multiple Range Test ($p < 0.05$).

5. Conclusions

As a general conclusion, in this work we have demonstrated that the plum line J8-1 is tolerant to NaCl in terms of plant growth and plant performance (chlorophyll content and chlorophyll fluorescence parameters, shoot biomass and leaf and root water contents) under the tested conditions. In the presence of NaCl, the MD treatment produced the highest SA and JA increases, but it also induced the expression of *NPR1* and *TrxH* transcripts. These results, similar to those reported by other authors, suggest that the *NPR1/TrxH* interaction, along with SA and JA accumulation, may play an important role in the tolerant response of the J8-1 plum line to salt stress. The biosynthetic pathways of SA, JA and ABA take place in the chloroplast [43], and this organelle is rapidly affected by salt stress [2]. Therefore, a connection of the SA, JA, and ABA pathways and qN with the expression of *NPR1* and *TrxH*, mediated by the redox state of the chloroplast can be suggested.

Finally, the results led us to think that the new SA synthesis pathway demonstrated in peach seemed not to be operative in plum under in vitro conditions. However, MD could be involved in the SA biosynthesis under NaCl stress conditions in plum plants under greenhouse conditions. In the transgenic plum line J8-1 a crosstalk between different signalling pathways (NPR1/Trx and SA/JA) leading to salinity tolerance is suggested.

Author Contributions: A.B.V. and P.D.V performed and analyzed the metabolomics, plant growth, chlorophyll content and ascorbate and glutathione determination assays. D.C. performed and analyzed the gene expression and nutrient content experiments. C.P. performed the micropropagation and MD and Phe treatments of micropropagated shoots. J.A.H. performed and analyzed the chlorophyll fluorescence assay. P.D.V. and J.A.H. designed the experiments, performed data curation and write, review and edit the manuscript. Supervision, project administration and funding acquisition by P.D.V. and J.A.H. All authors discussed and commented on the content of the paper.

Acknowledgments: PDV and CP thank CSIC and UPCT, respectively, as well as the Spanish Ministry of Economy and Competitiveness for their 'Ramon and Cajal' research contract, co-financed by FEDER funds. This work was supported by the Spanish Ministry of Economy and Competitiveness (Projects AGL2014-52563-R and INIA-RTA2013-00026-C03-00).

References

1. Hossain, M.S.; Dietz, K.J. Tuning of Redox Regulatory Mechanisms, Reactive Oxygen Species and Redox Homeostasis under Salinity Stress. *Front. Plant Sci.* **2016**, *7*, 548. [CrossRef] [PubMed]

2. Acosta-Motos, J.; Ortuño, M.; Bernal-Vicente, A.; Diaz-Vivancos, P.; Sanchez-Blanco, M.; Hernandez, J. Plant Responses to Salt Stress: Adaptive Mechanisms. *Agronomy* **2017**, *7*, 18. [CrossRef]

3. Miura, K.; Tada, Y. Regulation of water, salinity, and cold stress responses by salicylic acid. *Front. Plant Sci.* **2014**, *5*, 4. [CrossRef] [PubMed]

4. Khan, M.I.; Fatma, M.; Per, T.S.; Anjum, N.A.; Khan, N.A. Salicylic acid-induced abiotic stress tolerance and underlying mechanisms in plants. *Front. Plant Sci.* **2015**, *6*, 462. [CrossRef] [PubMed]

5. Diaz-Vivancos, P.; Bernal-Vicente, A.; Cantabella, D.; Petri, C.; Hernández, J.A. Metabolomic and Biochemical Approaches Link Salicylic Acid Biosynthesis to Cyanogenesis in Peach Plants. *Plant Cell Physiol.* **2017**, *58*, 2057–2066. [CrossRef] [PubMed]

6. Gleadow, R.M.; Møller, B.L. Cyanogenic Glycosides: Synthesis, Physiology, and Phenotypic Plasticity. *Annu. Rev. Plant Biol.* **2014**, *65*, 155–185. [CrossRef] [PubMed]

7. Catinot, J.; Buchala, A.; Abou-Mansour, E.; Metraux, J.P. Salicylic acid production in response to biotic and abiotic stress depends on isochorismate in *Nicotiana benthamiana*. *FEBS Lett.* **2008**, *582*, 473–478. [CrossRef] [PubMed]

8. Ogawa, D.; Nakajima, N.; Seo, S.; Mitsuhara, I.; Kamada, H.; Ohashi, Y. The phenylalanine pathway is the main route of salicylic acid biosynthesis in *Tobacco mosaic virus* infected tobacco leaves. *Plant Biotechnol.* **2006**, *23*, 395–398. [CrossRef]

9. Liu, X.; Rockett, K.S.; Kørner, C.J.; Pajerowska-Mukhtar, K.M. Salicylic acid signalling: New insights and prospects at a quarter-century milestone. *Essays Biochem.* **2015**, *58*, 101–113. [CrossRef] [PubMed]

10. Foyer, C.H.; Noctor, G. Oxidant and antioxidant signalling in plants: A re-evaluation of the concept of oxidative stress in a physiological context. *Plant Cell Environ.* **2005**, *28*, 1056–1071. [CrossRef]

11. Foyer, C.H.; Noctor, G. Ascorbate and glutathione: The heart of the redox hub. *Plant Physiol.* **2011**, *155*, 2–18. [CrossRef] [PubMed]

12. Vieira Dos Santos, C.; Rey, P. Plant thioredoxins are key actors in the oxidative stress response. *Trends Plant Sci.* **2006**, *11*, 329–334. [CrossRef] [PubMed]

13. Tada, Y.; Spoel, S.H.; Pajerowska-Mukhtar, K.; Mou, Z.; Song, J.; Wang, C.; Zuo, J.; Dong, X. Plant Immunity Requires Conformational Charges of NPR1 via S-Nitrosylation and Thioredoxins. *Science* **2008**, *321*, 952–956. [CrossRef] [PubMed]

14. Brosché, M.; Kangasjärvi, J. Low antioxidant concentrations impact on multiple signalling pathways in *Arabidopsis thaliana* partly through NPR1. *J. Exp. Bot.* **2012**, *63*, 1849–1861. [CrossRef] [PubMed]

15. Faize, M.; Faize, L.; Petri, C.; Barba-Espin, G.; Diaz-Vivancos, P.; Clemente-Moreno, M.J.; Koussa, T.; Rifai, L.A.; Burgos, L.; Hernandez, J.A. Cu/Zn superoxide dismutase and ascorbate peroxidase enhance in vitro shoot multiplication in transgenic plum. *J. Plant Physiol.* **2013**, *170*, 625–632. [CrossRef] [PubMed]

16. Diaz-Vivancos, P.; Faize, M.; Barba-Espin, G.; Faize, L.; Petri, C.; Hernández, J.A.; Burgos, L. Ectopic expression of cytosolic superoxide dismutase and ascorbate peroxidase leads to salt stress tolerance in transgenic plums. *Plant Biotechnol. J.* **2013**, *11*, 976–985. [CrossRef] [PubMed]

17. Diaz-Vivancos, P.; Faize, L.; Nicolas, E.; Clemente-Moreno, M.J.; Bru-Martinez, R.; Burgos, L.; Hernandez, J.A. Transformation of plum plants with a cytosolic ascorbate peroxidase transgene leads to enhanced water stress tolerance. *Ann. Bot.* **2016**, *117*, 1121–1131. [CrossRef] [PubMed]

18. Bernal-Vicente, A.; Petri, C.; Hernández, J.A.; Diaz-Vivancos, P. The effect of abiotic and biotic stress on the salicylic acid biosynthetic pathway from mandelonitrile in peach. *J. Plant Physiol.* (under review).

19. Acosta-Motos, J.-R.; Diaz-Vivancos, P.; Alvarez, S.; Fernandez-Garcia, N.; Jesus Sanchez-Blanco, M.; Antonio Hernandez, J. Physiological and biochemical mechanisms of the ornamental *Eugenia myrtifolia* L. plants for coping with NaCl stress and recovery. *Planta* **2015**, *242*, 829–846. [CrossRef] [PubMed]

20. Acosta-Motos, J.R.; Diaz-Vivancos, P.; Alvarez, S.; Fernandez-Garcia, N.; Jesus Sanchez-Blanco, M.; Hernandez, J.A. NaCl-induced physiological and biochemical adaptative mechanisms in the ornamental *Myrtus communis* L. plants. *J. Plant Physiol.* **2015**, *183*, 41–51. [CrossRef] [PubMed]

21. Cantabella, D.; Piqueras, A.; Acosta-Motos, J.R.; Bernal-Vicente, A.; Hernandez, J.A.; Diaz-Vivancos, P. Salt-tolerance mechanisms induced in *Stevia rebaudiana* Bertoni: Effects on mineral nutrition, antioxidative metabolism and steviol glycoside content. *Plant Physiol. Biochem.* **2017**, *115*, 484–496. [CrossRef] [PubMed]

22. Boatwright, J.L.; Pajerowska-Mukhtar, K. Salicylic acid: An old hormone up to new tricks. *Mol. Plant Pathol.* **2013**, *14*, 623–634. [CrossRef] [PubMed]

23. Qiu, Z.; Guo, J.; Zhu, A.; Zhang, L.; Zhang, M. Exogenous jasmonic acid can enhance tolerance of wheat seedlings to salt stress. *Ecotoxicol. Environ. Saf.* **2014**, *104* (Suppl. C), 202–208. [CrossRef] [PubMed]

24. He, Y.; Zhu, Z.J. Exogenous salicylic acid alleviates NaCl toxicity and increases antioxidative enzyme activity in *Lycopersicon esculentum*. *Biol. Plant.* **2008**, *52*, 792. [CrossRef]

25. Acosta-Motos, J.R.; Ortuño, M.F.; Álvarez, S.; López-Climent, M.F.; Gómez-Cadenas, A.; Sánchez-Blanco, M.J. Changes in growth, physiological parameters and the hormonal status of *Myrtus communis* L. plants irrigated with water with different chemical compositions. *J. Plant Physiol.* **2016**, *191* (Suppl. C), 12–21. [CrossRef] [PubMed]

26. Zhang, H.; Zhang, Q.; Zhai, H.; Li, Y.; Wang, X.; Liu, Q.; He, S. Transcript profile analysis reveals important roles of jasmonic acid signalling pathway in the response of sweet potato to salt stress. *Sci. Rep.* **2017**, *7*, 40819. [CrossRef] [PubMed]

27. Khokon, A.R.; Okuma, E.; Hossain, M.A.; Munemasa, S.; Uraji, M.; Nakamura, Y.; Mori, I.C.; Murata, Y. Involvement of extracellular oxidative burst in salicylic acid-induced stomatal closure in *Arabidopsis*. *Plant Cell Environ.* **2011**, *34*, 434–443. [CrossRef] [PubMed]

28. Clemente-Moreno, M.J.; Diaz-Vivancos, P.; Piqueras, A.; Antonio Hernandez, J. Plant growth stimulation in *Prunus* species plantlets by BTH or OTC treatments under in vitro conditions. *J. Plant Physiol.* **2012**, *169*, 1074–1083. [CrossRef] [PubMed]

29. Yang, Y.; Qi, M.; Mei, C. Endogenous salicylic acid protects rice plants from oxidative damage caused by aging as well as biotic and abiotic stress. *Plant J.* **2004**, *40*, 909–919. [CrossRef] [PubMed]

30. Herrera-Vasquez, A.; Salinas, P.; Holuigue, L. Salicylic acid and reactive oxygen species interplay in the transcriptional control of defense genes expression. *Front. Plant Sci.* **2015**, *6*, 171. [CrossRef] [PubMed]

31. Vlot, A.C.; Dempsey, D.A.; Klessig, D.F. Salicylic acid, a multifaceted hormone to combat disease. *Annu. Rev. Phytopathol.* **2009**, *47*, 177–206. [CrossRef] [PubMed]

32. Gelhaye, E.; Rouhier, N.; Jacquot, J.P. The thioredoxin *h* system of higher plants. *Plant Physiol. Biochem.* **2004**, *42*, 265–271. [CrossRef] [PubMed]

33. Tsukamoto, S.; Morita, S.; Hirano, E.; Yokoi, H.; Masumura, T.; Tanaka, K. A Novel cis-Element That Is Responsive to Oxidative Stress Regulates Three Antioxidant Defense Genes in Rice. *Plant Physiol.* **2005**, *137*, 317–327. [CrossRef] [PubMed]

34. Issakidis-Bourguet, E.; Mouaheb, N.; Meyer, Y.; Miginiac-Maslow, M. Heterologous complementation of yeast reveals a new putative function for chloroplast m-type thioredoxin. *Plant J.* **2001**, *25*, 127–135. [CrossRef] [PubMed]

35. Zhang, C.-J.; Zhao, B.-C.; Ge, W.-N.; Zhang, Y.-F.; Song, Y.; Sun, D.-Y.; Guo, Y. An Apoplastic H-Type Thioredoxin Is Involved in the Stress Response through Regulation of the Apoplastic Reactive Oxygen Species in Rice. *Plant Physiol.* **2011**, *157*, 1884–1899. [CrossRef] [PubMed]

36. Jayakannan, M.; Bose, J.; Babourina, O.; Rengel, Z.; Shabala, S. Salicylic acid in plant salinity stress signalling and tolerance. *Plant Growth Regul.* **2015**, *76*, 25–40. [CrossRef]

37. Barba-Espin, G.; Clemente-Moreno, M.J.; Alvarez, S.; Garcia-Legaz, M.F.; Hernandez, J.A.; Diaz-Vivancos, P. Salicylic acid negatively affects the response to salt stress in pea plants. *Plant Biol.* **2011**, *13*, 909–917. [CrossRef] [PubMed]

38. Bernal-Vicente, A.; Cantabella, D.; Hernández, J.A.; Diaz-Vivancos, P. The effect of mandelonitrile, a recently described salicylic acid precursor, on peach plant response against abiotic and biotic stresses. *Plant Biol.* **2018**, *20*, 986–994. [CrossRef] [PubMed]

39. Liu, X.; Zhang, S.; Lou, C. Involvement of nitric oxide in the signal transduction of salicylic acid regulating stomatal movement. *Chin. Sci. Bull.* **2003**, *48*, 449–452. [CrossRef]

40. Arnon, D.I. Copper enzymes in isolated chloroplasts. *Polyphenoloxidase in Beta vulgaris. Plant Physiol.* **1949**, *24*, 1–15. [PubMed]

41. Vivancos, P.D.; Dong, Y.P.; Ziegler, K.; Markovic, J.; Pallardo, F.V.; Pellny, T.K.; Verrier, P.J.; Foyer, C.H. Recruitment of glutathione into the nucleus during cell proliferation adjusts whole-cell redox homeostasis in *Arabidopsis thaliana* and lowers the oxidative defence shield. *Plant J.* **2010**, *64*, 825–838. [CrossRef] [PubMed]

42. Pellny, T.K.; Locato, V.; Vivancos, P.D.; Markovic, J.; De Gara, L.; Pallardo, F.V.; Foyer, C.H. Pyridine Nucleotide Cycling and Control of Intracellular Redox State in Relation to Poly (ADP-Ribose) Polymerase Activity and Nuclear Localization of Glutathione during Exponential Growth of *Arabidopsis* Cells in Culture. *Mol. Plant* **2009**, *2*, 442–456. [CrossRef] [PubMed]

43. Czarnocka, W.; Karpiński, S. Friend or foe? Reactive oxygen species production, scavenging and signaling in plant response to environmental stresses. *Free Radic. Biol. Med.* **2018**, *122*, 4–20. [CrossRef] [PubMed]

10

The *Arabidopsis* Ca^{2+}-Dependent Protein Kinase CPK12 is Involved in Plant Response to Salt Stress

Huilong Zhang [1,†], **Yinan Zhang** [1,†], **Chen Deng** [1,†], **Shurong Deng** [1], **Nianfei Li** [1], **Chenjing Zhao** [1], **Rui Zhao** [1,*], **Shan Liang** [2,*] and **Shaoliang Chen** [1]

[1] Beijing Advanced Innovation Center for Tree Breeding by Molecular Design, College of Biological Sciences and Technology, Beijing Forestry University, Beijing 100083, China; hlzhang2018@126.com (H.Z.); xhzyn007@163.com (Y.Z.); ced501@163.com (C.D.); danceon@126.com (S.D.); nl1669@nyu.edu (N.L.); 1120170396@mail.nankai.edu.cn (C.Z.); lschen@bjfu.edu.cn (S.C.)

[2] Beijing Advanced Innovation Center for Food Nutrition and Human Health, School of Food and Chemical Engineering, Beijing Technology and Business University, Beijing 100048, China

* Correspondence: ruizhao926@126.com (R.Z.); liangshan@btbu.edu.cn (S.L.)

† These authors contributed equally to this work.

Abstract: CDPKs (Ca^{2+}-Dependent Protein Kinases) are very important regulators in plant response to abiotic stress. The molecular regulatory mechanism of CDPKs involved in salt stress tolerance remains unclear, although some CDPKs have been identified in salt-stress signaling. Here, we investigated the function of an *Arabidopsis* CDPK, CPK12, in salt-stress signaling. The *CPK12*-RNA interference (RNAi) mutant was much more sensitive to salt stress than the wild-type plant GL1 in terms of seedling growth. Under NaCl treatment, Na$^+$ levels in the roots of *CPK12*-RNAi plants increased and were higher than levels in GL1 plants. In addition, the level of salt-elicited H$_2$O$_2$ production was higher in *CPK12*-RNAi mutants than in wild-type GL1 plants after NaCl treatment. Collectively, our results suggest that CPK12 is required for plant adaptation to salt stress.

Keywords: *Arabidopsis*; CDPK; ion homeostasis; NMT; ROS; salt stress

1. Introduction

Saline soil cannot be used for agriculture and forestry production [1], and soil salinity is a major abiotic stress for plants worldwide [2,3]. When plants suffer from salt environments, the accumulation of sodium and chloride ions breaks the ion balance and causes secondary stress, such as oxidative bursts [4,5].

Plants have evolved sophisticated regulatory mechanisms to avoid and acclimate to salt stress and repair related damage, processes based on morphological, physiological, biochemical and molecular changes [6]. Salt overly sensitive (SOS) signaling is the most important pathway for regulating plant adaptation to salt stress [4,7]. In *Arabidopsis*, salt-induced increases in cytoplasmic calcium (Ca^{2+}) are sensed by the EF-hand–type Ca^{2+}-binding protein SOS3. Ca^{2+} together with SOS3 activates SOS2, a serine/threonine protein kinase. Activated SOS2 phosphorylates and stimulates the activity of SOS1, a plasma membrane–localized Na$^+$/H$^+$ antiporter, leading to regulation of ion homeostasis during salt stress [8–11]. A Na$^+$/H$^+$ exchanger, which is localized to plasma membrane, also plays an important role in *Populus euphratica*, the roots of which exhibit a strong capacity to extrude Na$^+$ under salt stress; furthermore, the protoplasts from root display enhanced Na$^+$/H$^+$ transport activity [12]. In addition, wheat *Nax1* and *Nax2* affect activity and expression levels of the SOS1-like Na$^+$/H$^+$ exchanger in both root cortical and stellar tissues [13].

Salt stress increases the production of reactive oxygen species (ROS), which plays a dual role in plants: they function as toxic byproducts of metabolism and as important signal transduction molecules [14–16]. Peroxisomes and chloroplasts are the major organelles of ROS generation [17–19], and plants eliminate ROS through non-enzymatic and enzymatic scavenging mechanisms [20]. Non-enzymatic antioxidants include the major cellular redox buffers glutathione and ascorbate, as well as flavonoids, carotenoids, alkaloids, and tocopherol [21]. Enzymatic ROS scavenging pathways in plants include superoxide dismutase (SOD), ascorbate peroxidase (APX), glutathione peroxidase (GPX), and catalase (CAT) [20]. H_2O_2 is the end-product of SOD, which is harmful to DNA, proteins, and lipids [20]. Halophytes can send stress signals quickly through H_2O_2, and have an efficient antioxidant ability to scavenge H_2O_2 upon completion of signaling [22]. In addition, H_2O_2 is a signaling molecule in the plant response to salt stress [23,24]. *P. euphratica* responds to salt stress with rapid H_2O_2 production, and exogenous H_2O_2 application enhances the Na^+/H^+ exchange [25]. Pharmacological experiments have strongly indicated that NaCl-induced Na^+/H^+ antiport is inhibited when H_2O_2 is absent [25], and H_2O_2-regulated K^+/Na^+ homeostasis in the salt-stressed plant is Ca^{2+}-dependent [25]. Exogenous H_2O_2 causes elevated cytosolic Ca^{2+} [25], which stimulates plasma membrane–localized Na^+/H^+ antiporters through the SOS signaling pathway [2,4,5]. Furthermore, H_2O_2 mediates increased *SOS1* mRNA stability in *Arabidopsis* and may therefore contribute to cellular Na^+ protection [26].

Ca^{2+} is a conserved second messenger in plant growth and development pathways and contributes to plant adaptations to environmental challenges [27,28]. In plants, calmodulin (CAM), calcineurin B-like proteins (CBL), and Ca^{2+}-dependent protein kinases (CDPKs) are important Ca^{2+} sensors [29–33]. For CDPKs, *Arabidopsis* has 34 members, rice (*Oryza sativa*) has 29 members, wheat (*Triticum aestivum*) has 20 members, and poplar (*Populus trichocarpa*) has 30 members [34–37]. In recent years, CDPKs have been characterized as playing an important function in mediating stress-signaling networks [38–40].

Genetic and biochemical evidence implicates several CDPKs in plant adaptations to environmental stress. *Arabidopsis* CPK32 (Ca^{2+}-Dependent Protein Kinase 32) phosphorylates ABF4 (ABRE Binding Factor 4) to participate in abscisic acid (ABA) signaling [41]. CPK4 and CPK11 are important positive regulators mediating ABA signaling pathways [42], but their homolog, CPK12, plays a negative role in this signaling [43,44]. CPK10 interacts with HSP1, which contributes to plant drought responses by modulating signaling through ABA and Ca^{2+} [45]. CPK23 responds to drought and salt stresses, and together with CPK21 constitutes a pair of critical Ca^{2+}-dependent regulators of the guard cell anion channel SLAC1 (Slow Anion Channel-Associated 1) in ABA signaling [46–48]. CPK3 and CPK6 positively regulate ABA signaling in stomatal movement [49–51], and CPK6 functions as a positive regulator of methyl jasmonate signaling in guard cells [52]. CPK13 inhibits opening of the stomata through its inhibition of guard cell–expressed KAT2 (K^+ transporter 2) and KAT1 (K^+ transporter 1) channels [53]. The expression of *CPK27* is induced by NaCl, and the *cpk27-1* mutant is much more sensitive to salt stress than wild-type plants in terms of seed germination and post-germination seedling growth [54].

Overexpression of rice *CPK21* enhances rice's capacity to tolerate high salinity [55]. OsCPK12 reduces ROS levels to regulate salt tolerance [56] and plays a positive role in plant responses to drought, osmotic stress, and dehydration [57]. In maize (*Zea mays*), ZmCCaMK is required for ABA-induced antioxidant defense, and the ABA-induced activation of ZmCCaMK is required for H_2O_2-dependent nitric oxide production [58]. ZmCPK4 positively regulates ABA signaling and enhances drought stress tolerance in *Arabidopsis* [59], while ZmCPK11 functions upstream of ZmMPK5 and regulates ABA-induced antioxidant defense [60]. The expression of *PeCPK10*, a gene cloned from *P. euphratica*, is induced by salt, drought and cold treatments, overexpression of *PeCPK10* in *Arabidopsis* improves the plant's tolerance of freezing [61]. In grape berry, ABA stimulates ACPK1 (ABA-stimulated calcium-dependent protein kinase1), which is potentially involved in ABA signaling [62]. Heterologous overexpression of ACPK1 in *Arabidopsis* promotes significant plant growth and enhances ABA sensitivity in seed germination, early seedling growth, and stomatal movement, providing evidence that ACPK1 is involved in ABA signal transduction as a positive regulator [63].

Although the functions of CDPKs in plant response to environmental stress have been demonstrated, the molecular biological mechanisms of CDPKs remain unclear. Previously, we reported that *Arabidopsis* CPK12 negatively regulates ABA signaling [43,44]. Here, we show that CPK12 mediates salt stress tolerance by regulating ion homeostasis and H_2O_2 production. Down-regulation of *CPK12* results in salt hypersensitivity in seedling growth and accumulation of higher levels of Na^+ and H_2O_2. Our results show that CPK12 may modulate salt stress tolerance in *Arabidopsis*.

2. Results

2.1. Identification of RNA Interference (RNAi) Mutants of CPK12

We previously identified the function of *Arabidopsis CPK12*, generated *CPK12*-RNAi lines, and observed that down-regulation of *CPK12* results in ABA hypersensitivity in seed germination and post-germination growth. CPK12 interacted with and phosphorylated and stimulated the type 2C protein phosphatase ABI2. In addition, CPK12 together with ABI2 negatively regulates ABA signal transduction [43]. Thus, we wondered whether *CPK12* was involved in salt-stress signaling. To address this question, we re-generated *CPK12*-RNAi lines and selected four (lines *R1, R4, R7, R8*) as examples for this study. Expression of *CPK12* was down-regulated in these RNAi lines, and the level of *CPK12* mRNA gradually decreased from line *R8* to line *R1*, creating a gradient of *CPK12* expression levels (Figure 1). In addition, the expression of a control gene *EF-1α* (*Elongation Factor-1α*), which is not related to salt-stress, was not affected in those *CPK12*-RNAi lines (Figure 1).

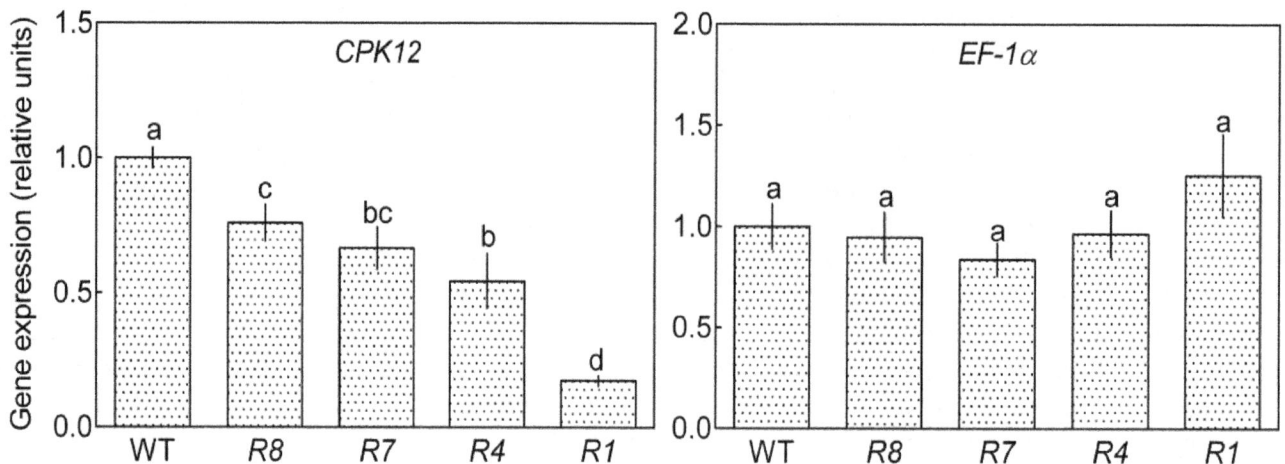

Figure 1. The expression of *CPK12* (*Ca²⁺-Dependent Protein Kinase 12*) and *EF-1α* (*Elongation Factor-1α*) in WT and *CPK12*-RNAi mutant. The mRNA levels (relative units, normalized relative to the mRNA level of the wild-type GL1 taken as 100%) of *CPK12* and *EF-1α*, estimated by qRT-PCR, in the non-transgenic GL1 (WT) and four different transgenic *CPK12*-RNAi lines (indicated by *R8, R7, R4,* and *R1*). Values are mean ± standard error from three independent experiments. Columns labeled with different letters indicate significant differences at $p < 0.05$.

2.2. Down-Regulation of CPK12 Results in NaCl Hypersensitivity in Seedling Growth

We next examined whether CPK12 affects seedling growth under salt stress. Seeds from the *CPK12*-RNAi mutant and GL1 plants were sown on medium containing various concentrations of NaCl. The presence of 110–150 mM NaCl inhibited *CPK12*-RNAi mutant growth. In addition, the cotyledons of the *CPK12*-RNAi mutants were chlorotic compared with GL1 seedlings (Figure 2).

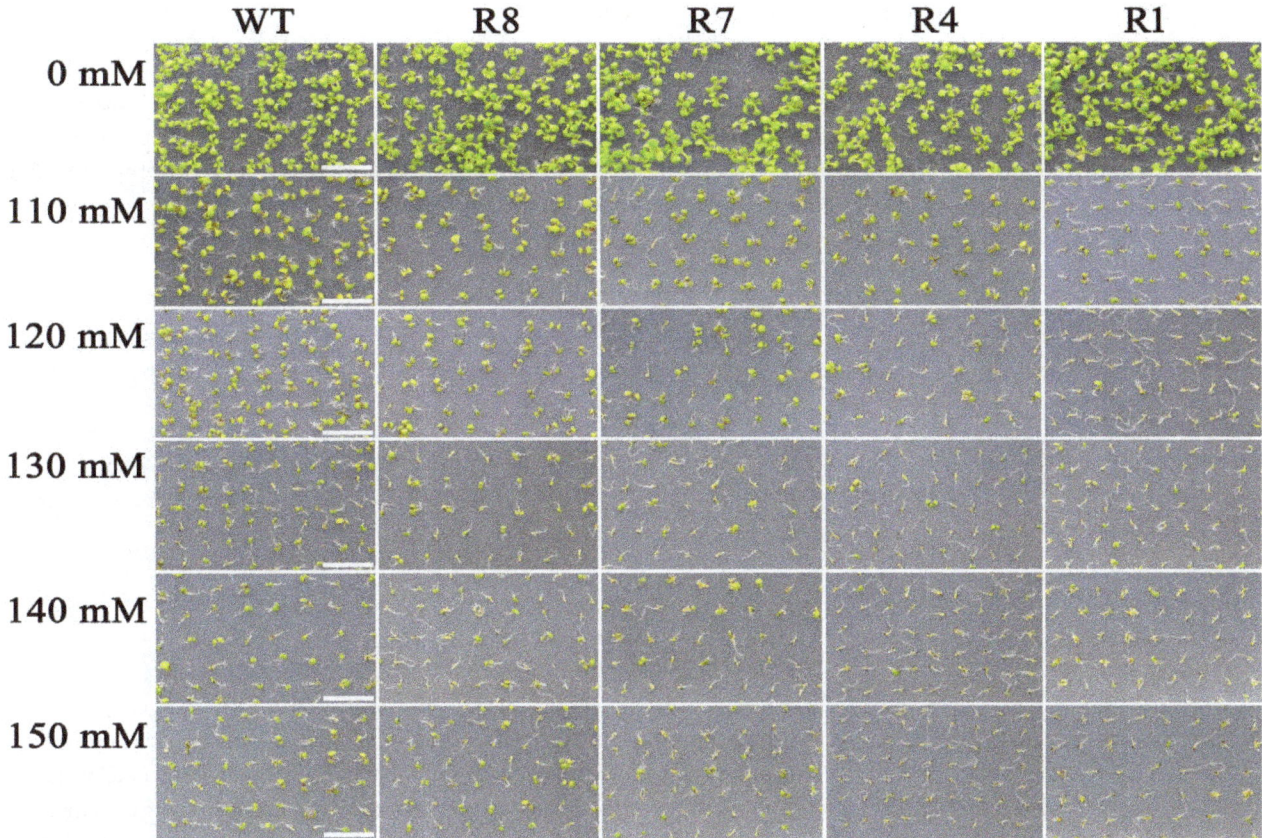

Figure 2. Down-regulation of *CPK12* results in NaCl-hypersensitive seedling growth. Seeds were planted in NaCl-free (0 mM) medium or media containing 110, 120, 130, 140, or 150 mM NaCl, and seedling growth was investigated 10 days after stratification. Scale bars, 1 cm.

CPK12-RNAi mutants exhibited the same post-germination seedling growth status as GL1 plants in the free of NaCl medium; however, compared with GL1 plants, NaCl suppressed root growth of *CPK12*-RNAi mutants more strongly, the reduction in the growth of salt-stressed *CPK12*-RNAi seedlings was more pronounced than for GL1 seedling (Figure 3). Taken together, these results suggest that *CPK12* is involved in salt-stress tolerance in *Arabidopsis*.

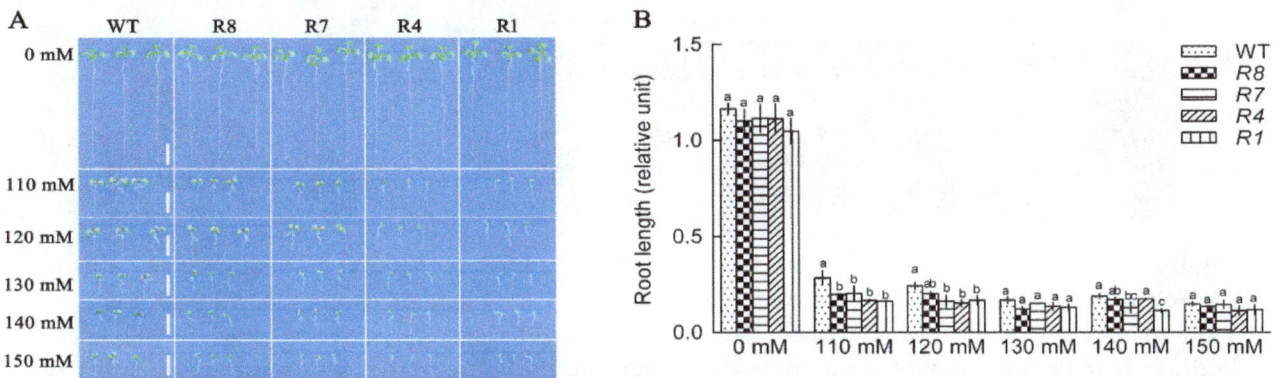

Figure 3. Down-regulation of *CPK12* results in NaCl-hypersensitive root growth. (**A**) Seeds were planted in NaCl-free (0 mM) medium or media containing 110, 120, 130, 140, 150, or mM NaCl, and root growth was detected 10 days after stratification. Scale bars, 0.5 cm. (**B**) Root lengths are mean ± standard error from three independent experiments. Thirty plants were measured for each genotype in each treatment. The mean values of root lengths are labeled with letters in the same group to denote significant differences ($p < 0.05$).

2.3. Salt Stress Induced the Ca²⁺ Elevation in Root Tissue

We examined the Ca^{2+} level in the *CPK12*-RNAi plants and wild type plants GL1 after salt stress using the Ca^{2+} specific probe, Rhod-2 AM. In the absence of salt treatment, the relative fluorescence intensity was not significantly different between GL1 and *CPK12*-RNAi plants, except the R1 line, probably due to the lowest expression of *CPK12* in R1 line. Under salt treatment, as expected, the Ca^{2+} levels in the roots of *CPK12*-RNAi plants and GL1 increased (Figure 4).

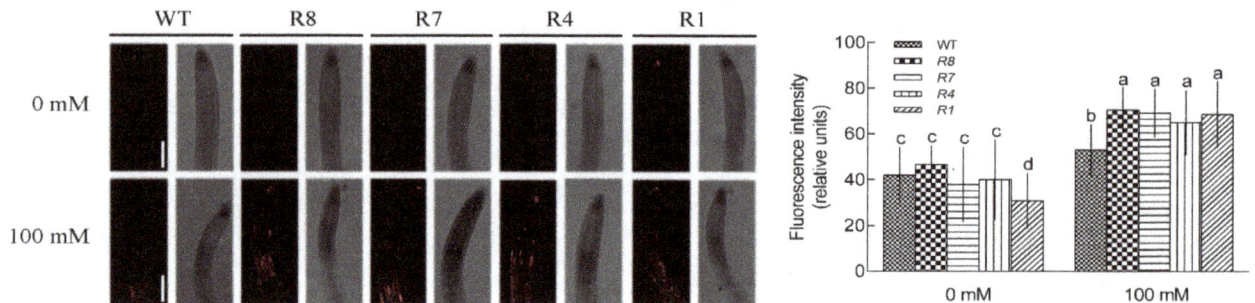

Figure 4. Ca^{2+} levels within roots of GL1 and *CPK12*-RNAi plants. (**A**) Seven-day-old seedlings were transferred to MS medium supplemented with (100 mM) or without NaCl (0 mM) for 12 h, then stained with the Ca^{2+}-specific fluorescent probe Rhod-2 AM for 1 h at room temperature. Orange-red fluorescence within cells was detected at the apical region of roots under a Leica confocal microscope. Representative confocal images show cytosolic Ca^{2+} content in plant roots. Scale bars, 100 μm. (**B**) The relative fluorescence intensity (±SD) represents the mean of 10 independent seedlings. The mean values of Ca^{2+} fluorescence are labeled with letters in the same group to denote significant differences ($p < 0.05$).

2.4. Down-Regulation of CPK12 Leads to Na⁺ Accumulation in Root Tissue

To investigate the cause of the observed hypersensitivity of *CPK12*-RNAi mutants to salt stress, sodium accumulation in root cells was examined using the sodium-specific dye CoroNa-Green. Under no salt stress, CoroNa-Green fluorescence was almost undetectable in the root cells of *CPK12*-RNAi and wild-type plants because of low Na^+ content in root cells. Under NaCl treatment, Na^+ levels in the roots of *CPK12*-RNAi plants increased and were higher than levels in GL1 plants (Figure 5).

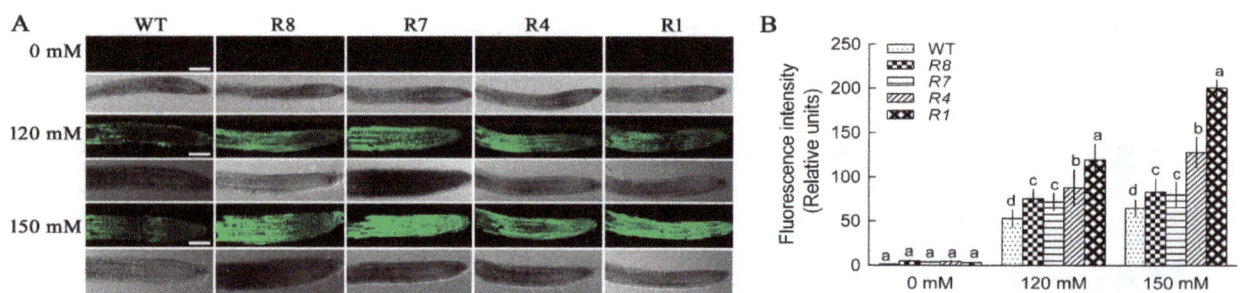

Figure 5. Na^+ levels in root cells of wild-type (WT) GL1 and *CPK12*-RNAi plants under salt stress. (**A**) Seven-day-old seedlings were transferred to MS medium supplemented with (120 mM, 150 mM) or without NaCl (0 mM) for 12 h, then seedlings were treated with CoroNa-Green AM (green fluorescence, sodium-specific) for 1 h. Green fluorescence in root cells was observed at the apical region of roots using a Leica confocal microscope. Typical images show Na^+ content in plant roots. Scale bars, 100 μm. (**B**) The mean relative fluorescence values marked with letters in the same group represent significant differences ($p < 0.05$).

To determine whether CPK12 contributed to the regulation of ion homeostasis, NMT (Non-invasive Micro-test Technique) was used to record root Na^+ fluxes in the *CPK12*-RNAi plants and GL1 plants under a long-term NaCl treatment (0, 110, 120, 130 mM; 7 d). In the absence of salt stress, Na$^+$ efflux in the apical region of roots was not significant between *CPK12*-RNAi plants and wild type GL1 plants. However, long-term salt treatment caused a rise in Na$^+$ efflux in GL1 plants and *CPK12*-RNAi mutants, but this was more pronounced in GL1 plants than in *CPK12*-RNAi plants (Figure 6). These observations indicate that CPK12 participates in salt-stress tolerance by regulating ion balance in root tissue.

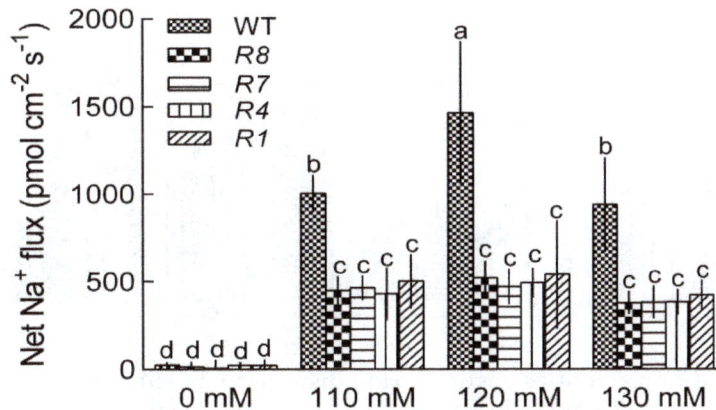

Figure 6. Na$^+$ flux in GL1 and *CPK12*-RNAi plants. Seeds were germinated for one week in a vertical direction on MS agar medium containing 0, 110, 120, 130 mM NaCl. Continuous NMT recording were applied at the meristem region of the root tips. Each column is the mean of six independent seedlings; bars show the standard error of the mean. Columns marked with letters in the same group indicate significant differences at $p < 0.05$.

2.5. Down-Regulation of CPK12 Results in H$_2$O$_2$ Burst and Accumulation

ROS accumulates when plants are exposed to salt stress, so we investigated H$_2$O$_2$ levels in *CPK12*-RNAi plants using a H$_2$O$_2$-specific fluorescent probe, H$_2$DCF-DA. In the NaCl shock condition, the levels of H$_2$O$_2$ in *CPK12*-RNAi plants were higher than GL1 plants (Figure 7). The salt stress-induced H$_2$O$_2$ accumulation in *Arabidopsis* was also detected after 12 h or 24 h treatment; compared with GL1 plants, the level of H$_2$O$_2$ in *CPK12*-RNAi plants was significantly higher after exposure to high NaCl concentrations (Figure 7).

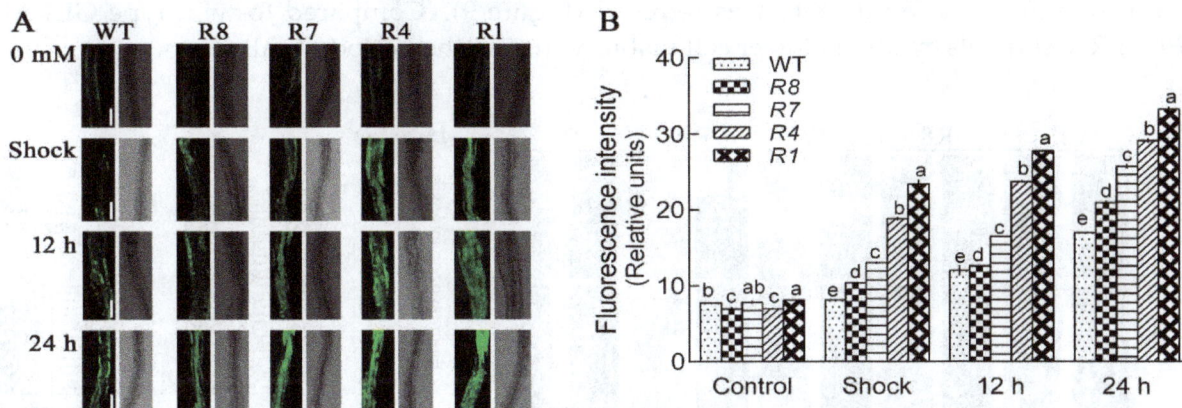

Figure 7. Accumulation of H$_2$O$_2$ in the root tips of GL1 and *CPK12*-RNAi plants exposed to salt stress. (**A**) After germinating seven days, the *Arabidopsis* seedlings were transferred to MS medium containing 0 or 100 mM NaCl for 10 min, 12 h, or 24 h. These seedlings were incubated with H$_2$DCF-DA for 5 min. The green fluorescence within cells at the apical region of roots was detected using a Leica confocal microscope. Scale bars, 100 μm. (**B**) The relative fluorescence intensity ($\pm$SD) represents the mean of 10 *Arabidopsis* seedlings. The mean values of H$_2$O$_2$ fluorescence are labeled with letters in the same group to denote significant differences ($p < 0.05$).

We measured the activities of antioxidant enzymes, such as SOD, CAT, and APX, in GL1 and CPK12-RNAi plants. In the absence of salt treatment, the activities of SOD and CAT were not significantly different between GL1 and CPK12-RNAi plants, but the activity of APX in CPK12-RNAi plants was lower than GL1. Under salt stress conditions, the activity of SOD in GL1 was higher than CPK12-RNAi plants, but CAT and APX were lower in CPK12-RNAi plants, when compared with GL1 (Figure 8). Taken together, these data imply that CPK12 is involved in the elimination of H_2O_2 under salt stress.

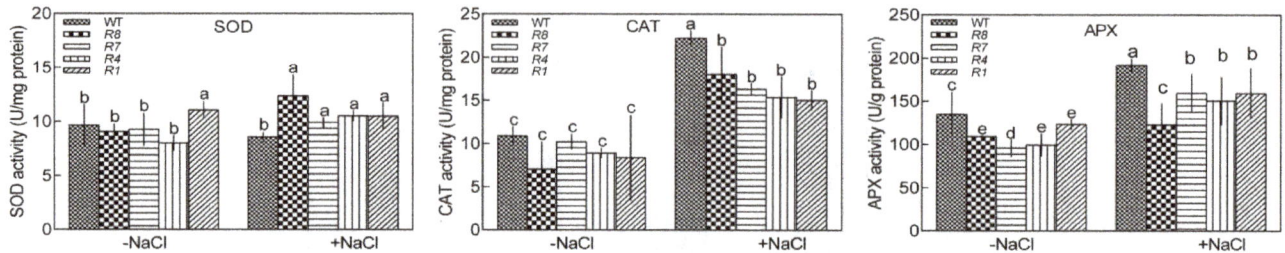

Figure 8. Effect of NaCl on activities of superoxide dismutase (SOD), catalase (CAT), and ascorbate peroxidase (APX) in wild-type (GL1) and CPK12-RNAi lines. Seven-day-old seedlings were transferred to MS medium supplemented with or without 100 mM NaCl for 10 d. The activities of antioxidant enzymes were analyzed. Each column shows the mean of three replicated experiments and bars represent the standard error of the mean. Columns labeled with letters in the same group denote significant difference at $p < 0.05$.

2.6. Down-Regulation of CPK12 Suppressed Cell Viability in Arabidopsis Roots

Previous studies showed that a high level of NaCl could reduce viability and increase programmed cell death in plants [64–67]. Cell viability was assayed with fluorescein diacetate (FDA) to determine whether salt stress could induce cell death in CPK12-RNAi plants. FDA staining showed an effect of salt treatment on cell viability in the elongation zone of roots. Wild-type GL1 and CPK12-RNAi plants grown in control conditions (NaCl-free Murashige–Skoog (MS) medium) showed clear FDA fluorescence with the cytoplasm of root cells, features indicating that the cells were viable. However, in salt-stressed CPK12-RNAi plants, the FDA fluorescence was undetectable in a number of root cells, and the fluorescence intensity was reduced (Figure 9). Compared to wild-type GL1 plants, the CPK12-RNAi plants exhibited lower cell viability during the period of salt stress.

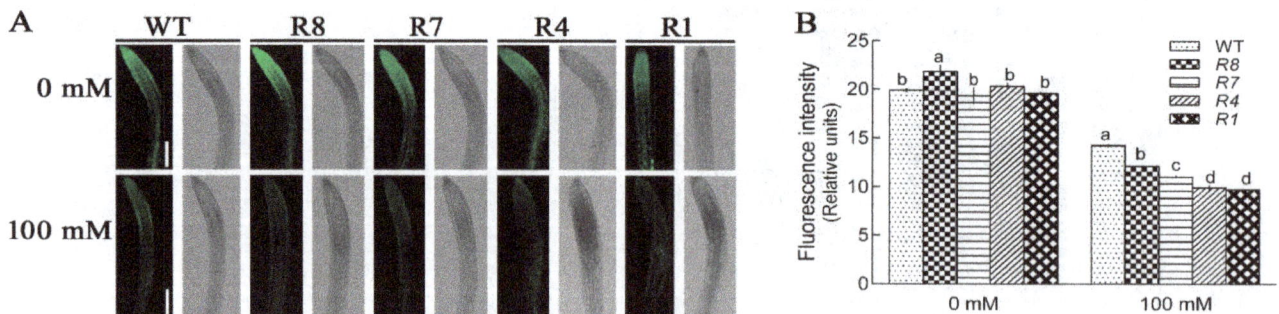

Figure 9. Effect of salt stress on cell viability in wild-type (GL1) and CPK12-RNAi lines. (**A**) Seven-day-old seedlings were transferred to MS medium supplemented with or without 100 mM NaCl for 12 h. Cell viability was assayed with fluorescein diacetate (FDA, green) stain. Representative images of apical region of roots are shown. Scale bars, 100 μm. (**B**) The fluorescence intensity (±SD) represents the mean of 10 independent seedlings. The mean values of FDA fluorescence are labeled with letters in the same group to denote significant differences ($p < 0.05$).

2.7. Down-Regulation of CPK12 Alters the Expression of Some Salt-Responsive Genes

We tested the expression of the following salt-related genes in the GL1 and *CPK12*-RNAi plants: *SOS1*, *SOS2*, and *SOS3* [4,7–11], *AHA1* and *AHA2* [68], *PER1* [69,70], *SOD*, *CAT*, and *APX*. In the absence of salt treatment, the expression of *AHA1* was not significantly difference between wild-type GL1 plants and *CPK12*-RNAi plants, but the expression of *SOS1*, *SOS2*, *SOS3 AHA2*, and *PER1* was down-regulated in *CPK12*-RNAi plants (Figure 10). Under salt stress, down-regulation of *CPK12* did not affect expression of *AHA2* and *PER1*, but significantly reduced expression of *SOS1* in *R8*, *R7*, and *R4* plants, *SOS2* in *R7*, *R4*, and *R1* plants, *AHA1* in *R8*, *R7*, *R4*, and *R1* plants (Figure 10). In the absence of salt treatment, the expression of *SOD* and *CAT* was down-regulated in *CPK12*-RNAi plants, and *APX* was down-regulated in *R8*, *R7*, and *R4* plants. Under salt stress, the expression of *APX* was down-regulated in *R1*. It is interesting that the expression of *SOD*, *CAT*, and *APX* was nearly undetectable in *R8* and *R7* plants, whether under salt or no-salt stress (Figure 10).

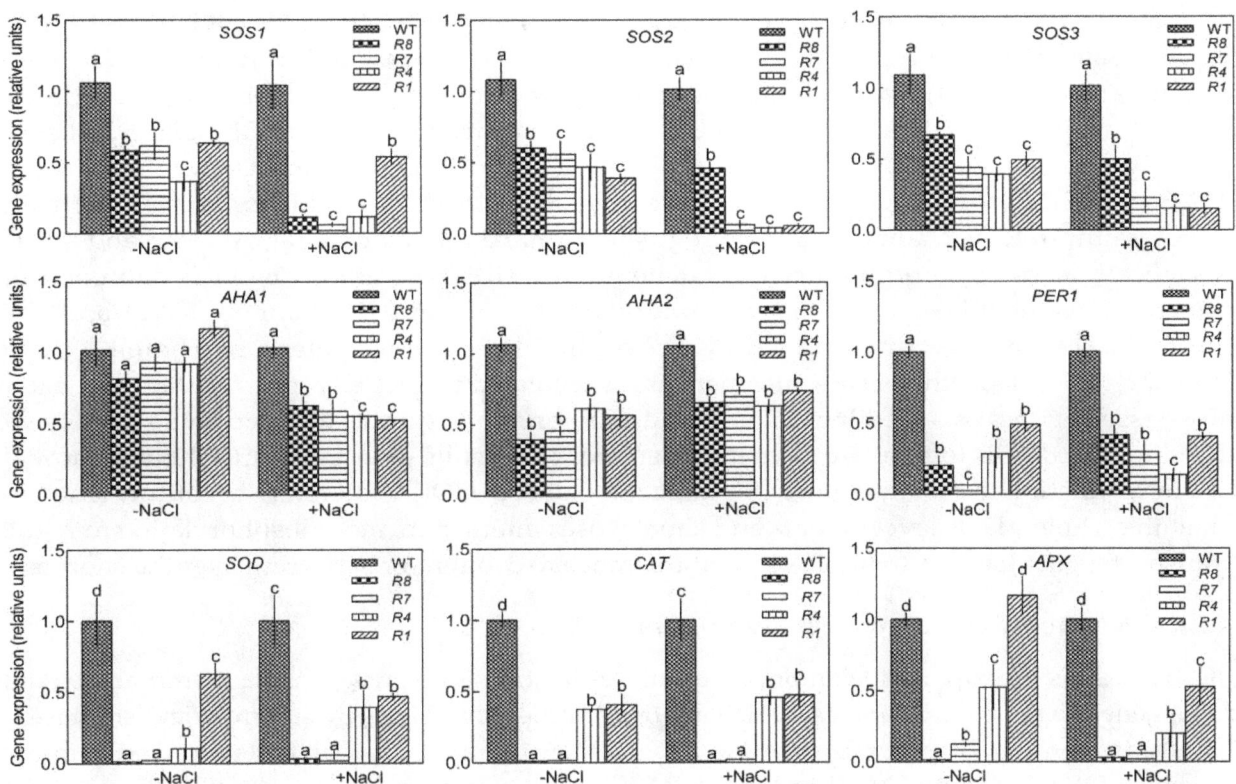

Figure 10. Changes in *CPK12* expression alter expression of a subset of genes involved in salt stress responses. The mRNA levels in the seedling of wild-type GL1, *CPK12*-RNAi mutants were determined by qRT-PCR. One-week-old seedlings were transferred to MS medium with and without the addition of 100 mM NaCl for ten days. The expression of salt stress responsive genes was analyzed. The gene expression levels were normalized relative to the value of the GL1 plants. Each value is the mean of the three independent determinations; columns labeled with letters in the same group indicate significant differences ($p < 0.05$).

3. Discussion

3.1. CPK12 Is Involved in Salt Stress Tolerance in Plants

In this investigation, *Arabidopsis* CPK12 was identified and characterized as a regulatory component involved in salt tolerance in terms of seedling growth. Previously we reported that CPK12 negatively regulates ABA signal transduction [43,44]. These results together imply an important function of CPK12 in regulating plant salt stress tolerance. Of note, although the expression of some salt-related genes were down-regulated in *CPK12*-RNAi plants without salt stress treatment (except *AHA1*), down-regulated *CPK12* did not influence post-germination seedling growth, and the expression of a control gene *EF-1α* that is not related to salt-stress, indicating that CPK12 is involved in salt stress signal transduction, but is not related to seedling development. Previously we reported that CPK27's function in salt stress tolerance [54], although CPK12 and CPK27 had similar functions. Our results suggest that the function of CPK12 may not be redundant for CPK27, because independent down-regulation of CPK12 or CPK27 can change the responses of *Arabidopsis* seedlings to salt stress. Current evidence suggests that CDPKs regulate plant tolerance to abiotic stress through ABA and Ca^{2+} pathways. For example, CPK10 regulates plants response to drought stress through ABA- and Ca^{2+}-mediated stomatal movements [45]; CPK4 and CPK11 are positive regulators in Ca^{2+}-mediated ABA signaling [42]. Thus, CPK12 and other CDPK members constitute a complicated regulation network, which functions in plant adaptation to salt and drought stresses.

3.2. CPK12 Regulates Na⁺ Balance in Salt-Stressed Plants

The ability to retain ion balance is very important for plant survival in saline environments [1,5,71]. Here, under salt stress, wild-type plants GL1 and *CPK12*-RNAi absorbed and accumulated Na^+ in roots; when compared with GL1, Na^+ accumulation was significantly higher in the roots of *CPK12*-RNAi plants, and net Na^+ efflux was reduced in *CPK12*-RNAi roots compared to GL1 plants. There are many CDPKs that can interact and regulate the activity of ion transporters. Under drought stress, AtCPK23 phosphorylates the guard cell anion channel SLAC1, which is collaborated with activation of the potassium-release channel GORK to regulate stomatal movement [47]. AtCPK3 and AtCPK6 are specifically expressed in guard-cell, regulate guard cell S-type anion channels and contribute to stomatal movement [49]. AtCPK13 phosphorylates two inward K^+ channels, KAT1 and KAT2, to restrict the stomatal aperture, and AtCPK3 phosphorylates and activates a two pore K^+ channel TPK1 [53,72]. Thus, like other CDPK members, it is speculated that CPK12 may regulate Na^+ balance in salt-stressed *Arabidopsis* seedlings. Compared with previous studies wherein some CDPKs were localized in guard cells to regulate stomatal aperture, our results give a new insight and show that CPK12 may function in root systems to regulate ion balance. CDPKs therefore constitute a network, which at the whole-plants level in roots and shoots, uses interaction and phosphorylation to regulate ion transporters or channels to improve plant tolerance to drought or salt stress over the short term.

3.3. CPK12 Regulates ROS Homeostasis in Salt-Stressed Plants

ROS, such as the superoxide anion, accumulates in stressed plants. Plasma membrane NADPH oxidases generate superoxide anions, which are transformed into H_2O_2 by superoxide dismutase [20]. In this study, compared with GL1 plants, *CPK12*-RNAi plants accumulated more H_2O_2 in roots, irrespective of the duration of NaCl treatment. Without salt-stress, compared with GL1, the expression of *CAT* gene was down-regulated in *CPK12*-RNAi plants, but the activity of CAT was not reduced, and the level of H_2O_2 was not significantly different between GL1 and *CPK12*-RNAi plants. Furthermore, under salt stress treatment, the activity of SOD was higher in *CPK12*-RNAi plants than GL1, but the activity of CAT was lower in *CPK12*-RNAi plants than GL1. The enhanced activity of SOD in *CPK12*-RNAi plants results in H_2O_2 production, but the reduced activity of CAT lead to H_2O_2 accumulation; these results suggest that down-regulated *CPK12* cannot scavenge NaCl-induced H_2O_2 bursts. Previous studies showed that down-regulated expression of CDPKs, which are related to antioxidases, may cause the accumulation of H_2O_2. *Arabidopsis cpk27-1* mutants accumulate more H_2O_2 in roots [54]. Potato StCDPK4 and StCDPK5 regulate the production of ROS [73]. In rice, *OsCPK12*-OX plants accumulate less H_2O_2 under conditions of high salinity, and this accumulation is more pronounced in *oscpk12* mutants and *OsCPK12* RNAi plants [56]. Similarly, excess H_2O_2 leads to oxidative damage and growth inhibition in *CPK12*-RNAi plants under salinity conditions. In contrast, *Arabidopsis cpk5 cpk6 cpk11 cpk4* quadruple mutants harbor decreased ROS content, suggesting that these CDPKs regulate ROS production potentially by phosphorylating the NADPH oxidase RBOHB [74]. Therefore, CDPKs play key roles in regulating ROS production and accumulation in plants [75].

3.4. CPK12 Functions with Potential Substrates in Salt Stress Signaling

In recent years, many substrates of CDPKs were identified. The transcription factor ABF4 is a substrate of CPK4/11 in *Arabidopsis* [42]. CPK12 can phosphorylate type-2C protein phosphatase ABI2 [43]. HSP1 interacts with CPK10 [45], SLAC1 is an interacting partner of CPK21 and CPK23 [47], and CPK13 specifically phosphorylates KAT2 and KAT1 [53]. In nutrient signaling, CPK10, CPK30, and CPK32 could potentially phosphorylate and activate all NLPs and possibly other transcription factors with overlapping or distinct target genes to support transcriptional, metabolic, and system-wide nutrient-growth regulations [76]. Potato CDPK5 phosphorylates the N-terminal region of plasma membrane RBOH (respiratory burst oxidase homolog) protein and participates in RBOHB-mediated ROS bursts, conferring resistance to near-obligate pathogens but increasing susceptibility to necrotrophic pathogens [77]. In this work, CPK12 was involved in plant adaptation to salt stress by regulating Na^+ and H_2O_2 homeostasis. These results indicate that CPK12 may interact with and phosphorylate several salt stress-related proteins as potential substrates in its regulatory function. To deeply demonstrate the regulatory mechanism of CPK12, the downstream components of CPK12 need to be identified, and their relationship with the whole complex CDPK regulation network need to be elucidated. Although the functions of CDPKs are widely identified in recent years, the complete CDPK signal transduction pathway is still not clearly illustrated, as the CDPK transduction network is very complex. Future progress is likely to identify sensors, channels, and other regulators involved in generating complex Ca^{2+} signatures in plant responses to hormones and environmental cues.

4. Materials and Methods

4.1. Plant Materials, Constructs, and Arabidopsis Transformation

Arabidopsis thaliana GL1 (Col-5) was used in this work for generating the *CPK12*-RNAi plants. A specific 242-bp fragment of *CPK12* (At5g23580) corresponding to the region of nt 6 to 247 was amplified with forward primer 5′-ACGCGTCGACGAACAAACCAAGAACCAGATGGGTT-3′ and reverse primer 5′-CCGCTCGAGCGTTGGGGTATTCAGACAAGTGATG-3′. This fragment was inserted into the pSK-int vector, which was digested with the *Xho*I and *Sal*I. The same fragment, amplified with forward primer 5′-AACTGCAGGAACAAACCAAGAACCAGATGGGTT-3′ and reverse primer 5′-GGACTAGTCGTTGGGGTATTCAGACAAGTGATG-3′, was inserted into the pSK-int carrying the previous fragment. The entire RNAi cassette linked with actin 11 intron was excised from pSK-int vector, and inserted into the *Sac*I *Apa*I digested vector pSUPER1300(+) [78]. The construct was introduced into *Agrobacterium tumefaciens* GV3101 and transformed into GL1 by the floral dip method [79]. Transgenic plants were grown on MS agar plates containing hygromycin (50 µg/mL) to screen for positive seedlings. The homozygous T3 seeds of the transgenic plants were used for further analysis. Plants were grown in a growth chamber at 20–21 °C on MS medium at about 80 µmoL photons $m^{-2} \cdot s^{-1}$, or in compost soil at about 120 µmoL photons $m^{-2} \cdot s^{-1}$ over a 16-h photoperiod.

4.2. qRT-PCR Analysis

To assay the gene expression in the transgenic plants, quantitative real-time PCR analysis was performed with the RNA samples isolated from 10-day-old seedlings. For *CPK12*, qRT-PCR amplification was performed with forward primer 5′-CGAAACCCTCAAAGAAATAA-3′ and reverse primer 5′-TGGTGTCCTCGTACGCACTCTC-3′. The primers specific for salt-related genes were: forward 5′-CACAAACATTTACCGAAAACCA-3′ and reverse 5′-CAAATTTGCAAAGCTCATATCG-3′ for *AHA1* (At2g18960); forward 5′-TGACTGATCTTCGATCCTCTCA-3′ and reverse 5′-GAGAATGT GCATGTGCCAAA-3′ for *AHA2* (At4g30190); forward 5′-CGTGCCCTTCATATTGTTGG-3′ and reverse 5′-GACGCCATCAACAACGAGTC-3′ for *PER1* (At1g48130); forward 5′-GTGAAGCAATCAAGC GGAAA-3′ and reverse 5′-TGCGAAGAAGGCGTAGAACA-3′ for *SOS1* (At2g01980); forward 5′-GCGAACTCAATGGGTTTTAAGT-3′ and reverse 5′-CTTACGTCTACCATGAAAAGCG-3′ for *SOS2*

(At5g35410); forward 5′-CCGGTCCATGAAAAAGTCAAAT-3′ and reverse 5′-CTCTTTCAATTCTT
CTCGCTCG-3′ for *SOS3* (At5g24270); forward 5′-AGGAAACATCACTGTTGGAGAT-3′ and reverse
5′-GAGTTTGGTCCAGTAAGAGGAA-3′ for *SOD* (At1g08830); forward 5′-AGGATCAAACTTT
GAGGGGTAG-3′ and reverse 5′-CTTGTGGTTCCTGGAATCTACT-3′ for *CAT* (At1g20620); forward
5′-GATGTCTTTGCTAAGCAGATGG-3′ and reverse 5′-GAGTTGTCGAAGATTAGAGGGT-3′ for *APX*
(At1g07890); forward 5′-CACCACTGGAGGTTTTGAGG-3′ and reverse 5′-TGGAGTATTTGGGG
GTGGT-3′ for *EF-1α* (At5g60390). Amplification of *ACTIN2/8* (forward primer 5′-GGTAACATTGTG
CTCAGTGGTGG-3′, reverse primer 5′-AACGACCTTAATCTTCATGCTGC-3′) gene was used as an
internal control.

4.3. Phenotype Identification

For the seedling growth experiment, seeds were germinated after stratification on MS medium
supplemented with various concentrations of NaCl. Seedling growth was examined 10 days
after stratification.

4.4. Measurement of Cytosolic Ca^{2+} Concentrations

For cytosolic Ca^{2+} concentration analysis, the Ca^{2+}-specific fluorescent probe Rhod-2 AM
(Invitrogen, Carlsbad, CA, USA) was used to measure the concentration of Ca^{2+} as previously
described [80]. Briefly, CPK12-RNAi mutants and GL1 seedlings were treated with MS liquid solution
supplemented with or without 100 mM NaCl for 12 h. Then, control and salinized plants roots
were 2 μM Rhod-2 AM (prepared in MS liquid solution, pH 5.8) incubated in the dark for 1 h at
room temperature. Then, the Arabidopsis plants were washed four to five times with distilled water.
The image of Ca^{2+} fluorescence in the probe-loaded roots was measured with a Leica SP5 confocal
microscope (Leica. Microsystems GmbH, Wetzlar, Germany), with emission at 570–590 nm and
excitation at 543 nm [80].

4.5. Detection of Cytosolic Na^+ Concentrations

Root cellular Na^+ levels were detected with Na^+-specific fluorescent probe, CoroNa™ Green,
AM (Invitrogen, Carlsbad, CA, USA). Seedlings of wild-type GL1 and *CPK12*-RNAi mutants were
treated with 0, 120, or 150 mM NaCl in MS liquid solution for 12 h. Then, control and salinized
seedlings were incubated with CoroNa in the dark for 1 h, and washed 3–4 times with distilled
water subsequently. Na^+ fluorescence was observed with a Leica SP5 confocal microscope (excitation:
488 nm; emission: 510–530 nm, Microsystems GmbH, Wetzlar, Germany) [25,80,81]. ImageJ software
(Version 1.48, National Institutes of Health, Bethesda, MD, USA) was used to quantify relative
fluorescence intensity. It is worth noting that there are several commercial Na^+ specific probes;
for example, SBFI, Sodium Green, CoroNa etc., Sodium Green displays a modest fluorescence increase
in response to Na^+ binding [82,83], while CoroNa is more suitable for detecting Na^+ in a wider range
of concentrations; and the selectivity of CoroNa is 4 times higher to Na^+ than to K^+ binding [82], but
CoroNa is not suitable for the detection of relatively low Na^+ changes in cells [83]. In this work, after
NaCl treatment, the roots absorbed and accumulated high levels of Na^+, and Na^+ efflux was inhibited.
Our previous studies showed that CoroNa is suitable for detecting Na^+ level after NaCl treatment in
tobacco, *Arabidopsis*, and *Glycyrrhiza uralensis* [84–86]; thus, we selected CoroNa Green to detect the
cytosolic Na^+ level in this work.

4.6. Net Fluxes Measurements of Na^+

Net Na^+ flux was measured using the NMT technique (NMT-YG-100, Younger, Amherst,
Massachusetts, USA) as described previously [12,80]. One-week-old seedling grown on MS medium
containing 0, 110, 120, 130 mM NaCl was washed 4–5 times with ddH_2O and transferred to the
measuring chamber containing 10–15 mL measuring solution, which included 0.1 mM NaCl, 0.5 mM
KCl, 0.1 mM $CaCl_2$, 0.1 mM $MgCl_2$, and 2.5% sucrose, pH 5.8. After the roots were immobilized to

the bottom of the chamber, Na flux measurements were started at 200–300 μm from the root apex. The Na^+ flux was continuously recorded for 17–20 min. The Na^+ flux was detected by shifting the ion-selective microelectrode between two sites close the roots over a preset length (30 μm for intact roots in this experiment) at a frequency in the range of 0.3–0.5 Hz. The electrode was stepped from one site to another in a predesigned sampling routine, while the sample was also scanned with a 3-D microstepper motor manipulator (CMC-4). Pre-pulled and salinized glass micropipettes (4–5 μm aperture, XYPG120-2; Xuyue Sci. and Tech. Co., Ltd., Beijing, China) were processed with a backfilling solution (Na: 250 mM NaCl) to a length of approximately 1 cm from the tip, then front-filled with about 10 μm columns of selective liquid ion-exchange cocktails (Na: Fluka 71178). An Ag/AgCl wire electrode holder (XYEH01-1; Xuyue Sci. and Tech. Co., Ltd., Beijing, China) was used to make electrical contact with the electrolyte solution. The reference electrode was DRIREF-2 (World Precision Instruments, www.wpiinc.com). Ion-selective electrodes were calibrated prior to flux measurements (The concentration of Na^+ was usually 0.1 mM in the measuring buffer for root samples). Na^+ flux was calculated by Fick's law of diffuseon: $J = -D(dc/dx)$, where J represents the Na^+ flux in the x direction, dc/dx is the Na^+-concentration gradient, and D is the Na^+ diffusion constant.

4.7. H_2O_2 Production with Root Cells

A specific fluorescent probe, 2′,7′-dichlorodihydrofluorescein diacetate (H_2DCF-DA; Molecular Probes) was used for H_2O_2 detection in the roots of GL1 plants and *CPK12*-RNAi plants. Shock and short-term responses of H_2O_2 to NaCl exposure were examined in this study.

Seven-day-old seedlings (GL1 and *CPK12*-RNAi) grown on MS medium were exposed to 0 or 100 mM NaCl for 10 min, 12 h, and 24 h and then incubated with 10 μM H_2DCF-DA (prepared in liquid MS medium, pH 5.7) for 5 min at room temperature in the dark. The H_2DCF-DA-loaded seedlings were washed 3–4 times with liquid MS solution. DCF-specific fluorescence was examined under a Leica SP5 confocal microscope (Leica Microsystem GmbH, Wetzlar, Germany), with excitation at 488 nm and emission at 510–530 nm. Relative H_2DCF-DA fluorescence intensities in root cells were measured with ImageJ 1.48 (National Institutes of Health, Bethesda, MD, USA).

4.8. Activity Analyses of Antioxidant Enzymes

For the detection of antioxidant enzyme activities, after salt treatment for ten days, the *Arabidopsis* seedling (0.2 g) was ground to a fine powder in liquid nitrogen, and then 2 mL ice-cold 50 mM potassium phosphate buffer (pH 7.0) was added. After centrifugation at 12,000 g for 20 min, the supernatant was used to detect the enzymatic activities of the antioxidant enzymes SOD, CAT, and APX. The activities of antioxidant enzymes were determined using commercial kits (Nanjing Jiancheng Bioengineering Institute, Nanjing, China). SOD activity was measured using the Superoxide Dismutase WST-1 Assay Kit, which, based on the xanthine/xanthine oxidase method, depended on the production of O^{2-} anions. CAT activity was measured by analyzing the yellowish complex produced by the reaction between H_2O_2 and ammonium molybdate and calculating CAT activity by measuring OD value at 405 nm. APX activity was estimated based on the reaction of ASA with H_2O_2 to oxidize ASA to MDASA; APX activity was calculated by measuring the reduced OD value at 290 nm. Activities of SOD and CAT are expressed as units per milligrams of protein (U/mg protein). The activity of APX is expressed as (U/g protein).

4.9. Cell Viability Analyses

Seven-day-old seedlings (GL1 and *CPK12*-RNAi) grown on MS medium were exposed to 0 or 100 mM NaCl for 12 h, and cell viability was measured by staining seedlings with FDA (Invitrogen, Carlsbad, CA, USA). The confocal parameters were set as described in previous studies: the excitation wavelength was 488 nm, and the emission wavelengths were 505 to 525 nm. Relative FDA fluorescence intensities in root cells were measured with ImageJ 1.48 (National Institutes of Health, Bethesda, MD, USA).

4.10. Data Analysis

All experimental data were analyzed with SPSS version 17.0 software (IBM China Company Ltd., Beijing, China) for statistical evaluations. Statistical analysis were performed using one-way ANOVA. Differences were considered significant at $p < 0.05$, unless otherwise stated.

Author Contributions: Conceptualization, R.Z., S.L. and S.C.; Investigation, H.Z., Y.Z., C.D., N.L. and C.Z.; Resources, R.Z. and S.D.; Writing—original draft preparation, H.Z., R.Z. and S.L.; Writing—review and editing, H.Z., R.Z. and S.L.; All authors discussed the results and comments on the manuscript.

Abbreviations

ABA	Abscisic Acid
APX	Ascorbate Peroxidase
CAT	Catalase
CDPK	Ca^{2+}-Dependent Protein Kinases
FDA	Fluorescein Diacetate
H_2DCF-DA	$2',7'$-Dichlorodihydrofluorescein Diacetate
HSP	Heat Shock Protein
MS	Murashige–Skoog Medium
NADPH	Nicotinamide Adenine Dinucleotide Phosphate
NLP	NIN-Like Protein
NMT	Non-Invasive Micro-Test Technique
qRT-PCR	Quantitative Reverse Transcription PCR
RBOHB	Respiratory Burst Oxidase Homolog Protein B
ROS	Reactive Oxygen Species
SOD	Superoxide Dismutase
SOS	Salt overly sensitive

References

1. Polle, A.; Chen, S. On the salty side of life: Molecular physiological and anatomical adaptation and acclimation of trees to extreme habitats. *Plant Cell Environ.* **2015**, *38*, 1794–1816. [CrossRef] [PubMed]
2. Zhu, J.K. Plant salt tolerance. *Trends Plant Sci.* **2001**, *6*, 66–71. [CrossRef]
3. Janz, D.; Polle, A. Harnessing salt for woody biomass production. *Tree Physiol.* **2012**, *32*, 1–3. [CrossRef] [PubMed]
4. Zhu, J.K. Salt and drought stress signal transduction in plants. *Annu. Rev. Plant Biol.* **2002**, *53*, 247–273. [CrossRef] [PubMed]
5. Zhu, J.K. Regulation of ion homeostasis under salt stress. *Curr. Opin. Plant Biol.* **2003**, *6*, 441–445. [CrossRef]
6. Acosta-Motos, J.R.; Ortuño, M.F.; Bernal-Vicente, A.; Diaz-Vivancos, P.; Sanchez-Blanco, M.J.; Hernandez, J.A. Plant responses to salt stress: Adaptive mechanisms. *Agronomy* **2017**, *7*, 18. [CrossRef]
7. Xiong, L.; Zhu, J.K. Salt tolerance. *Arabidopsis Book* **2002**, *1*, e0048. [CrossRef]
8. Qiu, Q.S.; Guo, Y.; Dietrich, M.A.; Schumaker, K.S.; Zhu, J.K. Regulation of SOS1 a plasma membrane Na^+/H^+ exchanger in *Arabidopsis thaliana* by SOS2 and SOS3. *Proc. Natl. Acad. Sci. USA* **2002**, *99*, 8436–8441. [CrossRef]
9. Quan, R.; Lin, H.; Mendoza, I.; Zhang, Y.; Cao, W.; Yang, Y.; Shang, M.; Chen, S.; Pardo, J.M.; Guo, Y. SCABP8/CBL10 a putative calcium sensor interacts with the protein kinase SOS2 to protect *Arabidopsis* shoots from salt stress. *Plant Cell* **2007**, *19*, 1415–1431. [CrossRef]
10. Zhu, J.; Fu, X.; Koo, Y.D.; Zhu, J.K.; Jenney, F.E., Jr.; Adams, M.W.; Zhu, Y.; Shi, H.; Yun, D.J.; Hasegawa, P.M.; et al. An enhancer mutant of *Arabidopsis* salt overly sensitive 3 mediates both ion homeostasis and the oxidative stress response. *Mol. Cell Biol.* **2007**, *27*, 5214–5224. [CrossRef]
11. Yang, Q.; Chen, Z.Z.; Zhou, X.F.; Yin, H.B.; Li, X.; Xin, X.F.; Hong, X.H.; Zhu, J.K.; Gong, Z. Overexpression of SOS (Salt Overly Sensitive) genes increases salt tolerance in transgenic *Arabidopsis*. *Mol. Plant* **2009**, *2*, 22–31. [CrossRef] [PubMed]

12. Sun, J.; Chen, S.; Dai, S.; Wang, R.; Li, N.; Shen, X.; Zhou, X.; Lu, C.; Zheng, X.; Hu, Z.; et al. NaCl-induced alternations of cellular and tissue ion fluxes in roots of salt-resistant and salt-sensitive poplar species. *Plant Physiol.* **2009**, *149*, 1141–1153. [CrossRef] [PubMed]

13. Zhu, M.; Shabala, L.; Cuin, T.A.; Huang, X.; Zhou, M.; Munns, R.; Shabala, S. *Nax* loci affect SOS1-like Na$^+$/H$^+$ exchanger expression and activity in wheat. *J. Exp. Bot.* **2016**, *67*, 835–844. [CrossRef] [PubMed]

14. Mittler, R. Oxidative stress antioxidants and stress tolerance. *Trends Plant Sci.* **2002**, *7*, 405–410. [CrossRef]

15. Miller, G.; Shulaev, V.; Mittler, R. Reactive oxygen signaling and abiotic stress. *Physiol. Plant* **2008**, *133*, 481–489. [CrossRef] [PubMed]

16. Miller, G.; Suzuki, N.; Ciftci-Yilmaz, S.; Mittler, R. Reactive oxygen species homeostasis and signalling during drought and salinity stresses. *Plant Cell Environ.* **2010**, *33*, 453–467. [CrossRef] [PubMed]

17. Corpas, F.J.; Barroso, J.B.; del Rio, L.A. Peroxisomes as a source of reactive oxygen species and nitric oxide signal molecules in plant cells. *Trends Plant Sci.* **2001**, *6*, 145–150. [CrossRef]

18. Asada, K. Production and scavenging of reactive oxygen species in chloroplasts and their functions. *Plant Physiol.* **2006**, *141*, 391–396. [CrossRef]

19. Palma, J.M.; Corpas, F.J.; del Rio, L.A. Proteome of plant peroxisomes: New perspectives on the role of these organelles in cell biology. *Proteomics* **2009**, *9*, 2301–2312. [CrossRef]

20. Apel, K.; Hirt, H. Reactive oxygen species: Metabolism, oxidative stress, and signal transduction. *Annu. Rev. Plant Biol.* **2004**, *55*, 373–399. [CrossRef]

21. Creissen, G.; Firmin, J.; Fryer, M.; Kular, B.; Leyland, N.; Reynolds, H.; Pastori, G.; Wellburn, F.; Baker, N.; Wellburn, A.; et al. Elevated glutathione biosynthetic capacity in the chloroplasts of transgenic tobacco plants paradoxically causes increased oxidative stress. *Plant Cell* **1999**, *11*, 1277–1292. [CrossRef] [PubMed]

22. Bose, J.; Rodrigo-Moreno, A.; Shabala, S. ROS homeostasis in halophytes in the context of salinity stress tolerance. *J. Exp. Bot.* **2014**, *65*, 1241–1257. [CrossRef] [PubMed]

23. Rentel, M.C.; Knight, M.R. Oxidative stress-induced calcium signaling in *Arabidopsis*. *Plant Physiol.* **2004**, *135*, 1471–1479. [CrossRef] [PubMed]

24. Chen, S.; Polle, A. Salinity tolerance of *Populus*. *Plant Biol.* **2010**, *12*, 317–333. [CrossRef] [PubMed]

25. Sun, J.; Wang, M.J.; Ding, M.Q.; Deng, S.R.; Liu, M.Q.; Lu, C.F.; Zhou, X.Y.; Shen, X.; Zheng, X.J.; Zhang, Z.K.; et al. H$_2$O$_2$ and cytosolic Ca^{2+} signals triggered by the PM H-coupled transport system mediate K$^+$/Na$^+$ homeostasis in NaCl-stressed *Populus euphratica* cells. *Plant Cell Environ.* **2010**, *33*, 943–958. [CrossRef] [PubMed]

26. Chung, J.S.; Zhu, J.K.; Bressan, R.A.; Hasegawa, P.M.; Shi, H. Reactive oxygen species mediate Na$^+$-induced SOS1 mRNA stability in *Arabidopsis*. *Plant J.* **2008**, *53*, 554–565. [CrossRef] [PubMed]

27. Sanders, D.; Pelloux, J.; Brownlee, C.; Harper, J.F. Calcium at the crossroads of signaling. *Plant Cell* **2002**, *14* (Suppl. S1), 401–417. [CrossRef]

28. Hepler, P.K. Calcium: A central regulator of plant growth and development. *Plant Cell* **2005**, *17*, 2142–2155. [CrossRef]

29. Zielinski, R.E. Calmodulin and calmodulin-binding proteins in plants. *Annu. Rev. Plant Physiol. Plant Mol. Biol.* **1998**, *49*, 697–725. [CrossRef]

30. Cheng, S.H.; Willmann, M.R.; Chen, H.C.; Sheen, J. Calcium signaling through protein kinases. The *Arabidopsis* calcium-dependent protein kinase gene family. *Plant Physiol.* **2002**, *129*, 469–485. [CrossRef]

31. Luan, S.; Kudla, J.; Rodriguez-Concepcion, M.; Yalovsky, S.; Gruissem, W. Calmodulins and calcineurin B-like proteins: Calcium sensors for specific signal response coupling in plants. *Plant Cell* **2002**, *14* (Suppl. 1), 389–400. [CrossRef]

32. Harper, J.F.; Breton, G.; Harmon, A. Decoding Ca^{2+} signals through plant protein kinases. *Annu. Rev. Plant Biol.* **2004**, *55*, 263–288. [CrossRef] [PubMed]

33. Bouche, N.; Yellin, A.; Snedden, W.A.; Fromm, H. Plant-specific calmodulin-binding proteins. *Annu. Rev. Plant Biol.* **2005**, *56*, 435–466. [CrossRef] [PubMed]

34. Hrabak, E.M.; Chan, C.W.; Gribskov, M.; Harper, J.F.; Choi, J.H.; Halford, N.; Kudla, J.; Luan, S.; Nimmo, H.G.; Sussman, M.R.; et al. The *Arabidopsis* CDPK-SnRK superfamily of protein kinases. *Plant Physiol.* **2003**, *132*, 666–680. [CrossRef] [PubMed]

35. Asano, T.; Tanaka, N.; Yang, G.; Hayashi, N.; Komatsu, S. Genome-wide identification of the rice calcium-dependent protein kinase and its closely related kinase gene families: Comprehensive analysis of the CDPKs gene family in rice. *Plant Cell Physiol.* **2005**, *46*, 356–366. [CrossRef]

36. Li, A.L.; Zhu, Y.F.; Tan, X.M.; Wang, X.; Wei, B.; Guo, H.Z.; Zhang, Z.L.; Chen, X.B.; Zhao, G.Y.; Kong, X.Y.; et al. Evolutionary and functional study of the CDPK gene family in wheat (*Triticum aestivum* L.). *Plant Mol. Biol.* **2008**, *66*, 429–443. [CrossRef]

37. Zuo, R.; Hu, R.; Chai, G.; Xu, M.; Qi, G.; Kong, Y.; Zhou, G. Genome-wide identification classification and expression analysis of CDPK and its closely related gene families in poplar (*Populus trichocarpa*). *Mol. Biol. Rep.* **2013**, *40*, 2645–2662. [CrossRef]

38. Boudsocq, M.; Sheen, J. CDPKs in immune and stress signaling. *Trends Plant Sci.* **2013**, *18*, 30–40. [CrossRef]

39. Schulz, P.; Herde, M.; Romeis, T. Calcium-dependent protein kinases: Hubs in plant stress signaling and development. *Plant Physiol.* **2013**, *163*, 523–530. [CrossRef]

40. Hamel, L.P.; Sheen, J.; Seguin, A. Ancient signals: Comparative genomics of green plant CDPKs. *Trends Plant Sci.* **2014**, *19*, 79–89. [CrossRef]

41. Choi, H.I.; Park, H.J.; Park, J.H.; Kim, S.; Im, M.Y.; Seo, H.H.; Kim, Y.W.; Hwang, I.; Kim, S.Y. *Arabidopsis* calcium-dependent protein kinase AtCPK32 interacts with ABF4, a transcriptional regulator of abscisic acid-responsive gene expression, and modulates its activity. *Plant Physiol.* **2005**, *139*, 1750–1761. [CrossRef] [PubMed]

42. Zhu, S.Y.; Yu, X.C.; Wang, X.J.; Zhao, R.; Li, Y.; Fan, R.C.; Shang, Y.; Du, S.Y.; Wang, X.F.; Wu, F.Q.; et al. Two calcium-dependent protein kinases CPK4 and CPK11 regulate abscisic acid signal transduction in *Arabidopsis*. *Plant Cell* **2007**, *19*, 3019–3036. [CrossRef] [PubMed]

43. Zhao, R.; Sun, H.L.; Mei, C.; Wang, X.J.; Yan, L.; Liu, R.; Zhang, X.F.; Wang, X.F.; Zhang, D.P. The *Arabidopsis* Ca^{2+}-dependent protein kinase CPK12 negatively regulates abscisic acid signaling in seed germination and post-germination growth. *New Phytol.* **2011**, *192*, 61–73. [CrossRef] [PubMed]

44. Zhao, R.; Wang, X.F.; Zhang, D.P. CPK12: A Ca^{2+}-dependent protein kinase balancer in abscisic acid signaling. *Plant Signal. Behav.* **2011**, *6*, 1687–1690. [CrossRef] [PubMed]

45. Zou, J.J.; Wei, F.J.; Wang, C.; Wu, J.J.; Ratnasekera, D.; Liu, W.X.; Wu, W.H. *Arabidopsis* calcium-dependent protein kinase CPK10 functions in abscisic acid- and Ca^{2+}-mediated stomatal regulation in response to drought stress. *Plant Physiol.* **2010**, *154*, 1232–1243. [CrossRef] [PubMed]

46. Ma, S.Y.; Wu, W.H. AtCPK23 functions in *Arabidopsis* responses to drought and salt stresses. *Plant Mol. Biol.* **2007**, *65*, 511–518. [CrossRef] [PubMed]

47. Geiger, D.; Scherzer, S.; Mumm, P.; Marten, I.; Ache, P.; Matschi, S.; Liese, A.; Wellmann, C.; Al-Rasheid, K.A.; Grill, E.; et al. Guard cell anion channel SLAC1 is regulated by CDPK protein kinases with distinct Ca^{2+} affinities. *Proc. Natl. Acad. Sci. USA* **2010**, *107*, 8023–8028. [CrossRef]

48. Franz, S.; Ehlert, B.; Liese, A.; Kurth, J.; Cazale, A.C.; Romeis, T. Calcium-dependent protein kinase CPK21 functions in abiotic stress response in *Arabidopsis thaliana*. *Mol. Plant* **2011**, *4*, 83–96. [CrossRef]

49. Mori, I.C.; Murata, Y.; Yang, Y.; Munemasa, S.; Wang, Y.F.; Andreoli, S.; Tiriac, H.; Alonso, J.M.; Harper, J.F.; Ecker, J.R.; et al. CDPKs CPK6 and CPK3 function in ABA regulation of guard cell S-type anion- and Ca^{2+}-permeable channels and stomatal closure. *PLoS Biol.* **2006**, *4*, e327. [CrossRef]

50. Mehlmer, N.; Wurzinger, B.; Stael, S.; Hofmann-Rodrigues, D.; Csaszar, E.; Pfister, B.; Bayer, R.; Teige, M. The Ca^{2+}-dependent protein kinase CPK3 is required for MAPK-independent salt-stress acclimation in *Arabidopsis*. *Plant J.* **2010**, *63*, 484–498. [CrossRef]

51. Xu, J.; Tian, Y.S.; Peng, R.H.; Xiong, A.S.; Zhu, B.; Jin, X.F.; Gao, F.; Fu, X.Y.; Hou, X.L.; Yao, Q.H. AtCPK6 a functionally redundant and positive regulator involved in salt/drought stress tolerance in *Arabidopsis*. *Planta* **2010**, *231*, 1251–1260. [CrossRef] [PubMed]

52. Ye, W.; Muroyama, D.; Munemasa, S.; Nakamura, Y.; Mori, I.C.; Murata, Y. Calcium-dependent protein kinase CPK6 positively functions in induction by yeast elicitor of stomatal closure and inhibition by yeast elicitor of light-induced stomatal opening in *Arabidopsis*. *Plant Physiol.* **2013**, *163*, 591–599. [CrossRef] [PubMed]

53. Ronzier, E.; Corratge-Faillie, C.; Sanchez, F.; Prado, K.; Briere, C.; Leonhardt, N.; Thibaud, J.B.; Xiong, T.C. CPK13 a noncanonical Ca^{2+}-dependent protein kinase specifically inhibits KAT2 and KAT1 shaker K^+ channels and reduces stomatal opening. *Plant Physiol.* **2014**, *166*, 314–326. [CrossRef] [PubMed]

54. Zhao, R.; Sun, H.M.; Zhao, N.; Jing, X.S.; Shen, X.; Chen, S.L. The *Arabidopsis* Ca^{2+}-dependent protein kinase CPK27 is required for plant response to salt-stress. *Gene* **2015**, *563*, 203–214. [CrossRef] [PubMed]

55. Asano, T.; Hakata, M.; Nakamura, H.; Aoki, N.; Komatsu, S.; Ichikawa, H.; Hirochika, H.; Ohsugi, R. Functional characterisation of OsCPK21, a calcium-dependent protein kinase that confers salt tolerance in rice. *Plant Mol. Biol.* **2011**, *75*, 179–191. [CrossRef] [PubMed]

56. Asano, T.; Hayashi, N.; Kobayashi, M.; Aoki, N.; Miyao, A.; Mitsuhara, I.; Ichikawa, H.; Komatsu, S.; Hirochika, H.; Kikuchi, S.; et al. A rice calcium-dependent protein kinase OsCPK12 oppositely modulates salt-stress tolerance and blast disease resistance. *Plant J.* **2012**, *69*, 26–36. [CrossRef] [PubMed]

57. Wei, S.; Hu, W.; Deng, X.; Zhang, Y.; Liu, X.; Zhao, X.; Luo, Q.; Jin, Z.; Li, Y.; Zhou, S.; et al. A rice calcium-dependent protein kinase OsCPK9 positively regulates drought stress tolerance and spikelet fertility. *BMC Plant Biol.* **2014**, *14*, 133. [CrossRef]

58. Ma, F.; Lu, R.; Liu, H.; Shi, B.; Zhang, J.; Tan, M.; Zhang, A.; Jiang, M. Nitric oxide-activated calcium/calmodulin-dependent protein kinase regulates the abscisic acid-induced antioxidant defence in maize. *J. Exp. Bot.* **2012**, *63*, 4835–4847. [CrossRef]

59. Jiang, S.; Zhang, D.; Wang, L.; Pan, J.; Liu, Y.; Kong, X.; Zhou, Y.; Li, D. A maize calcium-dependent protein kinase gene, ZmCPK4, positively regulated abscisic acid signaling and enhanced drought stress tolerance in transgenic *Arabidopsis*. *Plant Physiol. Biochem.* **2013**, *71*, 112–120. [CrossRef]

60. Ding, Y.; Cao, J.; Ni, L.; Zhu, Y.; Zhang, A.; Tan, M.; Jiang, M. ZmCPK11 is involved in abscisic acid-induced antioxidant defence and functions upstream of ZmMPK5 in abscisic acid signalling in maize. *J. Exp. Bot.* **2013**, *64*, 871–884. [CrossRef]

61. Chen, J.; Xue, B.; Xia, X.; Yin, W. A novel calcium-dependent protein kinase gene from *Populus euphratica*, confers both drought and cold stress tolerance. *Biochem. Biophys. Res. Commun.* **2013**, *441*, 630–636. [CrossRef] [PubMed]

62. Yu, X.C.; Li, M.J.; Gao, G.F.; Feng, H.Z.; Geng, X.Q.; Peng, C.C.; Zhu, S.Y.; Wang, X.J.; Shen, Y.Y.; Zhang, D.P. Abscisic acid stimulates a calcium-dependent protein kinase in grape berry. *Plant Physiol.* **2006**, *140*, 558–579. [CrossRef]

63. Yu, X.C.; Zhu, S.Y.; Gao, G.F.; Wang, X.J.; Zhao, R.; Zou, K.Q.; Wang, X.F.; Zhang, X.Y.; Wu, F.Q.; Peng, C.C.; et al. Expression of a grape calcium-dependent protein kinase ACPK1 in *Arabidopsis thaliana* promotes plant growth and confers abscisic acid-hypersensitivity in germination postgermination growth and stomatal movement. *Plant Mol. Biol.* **2007**, *64*, 531–538. [CrossRef] [PubMed]

64. Lin, J.S.; Wang, T.; Wang, G.X. Salt stress-induced programmed cell death via Ca^{2+}-mediated mitochondrial permeability transition in tobacco protoplasts. *Plant Growth Regul.* **2005**, *45*, 243–250. [CrossRef]

65. Lin, J.S.; Wang, Y.; Wang, G.X. Salt stress-induced programmed cell death in tobacco protaplasts is mediated by reactive oxygen species and mitochondrial permeability transition pore status. *J. Plant Physiol.* **2006**, *163*, 731–739. [CrossRef] [PubMed]

66. Li, J.Y.; Jiang, A.; Zhang, W. Salt stress-induced programmed cell death in rice root tip cells. *J. Integr. Plant Biol.* **2007**, *49*, 481–486. [CrossRef]

67. Shabala, S. Salinity and programmed cell death: Untavelling mechanisms for ion specific signalling. *J. Exp. Bot.* **2009**, *60*, 709–712. [CrossRef]

68. Bose, J.; Xie, Y.; Shen, W.; Shabala, S. Haem oxygenase modifies salinity tolerance in Arabidopsis by controlling K$^+$ retention via regulation of the plasma membrane H$^+$-ATPase and by altering SOS1 transcript levels in roots. *J. Exp. Bot.* **2013**, *64*, 471–481. [CrossRef]

69. Lee, S.; Lee, H.J.; Jung, J.H.; Park, C.M. The *Arabidopsis thaliana* RNA-binding protein FCA regulates thermotolerance by modulating the detoxification of reactive oxygen species. *New Phytol.* **2015**, *205*, 555–569. [CrossRef]

70. Ha, J.H.; Kim, J.H.; Kim, S.G.; Sim, H.J.; Lee, G.; Halitschke, R.; Baldwin, I.T.; Kim, J.I.; Park, C.M. Shoot phytochrome B modulates reactive oxygen species homeostasis in roots via abscisic acid signaling in *Arabidopsis*. *Plant J.* **2018**, *94*, 790–798. [CrossRef]

71. Adams, E.; Shin, R. Transport, signaling, and homeostasis of potassium and sodium in plants. *J. Integr. Plant Biol.* **2014**, *56*, 231–249. [CrossRef] [PubMed]

72. Latz, A.; Mehlmer, N.; Zapf, S.; Mueller, T.D.; Wurzinger, B.; Pfister, B.; Csaszar, E.; Hedrich, R.; Teige, M.; Becker, D. Salt stress triggers phosphorylation of the *Arabidopsis* vacuolar K$^+$ channel TPK1 by calcium-dependent protein kinases (CDPKs). *Mol. Plant* **2013**, *6*, 1274–1289. [CrossRef] [PubMed]

73. Kobayashi, M.; Ohura, I.; Kawakita, K.; Yokota, N.; Fujiwara, M.; Shimamoto, K.; Doke, N.; Yoshioka, H. Calcium-dependent protein kinases regulate the production of reactive oxygen species by potato NADPH oxidase. *Plant Cell* **2007**, *19*, 1065–1080. [CrossRef] [PubMed]

74. Boudsocq, M.; Willmann, M.R.; McCormack, M.; Lee, H.; Shan, L.; He, P.; Bush, J.; Cheng, S.H.; Sheen, J. Differential innate immune signalling via Ca^{2+} sensor protein kinases. *Nature* **2010**, *464*, 418–422. [CrossRef] [PubMed]

75. Romeis, T.; Herde, M. From local to global: CDPKs in systemic defense signaling upon microbial and herbivore attack. *Curr. Opin. Plant Biol.* **2014**, *20c*, 1–10. [CrossRef] [PubMed]

76. Liu, K.H.; Niu, Y.; Konishi, M.; Wu, Y.; Du, H.; Sun Chung, H.; Li, L.; Boudsocq, M.; McCormack, M.; Maekawa, S.; et al. Discovery of nitrate-CPK-NLP signalling in central nutrient-growth networks. *Nature* **2017**, *545*, 311–316. [CrossRef]

77. Kobayashi, M.; Yoshioka, M.; Asai, S.; Nomura, H.; Kuchimura, K.; Mori, H.; Doke, N.; Yoshioka, H. StCDPK5 confers resistance to late blight pathogen but increases susceptibility to early blight pathogen in potato via reactive oxygen species burst. *New Phytol.* **2012**, *196*, 223–237. [CrossRef]

78. Ni, M.; Cui, D.; Einstein, J.; Narasimhulu, S.; Vergara, C.E.; Gelvin, S.B. Strength and tissue specificity of chimeric promoters derived from the octopine and mannopine synthase genes. *Plant J.* **1995**, *7*, 661–676. [CrossRef]

79. Clough, S.J.; Bent, A.F. Floral dip: A simplified method for *Agrobacterium*-mediated transformation of *Arabidopsis thaliana*. *Plant J.* **1998**, *16*, 735–743. [CrossRef]

80. Sun, J.; Zhang, X.; Deng, S.; Zhang, C.; Wang, M.; Ding, M.; Zhao, R.; Shen, X.; Zhou, X.; Lu, C.; et al. Extracellular ATP signaling is mediated by H_2O_2 and cytosolic Ca^{2+} in the salt response of *Populus euphratica* cell. *PLoS ONE* **2012**, *7*, e53136. [CrossRef]

81. Sun, J.; Li, L.; Liu, M.; Wang, M.; Ding, M.; Deng, S.; Lu, C.; Zhou, X.; Chen, X.; Zheng, X.; et al. Hydrogen peroxide and nitric oxide mediate K^+/Na^+ homeostasis and antioxidant defense in NaCl-stressed callus cells of two contrasting poplars. *Plant Cell Tiss. Organ. Cult.* **2010**, *103*, 205–215. [CrossRef]

82. Martin, V.V.; Rothe, A.; Gee, K.R. Fluorescent metal ion indicators based on benzoannelated crown systems: A green fluorescent indicator for intracellular sodium ions. *Bioorg. Med. Chem. Lett.* **2005**, *15*, 1851–1855. [CrossRef]

83. Iamshanova, O.; Mariot, P.; Lehen'kyi, V.; Prevarskaya, N. Comparison of fluorescence probes for intracellular sodium imaging in prostate cancer cell lines. *Eur. Biophys. J.* **2016**, *45*, 765–777. [CrossRef] [PubMed]

84. Han, Y.; Wang, W.; Sun, J.; Ding, M.; Zhao, R.; Deng, S.; Wang, F.; Hu, Y.; Wang, Y.; Lu, Y.; et al. *Populus euphratica* XTH overexpression enhances salinity tolerance by the development of leaf succulence in transgenic tobacco plants. *J. Exp. Bot.* **2013**, *64*, 4225–4238. [CrossRef] [PubMed]

85. Zhang, Y.N.; Wang, Y.; Sa, G.; Zhang, Y.H.; Deng, J.Y.; Deng, S.R.; Wang, M.J.; Zhang, H.L.; Yao, J.; Ma, X.Y.; et al. *Populus euphratica* J3 mediates root K^+/Na^+ homeostasis by activating plasma membrane H^+-ATPase in transgenic Arabidopsis under NaCl salinity. *Plant Cell Tiss. Organ Cult.* **2017**, *131*, 75–88. [CrossRef]

86. Lang, T.; Deng, S.; Zhao, N.; Deng, C.; Zhang, Y.; Zhang, Y.; Zhang, H.; Sa, G.; Yao, J.; Wu, C.; et al. Salt-sensitive signaling networks in the mediation of K^+/Na^+ homeostasis gene expression in *Glycyrrhiza uralensis* roots. *Front. Plant Sci.* **2017**, *8*, 1403. [CrossRef] [PubMed]

11

Comprehensive Analysis of Differentially Expressed Unigenes under NaCl Stress in Flax (*Linum usitatissimum* L.) using RNA-Seq

Jianzhong Wu [1], Qian Zhao [2], Guangwen Wu [2], Hongmei Yuan [2], Yanhua Ma [1], Hong Lin [1], Liyan Pan [1], Suiyan Li [1] and Dequan Sun [1,*

[1] Institute of Forage and Grassland Sciences, Heilongjiang Academy of Agricultural Sciences, Harbin 150086, China; wujianzhong176@163.com (J.W.); mayanhua@163.com (Y.M.); linhong@163.com (H.L.); panliyan@163.com (L.P.); lisuiyan@163.com (S.L.)

[2] Institute of Industrial Crop, Heilongjiang Academy of Agricultural Sciences, Harbin 150086, China; zhaoqian0401@sina.com (Q.Z.); wuguangwenflax@163.com (G.W.); yuanhm@163.com (H.Y.)

* Correspondence: sundequan0451@163.com (D.S.)

Abstract: Flax (*Linum usitatissimum* L.) is an important industrial crop that is often cultivated on marginal lands, where salt stress negatively affects yield and quality. High-throughput RNA sequencing (RNA-seq) using the powerful Illumina platform was employed for transcript analysis and gene discovery to reveal flax response mechanisms to salt stress. After cDNA libraries were constructed from flax exposed to water (negative control) or salt (100 mM NaCl) for 12 h, 24 h or 48 h, transcription expression profiles and cDNA sequences representing expressed mRNA were obtained. A total of 431,808,502 clean reads were assembled to form 75,961 unigenes. After ruling out short-length and low-quality sequences, 33,774 differentially expressed unigenes (DEUs) were identified between salt-stressed and unstressed control (C) flax. Of these DEUs, 3669, 8882 and 21,223 unigenes were obtained from flax exposed to salt for 12 h (N1), 24 h (N2) and 48 h (N4), respectively. Gene function classification and pathway assignments of 2842 DEUs were obtained by comparing unigene sequences to information within public data repositories. qRT-PCR of selected DEUs was used to validate flax cDNA libraries generated for various durations of salt exposure. Based on transcriptome sequences, 1777 EST-SSRs were identified of which trinucleotide and dinucleotide repeat microsatellite motifs were most abundant. The flax DEUs and EST-SSRs identified here will serve as a powerful resource to better understand flax response mechanisms to salt exposure for development of more salt-tolerant varieties of flax.

Keywords: RNA-seq; DEUs; flax; NaCl stress; EST-SSR

1. Introduction

Worldwide, flax (*Linum usitatissimum* L.) is an economically important fiber crop, with the flax fiber industry rapidly expanding to meet increasing demand. However, in China flax competes with higher priority food crops, such as grains, that require ever increasing acreage to meet increasing food demands [1]. Consequently, flax cultivation there is confined to barren or even high-salinity plots, where flax varieties with high salt stress tolerance are urgently needed to increase yields of high quality fiber [2]. Unfortunately, the current lack of suitable salt tolerant varieties awaits development of salt-tolerant and high-yield flax germplasm resources. Toward this end, a current research focus is to identify flax genes involved in salt tolerance and salt stress responses [3]. To date, studies of flax salt stress responses have mainly focused on physiological and biochemical aspects instead of molecular response mechanisms. With the completion of the flax genome sequence [4] and continuous development of powerful molecular biological techniques, tools are now available to study regulatory responses to salt stress at the molecular level. Because the neutral salt NaCl is the main source of harmful salt in saline-alkali soils found on marginal lands, flax varieties that are specifically tolerant to NaCl are desired in China [5]. Only after a better understanding of complex salt tolerance mechanisms is mastered will great strides me made toward successful breeding of flax varieties with greater resistance to salt stress.

Research on mechanisms of salt tolerance in plants other than flax has developed rapidly and has revealed that specific molecular mechanisms of salt and alkali tolerance tend to be very complicated [6]. Plants employ adaptations and morphological changes to cope with various abiotic stresses through molecular, cellular, physiological and biochemical responses to stressors [7]. For example, plants regulate and balance osmotic pressure inside and outside of cells by accumulating metabolites to reduce or eliminate stress damage caused by water loss [8]. Nitric oxide (NO) was one of the factors which was operating the melatonin downstream to promote salinity tolerance in rapeseed based on the pharmacological, molecular and genetic data [9]. Arabidopsis EARLY FLOWERING3 (ELF3) enhances plants' resilience to salt stress [10]. Under salt stress, ELF3 suppresses GIGANTEA (GI) at the post-translational level and PHYTOCHROME INTERACTING FACTOR4 (PIF4) at the transcriptional level and PIF4 directly up-regulates the transcription of ORESARA1 and SAG29, which were the two genes that are positive regulators of salt stress response pathways.

Effects of salt stress on plant morphological development are mainly observed as reductions in seed germination, seedling growth and altered growth and development of plant tissues and organs [11–13]. Under salt stress, plant growth is generally inhibited by a water deficit manifesting as water exosmosis from cells. Exosmosis decreases plant growth rate significantly, causing wilting of plants and cell membrane damage that leads to plant cell death [14–16]. Meanwhile, Na^+ competition with various nutrients prevents plants from absorbing other key mineral elements, causing nutrient deficits such as K^+ deficiency, the most common NaCl-induced nutrient deficiency observed [17,18]. So far, a large number of studies on salt tolerance have been carried out in *Arabidopsis thaliana* [19,20], Oryza sativa [21,22] and other crops [23–28] using RNA-Seq technology. The integration of spatiotemporal expression patterns and response characteristics of different genes helps to identify a large number of differentially expressed genes (DEGs) and mechanisms related to salt stress.

In recent years, salt damage has seriously affected flax production in northeast China. Therefore, it is of great significance to understand salt stress response mechanisms and signal pathways, a challenge which currently has important economic, environmental and scientific urgency. Exploring the functional molecules of salt stress signals is fundamental to understanding the mechanism of salt-tolerant crops, so as to conduct genetic engineering and accelerate the breeding of salt-tolerant crops [29]. Ultimately, future agricultural breeding programs will benefit from enhancement of our understanding of resistance mechanisms to a variety of other stressors as well [30].

Transcriptome analysis, a recently developed tool, has greatly enhanced our understanding of plant stress resistance mechanisms [21,23,25–27,31]. However, this powerful technology has seldom been applied to the study of molecular mechanisms involved in flax tolerance, with only a few known resistance genes characterized to date. Here, we sequenced the flax transcriptome to identify differentially expressed unigenes (DEUs) for different NaCl stress exposure durations to better understand flax adaptive molecular responses mechanisms under NaCl stress. After flax transcriptome results were confirmed using qRT-PCR, large-scale analysis of EST-SSRs was conducted using public resources to understand the functions of genes involved in salt stress. The information obtained from this work lays the foundation for understanding molecular mechanisms that participate in the flax response to salt and other stressors, while also identifying relevant and useful genes and markers for future development of salt-tolerant flax varieties.

2. Results and Discussion

2.1. Response of Flax to NaCl Stress

Membrane systems are primary sites of salt stress injury, where such damage causes changes in or loss of plasma membrane semiperme ability, leading to increased electrolyte extravasation [8]. Because field experimentation is difficult to control and time-consuming, in its place laboratory studies have been conducted that measure plant leaf electrical conductivity to study salt stress. Based on previous preliminary conductivity results, three exposure times (N1 for 12 h, N2 for 24 h or N4 for 48 h) were selected to measure changes in gene expression profiles of flax exposed to 100 mM NaCl solution for each exposure duration in the laboratory.

2.2. Transcriptome Analysis

Illumina paired-end sequencing technology was employed to explore DEUs related to NaCl stress in flax using two biological replicates per time point. A total of 185,457,832 clean reads were generated after removing low quality regions and adapter sequences and sequences were mapped to the flax genome (ftp://climb.genomics.cn/pub/10.5524/100001_101000/100081/Flax.cds) using bowtie 2 (v2.1.0) (https://sourceforge.net/projects/bowtie-bio/files/bowtie2/2.1.0/) for short reads with default parameter settings (Table 1). Only 42.92% of reads could be mapped to the reference genome, possibly due to incomplete flax genome assembly. After RSEM (V1.2.4) (https://omictools. com/rsem-tool) [32] was used to evaluate the expression level of each gene, 304,809 of 422,316 unigenes remained with at least 1 fragment per kilobase of transcript per million mapped reads (FPKM) (Table 2). Sequence regions were assigned to exon, intron and intergenic types after comparing total mapped sequence reads with the reference genome. Since exon-type sequences accounted for more than 80% of genome-mapped sequences, as expected (Figure 1), these results confirmed a high degree of annotation accuracy. Moreover, comparison of FPKM box plots of gene expression levels of all genes for different experimental conditions demonstrated that sequence results were reliable, since each sample yielded equivalent reads and coverage depths between duplicates (Figure 2). Furthermore, comparison rates between sequence data and reference genes also demonstrated sequence reliability to some extent.

Table 1. Summary of data generated in the transcriptome sequencing of flax.

Sample *	Clean Reads	Total Mapped	Unique Mapped	Multi Mapped	Bases (Gb)	Q20 (%)	GC (%)
C1	60,125,334	25,807,774	9,037,386	16,770,388	7.48	99.36	48.35
C2	54,128,778	23,792,226	9,600,424	14,191,802	6.67	99.23	47.68
N11	62,422,020	26,632,880	9,637,906	16,994,974	7.77	99.33	47.67
N12	57,381,976	22,307,914	8,055,084	14,252,830	7.13	99.3	47.84
N21	60,066,230	25,957,230	9,698,022	16,259,208	7.48	99.37	47.87
N22	21,827,146	9,277,362	3,136,434	6,140,928	2.58	97.64	47.81
N41	56,385,726	25,227,542	9,202,726	16,024,816	7.01	99.35	47.96
N42	59,471,292	26,454,904	9,667,984	16,786,920	7.39	99.31	48.03

* Samples: Name of sequencing sample, C1/C2 for the two biological replicates of control, N11/N12, N21/N22 and N41/N42 represent for the two biological replicates of treatment with 100 mM NaCl for 12 h, 24 h and 48 h respectively; Clear reads: Number of clean reads participating in the comparison; Total mapped: Numbers of all reads compared to reference genes; Unique mapped: The number of reads compared to the unique location of the reference gene which was used for gene expression analysis; Multi mapped: Number of reads compared to multiple locations of reference genes; Bases (Gb): The total number of bases, Gb represent for billion base pairs; Q20: Percentage of sequencing error rate is smaller than 1% of base number; GC(%):The percentage of G + C of the total number bases.

Table 2. The number of genes in different expression levels.

FPKM Interval	0–0.1	0.1–1	1–3	3–15	15–60	>60
C1	1635 (3.03%)	9893 (18.31%)	11,852 (21.93%)	21,176 (39.19%)	7660 (14.17%)	1823 (3.37%)
C2	1805 (3.37%)	11,354 (21.21%)	11,228 (20.97%)	18,620 (34.78%)	8258 (15.42%)	2278 (4.25%)
N11	1805 (3.31%)	10,645 (19.52%)	11,210 (20.56%)	20,619 (37.82%)	8343 (15.30%)	1901 (3.49%)
N12	1784 (3.28%)	11,534 (21.18%)	11,789 (21.65%)	19,438 (35.70%)	7855 (14.43%)	2047 (3.76%)
N21	2226 (4.15%)	12,475 (23.27%)	11,564 (21.57%)	17,954 (33.49%)	7324 (13.66%)	2061 (3.84%)
N22	1015 (2.03%)	11,621 (23.24%)	11,373 (22.75%)	16,902 (33.80%)	7049 (14.10%)	2039 (4.08%)
N41	3221 (6.32%)	15,996 (31.38%)	10,872 (21.33%)	13,032 (25.57%)	5696 (11.17%)	2154 (4.23%)
N42	3804 (7.43%)	16,694 (32.61%)	10,687 (20.88%)	12,463 (24.35%)	5431 (10.61%)	2111 (4.12%)

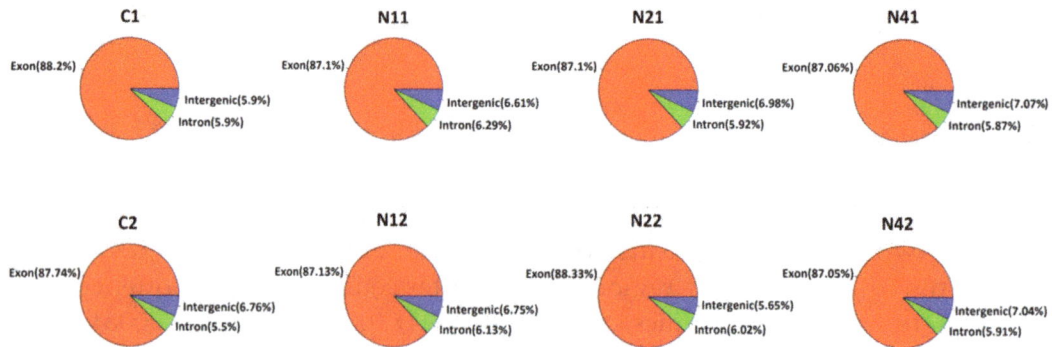

Figure 1. The comparison of clean reads with reference genome in different regions.

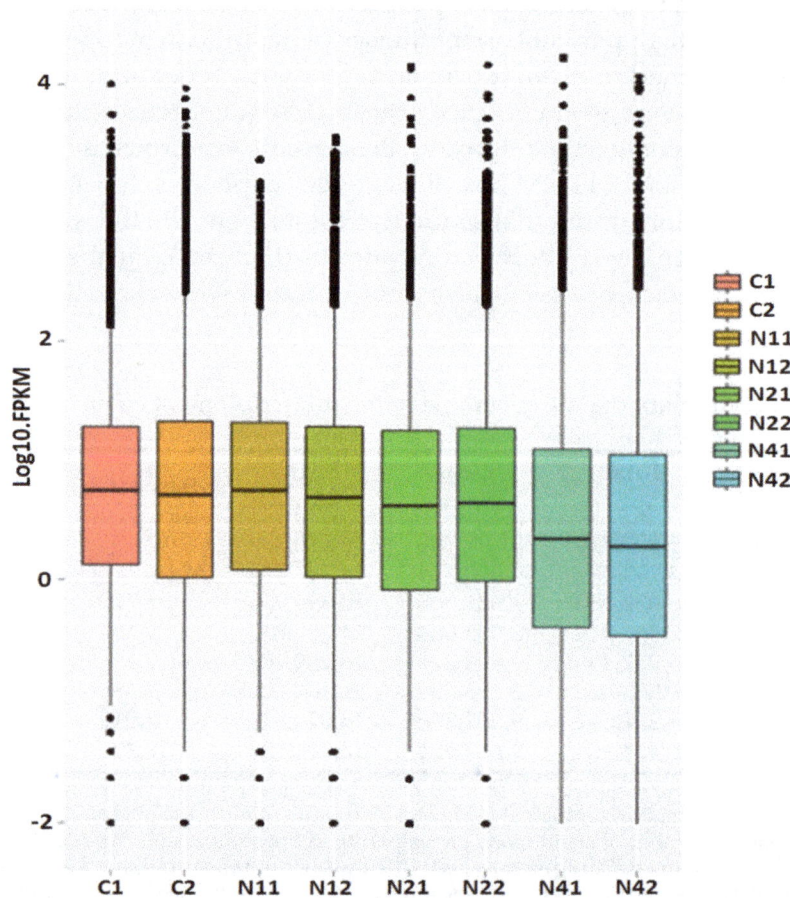

Figure 2. The box plot of FPKM in gene expression levels. Abscissa for sample name, ordinate for Log10.FPKM, box plot for each region for five statistics (From top to bottom: maximum, upper quartile, median, lower quartile and minimum), the outlier is shown in black dots.

2.3. Identification of Differentially Expressed Unigenes (DEUs)

Gene expression profiles in response to NaCl stress exposures of 12 h (N1), 24 h (N2) and 48 h (N4) were compared to the no-treatment control (water). DEUs were identified within the set of transcriptome sequences, with 33,774 significant DEUs identified (18,040 up-regulated and 15,734 down-regulated), with values of false discovery rate (FDR) $\leq$ 0.05 and 2× fold change significance cutoffs generated for various treatment time points.

At 12 h, 24 h and 48 h of stress exposure, 3669 DEUs (2219 up-regulated and 1450 down-regulated), 8882 DEUs (4865 up-regulated and 4017 down-regulated) and 21,223 DEUs (10,956 up-regulated and 10,267 down-regulated) were significantly differentially regulated in response to NaCl stress exposure, respectively (Figure 3). While different numbers of DEUs were observed for various pairings of stress exposure durations, 2581 DEUs (11.5%) were identified that were shared among all stress-exposed samples (Figure 4), in which 2576 co-expressed unigenes were detected (1322 up-regulated and 1254 down-regulated), with five unigenes not co-expressed (Supplementary Table S1). The proportion of DEUs common to paired exposure time points of 24 h/48 h (25.2%) was higher than corresponding proportions for 12 h/24 h (0.9%) and 12 h/48 h (1.9%). These results may be attributed to relatively greater effects of salt injury stress on flax from 24 h to 48 h than for other exposure windows, with involvement of a larger number of regulatory genes observed. This result is not surprising, since expression trends of common DEUs were not entirely consistent among different time periods. In general, DEUs analysis of flax under NaCl stress exposures should enhance our understanding of factors influencing gene expression during salt stress responses and provide clues to genes involved in salt tolerance.

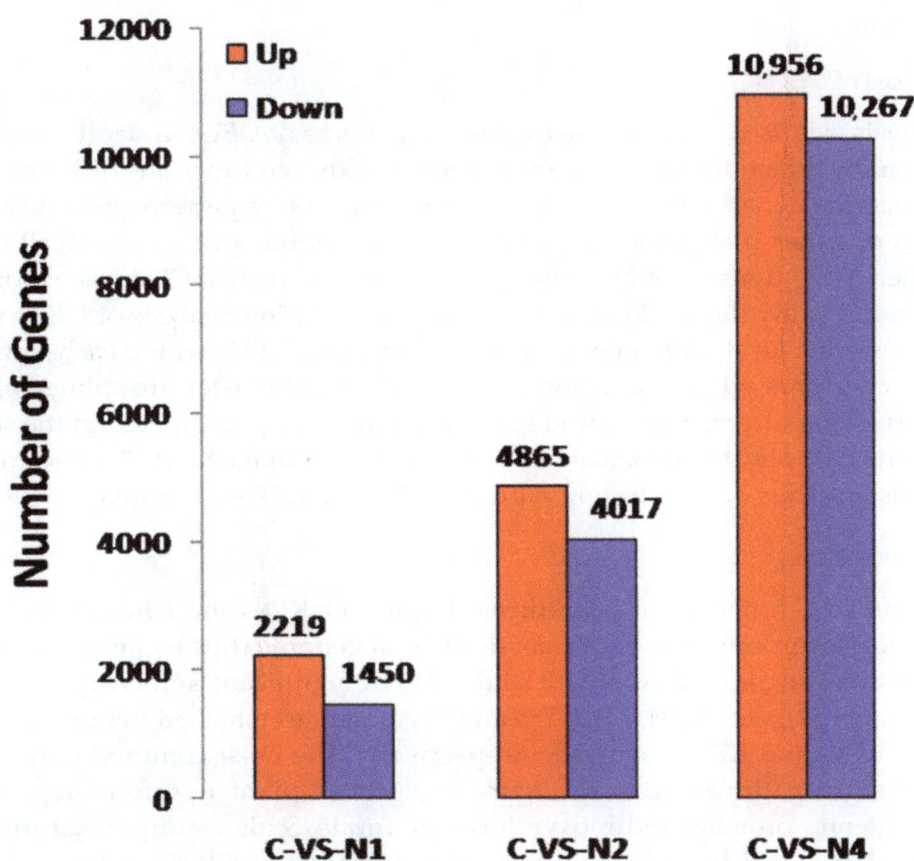

Figure 3. Comparison of up and down-regulation of DEGs. C-vs-N1, C-vs-N2 and C-vs-N4 representing the DEUs under the exposure time of 12 h, 24 h and 48 h in NaCl solution, respectively.

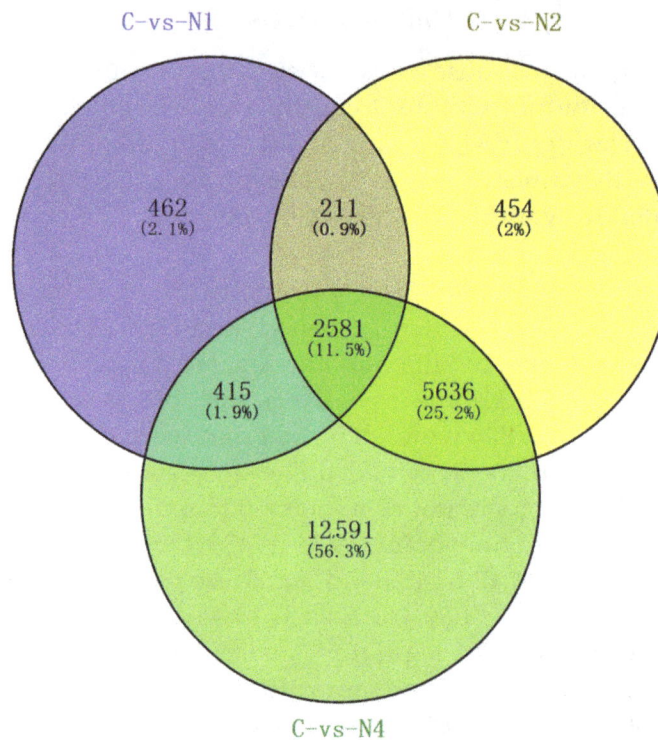

Figure 4. The Venn diagram of DEUs. Venn diagram representing the distribution of NaCl-responsive genes. The numbers in the Venn diagram indicated total numbers of regulated genes in the unique treatment.

2.4. Cluster Analysis of DEUs

Cluster analysis was used to determine expression patterns of DEUs under different experimental conditions. Generally, functions of unknown genes or unknown functions of known genes can be identified by clustering genes with the same or similar expression patterns into classes; genes with similar functions or genes that participate in the same metabolic processes or cell pathways tend to cluster together. DEU dynamic expression patterns for various NaCl stress exposure durations were identified in this study (Figure 5). After hierarchical clustering analysis of DEUs was conducted based on gene expression levels determined from FPKM values, DEUs within each single cluster were considered to be co-expressed genes. Color-coding of different cluster groupings highlights genes with similar expression patterns that shared similar functions or participated in the same biological processes. Moreover, duplicate biological samples for control or each NaCl stress treatment group were highly consistent, thus demonstrating reproducibility of RNA-seq results.

2.5. Functional Annotation

To understand DEU functions, we conducted Pfam, GO, KOG and KEGG enrichment analyses against the genetic background of the *Arabidopsis thaliana* genome (https://www.arabidopsis.org/). Of 2582 co-expressed unigenes, 2482 (96.13%) displayed significant similarity to known proteins (Supplementary Table S2), with 2104, 1473, 785 and 913 unigenes matched to homologous sequences using Pfam, GO, KOG and KEGG analyses, respectively. The most common enrichment analysis terms were related to pathway factors (Figure 6) such as plant hormone signal transduction, photosynthesis-antenna proteins and biosynthesis of amino acids as important in flax responses to NaCl exposure. Not surprisingly, here we identified a number of differentially expressed unigenes (DEUs), which were homologous to known stress regulating plant transcription factors, such as: bZIP (lus10033630), HD-ZIP (lus10025232), WRKY (Lus10020023, Lus10024380, Lus10022736, Lus10012870 and Lus10030517), NAC (Lus10042731, Lus10026617, Lus10036773, Lus10025118, Lus10041534,

Lus10003269 and Lus10042518), MYB (Lus10006740, Lus10019085 and Lus10006647), GRF (Lus10015651 and Lus10037668), GATA (Lus10031464, Lus10025829 and Lus10038273), ERF (Lus10005285, Lus10029987 and Lus10012226), CAMTA (Lus10024044) and B3 (Lus10018583 and Lus10039816). The basic leucine-region zipper (bZIP) transcription factors (TFs) act as crucial regulators in salt stress responses in plants [33]. In our study, Lus10033630 was homologous to AtbZIP34 which was required for Arabidopsis pollen wall patterning and the control of lipid metabolism and/or cellular transport in developing pollen.

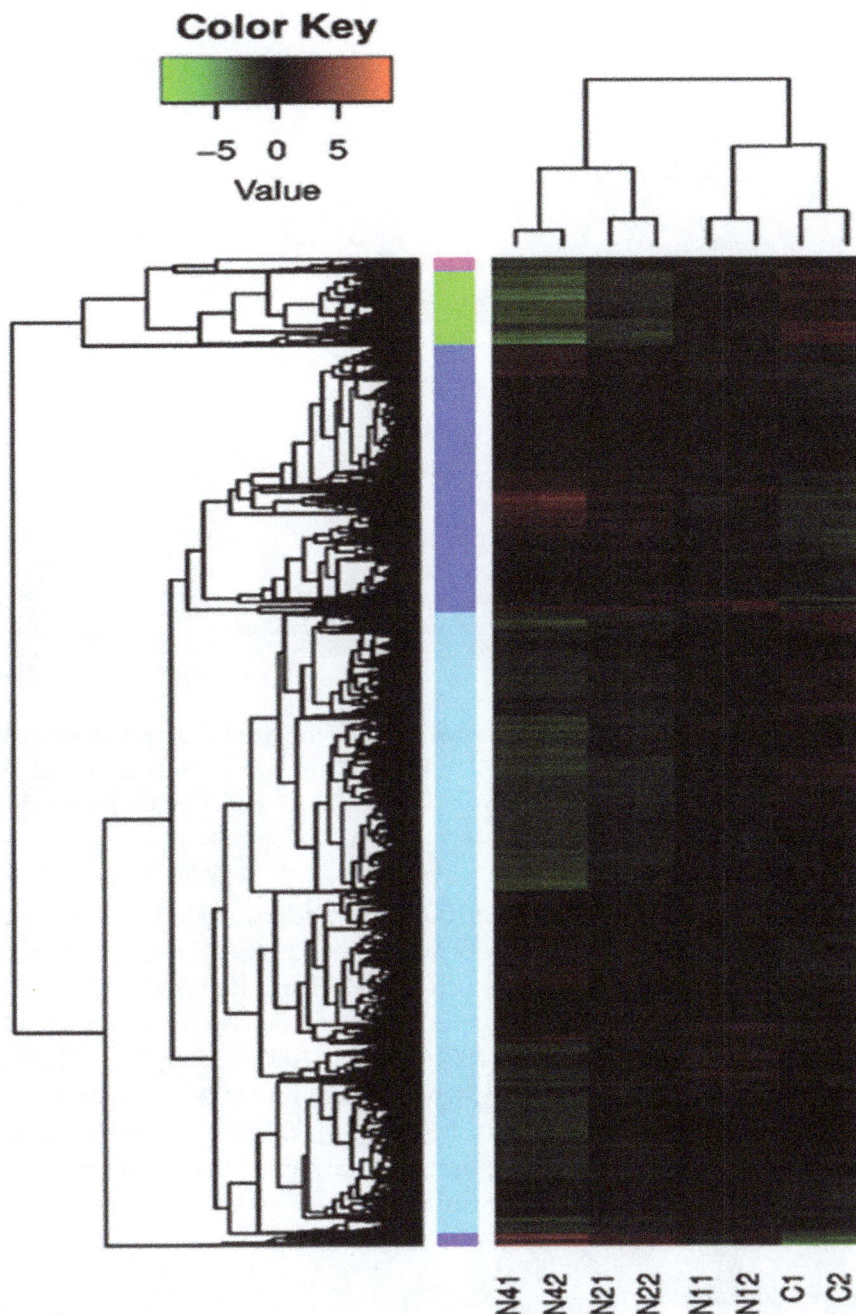

Figure 5. Cluster diagram of DEUs (FDR ≤ 0.05, fold change ≥ 2). The darker color represents the higher of the gene expression level. Each color block on the left represents a cluster of genes with similar expression levels. The Log2.FPKM value was used for clustering, with red for high expression gene and green for low expression gene. The color ranges from green to red, indicating higher gene expression.

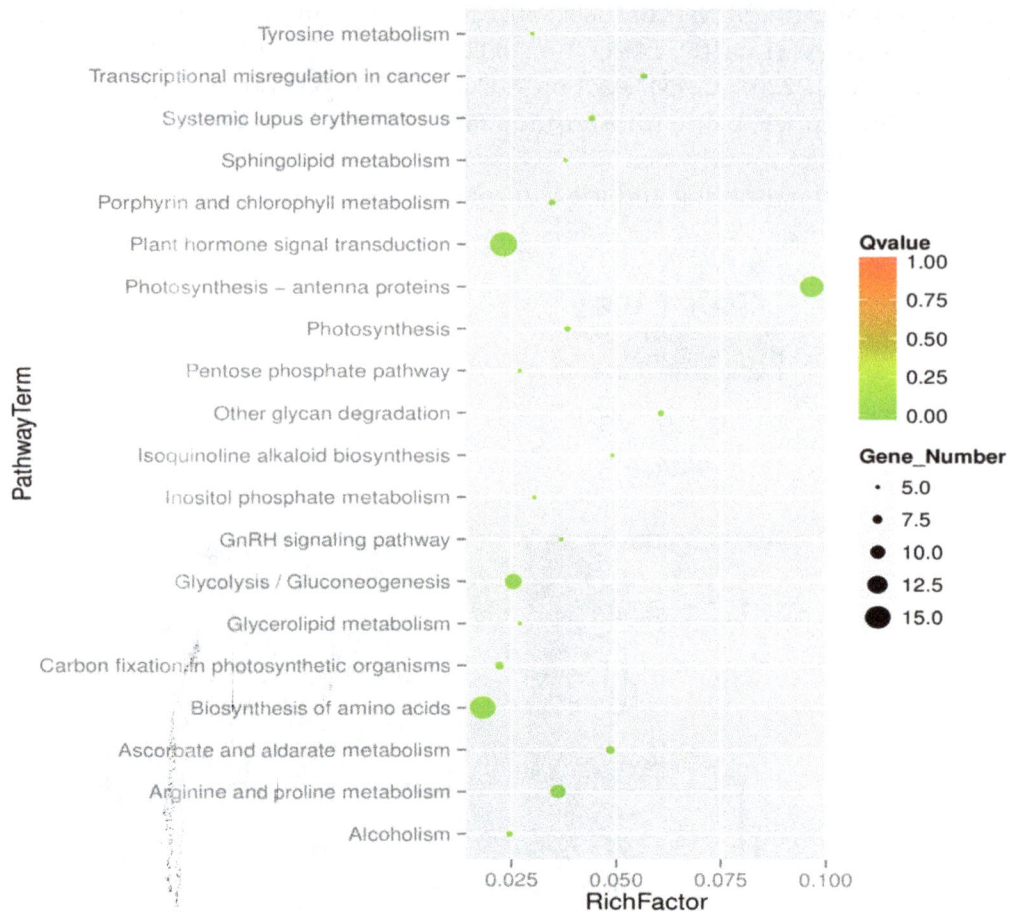

Figure 6. Gene enrichment of non-redundant unigenes. The longitudinal axis represents the different pathway and horizontal axis represents the Rich factor. The size of dots indicates the number of differentially expressed genes in this pathway and the color of the point corresponds to a different Qvalue range.

Interestingly, two salt-tolerant genes (lus10015754 and lus10000310) were obtained here to homologous with Arabidopsis Senescence-Associated Gene29 (SAG29), which was consistent with previous reports [10] and two salt-tolerant genes (lus10015285 and lus10025409) were homologous with SAG12 whose promoter control expression of isopentenyl transferase (ipt) gene in the decaying leaves of the lower part of the plant. Three genes (*Lus10040248*, *Lus10019786* and *Lus10037551*) belong to the Rho-like GTPases from plants (*ROP*) gene family, which might enhance salt tolerance by increasing root length, improving membrane injury and ion distribution [34]. Overall, all co-expressed unigenes could be aligned to the reference genome, suggesting that annotation and classification analyses performed here could be used to reliably predict flax gene functions.

2.6. RNA-Seq Expression Validation

To quantitatively assess the reliability of transcriptome data, six candidate DEUs were selected for analysis using real-time reverse transcription quantitative PCR (qRT-PCR) of biological duplicate samples. Consistent with RNA-sequencing analysis results, qRT-PCR showed significant log2-fold expression changes of DEUs among different salt-stress exposure treatments (Figure 7), thus demonstrating the reliability and accuracy of the transcriptome analysis of NaCl stress in flax conducted here.

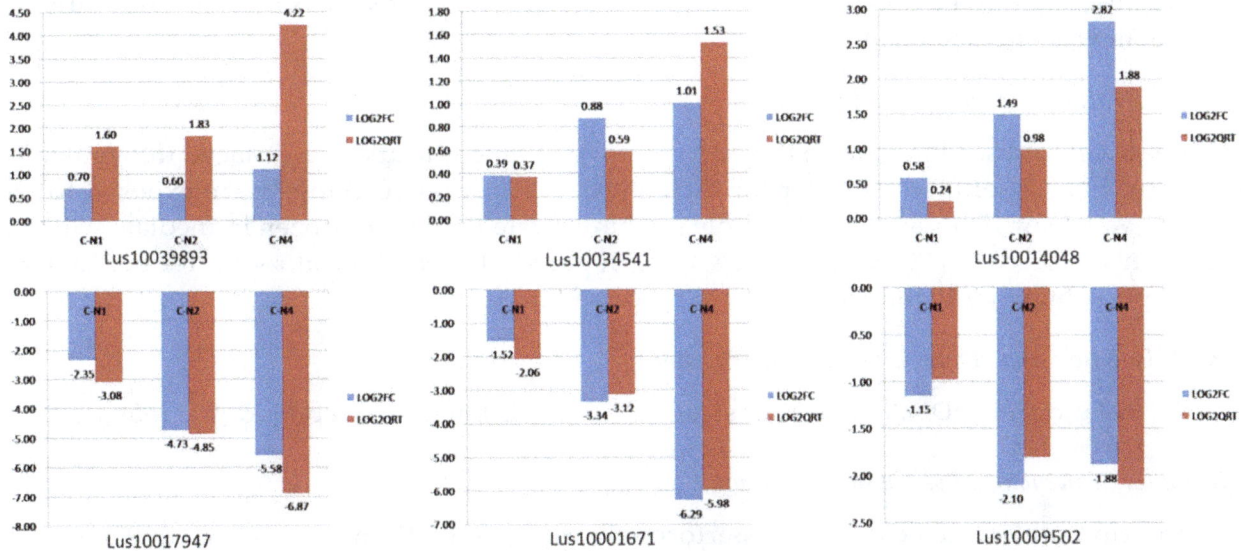

Figure 7. Validation of the RNA-seq data expression profile by qRT-PCR. The relative expression levels of 6 DEUs were calculated according to the $2^{-\Delta\Delta Ct}$ method using the *Actin* gene as an internal reference gene. The *x*-axis indicates the different exposure of 100 mM NaCl solution with 12 h (C-N1), 24 h (C-N2) and 48 h (C-N4). LOG2FC and LOG2QRT represent for the binary logarithm of fold changes of differentially expressed genes in RNA-seq and qRT-PCR, respectively.

2.7. Distribution Characteristics of EST-SSRs

A total of 1777 EST-SSRs with 2–6 bp repeat numbers were identified from sequence data (Table 3). Among SSR loci, trinucleotide microsatellites were the most abundant repeat motif (1002 SSRs, accounting for 56.39%), followed by dinucleotide microsatellites (623 SSRs, accounting for 35.06%), tetranucleotide microsatellites (68 SSRs, accounting for 3.83%), pentanucleotide microsatellites (58 SSRs, accounting for 3.26%) and hexanucleotide microsatellites (26 SSRs, accounting for 1.46%). A repeat iteration number of five (621) was the most common repeat iteration number observed, followed by six (502) and seven (258) repeat iteration numbers. AT/TA motifs were the dinucleotide microsatellite motifs most frequently observed of the six possible motifs (AC/TG, AG/TC, AT/TA, CT/GA, CG/GC and GT/CA), while CTT/GAA motifs were the most frequently represented trinucleotide microsatellites of the 30 possible motifs. The details of all EST-SSRs with the primer pairs are shown in Supplementary Table S3.

Table 3. Distribution of EST-SSR types.

Type	Repeat Number									Percentage (%)
	4	5	6	7	8	9	10	>10	Total	
Dinucleotide			238	148	82	57	38	60	623	35.06
Trinucleotide		558	250	107	49	36	2		1002	56.39
Quadnucleotide		53	12	3					68	3.83
Pentanucleotide	49	7	2						58	3.26
Hexanucleotide	23	3							26	1.46
Total	72	621	502	258	131	93	40	60	1777	

3. Materials and Methods

3.1. Material Planting and Processing

The fiber flax variety (Agatha), provided by Heilongjiang Academy of Agricultural Sciences in Harbin, China, provided material for high-throughput RNA-seq. Flaxseeds were placed in cups filled with sterilized vermiculite and maintained at 28 °C during the day and 22 °C at night in a growing room on a 16-h light/8-h dark cycle. Plants were irrigated every three days and the humidity in the growth room was maintained at 70%.

3.2. Stress Treatments and Sample Preparation

Three-week-old seedlings were rinsed to remove vermiculite and were placed into tanks filled with 100 mM NaCl solution or water for the control (CK). After seedlings were exposed to NaCl solution or water for 12 h, 24 h or 48 h, whole seedlings were harvested, frozen immediately in liquid nitrogen, then stored at $-80\ ^{\circ}$C until RNA was prepared. Two biological replicates per treatment (each sample containing 10 plants each) were processed in parallel.

3.3. RNA Isolation and cDNA Library Construction

RNA isolation and cDNA library construction methods have been reported previously [35].

3.4. Illumina Sequencing, Assembly and Annotation

Transcriptome sequencing was performed using the Illumina HiSeq 2500 platform (https://www.illumina.com/systems/sequencing-platforms/hiseq-2500.html) to generate paired-end (PE) raw reads, each of ~100 bp in length. Clean reads were generated from raw reads by removal of adaptor sequences, ambiguous 'N' nucleotides (with a ratio of 'N' greater than 10%) and low quality sequences (with quality scores less than 5) and were assembled as described by using bowtie2 (2.1.0) software (http://bowtie-bio.sourceforge.net/bowtie2/index.shtml) against a reference genome (ftp://climb.genomics.cn/pub/10.5524/100001_101000/100081/). Only clean, high quality sequence data was used in subsequent analyses.

For homology-based annotation, the model plant genome of *Arabidopsis thaliana* (https://www.arabidopsis.org/) was selected for use as genetic background, with non-redundant sequences subjected to Pfam (http://pfam.xfam.org/), Gene Ontology (GO) (http://www.geneontology.org/), Eukaryotic Clusters of Orthologous Groups (KOG) (https://www.ncbi.nlm.nih.gov/COG/) and Kyoto Encyclopedia of Genes and Genomes (KEGG) (http://www.genome.jp/kegg/). For gene expression profiling analysis, functional assignments were mapped to GO terms [36]. Significantly enriched pathways were identified according to p values and enrichment factors [37].

3.5. Identification of DEUs and Cluster Analysis

DEUs were identified based on a negative binomial distribution using the edgeR package (https://www.rdocumentation.org/packages/edgeR/versions/3.14.0) [38]. Candidate genes exhibited false discovery rate (FDR) values ≤ 0.01, which were calculated from numbers of mapped reads. In addition, the fragments per kilobase of transcript per million mapped reads (FPKM) values of candidate genes were calculated using RSEM (http://deweylab.github.io/RSEM/; RNA-Seq by Expectation-Maximization) [32]. All expressed genes were divided into six categories: $0 < FPKM \leq 0.1$, $0.1 < FPKM \leq 1$, $1 < FPKM \leq 3$, $3 < FPKM \leq 15$, $15 < FPKM \leq 60$ and $FPKM > 60$. The formula for calculating expression level reflected by FPKM for each target gene is:

$$FPKM = 10C^9/NL$$

where C is the number of fragments of the target gene, N is the number of fragments for all genes and L is the base length of the target gene.

Finally, the fold change of FPKM for each sequence was calculated and genes with FPKM fold changes greater or equal to 2 were classified as DEUs. Common DEUs among different time points that exhibited FPKM fold change differences greater than 2 between biological replicates were eliminated. For each gene, normalized FPKM values for each transcript were clustered using the hclust function in R (http://www.r-project.org/) using a distance matrix representing FPKM level profiles of genes across the four sampled time points. The tree produced by the clustering process was split into two branches using the cutree function.

3.6. Real-Time qRT-PCR Validation

DEUs identified in transcriptome sequencing analysis were validated using qRT-PCR (quantitative RT-PCR) to further confirm RNA-Seq analysis gene expression results. Six genes (*Lus10017947*, *Lus10001671*, *Lus10009502*, *Lus10039893*, *Lus10034541* and *Lus10014048*) were selected for analysis of expression levels in flax treated with NaCl solution exposure times of 12 h, 24 h and 48 h using *L. ussitatissimum Act1* (GenBank accession no. AY857865) as the internal reference gene. Primer sets were designed from Illumina sequencing data using Primer Premier v6.24 (http://www.premierbiosoft.com/crm/jsp/com/pbi/crm/clientside/ProductList.jsp) (listed in Supplementary Table S4). Primers were synthesized by GENEWIZ, Inc. (Suzhou, China). Quantitative RT-PCR was performed with a SYBR® Premix Ex Taq™ II Kit (TaKaRa, Dalian, China) using an ABI 7500 Real-Time PCR System (Applied Biosystems, Foster City, CA, USA). Data for each experimental sample was corrected for sample loading differences using results of flax *Act1* qRT-PCR using the $2^{-\Delta\Delta Ct}$ method [39]. PCR amplification was performed using thermal cycling conditions of denaturation at 95 °C for 30 s followed by 40 cycles of amplification (95 °C for 5 s, 60 °C for 34 s). The control reaction (to normalize expression levels) and all samples were tested in triplicate.

3.7. Development of EST-SSR

All DEUs identified by transcriptome sequencing were analyzed to identify SSRs with dimer, trimer, tetramer, pentamer and hexamer motifs using SSRIT (http://archive.gramene.org/db/markers/ssrtool). SSR primer pairs were designed with highest primer scores using Primer-BLAST (https://www.ncbi.nlm.nih.gov/tools/primer-blast/index.cgi?LINK_LOC=BlastHomeAd) with the following standard parameters: target amplicon length of 100–300 bp, annealing temperature variation from 55 to 65 °C and primer size of 18–28 bp.

4. Conclusions

As an important industrial crop, flax is currently cultivated on marginal lands in the northeast region of China. By analyzing the genome-wide transcriptome of flax during exposure to 100 mM NaCl solution, we identified 3669, 8882 and 21,223 potential salt stress-responsive DEUs after salt exposure for 12 h, 24 h and 48 h, respectively, compared with untreated control. All of these unigenes were identified as part of an extensive investigation of flax genes involved in the response to salt exposure to collectively provide high-resolution gene expression profiles for flax under control and salt stress conditions. Of the 2582 co-expressed unigenes, 2482 were annotated using at least one public database (GO, KOG, KO, Pfam, KEGG) and pathways linked to flax response to NaCl exposure were mainly associated with plant hormone signal transduction, photosynthesis-antenna proteins and biosynthesis of amino acids. Number of the genes are homologous to known stress regulators. Transcriptome sequencing results were verified by qRT-PCR. Ultimately, 1777 EST-SSRs based on transcriptome results were identified as important in flax response to NaCl exposure. The results described here thus lay the groundwork for further elucidation of molecular mechanisms of flax stress responses. Ultimately, this information will guide the development of flax varieties that are more tolerant to a wide range of environmental stresses.

Author Contributions: J.W. conceived and designed the study, interpreted the data, performed the experimental work and wrote the manuscript; Q.Z. drafted, wrote the manuscript, did the bioinformatics analysis and revised the manuscript; G.W. contributed to materials and revised the manuscript; H.Y., Y.M., H.L. and L.P. performed the experimental work and interpreted the data; S.L. and D.S. contributed to the experiments and discussion of results. All authors read and approved the final manuscript.

Acknowledgments: The authors would also like to thank Tingbo Jiang for helpful critical comments on the manuscript.

References

1. Yu, Y.; Huang, W.G.; Chen, H.Y.; Wu, G.W.; Yuan, H.M.; Song, X.X.; Kang, Q.H.; Zhao, D.S.; Jiang, W.D.; Liu, Y.; et al. Identification of differentially expressed genes in flax (*Linum usitatissimum* L.) under saline-alkaline stress by digital gene expression. *Gene* **2014**, *549*, 113–122. [CrossRef] [PubMed]

2. El-Hariri, D.M.; Al-Kordy, M.A.; Hassanein, M.S.; Ahmed, M.A. Partition of photosynthates and energy production in different flax cultivars. *J. Nat. Fibers* **2005**. [CrossRef]

3. Pecenka, R.; Fürll, C.; Ola, D.C.; Budde, J.; Gusovius, H.J. Efficient use of agricultural land in production of energy: Natural insulation versus bio-energy. In Proceedings of the International Conference of Agricultural Engineering CIGR-AgEng, Agriculture and Engineering for a Healthier Life, Valencia, Spain, 8–12 July 2012.

4. Wang, Z.; Hobson, N.; Galindo, L.; Zhu, S.; Shi, D.; McDill, J.; Yang, L.; Hawkins, S.; Neutelings, G.; Datla, R.; et al. The genome of flax (*Linum usitatissimum*) assembled de novo from short shotgun sequence reads. *Plant J.* **2012**, *72*, 461–473. [CrossRef] [PubMed]

5. Marshall, G. Flax: Breeding and utilisation. *Plant Growth Regul.* **1991**, *10*, 171–172. [CrossRef]

6. Abdel, W.W.; Ahmed, S.A. Response surface methodology for production, characterization and application of solvent, salt and alkali-tolerant alkaline protease from isolated fungal strain, *Aspergillus niger*, WA 2017. *Int. J. Biol. Macromol.* **2018**, *115*, 447–458. [CrossRef]

7. Yamaguchi-Shinozaki, K.; Shinozaki, K. Transcriptional regulatory networks in cellular responses and tolerance to dehydration and cold stresses. *Annu. Rev. Plant Biol.* **2006**, *57*, 781–803. [CrossRef] [PubMed]

8. Zhu, J.K. Salt and Drought Stress Signal Transduction in Plants. *Annu. Rev. Plant Biol.* **2002**, *53*, 247–273. [CrossRef]

9. Zhao, G.; Zhao, Y.Y.; Yu, X.L.; Kiprotich, F.; Han, H.; Guan, R.Z.; Wang, R.; Shen, W.B. Nitric Oxide Is Required for Melatonin-Enhanced Tolerance against Salinity Stress in Rapeseed (*Brassica napus* L.) Seedlings. *Int. J. Mol. Sci.* **2018**, *19*, 1912. [CrossRef]

10. Sakuraba, Y.; Bulbul, S.; Piao, W.; Choi, G.; Paek, N.C. Arabidopsis EARLY FLOWERING3 increases salt tolerance by suppressing salt stress response pathways. *Plant J.* **2017**, *92*, 1106–1120. [CrossRef]

11. Li, Y. Effect of salt stress on seed germination and seedling growth of three salinity plants. *Pak. J. Biol. Sci.* **2008**, *11*, 1268–1272. [CrossRef]

12. Çavuşoğlu, K.; Kılıç, S.; Kabar, K. Some morphological and anatomical observations during alleviation of salinity (NaCl) stress on seed germination and seedling growth of barley by polyamines. *Acta Physiol. Plant.* **2007**, *29*, 551–557. [CrossRef]

13. Jiang, A.; Gan, L.; Tu, Y.; Ma, H.; Zhang, J.; Song, Z.; He, Y.C.; Cai, D.T.; Xue, X.D. The effect of genome duplication on seed germination and seedling growth of rice under salt stress. *Aust. J. Crop Sci.* **2013**, *7*, 1814–1821.

14. François, T.; Parent, B.; Caldeira, C.F.; Welcker, C. Genetic and physiological controls of growth under water deficit. *Plant Physiol.* **2014**, *164*, 1628–1635. [CrossRef]

15. Álvarez, S.; Rodríguez, P.; Broetto, F.; Sánchez-Blanco, M.J. Long term responses and adaptive strategies of pistacialentiscus under moderate and severe deficit irrigation and salinity: Osmotic and elastic adjustment, growth, ion uptake and photosynthetic activity. *Agric. Water Manag.* **2018**, *202*, 253–262. [CrossRef]

16. Li, H.; Chang, J.; Chen, H.; Wang, Z.; Gu, X.; Wei, C.; Zhang, Y.; Ma, J.; Yang, J.; Zhang, X. Exogenous melatonin confers salt stress tolerance to watermelon by improving photosynthesis and redox homeostasis. *Front. Plant Sci.* **2017**, *8*, 295. [CrossRef]

17. Nguyen, M.T.; Yang, L.E.; Fletcher, N.K.; Lee, D.H.; Kocinsky, H.; Bachmann, S.; Delpire, E.; McDonough, A. Effects of K^+-deficient diets with and without nacl supplementation on Na^+, K^+, and H_2O transporters' abundance along the nephron. *Am. J. Physiol. Renal Physiol.* **2012**, *303*, F92–F104. [CrossRef] [PubMed]

18. Rus, A.; Yokoi, S.; Sharkhuu, A.; Reddy, M.; Lee, B.H.; Matsumoto, T.K.; Koiwa, H.; Zhu, J.K.; Bressan, R.A.; Hasegawa, P.M. Athkt1 is a salt tolerance determinant that controls Na⁺ entry into plant roots. *Proc. Natl. Acad. Sci. USA* **2001**, *98*, 14150–14155. [CrossRef]

19. Atkinson, N.J.; Lilley, C.J.; Urwin, P.E. Identification of genes involved in the response of Arabidopsis to simultaneous biotic and abiotic stresses. *Plant Physiol.* **2013**, *162*, 2028–2041. [CrossRef]

20. Han, N.; Lan, W.J.; He, X.; Shao, Q.; Wang, B.S.; Zhao, X.J. Expression of a Suaeda salsa Vacuolar H+/Ca2+ Transporter Gene in Arabidopsis Contributes to Physiological Changes in Salinity. *Plant Mol. Boil. Rep.* **2012**, *30*, 470–477. [CrossRef]

21. Lu, T.; Lu, G.; Fan, D.; Zhu, C.; Wei, L.; Qiang, Z.; Qi, F.; Yan, Z.; Guo, Y.; Li, W. Function annotation of the rice transcriptome at single-nucleotide resolution by RNA-seq. *Genome Res.* **2010**, *20*, 1238–1249. [CrossRef]

22. Zhou, Y.; Yang, P.; Cui, F.; Zhang, F.; Luo, X.; Xie, J. Transcriptome Analysis of Salt Stress Responsiveness in the Seedlings of Dongxiang Wild Rice (Oryza rufipogon Griff.). *PLoS ONE* **2016**. [CrossRef] [PubMed]

23. Liu, A.; Xiao, Z.; Li, M.W.; Wong, F.L.; Yung, W.S.; Ku, Y.S.; Wang, Q.; Wang, X.; Xie, M.; Yim, A.K. Transcriptomic reprogramming in soybean seedlings under salt stress. *Plant Cell Environ.* **2018**, *12*, e0189159. [CrossRef]

24. Guo, J.; Li, Y.; Han, G.; Song, J.; Wang, B. NaCl markedly improved the reproductive capacity of the euhalophyte Suaeda Salsa. *Funct. Plant Boil.* **2018**, *45*, 350. [CrossRef]

25. Cui, F.; Sui, N.; Duan, G.; Liu, Y.; Han, Y.; Liu, S.; Wan, S.; Li, G. Identification of Metabolites and Transcripts Involved in Salt Stress and Recovery in Peanut. *Front. Plant Sci.* **2018**, *9*, 217. [CrossRef]

26. Yuan, F.; Lyu, M.J.A.; Leng, B.Y.; Zhu, X.G.; Wang, B.S. The transcriptome of NaCl-treated *Limonium bicolor* leaves reveals the genes controlling salt secretion of salt gland. *Plant Mol. Boil.* **2016**, *91*, 241–256. [CrossRef] [PubMed]

27. Wang, N.; Qian, Z.; Luo, M.; Fan, S.; Zhang, X.; Zhang, L. Identification of salt stress responding genes using transcriptome analysis in green alga chlamydomonas reinhardtii. *Int. J. Mol. Sci.* **2018**, *19*, 3359. [CrossRef]

28. Li, H.; Lin, J.; Yang, Q.S.; Li, X.G.; Chang, Y.H. Comprehensive analysis of differentially expressed genes under salt stress in pear (*Pyrus betulaefolia*) using RNA-seq. *Plant Growth Regul.* **2017**, *82*, 409–420. [CrossRef]

29. Xu, Y.Y.; Li, X.G.; Lin, J.; Wang, Z.H.; Yang, Q.S.; Chang, Y.H. Transcriptome sequencing and analysis of major genes involved in calcium signaling pathways in pear plants (*Pyrus calleryana* Decne.). *BMC Genom.* **2015**, *16*, 738. [CrossRef]

30. Hamed, K.B.; Ellouzi, H.; Talbi, O.Z.; Hessini, K.; Slama, I.; Ghnaya, T.; Bosch, S.M.; Savour, A.; Abdelly, C. Physiological response of halophytes to multiple stresses. *Funct. Plant Biol.* **2013**, *40*, 883–896. [CrossRef]

31. Dash, P.K.; Cao, Y.; Jailani, A.K.; Gupta, P.; Venglat, P.; Xiang, D.; Rai, R.; Sharma, R.; Thirunavukkarasu, N.; Abdin, M.Z.; et al. Genome-wide analysis of drought induced gene expression changes in flax (*Linum usitatissimum*). *Gm Crop. Food* **2014**, *5*, 106–119. [CrossRef]

32. Dewey, C.N.; Bo, L. RSEM: Accurate transcript quantification from RNA-Seq data with or without a reference genome. *BMC Bioinform.* **2011**, *12*, 323. [CrossRef]

33. Antónia, G.; David, R.; Matczuk, K.; Nikoleta, D.; David, C.; Twell, D.; Honys, D. AtbZIP34 is required for Arabidopsis pollen wall patterning and the control of several metabolic pathways in developing pollen. *Plant Mol. Biol.* **2009**, *70*, 581–601. [CrossRef]

34. Miao, H.X.; Sun, P.G.; Liu, J.H.; Wang, J.Y.; Xu, B.Y.; Jin, Z.Q. Overexpression of a Novel ROP Gene from the Banana (MaROP5g) Confers Increased Salt Stress Tolerance. *Int. J. Mol. Sci.* **2018**, *19*, 3108. [CrossRef] [PubMed]

35. Wu, J.; Zhao, Q.; Sun, D.; Wu, G.; Zhang, L.; Yuan, H.; Yu, Y.; Zhang, S.; Yang, X.; Li, Z.; et al. Transcriptome analysis of flax (*Linum usitatissimum* L.) undergoing osmotic stress. *Ind. Crop. Prod.* **2018**, *116*, 215–223. [CrossRef]

36. Harris, M.A.; Clark, J.; Ireland, A.; Lomax, J.; Ashburner, M.; Foulger, R.; Eilbeck, K.; Lewis, S.; Marshall, B.; Mungall, C.; et al. The Gene Ontology (GO) database and informatics resource. *Nucleic Acids Res.* **2014**, *32*, D258–D261. [CrossRef]

37. Kanehisa, M.; Goto, S.; Hattori, M.; Aoki-Kinoshita, K.F.; Itoh, M.; Kawashima, S.; Katayama, T.; Araki, M.; Hirakawa, M. From genomics to chemical genomics: New developments in KEGG. *Nucleic Acids Res.* **2006**, *34*, D354–D357. [CrossRef] [PubMed]

38. Robinson, M.D.; McCarthy, D.J.; Smyth, G.K. edgeR: A Bioconductor package for differential expression analysis of digital gene expression data. *Bioinformatics* **2010**, *26*, 139–140. [CrossRef] [PubMed]

39. Livak, K.J.; Schmittgen, T.D. Analysis of relative gene expression data using real-time quantitative PCR and the $2^{-\Delta\Delta CT}$ method. *Methods* **2001**, *25*, 402–408. [CrossRef]

12

Role and Functional Differences of HKT1-Type Transporters in Plants under Salt Stress

Akhtar Ali [1], Albino Maggio [2], Ray A. Bressan [3] and Dae-Jin Yun [1,*]

[1] Department of Biomedical Science & Engineering, Konkuk University, Seoul 05029, Korea;
    gultkr@yahoo.com
[2] Department of Agriculture, University of Naples Federico II, Via Universita 100, I-80055 Portici, Italy;
    almaggio@unina.it
[3] Department of Horticulture and Landscape Architecture, Purdue University, West Lafayette, IN 47907-2010,
    USA; bressan@purdue.edu
*   Correspondence: djyun@konkuk.ac.kr

Abstract: Abiotic stresses generally cause a series of morphological, biochemical and molecular changes that unfavorably affect plant growth and productivity. Among these stresses, soil salinity is a major threat that can seriously impair crop yield. To cope with the effects of high salinity on plants, it is important to understand the mechanisms that plants use to deal with it, including those activated in response to disturbed Na^+ and K^+ homeostasis at cellular and molecular levels. HKT1-type transporters are key determinants of Na^+ and K^+ homeostasis under salt stress and they contribute to reduce Na^+-specific toxicity in plants. In this review, we provide a brief overview of the function of HKT1-type transporters and their importance in different plant species under salt stress. Comparison between HKT1 homologs in different plant species will shed light on different approaches plants may use to cope with salinity.

Keywords: abiotic stresses; high salinity; HKT1; halophytes; glycophytes

1. Introduction

Plants are sessile organisms, which are continuously challenged by various biotic and abiotic environmental stresses, such as soil salinity, extreme temperatures, drought, nutrients deficiency or pathogen attack. These stresses have a tremendous impact on agricultural crops, reducing their potential yields by more than half [1]. Soil salinization is one of the most serious causes of stress for world's agriculture, and it is progressing in most agricultural regions [2–4]. In saline soils, the ability of plants to grow and complete their life cycle can be severely compromised. Increased salinity leads to cytosolic osmotic stress and sodium ion specific toxicity that exert a combined inhibitory effect on physiological, biochemical and developmental pathways [5–7]. To deal with toxic levels of Na^+, plants may restrict Na^+ influx, compartmentalize Na^+ to vacuoles and/or mobilize the un-avoidable influx of Na^+ outside the cell and/or in different regions/organs of the plants [8–11]. In addition, the ability to take up K^+, a plant essential nutrient, is also crucial under salinity stress [12–14].

In dealing with the potentially detrimental effects of Na^+, sodium transporters play a pivotal role in plant protection in saline environments. These include antiporters that extrude Na^+ from root cells and/or re-distribute Na^+ throughout different tissues (Salt-Overly-Sensitive or SOS pathway) so as to reduce toxicity in critical cellular regions and reestablish to some degree water homeostasis [15–17]. Symporters, known as HKT1-type transporters (high-affinity potassium transporter1) also contribute to Na^+ detoxification by retrieving/diverting Na^+ from the xylem stream to protect the shoots from Na^+ toxicity [18–21]. the function of this mechanism is to confine toxic ions to the roots, thus protecting above ground tissues from damage [8,9,22]. the critical role of HKT1 transporters under salt stress has been well characterized in a number of plant species including the model plant *Arabidopsis*, wheat, rice, sorghum, tomato, as well as in extremophile models such as *Eutrema parvula* and *Eutrema salsuginea* [8,23–28]. HKT1-type transporters mediate the balance between Na^+ and K^+ ions under salt stress, a function that has recently been reported also for HKT1 transporters in monocots [12–14,29]. In cereal crops such as wheat and rice, which contain multiple *HKT1*-type genes, some members of this transporters family have been identified as key components of plant salt stress tolerance [12,30,31].

2. Na^+ Homeostasis

There are two main drawbacks of salt stress: osmotic stress and ion imbalance. Osmotic stress is caused by a decrease of the available water in the soil due to a reduced osmotic potential which makes more difficult for a plant to extract water [32]. Ion imbalance is mostly caused by excessive accumulation of Na^+ ions, which can inhibit normal cellular functions [33]. To achieve protection against high salinity, plants need to activate mechanisms that regulate both Na^+ uptake and homoeostasis [34,35]. the orchestrated distribution of Na^+ ions throughout the entire plant body represents one crucial activity to keep Na^+ away from sites of metabolism [36]. Moreover, the management of Na^+ must be balanced against the control of specific ion toxicity and the uptake of K^+, which is essential for normal plant growth and development [36]. High Na^+/K^+ ratios in the cytosol are toxic to plants, inhibiting various processes such as K^+ absorption, vital enzyme reactions, protein synthesis and photosynthesis [5,37]. In order to control the adverse effects of salt stress, plants have evolved various adaptive mechanisms to control ion homeostasis regulated by several proteins, working alone or in a group. Among them, HKT1-type transporters regulate sodium homeostasis by keeping a balance between Na^+ and K^+ in the cytoplasm [14,19,21,22,38]. To further explain and sheds light on the role of HKT1-type transporters, we will briefly discuss their discovery and classification based on their cation selectivity.

3. Discovery of HKT1-Type Transporters

HKT1-type transporters have an important role in mediating the distribution of Na^+ within the plant by a repeated pattern of Na^+ removal from the xylem, particularly in the roots, so that the amount of Na^+ reaching the shoot becomes more easily manageable [20,21,39]. Since the discovery of HKT1 in the early 90s [8,40], many more HKT transporter homologs from other species and with different cation transport properties have been isolated, which opened a new area of salt stress signaling in plants [7,23,25,27,28,41–48]. Interestingly, HKT1-type transporters from various species received the same name independently of their specific transport characteristics. For example, HKT1 from wheat was named TaHKT2;1 whereas its homolog from *Arabidopsis* was named AtHKT1. Nevertheless, the cation transport properties of these two HKT1 transporters are different from each other, AtHKT1 is a Na^+ transporter whereas TaHKT2;1 is a K^+/Na^+ symporter [8,23]. Subsequently, due to their different cation selectivity, HKT1-transporters have been divided into different classes based on their cation transport properties [30,49].

4. Classification HKT1-Type Transporters

Homologs of *HKT1* genes and proteins have been identified in a number of plant species, including *Arabidopsis*. Typically, their ion selectivity has been characterized in yeast and/or *Xenopus oocytes* [23,41,43]. Based on protein structure and ion selectivity, HKT1-type transporters have been divided into two sub-classes with differences in the amino acids of the first pore domain (PD) of the protein as the main distinguishing feature. This was the basis of an international agreement for the nomenclature of HKT1-type transporters established back in 2006 [30]. Accordingly, members of class-1 contain a Ser (serine) residue at the first pore-loop domain (pA) and show higher selectivity for Na^+ than K^+ (Figure 1). In contrast, members of class-2 contain Gly (glycine) residues at the same position and are considered to function as Na^+/K^+ co-transporters (Figure 1) [24,30,50]. *Arabidopsis* contains a single copy *HKT1* gene, *AtHKT1*, that encodes for a member of class-1 and shows highly specific sodium influx when expressed in *Xenopus laevis oocytes* and *Saccharomyces cerevisiae* [23]. Monocots such as rice and wheat contain more than one copy of the *HKT1* gene and their coding proteins belong to both class-1 as well as class-2 [3,12,24,25].

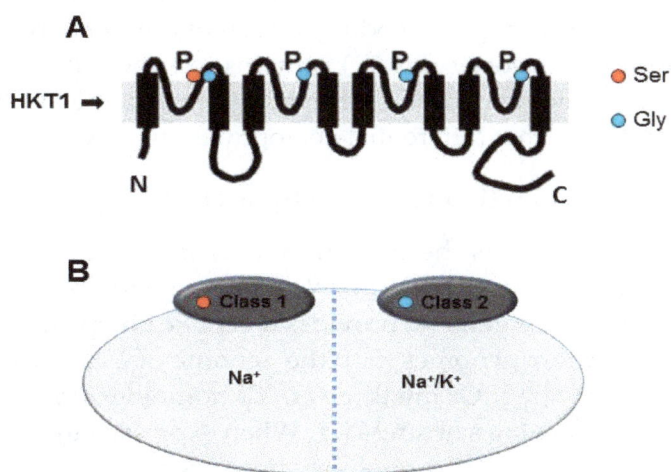

Figure 1. Classification and structure of HKT1 proteins. (**A**) Structural analysis of HKT1 protein containing Ser or Gly residues in their conserved regions. Members of class-1 transporters carry a Ser in the selectivity filter position while on the same position class-2 transporters contain a Gly residue. P denotes pore-loop domain while N and C indicate N-terminus and C-terminus of HKT-proteins. (**B**) Members of class-1 HKT1 that carry a Ser residue transport Na^+ while members of class-2 that carry a Gly residue can transport both Na^+ as well as K^+

5. Role of HKT1-Type Transporters in Glycophytes and Halophytes

HKT1-type transporters play a crucial role in Na^+ homeostasis. Knock-out of *HKT1* leads to NaCl sensitivity in *Arabidopsis* [19,20]. Homologs of *HKT1* have been isolated from different glycophytic species including wheat [8,40], *Arabidopsis* [19,21–23], rice [12,38,48], eucalyptus [43], barley [51], tomato [26], sorghum [27], strawberry [52], pumpkin [53], poplar [54]. the *Arabidopsis* genome contains a single *HKT1* gene that encodes for AtHKT1, a member of class-1 transporters. AtHKT1 acts as a high-affinity selective Na^+ transporter in heterologous systems such as *Xenopus oocytes* and yeast [19,23]. AtHKT1 has been shown to retrieve Na^+ from the xylem stream to reduce its transport and accumulation to the shoots [19,20]. This process prevents Na^+ toxicity in the shoots through recirculation of Na^+ to the roots from which it could be exported again [8,9,22]. On the other hand, members of class-2 transporters contribute to maintain a balanced Na^+/K^+ ratio in the cytoplasm under salt stress. the mode of action of class-2 transporters depends on the external Na^+ concentration. *TaHKT1* from wheat, a member of subclass-2, at low concentrations of Na^+ works as Na^+/K^+ symporter, but at high concentration of Na^+ TaHKT1 act as a Na^+ uniporter [8]. It has been recently demonstrated, via transgenic analysis, that over-expression of *AtHKT1* contributes to maintain optimal K^+/Na^+ in tobacco plants and to improve plant biomass under salt stress [55]. However, different modifications of HKT1 transporters may also cause variations of leaf Na^+ exclusion and salt tolerance in maize. Therefore, the exact mechanism through which HKT1 transporters confer salt tolerance deserves further attention. New insights in glycophytes have been obtained by comparative analysis of cultivated plants with wild relatives. In rice, different salt-tolerance during seed germination and seedling vegetative growth in weedy and cultivated plants has been associated to variants of *HKT1*-mediated transport and regulatory mechanisms that affect Na^+/K^+ balance [56].

Considering the activity of *HKT1* genes in glycophytic species, their functions in naturally salt-tolerant plants (halophytes) was also investigated [6,13,57]. Halophytes also control Na^+ toxicity based on efflux and re-distribution of Na^+ ions into various tissues to reduce its toxicity in specific plant organs and on Na^+ sequestration in the vacuole [13,58]. the *Arabidopsis* close relative *Eutrema salsuginea* (previously *Thellungiella halophila*) is a model halophyte [6,7,13]. *E. salsuginea*'s genome has been recently sequenced and it provides a resource to characterize the function of different genes in this species [11,59]. Although the precise nature of mechanisms that regulate halophytism is not fully understood [6,13,60,61], much progress has been made based on a comparative analysis of halophyte with glycophytes. Similar to glycophytes, halophytes also rely on genes coding for salt overly sensitive (SOS), vacuolar Na^+/H^+ antiporter (NHX) and sodium transporter (HKT1) proteins to cope with high salinity [7,28,59,60,62]. However, growing evidence indicates that these genes may have temporal and spatial expression patterns under normal and stress conditions that differentiate halophytes vs. glycophytes [7,11,28,58].

6. Functional Differences in Halophytic and Glycophytic HKTs

One close relative of *Arabidopsis* is the halophyte *Eutrema salsuginea* (previously *Thellungiella halophile* or *Thulengiella salsuginea*) [6,11,63]. *E. salsuginea*'s genome sequence is known and its juxta-positioning with the *Arabidopsis* genome provides a genetic blueprint that highlights similarities as well as differences between these genomes [59]. the genome of *E. salsuginea* includes three copies of *HKT1* genes in a tandem array [59]. Of the three *HKT1* homologs only *EsHKT1;2* is dramatically induced at the transcript level following salt stress [7]. When expressed in yeast, EsHKT1;2 shows high affinity for potassium whereas EsHKT1;1 more likely behaves as AtHKT1, with high specificity for sodium uptake [7]. Another *Arabidopsis* halophytic relative is *Eutrema parvula* (now *Schrenkiella parvula*). the genome of *Eutrema parvula* contains two *HKT1* genes, *EpHKT1;1* and *EpHKT1;2* [62]. *EpHKT1;2* is induced very rapidly upon salt stress [28]. All of these halophytic *HKT1* genes (three from *E. salsuginea* and two from *E. parvula*) contain a Ser residue at the selectivity filter in the first pore-loop domain and have therefore been grouped as class-1 transporters [30]. Members of class-1 HKT1-transporters lack the ability to uptake K^+. However, both EsHKT1;2 and EpHKT1;2 possess conserved Asp (aspartate) residue in the second pore-loop domain [7] and show K^+ uptake ability which makes them functionally different from other members of this class such as AtHKT1 (Figure 2).

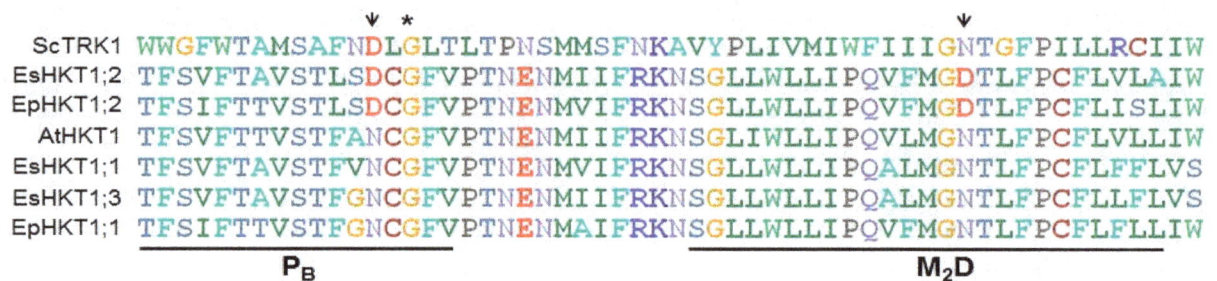

Figure 2. Sequence comparison of HKT homologs from *Arabidopsis*, *E. salsuginea* and *E. parvula*. Amino acid sequences in the second pore loop region (PB) and the adjacent transmembrane domain (M2B) are aligned by clustalw (http://www.ebi.ac.uk/Tools/msa/clustalw2/). the conserved Gly residues in the PB region [49] are indicated by asterisks. the Asp residues specific for EsHKT1;2 (D207) and EpHKT1;2 (D205) are indicated by arrows.

Excess of $[Na^+]$ in the cytosol impairs the optimal cytosolic Na^+/K^+ ratio, which is recognized by the plant as K^+ deficiency indicating that potassium homeostasis is important for plants during salt stress [7,13,38,64]. Induction of *HKT1* under K^+ shortage would be detrimental if HKT1 was Na^+ specific [7,24,28,65]. Under salt stress, when the cytosolic sodium concentration reaches a toxic level, plants activate high-affinity potassium transporters to re-establish an optimal $[Na^+]/[K^+]$ cellular balance [12,14,27]. One example of high-affinity potassium transporters is EsHKT1;2 (and possibly EpHKT1;2). Down-regulation of *EsHKT1;2* in *E. salsuginea* leads to hypersensitive phenotypes under K^+-deficient conditions (Figures 3 and 4). Based on these findings and on the activation of EsHKT1;2 and EpHKT1;2 in response to high salinity, these genes can be considered major contributors to the halophytic nature of *E. salsuginea* and *E. parvula* [28].

Figure 3. *EsHKT-RNAi* plants are sensitive to low K$^+$-limiting conditions. Wild type and knock-down lines of *EsHKT1;2* (*EsHKT1;2-RNAi*) were grown on MS-medium for 10-days and then transferred to K$^+$-deficient media with 0, 1 and 10 mM KCl (see Ali et al. 2012 for the detailed methodology) and allowed to grow for further 10-days. *EsHKT1;2-RNAi* lines were more sensitive to K$^+$-limiting conditions as compared with the wild type Control. A gradual increase of K$^+$ concentration greatly promotes the growth of wild type plants, whereas *EsHKT1;2-RNAi* lines were still sensitive. This result demonstrates the crucial role of EsHKT1;2 for K$^+$-uptake.

Figure 4. *EsHKT-RNAi* plants are sensitive to salt stress. Wild type and knock-down lines of *EsHKT1;2* (*EsHKT-RNAi*) were grown on MS-medium for 2-weeks and then transferred to inert soil (porous soil, see Ali et al. 2012 for details) and grown for further 3-weeks. Plants were then treated with 300 mM NaCl for another 3-weeks period, twice a week (control represents untreated plants). *EsHKT-RNAi* lines were more sensitive to salt stress as compared with wild type Control.

7. Importance of Conserved Amino Acids in the 2nd Pore-Loop of HKT1 in Glycophytes and Halophytes

As discussed earlier, certain residues in the HKT transporters have a crucial role in the functioning of the transporter. Alignment of most published HKTs with ScTRK1 showed that both EsHKT1;2 and EpHKT1;2 contain conserved Asp residues in their second pore-loop domains (Asp207 and Asp205, respectively), (Figure 2) [14]. Yeast ScTRK1, a known high-affinity potassium transporter [66], also carries an Asp in the second pore-loop position (Figure 2).

However, in most HKT1 proteins an Asn (asparagine) is present at the above said position (Asn211 in AtHKT1), while SlHKT1;1 and SbHKT1-4 both carry Ser residues (Ser264 and Ser277, respectively) [12,14,26]. These considerations were confirmed by showing that Asp207 and Asp205, could impart selectivity to subclass-1 HKT1 transporters. Asp207 to Asn207 in EsHKT1;2 and Asp205 to Asn205 in EpHKT1;2 were able to abolish potassium uptake and generate canonical subclass-1 Na^+-selective transporters [7,28]. In addition, changing the Asn residue in the 2nd pore-loop domain of AtHKT1 to Asp (N211D) resulted in a transporter that resembled EsHKT1;2 with high affinity for potassium transport. More importantly, transgenic *Arabidopsis* plants expressing AtHKT1[N211D] tolerate salt stress more effectively than the wild type AtHKT1 and show exactly the same phenotype as *EsHKT1;2*- or *EpHKT1;2*-overexpressing Arabidopsis plants [14,28]. This means that HKTs from dicots can be differentiated from each other with respect to their monovalent cation selectivity by the presence of either Asp or Asn residues in the 2nd pore loop domain.

8. Substitution of Conserved Residues in the Pore-Region Affects the Cation Selectivity of HKT1-Transporters

The cation selectivity of HKT1 transporters is convertible by exchanging a single amino acid. Ser in the 1st pore loop domain appears not to be the only essential amino acid favoring K^+ uptake (at least in *Arabidopsis* and *Eutrema* species), but it possibly functions as a supporting residue. Nevertheless, Ser or Gly at the first pore-loop differentiates class-1 and class-2 HKT-transporters based on their Na^+ or Na^+/K^+ co-transport activity.

However, this hypothesis failed to differentiate EsHKT1;2 which, although it contains a Ser residue at 1st pore-loop region and is a member of class-1 transporters, it unexpectedly functions as a K^+ transporter. In addition, EsHKT1;2 and EpHKT1;2 both contain a conserved Asp residue in the 2nd pore-loop region which is the key residue for their cation selectivity [14,28]. In contrast, it has been shown as indispensable the presence of an Asp (D) replacing an Asn residue (N211) to convert the Na^+ uniporter AtHKT1 into a Na^+/K^+ symporter [14]. Furthermore, the replacement of Asp with Asn in the EsHKT1;2 protein abolishes its potassium transport properties and it converts EsHKT1;2 into a Na^+ uniporter. Substitution of the corresponding Asn in AtHKT1 to Asp (N211D) confirmed the importance of this residue by expression in yeast cells, *Xenopus oocytes* and *Arabidopsis* (Figure 5) [14]. More recently, EpHKT1;2 was also shown to carry an Asp (D205) in the 2nd pore-loop domain. When Asp205 was substituted by Asn, EpHKT1;2 lost its ability to tolerate sodium stress in the presence of potassium [28]. According to these reports, HKT-type transporters possess several key amino acids which define their transport properties. In this regard, the presence of only Ser or Gly residues might not be sufficient to assign HKT-transporters to a specific class.

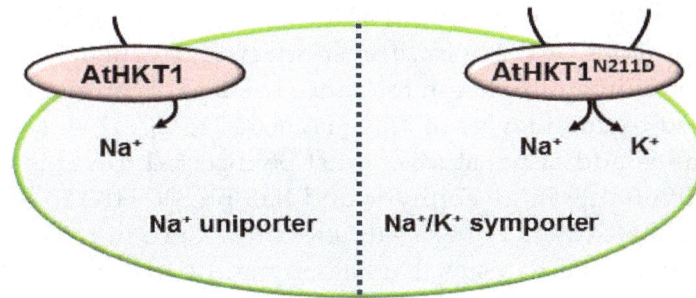

Figure 5. Functional properties of AtHKT1 and AtHKT1^{N211D} and differential selectivity for Na$^+$ and K$^+$ based on the Asn/Asp variance in the pore region. Wild type AtHKT1 is a sodium uniporter and does not confer salt stress tolerance. An altered version of AtHKT1 with a mutation of the Asn to Asp (AtHKT1^{N211D}) is also able to uptake potassium and confers salt stress tolerance. It has already been shown by Ali et al. 2016 that *athkt1-1* plants complemented by *AtHKT1^{N211D}* showed higher tolerance to salt stress than lines complemented by the wild type *AtHKT1*. Thus, the introduction of Asp, replacing Asn, in HKT1-type transporters established altered cation selectivity and uptake dynamics.

9. Contribution of HKT1 in Plant Na$^+$ Homeostasis and Salinity Tolerance

A balance between Na$^+$ and K$^+$ ions under salt stress is crucial for plant survival [64], but it is not clear how such balance can be established under conditions of (hyper-) accumulation of Na$^+$ (and to some degree Cl$^-$) leading to osmotic stress and ionic imbalance [67]. the localization of AtHKT1 in the xylem parenchyma cells appears to provide an answer because its activity can reduce the flux of Na$^+$ to the shoot tip in the (for most plants extremely rare) conditions of excess Na$^+$ in the root-zone. It is believed that high-affinity potassium transporters will be active during salt stress. the presence and stress-induced activity of Na$^+$/H$^+$ antiporters have also been shown [11]. Yet other transporters are active in partitioning Na$^+$ into vacuoles which can act as ultimate sinks for sodium ions [68].

For plants that are exposed to an excess of Na$^+$, the function of HKT1 isoforms seems to have changed from a distribution role that curtails Na$^+$ flux throughout the plant into a supporting role as K$^+$ transporters. the ability of *Thellungiella* species to maintain a low cytosolic Na$^+$/K$^+$ ratio in the presence of high salinity stress has been shown [69]. Suppression of *HKT1* expression in *E. salsuginea* by RNAi leads to hyper-accumulation of sodium in the shoots, reduced sodium content in the roots and, consequently, a disturbed Na$^+$/K$^+$ ratio in the plant. Shoot sodium hyper-accumulation brings salt sensitivity suggesting that *EsHKT1;2* in *E. salsuginea* is one of the major components of its halophytic behavior (Figures 3 and 4). While the RNAi targeted all *EsHKT1* copies, it is likely that the phenotype of the RNAi lines mirrored the function of *EsHKT1;2*, which is the most highly expressed copy among the three tandem duplicated *HKT1*paralogs in the *E. salsuginea* genome [59].

Our recent findings showed that AtHKT1^{N-D}, like the native EsHKT1;2 transporter, contributed to salt tolerance. This was demonstrated based on the phenotype of AtHKT1^{N-D} both in yeast and transgenic Arabidopsis lines [14]. More importantly the current generated by AtHKT1^{N211D} in *xenopus oocytes* were more similar to that of EsHKT1;2 rather than AtHKT1, indicating that enhanced uptake of K$^+$ can reduce Na$^+$ toxicity [26]. Molecular and structural studies of both AtHKT1 and EsHKT1;2 and their mutated versions further explained that HKT1-proteins contain different charge distributions at the pore site [14].

10. Concluding Remarks

Salinity tolerance in plants is a very complex process whose components are only partially known. A comparative analysis of halophytic and glycophytic systems has helped to understand the nature and function of critical genes in salt stress adaptation [11,19,29,70]. It has been shown that ThSOS1 and EsHKT1;2, and more recently EpHKT1;2 are essential determinants of the halophytic behavior of *E. salsuginea* and *E. pavula* [7,11,28]. Unlike their *Arabidopsis* counterparts, strong induction and activity of EsHKT1;2 [7] and EpHKT1;2 [28] under salt stress might suggest that the co-evolution of these ion transporters play a critical role in shaping halophytic lifestyles of *E. salsuginea* and *E. parvula*. Functional studies in *Arabidopsis* homologs of these ion transporters and genetic duplication in halophytes may help us to understand how multiple ion tolerance has been acquired to support survival in an environment characterized by high levels of Na^+ [11,59,62,71,72]. Therefore, to better understand salt tolerance in crop plants, additional studies must be directed towards defining the regulatory mechanisms operating differently in glycophytic and halophytic HKT1 so to reconcile expression and protein location with the extremely high salt stress tolerance of these species. Improving our understanding of key functional mechanisms that halophytes use to cope with high salinity could help us in designing applications and strategies to improve salt stress tolerance in crop plants in the future.

Author Contributions: D.-J.Y. designed research, A.A. performed experiments, A.A., A.M., R.A.B. and D.-J.Y. analyzed the data, collected information from literature and wrote the paper.

References

1. Bray, E.A.; Bailey-Serres, J.; Weretilnyk, E. Responses to abiotic stresses. In *Biochemistry and Molecular Biology of Plants*; Gruissem, W., Buchannan, B., Jones, R., Eds.; American Society of Plant Physiologists: Rockville, MD, USA, 2000; pp. 1158–1249.

2. Wang, W.; Vinocur, B.; Altman, A. Plant responses to drought, salinity and extreme temperatures: Towards genetic engineering for stress tolerance. *Planta* **2003**, *218*, 1–14. [CrossRef] [PubMed]

3. Huang, S.; Spielmeyer, W.; Lagudah, E.S.; Munns, R. Comparative mapping of HKT1 genes in wheat, barley and rice, key determinants of Na^+ transport, and salt tolerance. *J. Exp. Bot.* **2008**, *59*, 927–937. [CrossRef] [PubMed]

4. Cirillo, V.; Masin, R.; Maggio, A.; Zanin, G. Crop-weed interactions in saline environments. *Eur. J. Agron.* **2018**, *99*, 51–61. [CrossRef]

5. Murguía, J.R.; Bellés, J.M.; Serrano, R. A salt-sensitive 3′(2′),5′-bisphosphate nucleotidase involved in sulfate activation. *Science* **1995**, *267*, 232–234. [CrossRef] [PubMed]

6. Inan, G.; Zhang, Q.; Li, P.; Wang, Z.; Cao, Z.; Zhang, H.; Zhang, C.; Quist, T.M.; Goodwin, S.M.; Zhu, J.; et al. Salt cress. A halophyte and cryophyte *Arabidopsis* relative model system and its applicability to molecular genetic analyses of growth and development of extremophiles. *Plant Physiol.* **2004**, *135*, 1718–1737. [CrossRef] [PubMed]

7. Ali, Z.; Park, H.C.; Ali, A.; Oh, D.H.; Aman, R.; Kropornicka, A.; Hong, H.; Choi, W.; Chung, W.S.; Kim, W.Y.; et al. TsHKT1;2, a HKT1 Homolog from the Extremophile *Arabidopsis* Relative *Thellungiella salsuginea*, Shows K^+ Specificity in the Presence of NaCl. *Plant Physiol.* **2012**, *158*, 1463–1474. [CrossRef] [PubMed]

8. Rubio, F.; Gassmann, W.; Schroeder, J.I. Sodium-driven potassium uptake by the plant potassium transporter HKT1 and mutations conferring salt tolerance. *Science* **1995**, *270*, 1660–1663. [CrossRef] [PubMed]

9. Rus, A.; Yokoi, S.; Sharkhuu, A.; Reddy, M.; Lee, B.H.; Matsumoto, T.K.; Koiwa, H.; Zhu, J.K.; Bressan, R.A.; Hasegawa, P.M. AtHKT1 is a salt tolerance determinant that controls Na^+ entry into plant roots. *Proc. Natl. Acad. Sci. USA* **2001**, *98*, 14150–14155. [CrossRef] [PubMed]

10. Quintero, F.J.; Ohta, M.; Shi, H.Z.; Zhu, J.K.; Pardo, J.M. Reconstitution in yeast of the *Arabidopsis* SOS signaling pathway for Na^+ homeostasis. *Proc. Natl. Acad. Sci. USA* **2002**, *99*, 9061–9066. [CrossRef] [PubMed]

11. Oh, D.H.; Leidi, E.; Zhang, Q.; Hwang, S.M.; Li, Y.; Quintero, F.J.; Jiang, X.; D'Urzo, M.P.; Lee, S.Y.; Zhao, Y.; et al. Loss of halophytism by interference with SOS1 expression. *Plant Physiol.* **2009**, *151*, 210–222. [CrossRef] [PubMed]

12. Yao, X.; Horie, T.; Xue, S.; Leung, H.Y.; Katsuhara, M.; Brodsky, D.E.; Wu, Y.; Schroeder, J.I. Differential sodium and potassium transport selectivities of the rice OsHKT2;1 and OsHKT2;2 transporters in plant cells. *Plant Physiol.* **2010**, *152*, 341–355. [CrossRef] [PubMed]

13. Shao, Q.; Han, N.; Ding, T.; Zhou, F.; Wang, B. SsHKT1;1 is a potassium transporter of the C3 halophyte *Suaeda salsa* that is involved in salt tolerance. *Funct. Plant Biol.* **2014**, *41*, 790–802. [CrossRef]

14. Ali, A.; Raddatz, N.; Aman, R.; Kim, S.; Park, H.C.; Jan, M.; Baek, D.; Khan, I.U.; Oh, D.H.; Lee, S.Y.; et al. A single amino acid substitution in the sodium transporter HKT1 associated with plant salt tolerance. *Plant Physiol.* **2016**, *171*, 2112–2126. [CrossRef] [PubMed]

15. Qiu, Q.S.; Guo, Y.; Dietrich, M.A.; Schumaker, K.S.; Zhu, J.K. Regulation of SOS1, a plasma membrane Na⁺/H⁺ exchanger in *Arabidopsis thaliana*, by SOS2 and SOS3. *Proc. Natl. Acad. Sci. USA* **2002**, *99*, 8436–8441. [CrossRef] [PubMed]

16. Oh, D.H.; Lee, S.Y.; Bressan, R.A.; Yun, D.J.; Bohnert, H.J. Intracellular consequences of SOS1 deficiency during salt stress. *J. Exp. Bot.* **2010**, *61*, 1205–1213. [CrossRef] [PubMed]

17. Quintero, F.J.; Martinez-Atienza, J.; Villalta, I.; Jiang, X.; Kim, W.Y.; Ali, Z.; Hiroaki, F.; Imelda, M.; Yun, D.J.; Zhu, J.K.; et al. Activation of the plasma membrane Na/H antiporter Salt-Overly-Sensitive 1 (SOS1) by phosphorylation of an auto-inhibitory C-terminal domain. *Proc. Natl. Acad. Sci. USA* **2011**, *108*, 2611–2616. [CrossRef] [PubMed]

18. Maser, P.; Eckelman, B.; Vaidyanathan, R.; Horie, T.; Fairbairn, D.J.; Kubo, M.; Yamagami, M.; Yamaguchi, K.; Nishimura, M.; Uozumi, N.; et al. Altered shoot/root Na⁺ distribution and bifurcating salt sensitivity in *Arabidopsis* by genetic disruption of the Na⁺ transporter AtHKT1. *FEBS Lett.* **2002**, *531*, 157–161. [CrossRef]

19. Berthomieu, P.; Conejero, G.; Nublat, A.; Brackenbury, W.J.; Lambert, C.; Savio, C.; Uozumi, N.; Oiki, S.; Yamada, K.; Cellier, F.; et al. Functional analysis of AtHKT1 in *Arabidopsis* shows that Na⁺ recirculation by the phloem is phloem is crucial for salt tolerance. *EMBO J.* **2003**, *22*, 2004–2014. [CrossRef] [PubMed]

20. Sunarpi, H.T.; Horie, T.; Motoda, J.; Kubo, M.; Yang, H.; Yoda, K.; Horie, R.; Chan, W.Y.; Leung, H.Y.; Hattori, K.; et al. Enhanced salt tolerance mediated by AtHKT1 transporter-induced Na unloading from xylem vessels to xylem parenchyma cells. *Plant J.* **2005**, *44*, 928–938. [CrossRef] [PubMed]

21. Møller, I.S.; Gilliham, M.; Jha, D.; Mayo, G.M.; Roy, S.J.; Coates, J.C.; Haseloff, J.; Tester, M. Shoot Na⁺ exclusion and increased salinity tolerance engineered by cell type-specific alteration of Na⁺ transport in *Arabidopsis*. *Plant Cell* **2009**, *21*, 2163–2178. [CrossRef] [PubMed]

22. Davenport, R.J.; Muñoz-Mayor, A.; Jha, D.; Essah, P.A.; Rus, A.; Tester, M. the Na⁺ transporter AtHKT1;1 controls retrieval of Na⁺ from the xylem in *Arabidopsis*. *Plant Cell Environ.* **2007**, *30*, 497–507. [CrossRef] [PubMed]

23. Uozumi, N.; Kim, E.J.; Rubio, F.; Yamaguchi, T.; Muto, S.; Tsuboi, A.; Bakker, E.P.; Nakamura, T.; Schroeder, J.I. the *Arabidopsis* HKT1 gene homolog mediates inward Na⁺ currents in *Xenopus laevis oocytes* and Na⁺ uptake in *Saccharomyces cerevisiae*. *Plant Physiol.* **2000**, *122*, 1249–1259. [CrossRef] [PubMed]

24. Horie, T.; Yoshida, K.; Nakayama, H.; Yamada, K.; Oiki, S.; Shinmyo, A. Two types of HKT transporters with different properties of Na⁺ and K⁺ transport in *Oryza sativa*. *Plant J.* **2001**, *27*, 129–138. [CrossRef] [PubMed]

25. Golldack, D.; Su, H.; Quigley, F.; Kamasani, U.R.; Munoz-Garay, C.; Balderas, E.; Popova, O.V.; Bennett, J.; Bohnert, H.J.; Pantoja, O. Characterization of a HKT-type transporter in rice as a general alkali cation transporter. *Plant J.* **2002**, *31*, 529–542. [CrossRef] [PubMed]

26. Asins, M.J.; Villalta, I.; Aly, M.M.; Olias, R.; Morales, P.A.D.; Huertas, R.; Li, J.; Jaime-perez, N.; Haro, R.; Raga, V.; et al. Two closely linked tomato HKT coding genes are positional candidates for the major tomato QTL involved in Na⁺/K⁺ homeostasis. *Plant Cell Environ.* **2012**, *36*, 1171–1191. [CrossRef] [PubMed]

27. Wang, T.T.; Ren, Z.J.; Liu, Z.Q.; Feng, X.; Guo, R.Q.; Li, B.G.; Li, L.G.; Jing, H.-C. SbHKT1;4, a member of the high-affinity potassium transporter gene family from *Sorghum bicolor*, functions to maintain optimal Na⁺/K⁺ balance under Na⁺ stress. *J. Integr. Plant Biol.* **2014**, *56*, 315–332. [CrossRef] [PubMed]

28. Ali, A.; Khan, I.U.; Jan, M.; Khan, H.A.; Hussain, S.; Nisar, M.; Chung, W.S.; Yun, D.-J. the High-Affinity Potassium Transporter EpHKT1;2 From the Extremophile Eutrema parvula Mediates Salt Tolerance. *Front. Plant Sci.* **2018**, *9*, 1108. [CrossRef] [PubMed]

29. Ali, A.; Park, H.C.; Aman, R.; Ali, Z.; Yun, D.J. Role of HKT1 in *Thellungiella salsuginea*, a model extremophile plant. *Plant Signal. Behav.* **2013**, *8*, e25196. [CrossRef] [PubMed]

30. Platten, J.D.; Cotsaftis, O.; Berthomieu, P.; Bohnert, H.J.; Davenport, R.J.; Fairbairn, D.J.; Horie, T.; Leigh, R.A.; Lin, H.X.; Luan, S.; et al. Nomenclature for HKT transporters, key determinants of plant salinity tolerance. *Trends Plant Sci.* **2006**, *11*, 372–374. [CrossRef] [PubMed]

31. Munns, R.; James, R.A.; Xu, B.; Athman, A.; Conn, S.J.; Jordans, C.; Byrt, C.S.; Hare, R.A.; Tyerman, S.D.; Tester, M.; et al. Wheat grain yield on saline soils is improved by an ancestral Na⁺ transporter gene. *Nat. Biotechnol.* **2012**, *30*, 360–366. [CrossRef] [PubMed]

32. Cheong, M.S.; Yun, D.J. Salt-stress signaling. *J. Plant Biol.* **2008**, *50*, 148–155. [CrossRef]

33. Munns, R.; Tester, M. Mechanisms of salinity tolerance. *Annu. Rev. Plant Biol.* **2008**, *59*, 651–681. [CrossRef] [PubMed]

34. Hasegawa, P.M.; Bressan, R.A.; Zhu, J.K.; Bohnert, H.J. Plant Cellular and Molecular Responses to High Salinity. *Annu. Rev. Plant Physiol. Plant Mol. Biol.* **2000**, *51*, 463–499. [CrossRef] [PubMed]

35. Tester, M.; Davenport, R. Na$^+$ tolerance and Na$^+$ transport in higher plants. *Ann. Bot.* **2003**, *91*, 503–527. [CrossRef] [PubMed]

36. Hauser, F.; Horie, T. A conserved primary salt tolerance mechanism mediated by HKT transporters: A mechanism for sodium exclusion and maintenance of high K$^+$/Na$^+$ ratio in leaves during salinity stress. *Plant Cell Environ.* **2009**, *33*, 552–565. [CrossRef] [PubMed]

37. Tsugane, K.; Kobayashi, K.; Niwa, Y.; Ohba, Y.; Wada, K.; Kobayashi, H. A recessive *Arabidopsis* mutant that grows photo-autotrophically under salt stress shows enhanced active oxygen detoxification. *Plant Cell* **1999**, *11*, 1195–1206. [CrossRef] [PubMed]

38. Ren, Z.H.; Gao, J.P.; Li, L.G.; Cai, X.L.; Huang, W.; Chao, D.Y.; Zhu, M.Z.; Wang, Z.Y.; Luan, S.; Lin, H.X. A rice quantitative trait locus for salt tolerance encodes a sodium transporter. *Nat. Genet.* **2005**, *37*, 1141–1146. [CrossRef] [PubMed]

39. Chen, Z.H.; Zhou, M.X.; Newman, I.A.; Mendham, N.J.; Zhang, G.P.; Shabala, S. Potassium and sodium relations in salinised barley tissues as a basis of differential salt tolerance. *Funct. Plant Biol.* **2007**, *34*, 150–162. [CrossRef]

40. Schachtman, D.P.; Schroeder, J.I. Structure and transport mechanism of a high-affinity potassium uptake transporter from higher-plants. *Nature* **1994**, *370*, 655–658. [CrossRef] [PubMed]

41. Rubio, F.; Schwarz, M.; Gassmann, W.; Schroeder, J.I. Genetic selection of mutations in the high affinity K$^+$ transporter HKT1 that define functions of a loop site for reduced Na$^+$ permeability and increased Na$^+$ tolerance. *J. Biol. Chem.* **1999**, *274*, 6839–6847. [CrossRef] [PubMed]

42. Gassmann, W.; Rubio, F.; Schroeder, J.I. Alkali cation selectivity of the wheat root high-affinity potassium transporter HKT1. *Plant J.* **1996**, *10*, 869–882. [CrossRef]

43. Fairbairn, D.J.; Liu, W.H.; Schachtman, D.P.; Gomez-Gallego, S.; Day, S.R.; Teasdale, R.D. Characterization of two distinct HKT1-like potassium transporters from *Eucalyptus camaldulensis*. *Plant Mol. Biol.* **2000**, *43*, 515–525. [CrossRef] [PubMed]

44. Garciadeblás, B.; Senn, M.E.; Banuelos, M.A.; Rodriguez-Navarro, A. Sodium transport and HKT transporters: the rice model. *Plant J.* **2003**, *34*, 788–801. [CrossRef] [PubMed]

45. Su, H.; Balderas, E.; Vera-Estrella, R.; Golldack, D.; Quigley, F.; Zhao, C.S.; Pantoja, O.; Bohnert, J.H. Expression of the cation transporter McHKT1 in a halophyte. *Plant Mol. Biol.* **2003**, *52*, 967–980. [CrossRef] [PubMed]

46. Haro, R.; Banuelos, M.A.; Senn, M.A.E.; Barrero-Gil, J.; Rodríguez-Navarro, A. HKT1 mediates sodium uniport in roots: Pitfalls in the expression of HKT1 in yeast. *Plant Physiol.* **2005**, *139*, 1495–1506. [CrossRef] [PubMed]

47. Takahashi, R.; Liu, S.; Takano, T. Cloning and functional comparison of a high-affinity K$^+$ transporter gene PhaHKT1 of salt-tolerant and saltsensitive reed plants. *J. Exp. Bot.* **2007**, *58*, 4387–4395. [CrossRef] [PubMed]

48. Jabnoune, M.; Espeout, S.; Mieulet, D.; Fizames, C.; Verdeil, J.L.; Conejero, G.; Rodriguez-Navarro, A.; Sentenac, H.; Guiderdoni, E.; Abdelly, C.; et al. Diversity in expression patterns and functional properties in the rice HKT transporter family. *Plant Physiol.* **2009**, *150*, 1955–1971. [CrossRef] [PubMed]

49. Maser, P.; Hosoo, Y.; Goshima, S.; Horie, T.; Eckelman, B.; Yamada, K.; Yoshida, K.; Bakker, E.P.; Shinmyo, A.; Oiki, S.; et al. Glycine residues in potassium channel-like selectivity filters determine potassium selectivity in four-looper-subunit HKT transporters from plants. *Proc. Natl. Acad. Sci. USA* **2002**, *99*, 6428–6433. [CrossRef] [PubMed]

50. Kato, Y.; Sakaguchi, M.; Mori, Y.; Saito, K.; Nakamura, T.; Bakker, E.P.; Sato, Y.; Goshima, S.; Uozumi, N. Evidence in support of a four transmembrane-pore-transmembrane topology model for the *Arabidopsis thaliana* Na$^+$/K$^+$ translocating AtHKT1 protein, a member of the superfamily of K$^+$ transporters. *Proc. Natl. Acad. Sci. USA* **2001**, *98*, 6488–6493. [CrossRef] [PubMed]

51. Mian, A.; Oomen, R.J.; Isayenkov, S.; Sentenac, H.; Maathuis, F.J.; Very, A.A. Over-expression of an Na$^+$ -and K$^+$ -permeable HKT transporter in barley improves salt tolerance. *Plant J.* **2011**, *68*, 468–479. [CrossRef] [PubMed]

52. Garriga, M.; Raddatz, N.; Véry, A.A.; Sentenac, H.; Rubio-Meléndez, M.E.; González, W.; Dreyer, I. Cloning and functional characterization of HKT1 and AKT1 genes of Fragaria spp.—Relationship to plant response to salt stress. *J. Plant Physiol.* **2017**, *210*, 9–17. [CrossRef] [PubMed]

53. Sun, J.; Cao, H.; Cheng, J.; He, X.; Sohail, H.; Niu, M.; Huang, Y.; Bie, Z. Pumpkin CmHKT1;1 controls shoot Na$^+$ accumulation via limiting Na$^+$ transport from rootstock to scion in grafted cucumber. *Int. J. Mol. Sci.* **2018**, *19*, 2648. [CrossRef] [PubMed]

54. Xu, M.; Chen, C.; Cai, H.; Wu, L. Overexpression of pehkt1;1 improves salt tolerance in populous. *Genes* **2018**, *9*, 475. [CrossRef] [PubMed]

55. Wang, L.; Liu, Y.; Feng, S.; Wang, Z.; Zhang, J.; Zhang, J.; Wang, D.; Gan, Y. AtHKT1 gene regulating K$^+$ state in whole plant improves salt tolerance in transgenic tobacco plants. *Sci. Rep.* **2018**, *8*, 16585. [CrossRef] [PubMed]

56. Zhang, Y.; Fang, J.; Wu, X.; Dong, L. Na$^+$/K$^+$ balance and transport regulatory mechanisms in weedy and cultivated rice (*Oryza sativa* L.) under salt stress. *BMC Plant Biol.* **2018**, *18*, 375. [CrossRef] [PubMed]

57. Vinocur, B.; Altman, A. Recent advances in engineering plant tolerance to abiotic stress: Achievements and limitations. *Curr. Opin. Biotechnol.* **2005**, *16*, 123–132. [CrossRef] [PubMed]

58. Gong, Q.; Li, P.; Ma, S.; Indu Rupassara, S.; Bohnert, H.J. Salinity stress adaptation competence in the extremo-phile *Thellungiella halophila* in comparison with its relative *Arabidopsis thaliana*. *Plant J.* **2005**, *44*, 826–839. [CrossRef] [PubMed]

59. Wu, H.J.; Zhang, Z.; Wang, J.Y.; Oh, D.H.; Dassanayake, M.; Liu, B.; Huang, Q.; Sun, H.X.; Xia, R.; Wu, Y.; et al. Insights into salt tolerance from the genome of *Thellungiella salsuginea*. *Proc. Natl. Acad. Sci. USA* **2012**, *109*, 12219–12224. [CrossRef] [PubMed]

60. Oh, D.H.; Hong, H.; Lee, S.Y.; Yun, D.J.; Bohnert, H.J.; Dassanayake, M. Genome Structures and Transcriptomes Signify Niche Adaptation for the Multiple-Ion-Tolerant Extremophyte *Schrenkiella parvula* (*Thellungiella parvula*). *Plant Physiol.* **2014**, *164*, 2123–2138. [CrossRef] [PubMed]

61. Vera-Estrella, R.; Barkla, B.J.; Pantoja, O. Comparative 2D-DIGE analysis of salinity responsive microsomal proteins from leaves of salt-sensitive *Arabidopsis thaliana* and salt-tolerant *Thellungiella salsuginea*. *J. Proteom.* **2014**, *111*, 113–127. [CrossRef] [PubMed]

62. Dassanayake, M.; Oh, D.H.; Haas, J.S.; Hernandez, A.; Hong, H.; Ali, S.; Yun, D.J.; Bressan, R.A.; Zhu, J.K.; Bohnert, H.J.; et al. the genome of the extremophile crucifer *Thellungiella parvula*. *Nat. Genet.* **2011**, *43*, 913–918. [CrossRef] [PubMed]

63. Oh, D.H.; Dassanayake, M.; Haas, J.S.; Kropornika, A.; Wright, C.; d'Urzo, M.P.; Hong, H.; Ali, S.; Hernandez, A.; Lambert, G.M.; et al. Genome structures and halophyte-specific gene expression of the extremophile *Thellungiella parvula* in comparison with *Thellungiella salsuginea* (*Thellungiella halophila*) and *Arabidopsis*. *Plant Physiol.* **2010**, *154*, 1040–1052. [CrossRef] [PubMed]

64. Qi, Z.; Spalding, E.P. Protection of plasma membrane K$^+$ transport by the salt overly sensitive1 Na$^+$/H$^+$ antiporter during salinity stress. *Plant Physiol.* **2004**, *136*, 2548–2555. [CrossRef] [PubMed]

65. Kader, M.A.; Seidel, T.; Golldack, D.; Lindberg, S. Expressions of OsHKT1, OsHKT2, and OsVHA are differentially regulated under NaCl stress in salt-sensitive and salt-tolerant rice (*Oryza sativa* L.) cultivars. *J. Exp. Bot.* **2006**, *57*, 4257–4268. [CrossRef] [PubMed]

66. Ko, C.H.; Gaber, R.F. TRK1 and TRK2 encode structurally related K$^+$ transporters in *Saccharomyces cerevisiae*. *Mol. Cell. Biol.* **1991**, *11*, 4266–4273. [CrossRef] [PubMed]

67. Shabala, S. Ionic and osmotic components of salt stress specifically modulate net ion fluxes from bean leaf mesophyll. *Plant Cell Environ.* **2000**, *23*, 825–837. [CrossRef]

68. Kim, B.G.; Waadt, R.; Cheong, Y.H.; Pandey, G.K.; Dominguez-Solis, J.R.; Schultke, S.; Lee, S.C.; Kudla, J.; Luan, S. the calcium sensorCBL10 mediates salt tolerance by regulating ion homeostasis in *Arabidopsis*. *Plant J.* **2007**, *52*, 473–484. [CrossRef] [PubMed]

69. Orsini, F.; D'Urzo, M.P.; Inan, G.; Serra, S.; Oh, D.H.; Mickelbart, M.V.; Consiglio, F.; Li, X.; Jeong, J.C.; Yun, D.J.; et al. A comparative study of salt tolerance parameters in 11 wild relatives of *Arabidopsis thaliana*. *J. Exp. Bot.* **2010**, *61*, 3787–3798. [CrossRef] [PubMed]

70. Shi, H.; Ishitani, M.; Kim, C.; Zhu, J.K. the *Arabidopsis thaliana* salt tolerance gene SOS1 encodes a putative Na$^+$/H$^+$ antiporter. *Proc. Natl. Acad. Sci. USA* **2000**, *97*, 6896–6901. [CrossRef] [PubMed]

71. Volkov, V.; Amtmann, A. *Thellungiella halophila*, a salt-tolerant relative of *Arabidopsis thaliana*, has specific root ion-channel features supporting K$^+$/Na$^+$ homeostasis under salinity stress. *Plant J.* **2006**, *48*, 342–353. [CrossRef] [PubMed]

72. Volkov, V.; Wang, B.; Dominy, P.J.; Fricke, W.; Amtmann, A. *Thellungiella* halophila, a salt-tolerant relative of *Arabidopsis thaliana*, possesses effective mechanisms to discriminate between potassium and sodium. *Plant Cell Environ.* **2003**, *27*, 1–14. [CrossRef]

13

Progress in Understanding the Physiological and Molecular Responses of *Populus* to Salt Stress

Xiaoning Zhang [1,†], Lijun Liu [2,†], Bowen Chen [1], Zihai Qin [1], Yufei Xiao [1], Ye Zhang [1], Ruiling Yao [1], Hailong Liu [1,*] and Hong Yang [3,*]

[1] Guangxi Key Laboratory of Superior Timber Trees Resource Cultivation, Guangxi Forestry Research Institute, 23 Yongwu Road, Nanning 530002, China; sflzxn@163.com (X.Z.); gfri_bwchen@163.com (B.C.); qinzihai@126.com (Z.Q.); xiaoyufei33@163.com (Y.X.); elaine.ye@163.com (Y.Z.); jullyudi@163.com (R.Y.)

[2] Key Laboratory of State Forestry Administration for Silviculture of the lower Yellow River, College of Forestry, Shandong Agricultural University, Taian 271018, Shandong, China; lijunliu@sdau.edu.cn

[3] Key Laboratory of Economic Plants and Biotechnology, Kunming Institute of Botany, Academy of Sciences, Yunnan Key Laboratory for Wild Plant Resources, Kunming 650201, China

* Correspondence: hailon_liu@163.com (H.L.); yanghong@mail.kib.ac.cn (H.Y.)

† These authors contribute equally to this work.

Abstract: Salt stress (SS) has become an important factor limiting afforestation programs. Because of their salt tolerance and fully sequenced genomes, poplars (*Populus* spp.) are used as model species to study SS mechanisms in trees. Here, we review recent insights into the physiological and molecular responses of *Populus* to SS, including ion homeostasis and signaling pathways, such as the salt overly sensitive (SOS) and reactive oxygen species (ROS) pathways. We summarize the genes that can be targeted for the genetic improvement of salt tolerance and propose future research areas.

Keywords: poplars (*Populus*); salt tolerance; molecular mechanisms; SOS; ROS

1. Introduction

Poplars (*Populus* spp.), which include about 100 species [1], are widely distributed across a variety of climatic regions [2] and have become important species for global afforestation and shelterbelt projects because of their rapid growth and high biomass yields [3]. These traits, in combination with other characteristics, such as extensive and deep root systems, considerable genetic variation, small genome size, convenient asexual propagation, genetic transformability, and economic significance, have led to the use of poplars as model tree species [4–6].

The increasing salinization of soils has greatly limited the planting of salt-sensitive *Populus* species [7]. Salt stress (SS) induces water deficiency, osmotic stress, ion toxicity, and oxidative damage [8] and thereby reduces photosynthesis, respiration, transpiration, metabolism, and growth in poplars. Like most plants, poplars can adapt to SS by maintaining their cellular ion homeostasis, accumulating osmotic-adjustment substances, and activating scavengers of reactive oxygen species (ROS) via the initiation of an efficient signal transduction network [9]. Desert poplar (*Populus euphratica*) is one of the most salt-tolerant *Populus* species [10] and is often used to study the salt-response mechanisms of trees. *P. euphratica* was reported to be tolerant to up to 450 mM NaCl (about 2.63%) under hydroponic conditions and showed high recovery efficiency when NaCl was removed from the culture medium [11]. A previous study has shown that *P. euphratica* could grow in soils with up to 2.0% salinity and can survive in soils with up to 5.0% salinity [12]. As a non-halophyte, *P. euphratica* could activate salt secretion mechanisms when soil salinity concentrations are greater than 20%, which may be one of the reasons for its high salt-tolerance [13]. Most other *Populus* species are relatively salt-sensitive, including the grey poplar (*Populus* × *canescens*), *Populus* × *euramericana*, and *Populus popul003*, which are usually used as a salt-sensitive control in salt-response studies. *Populus alba* has a wide variation of salinity tolerance within the species: for example 'Guadalquivir F-21–38', 'Guadalquivir F-21–39', and 'Guadalquivir F-21–40' clones show salt tolerance, while most other clones have a common salt sensitivity. Considering the wide variation within the species, *P. alba* could be used as a model species to understand the mechanisms of SS [2].

Previous research primarily focused on the anatomical, physiological, and biochemical changes in poplars during SS; many recent studies have focused on the molecular mechanisms using new techniques, such as genome-scale transcript analysis [14], high-throughput sequencing [15], metabolite profiling [16], bioinformatic analyses [17–19], and a non-invasive micro-test technique (NMT) [20].

Here, we review the recent progress in understanding the physiological and molecular responses of *Populus* to SS, including SS injuries, the main mechanisms of salt tolerance, and the genes targeted for the genetic improvement of salt tolerance in *Populus*, with a major focus on ion homeostasis, osmotic adjustment, ROS scavenging, salt overly sensitive (SOS) signaling pathways, potential candidate genes, and transcription factor-mediated regulation of the SS response.

2. SS Injury

2.1. Inhibition of Poplar Growth by Salinity Stress

Salinity affects all stages of *Populus* growth, including germination [21], vegetation growth [15,22], and sexual reproduction [23]. The percentage of seeds that germinate and the extent of leaf expansion were both reported to decline as salt concentrations increase [21]. In addition, shoot growth is more sensitive to salt than root growth [8]. When salt-sensitive white poplar (*P. alba*) clones were exposed to SS (0.6% NaCl), striking reductions were observed in their leaf elongation rate and internode lengths, while significant increases were observed in the numbers of short branches and bud numbers, as well as in the levels of leaf epinasty, necrosis, and abscission [7].

Long-term SS also induces early leaf maturity, early flowering, and early tree maturation in *P. alba* clones, which display a greater architectural modification when exposed to high SS (0.6%) than lower SS (0.3%) [22]. When exposed to 0.6% NaCl, the heights, ground diameter, and leaf numbers of Poplar 107 were significantly reduced, while plants exposed to 0.3% NaCl stress had relatively minor phenotypic changes [15]. When *P. euphratica* was exposed to 300 mM (1.76%) NaCl stress, the three growth indexes (plant height, ground diameter, and leaf number) were reduced to 31%, 45.5%, and 20% of the control plants, respectively. The mean leaf area of these stressed trees was reduced by up to 60%, and the leaves began to wither and yellow after 10 days. By contrast, a treatment of 50 mM (about 0.29%) NaCl did not cause a significant reduction in these traits in *P. euphratica* [24].

2.2. Salt-Induced Physiological and Cellular Changes

The adverse effects of SS also result in physiological and microscopic anatomical changes. *P. euphratica* and *P. alba* trees exposed to SS have significantly reduced stomatal area, aperture, and conductance, but increased stomatal density and hydraulic conductance [25–27]. The salt-induced reduction of leaf area in *Populus* may be one of the reasons for the increased stomatal density and decreased stomatal area [24,27]. The percentage loss of hydraulic conductivity (PLC%) in *P. euphratica* increased from 31.81% at 0 mM NaCl to 83.83% at 150 mM NaCl (0.88%), causing a 40–80% decrease in hydraulic conductivity and ensuring high hydraulic efficiency [27]. *Populus* trees thus reduce their transpiration by decreasing their stomatal apertures and conductance and increasing their PLC% values to cope with the salt-induced water deficit.

Populus may adjust their xylems to adapt to salinity stress [28]; for example, when exposed to SS, salt-sensitive *P.* × *canescens* produces narrower xylem vessels in which it stores sodium ions (Na$^+$), reducing the effect of ion toxicity. In contrast, the salt-tolerant species *P. euphratica* produces narrow xylem vessels to reduce Na$^+$ uptake even under normal conditions, the abundance of which remains largely unaltered under moderate SS [29]. Overall, this may indicate an evolutionary adaptation of the xylem structure in *P. euphratica*.

2.3. Salt-Induced Damage to the Photosynthetic System

Chlorophyll (Chl) is the main pigment used for photosynthesis in plants and is often studied in the evaluation of salt damage. The contents of Chla and Chlb, as well as the relative electron transport rate, decreased significantly in poplar 107 (a superior variety selected from *Populus* × *euramericana* cv. '74/76' hybrids) under 0.6% NaCl but increased under 0.3% NaCl [15].

Almost all poplar trees show a decreased net photosynthetic rate (Pn) under high SS; for example, the Pn decreased by 48.3% in *P. euphratica* under a 300 mM NaCl treatment, in comparison with the unstressed plants [24]. SS has a negative effect on growth, possibly due to the Pn affecting the accumulation of biomass. The stomatal conductance (Gs), transpiration rate, and internal CO_2 (Ci) concentration of *P. euphratica* leaves also decreased by 48.5%, 42.1%, and 15.7% under 300 mM NaCl, indicating that the photosynthetic system was injured [24]. The fluorescence transient curve OJIP is highly sensitive to salinity stress; the OJ phase is the photochemical phase, leading to the reduction of QA to QA⁻. The I level is related to the heterogeneity in the filling up of the plastoquinone (PQ) pool. The P level is reached when all the PQ molecules are reduced to PQH2 [30]. While in poplars treated with 0.3% NaCl the OJIP curve follows the same trend as in the control plants, it significantly decreases at the J, I, and P phases in plants treated with 0.6% NaCl [15]. Large decreases in the J and I phases are also observed in *P. euphratica* treated with 300 mM NaCl [24]. As the electron transfer between Pheo and QA occurs during the J phase, these results indicate the higher salt concentration had a significant influence on electron transfer, which led to a great degree of salt damage [15]. Besides, F_0 (minimal constant fluorescence of dark-adapted plants) and Fv/Fm (a surrogate of the maximum quantum efficiency of PSII) are also used as fluorescence parameters to identify salt tolerance, for example, to detect genotypic differences in the sensitivity of white poplar clones to SS [31].

3. Primary Mechanism of Salt Tolerance in *Populus*

3.1. Maintaining an Optimal K^+/Na^+ Ratio

Salt tolerance can be determined by the net Na^+ efflux capacity in *Populus*. The net uptake of Na^+ depends on its influx, exclusion, and sequestration, as well as on other Na^+ regulation processes, such as xylem Na^+ loading and unloading, and phloem Na^+ recirculation [32]. Na^+ moves into cells through non-selective cation channels (NSCCs) and high-affinity K^+ transporter (HKT) [33], while excessive Na^+ can be extruded into the apoplast across the plasma membrane (PM) by the Na^+/H^+ antiporter salt overly sensitive1 (SOS1; see Section 3.4) [34], which may also involve Na^+ loading in the xylem [35]. The net Na^+ efflux increased significantly in *P. euphratica* after 0.5–12 h under 100 mM NaCl but was reduced at 0.5 h in *P. popularis*, which showed an overall Na^+ influx after 6–12 h of SS [36]. This is consistent with results of X-ray microanalysis showing that *P. euphratica* had lower concentration of Na^+ in all subcellular compartments than salt-sensitive poplar species [37]. Moreover, salt tolerance can be improved by increasing Na^+ efflux and the uptake of mineral nutrients via interactions between mycorrhizal fungi and the roots of salt-sensitive *Populus* [10,38].

It is crucial for *Populus* to maintain an optimal K^+/Na^+ ratio in the cytoplasm when exposed to high salinity. An excessive uptake of Na^+ ions not only increases their abundance in plant cells, but also induces the loss of potassium ions (K^+) by depolarizing the cellular membranes. Moreover, Na^+ can compete with K^+ for the binding sites of the uptake system, resulting in an imbalance of the K^+/Na^+ ratio that eventually causes ion toxicity [32]. Under high SS, *P. euphratica* maintained an optimal Na^+/K^+ ratio by restricting the net Na^+ uptake and transport from roots to shoots and by maintaining higher K^+ uptake and transport [39,40]. X-ray microanalysis showed that high salinity reduced the K^+/Na^+ ratio by 93% in *P. popularis* but only by 69% in *P. euphratica* [40].

Populus improves salt tolerance by accumulating Na^+/K^+ ions in the vacuoles. Na^+ is sequestered into plant cell vacuoles by the tonoplast Na^+/H^+ antiporter NHX1 [41]. Besides detoxifying the cytoplasm, the accumulation of Na^+ ions in the vacuoles is used as an osmoticum to draw water into the cells [42]. Vacuolar H^+-ATPase and vacuolar H^+-PPase generate an electrochemical gradient across the vacuolar membrane for the tonoplast Na^+/H^+ antiporters in *P. euphratica* cells [42]. In addition, expressing the *Arabidopsis thaliana* Na^+/H^+ antiporter gene *AtNHX1* in transgenic poplars improves their salt resistance by improving their Na^+/H^+ exchange activity [43–45]. AtNHX3 was previously shown to act as the tonoplast K^+/H^+ antiporter for the transportation of K^+ and the maintenance of ion homeostasis [46]; however, AtNHX1 was also recently found to enhance the accumulation of K^+ in the vacuoles of transgenic poplars [47]. Yang reported that the constitutive expression of either *AtNHX1* or *AtNHX3* in transgenic *Populus* increased the vacuolar accumulation of Na^+ and K^+, leading to improved salt tolerance and drought tolerance [47].

K^+ and Na^+ uptake and transport mediated by HKT1 facilitate the rapid response of *Populus* to SS. The Na^+/K^+ transporter HKT1 is located on the plasma membrane and mediates the uptake and transport of K^+ and Na^+ in poplar. Furthermore, HKT is also involved in xylem Na^+ unloading in tomato (*Solanum lycopersicum*) [48], as well as in the phloem-mediated recirculation of Na^+ from the shoots to the roots of rice (*Oryza sativa*) [49], avoiding the excessive accumulation of Na^+ in the leaves. The expression of *HKT1* in *P. euphratica* is three times higher after 1 h of a 1% NaCl treatment, which facilitates its rapid response to SS by taking a certain amount of Na^+ ions into the cells to maintain their osmotic balance [50].

Restricting the K^+ efflux is important for the salt resistance of *Populus*. Plasma membrane H^+-ATPase restricts K^+ efflux, thus improving salt resistance of *Populus*. K^+ can be transported into cells via an inward-rectifying K^+ channel and the high-affinity K^+ transporter HKT1, while K^+ efflux from the roots is mediated by the activation (by depolarization) of outward-rectifying K^+ channels (DA-KORCs) and NSCCs (DA-NSCCs), which can be induced by salt and inhibited by the plasma membrane H^+-ATPase [51,52]. The concentration of K^+ is markedly reduced in *P. euphratica* callus cells exposed to SS when pretreated with vanadate, an inhibitor of H^+-ATPase, because of the enhanced efflux of K^+ [53,54].

H^+-ATPase activity plays a large role in salt tolerance of *Populus*. In addition to restricting K^+ efflux, H^+-ATPases can also maintain a proton gradient across the membrane, used by the Na^+/H^+ antiporters [8,55,56], which is closely associated with the salt sensitivity of *Populus*. The activity of plasma membrane H^+-ATPase was higher in salt-tolerant genotypes than in salt-sensitive genotypes of *P. alba* [57]. Similarly, H^+-ATPase genes were more highly expressed in the salt-tolerant species *P. euphratica* than in the salt-sensitive *Populus trichocarpa* [55]. This is further supported by the work of Ma et al., who showed that the *P. euphratica* genome contains more copies of the P-type H^+-ATPase genes than *P. trichocarpa* [58].

Overall, except the above mechanisms, *P. euphratica* owns more effective mechanisms to respond to SS, for example, develop smaller vessel lumina than other salt sensitive poplars to limit ion loading into the xylem and develop leaf succulence after a long time of SS to dilute salt as a plastic morphological adaptation. Up to one-fifth of Na^+ and one-third of Cl^- are stored in foliage in the harvest season and are eliminated as leaves are ultimately shed. Excessive $Na^+(Cl^-)$ can also be extruded via phloem retranslocating into the roots [59]. Sequestering Cl^- in cortical vacuoles at high salinity is also important for restricting its transport into above-ground organs [60].

3.2. Accumulation of Osmotic-Adjustment Substances

The water potential of the soil and the availability of water to plant roots are lower in saline soils; therefore, salt-stressed plants first experience water deficiencies caused by osmotic stress [16]. Stress caused by 150 mM NaCl caused a drop of -0.68 MPa in the osmotic potential and a rapid decrease of 0.77 MPa in the shoot water potential in young *P. euphratica* trees [16]. Plant cells tend to accumulate soluble osmolytes to adjust their osmotic potential, such as proline, glycine betaine, soluble sugars, and proteins, which enable the plants to alleviate the osmotic stress and maintain cell turgor, water uptake, and metabolic activity [8,61]. Both salt-sensitive and salt-tolerant poplar species showed an accumulation of free amino acids under long-term SS [62]. Proline is an important osmotic-adjustment substance that exists in a free state in plant cells, has a low molecular weight, is highly soluble in water, is relatively non-toxic, and has no net charge in the physiological pH range [63]. Proline accumulation preserves the osmotic balance under salinity stress, and proline content can be used as a physiological index of plant resistance to SS [8,64]. Salt-tolerant *P. euphratica* increased proline accumulation by 50–90% when exposed to 150–300 mM NaCl [24], while salt-sensitive hybrid poplars, such as *P. alba* cv. *Pyramidalis* × *P. tomentosa*, showed no significant accumulation when exposed to 50 and 150 mM NaCl [65]. Sucrose and total soluble sugars increased with the elevation of foliar Na^+ and Cl^- concentrations in *P. euphratica* [59]. Except for Valine (Val) and Isoleucine (Ile), soluble carbohydrates, sugar alcohols, organic acids, and amino acids in *P. euphratica* leaves did not show significant changes after 24 h of SS. However, the changes of these amino acids were too low to significantly affect the total osmotic potential of leaves [16]. As a "cheap" osmolyte, the accumulation of sodium mainly contributes to osmotic recovery in *P. euphratica* [16,66].

3.3. ROS and Reactive Nitrogen Species (RNS)

ROS, including hydrogen peroxide (H_2O_2), superoxide anions ($O_2^{\cdot-}$), hydroxyl radicals ($^{\cdot}OH$), and singlet oxygen ($^1O_2$), accumulate when plants are exposed to high SS [67,68]. At moderate levels, functioning as signaling molecules, ROS trigger signal transduction events and elicit specific cellular responses thus regulating plant growth and stress responses [69]. Some ROS can react with almost all the components of living cells leading to severe damage to lipids, proteins, and nucleic acids [70]. Excessive ROS can induce oxidative damage and might be detoxified through enzymatic and non-enzymatic antioxidant systems. Poplars expressing *TaMnSOD* show greatly improved tolerance to NaCl, with higher superoxide dismutase (SOD) activities, lower malondialdehyde (MDA) contents, and lower relative electrical conductivity (REC) than the wild-type lines [71]. The peroxidase (POD) activity of *P. euphratica* was 61.8% higher under 200 mM NaCl stress relative to the control [26], while 3,3'-diaminobenzidine (DAB) staining and H_2O_2 measurement demonstrated a sharply increased level of H_2O_2 in Chinese white poplar *(Populus tomentosa)* exposed to 200 mM NaCl for 24 h [72]. Most of the genes encoding glutathione peroxidases (GSH-Px), glutathione S-transferases (GST), and glutaredoxins were more highly expressed in salt-stressed *P. tomentosa* than in the control [72], while the transcription levels of genes encoding antioxidant enzymes were upregulated [69] in plants exposed to 150 mM NaCl for 24 h [55]. Genes encoding enzymes involved in the glutathione metabolism pathway, including GSH-Px, glucose-6-phosphate Dehydrogenase (G6PD), glucose phosphate dehydrogenase (GPD), and isocitrate dehydrogenase (IDH), were significantly upregulated in plants treated with NaCl, which facilitated the detoxification of the salinity-induced ROS [15].

Non-enzymatic antioxidants include ascorbate (AsA), glutathione (GSH), and carotenoids (Car). Carotenoids not only protect against active oxygen species by quenching the excited states of photosensitizing molecules and singlet oxygen and by scavenging free radicals, but also protect biomembranes against oxidative damage by modifying the structural and dynamic properties of lipid membranes [73]. Recently, a carotenoid-deficient mutant of bacteria Pantoea sp. YR343 was found, showing reduced colonization on *Populus deltoids* roots [74].

In addition to the antioxidant enzymes, heat-shock transcription factors (HSFs) play a role in scavenging ROS in plants under SS. The transgenic expression of *PeHSF* in tobacco enhanced the activities of ascorbate peroxidase, GSH-Px, and glutathione reductase [75], and *PtHSP17.8* expression in *Arabidopsis* increased the activation levels of antioxidative enzymes under SS [76].

RNS includes nitric oxide ($NO^{\cdot}$), nitric dioxide ($NO_2^{\cdot}$), nitrous acid (HNO_2), and dinitrogen tetroxide (N_2O_4), which can be produced when plants are subjected to SS. Like ROS, RNS also function as signaling molecules in the response to abiotic stress. NO is involved in plant growth, development, senescence, as well as stress response [69]. NO was also reported to enhance salt tolerance in plants [77]. NO reacts with GSH, forming S-nitrosoglutathione (GSNO); NO, GSNO, and peroxynitrite (ONOO$^-$) can produce covalent post-translational modifications (PTMs), such as S-nitrosylation and the protein nitration [78]. However, ONOO$^-$, generated from nitric oxide NO and superoxide anion($O_2^{\cdot-}$), can produce tyrosine nitration of plant proteins and originate nitrosative damage in plant cells [69].

3.4. Poplar Salt Stress (SS) Signaling Pathways

Calcium ions (Ca^{2+}) are an important secondary messenger in plants and mediate poplar salt tolerance by enhancing Na^+ exclusion, restricting K^+ efflux, and sustaining the selectivity of the cell membrane [40]. In higher plants, the Ca^{2+}- dependent SOS signaling pathway helps maintain ion homoeostasis and thus confers salt tolerance under saline conditions [79,80]. Upon NaCl exposure, the resulting elevated cytosolic Ca^{2+} levels are sensed by SOS3, which activates SOS2 and stimulates the membrane-localized Na^+/H^+ antiporter SOS1, resulting in Na^+ efflux into the apoplast of the root [79,81]. Similarly, in *Populus*, although *SOS* gene expression is generally ubiquitous, some studies have indicated that SOS2 functions upstream of SOS1 and downstream of SOS3 [82].

In addition to extruding Na^+ from the roots, SOS1 controls the long-distance transport of Na^+ and affects its partitioning in plant organs [34,35]. *PeSOS1* (Salt overly sensitive 1 from *P. euphratica*) expression was upregulated 5- to 10-fold in *P. euphratica* leaves treated with 200 mM NaCl for 24 h relative to the untreated controls, and *PeSOS1* partially suppressed salt sensitivity when transgenically expressed in the *Escherichia coli* mutant strain EP432 [83]. Similarly, *PabSOS1* expression was about five times higher after 12 h of NaCl treatment [82]. SOS2 not only acts as a central regulator of Na^+ extrusion but also is involved in the signaling node between the SOS pathway and other signaling pathways [80].

Tang identified two *CBL10* homologs, *PtCBL10A* and *PtCBL10B*, in the western balsam poplar (*P. trichocarpa*) genome, which may interact with the salt tolerance component PtSOS2 and may help accumulate Na^+ in vacuoles [84]. Like PtSOS3, PtCBL10s also interacts with PtSOS2 to stimulate the activity of PtSOS1. Whereas PtCBL10s primarily functions in green tissues such as the shoots and targets the downstream component PtSOS2 to the tonoplast, PtSOS3 functions in the roots and targets PtSOS2 to the plasma membrane [84].

H^+-ATPases not only provide the proton-motive force used to enhance Na^+/H^+ antiporter activity, but also can restrict the NaCl-induced efflux of K^+ through DA-KORCs and DA-NSCCs (Figure 1). Genes encoding plasma membrane H^+-ATPases are upregulated in *P. euphratica* [37], likely enhancing the exchange of Na^+ and H^+ across the plasma membrane.

As a hinge signal molecule for sensing and responding to SS, H_2O_2 is vital for K^+/Na^+ homeostasis. In response to SS, the salt-resistant species *P. euphratica* rapidly produces H_2O_2 in a process triggered by proton-coupled ion transporters such as the H^+-pumps and the Na^+/H^+ antiporters in the plasma membrane [53]. This H_2O_2 accumulation causes a net Ca^{2+} influx by activating non-selective cation channels, which enhances Ca^{2+} concentration in the cytosol [54] and stimulates the SOS signaling pathway [81,85,86] (Figure 1). In addition, H_2O_2 signaling results in the upregulation of plasma membrane H^+-ATPases, whose activity limits the NaCl-induced efflux of K^+ [40,87]. Overall, H_2O_2 is involved in salt resistance in *P. euphratica* by controlling Na^+ extrusion via the H_2O_2–cytosolic [Ca^{2+}]–SOS pathway and by reducing K^+ efflux to maintain ion homeostasis via the H_2O_2–Ca^{2+}–PM H^+-ATPases pathway (Figure 1).

NADPH oxidases are the main source of H_2O_2. During SS, plasma membrane H^+-ATPases enhance H^+ efflux, decreasing the pH and contributing to the activation of NADPH oxidases, which leads to H_2O_2 production and triggers the Ca^{2+}-dependent SOS signaling pathway [53,54,87] (Figure 1). NaCl induces a transient increase in extracellular ATP (eATP), which is sensed by purinoceptors in the plasma membrane (e.g., $P2K_1$) and causes a rapid H_2O_2 burst that in turn increases the concentration of Ca^{2+} in the cytosol of *Populus* cells [36,88,89]. Consequently, the salt-elicited eATP-H_2O_2-cytosolic [Ca^{2+}] cascade contributes to enhancing Na^+ extrusion through the Ca^{2+}-dependent SOS pathways and to reducing K^+ efflux by activating H^+-ATPase, thus controlling cellular K^+/Na^+ homeostasis (Figure 1).

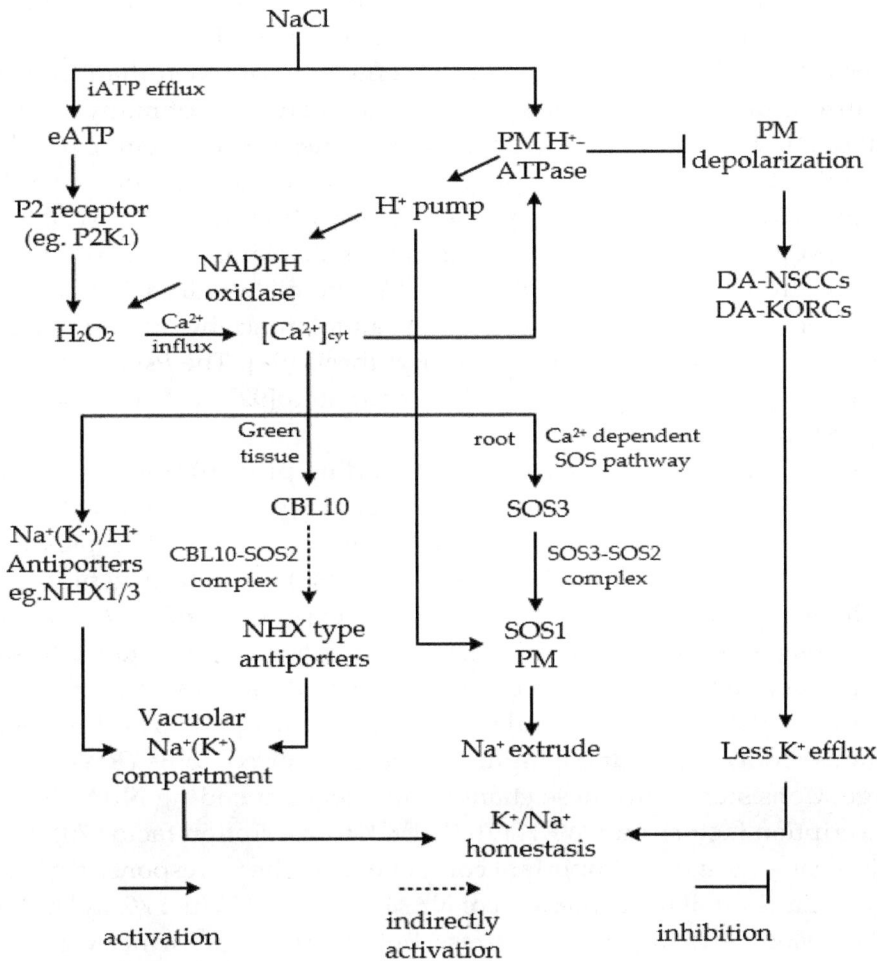

Figure 1. Schematic model showing multiple signaling networks active in *Populus* in response to NaCl stress. NaCl induces the efflux of intracellular ATP (iATP) and an increase in extracellular ATP (eATP), which is sensed by $P2K_1$ in the plasma membrane (PM) and leads to the induction of H_2O_2 production. This stimulates the movement of Ca^{2+} into the cells via Ca^{2+}-permeable channels. The elevated cytosolic Ca^{2+} concentration initiates the SOS pathway by stimulating Na^+/H^+ antiporters, such as SOS1, localized in the PM to extrude Na^+, or activates CBL10, forming the CBL10–SOS2 complex, which may indirectly target the NHX type antiporters to the tonoplast to compartmentalize Na^+ into vacuoles in green tissues. The elevated cytosolic Ca^{2+} also stimulates tonoplast-localized NHX1/3 to accumulate $Na^+(K^+)$ into vacuoles. Besides, the elevated cytosolic Ca^{2+} increases H^+-ATPase activity in the PM, which activates a H^+ pump to supply a proton gradient for the Na^+/H^+ antiporters, stimulating the extrusion of Na^+. A proton gradient supplied by the H^+ pump contributes to the activation of NADPH oxidases, which leads to H_2O_2 production. H^+-ATPases can also inhibit the efflux of K^+ by further polarizing the PM. All these signaling components help to maintain K^+/Na^+ homeostasis in *Populus* cells.

NaCl-induced expression of *HSF* (Heat shock transcription factor) in *P. euphratica* is markedly restricted by inhibitors of NADPH oxidase and Ca^{2+}-permeable channels, suggesting that salt-induced H_2O_2 and cytosolic Ca^{2+} enhance the transcription of *HSFs*, which in turn upregulate genes encoding antioxidant enzymes for scavenging ROS under saline conditions [75].

In conclusion, eATP, Ca^{2+}, H_2O_2, NADPH, H^+-ATPase, and $Na^+ (K^+)/H^+$ transporters play important roles in mediating salt tolerance in *Populus* trees.

4. Candidate Genes Used for the Genetic Improvement of Salt Tolerance

Currently, many studies are focused on introducing known salt-response signaling genes into *Populus* and testing the performance of the transgenic plants in high-salinity conditions. Many of these genes confer a significantly improved salt tolerance, as described below. Transferring transcription factors (TFs) genes into poplars is generally a more efficient approach than transferring structural genes, because transcription factors usually regulate the expression of many target genes in related pathways. A total of 59 *ERF* (Ethylene response factor) genes are associated with SS in *Populus* [90]. *ERF76* from dihaploid *P. simonii* × *P. nigra* plants was transferred into the same *Populus* clone and significantly upregulated 16 genes encoding other transcription factors, as well as 45 stress-related genes [91]. When exposed to SS, *ERF76*-expressing transgenic plants were significantly taller and had increased root lengths, fresh weights, abscisic acid (ABA) and gibberellic acid (GA) contents compared to the control. Transgenic *ERF76* expression enhanced salt tolerance by upregulating the expression of stress-related genes and increasing ABA and GA biosynthesis [91]. The *PsnERF75* gene from *P. simonii* × *P. nigra* is induced by salt, drought, and ABA treatments [92] and confers salt tolerance when transgenically expressed in *Arabidopsis* [91].

The DREB (for dehydration-responsive element-binding protein) transcription factors, members of the ERF family, are vital regulatory nodes in the signaling pathways involved in the salt-stress response [93]. *PeDREB2a*, encoding a DREB transcription factor in *P. euphratica*, improved salt tolerance in *Arabidopsis* or birdsfoot trefoil (*Lotus corniculatus*) when transgenically expressed under the stress-inducible *rd29A* promoter [94]. The transgenic expression of *LbDREB* (a *DREB* gene from the halophyte *Limonium bicolor*) in *Populus ussuriensis* enhanced its resistance to salt, increasing its SOD and POD activities and the expression of the genes encoding these enzymes, reducing its MDA content, and enhancing its proline accumulation in the leaves [93]. The transgenic *P. ussuriensis* plants also had higher root/shoot ratios, higher relative water contents (RWC), and lower relative electrolytic leakage. Consistent with these changes, the genes encoding NAM (no apical meristem), GT-1(trihelix transcription factor), and WRKY70 (WRKY transcription factor 70) displayed inducible temporal expression patterns and are important components in the SS response signaling networks [93]. The LbDREB protein may inhibit the expression of *NAM*, *GT-1*, and *WRKY70* and induce the expression of *SOD* and *POD* in response to high salinity stress, but this requires further verification.

The GTPase RabE is located in the Golgi apparatus and the plasma membrane, where it plays an important role in vesicle transport [95]. The overexpression of constitutively active *PtRabE1b* conferred salt tolerance in poplar [96]. This gene is directly co-expressed with many genes involved in salt tolerance, such as *HSFA4a*, *SOS2*, *MPK19*, and the Ca^{2+} signaling-related genes *CAM7*, *CKL6*, and calcium exchanger. *HSFA4a* expression is regulated by oxidative stress and MPK3/MPK6 and positively influenced salt tolerance in *Arabidopsis* [97]. *CmHSFA4*, a *Chrysanthemum* homologue of this gene, positively regulates salt tolerance by regulating the activities of SOS1, HKT2, and the ROS scavengers [98].

Recently, Yoon et al. identified a novel gene, *PagSAP1*(stress-associated proteins), from the hybrid poplar *P. alba* × *Populus glandulosa*. *PagSAP1* negatively mediates salt-stress responses, and SS can in turn suppress the expression of this gene in poplar roots [99]. *PagSAP1* overexpression resulted in enhanced sensitivity to SS, while *PagSAP1* silencing via RNA interference (RNAi) significantly increased cytosolic Ca^{2+} in the roots. This increased cytosolic Ca^{2+} activated SOS signal transduction, resulting in high *SOS3* transcript levels in the RNAi-derived plants. *HKT1* expression is significantly reduced in all poplar genotypes under salt treatment; however, the lowest level is observed in the *PagSAP1*-overexpressing lines. HKT1 is responsible for Na^+ influx and xylem-mediated Na^+ recirculation from the shoot to the root [100]; therefore, the low HKT1 activity levels in the *PagSAP1*-overexpressing lines may explain the higher Na^+ accumulation in the leaves and the lower Na^+ levels in the roots compared with the control and RNAi-derived lines. By contrast, the excess Na^+ in the roots of the *PagSAP1*-RNAi lines was eliminated by increased SOS1 activity, which resulted in lower Na^+ levels in both the roots and the leaves of these lines. As a result, the salt tolerance of the *PagSAP1*-RNAi lines was improved through the upregulation of *SOS3, SOS1, HKT1, H⁺-ATPase*, *AAA-type ATPase*, and Arabidopsis K+ channel 2 (*AKT2*), all of which are essential for maintaining Na^+/K^+ homeostasis [99].

The poplars SOS proteins share high functional conservation with their *Arabidopsis* homologues [82]. SOS2 interacts with or regulates the activity of several tonoplast-localized transporters, such as the Ca^{2+}/H^+ antiporter [101], the vacuolar H^+- ATPase [102], and the Na^+/H^+ exchanger [103,104]. *PtSOS2.1, PtSOS2.2,* and *PtSOS2.3* (the *PtSOS2* genes in *P. trichocarpa*) overexpression improves the salt tolerance of poplars and increases the concentrations of proline and photosynthetic pigments, relative water content, and the activity of their antioxidant enzymes, while significantly decreasing the levels of MDA [105].

The mutant SOS2 protein PtSOS2TD, generated by mutating the 169th amino acid in the activation loop of PtSOS2 from threonine (T) to aspartic acid (D), is more active than PtSOS2 and can sufficiently activate PtSOS1 in a PtSOS3-independent manner [80]. *PtSOS2TD* overexpression in poplars significantly increased salt tolerance, causing higher plasma membrane Na^+/H^+ exchange activity, greater Na^+ efflux, decreased Na^+ accumulation in the leaves, and improved ROS scavenging capacity [80].

The transgenic expression of the *PtCBL10s* (Calcineurin B-like from *P. trichocarpa*) conferred greater salt tolerance to poplars by maintaining shoot ion homeostasis under SS. The doubling of the *CBL10* genes in poplar may represent an evolutionary adaptation to the adverse environment [84].

The genes that have been shown to increase the salt tolerance of transgenic *Populus* are presented in Table 1.

Table 1. Candidate genes for improving salt tolerance in *Populus*.

Candidate Genes and Source	Transgenic Species	Effect of SS in Transgenic Species Compared with WT	Reference
AtNHX1/3 (Vacuolar Na⁺/H⁺ antiporter from *Arabidopsis thaliana*)	*Populus davidiana × Populus bolleana*	1 Normal growth and morphology; 2 Promoted vacuolar Na⁺ (K⁺)/H⁺ exchange activity; 3 Increased Na⁺ and K⁺ accumulation in the vacuoles; 4 Elevated the eaccumulation of proline.	[47]
AtNHX1 (See above)	*Populus × euramericana* 'Neva'	1 Enhanced plant growth and photosynthetic capacity; 2 Lowerd MDA and REC; 3 Increased Na⁺ (K⁺) accumulation in roots and leaves.	[44]
	Populus × euramericana 'Neva'	1 Reduced decrease in Chl, Car, PSII, Fv/Fm, and qP; 2 Smaller reduction of Pn, Gs, Ci, CE; 3 Greater increase of stem and leaf, smaller increase in root.	[43]
	Populus deltoides × P. euramericana CL 'NL895'	1 Higher content of sodium ions; 2 Decreased MDA content.	[45]
PtSOS2TD (Salt overly sensitive 2 from *Populus trichocarpa*)	*P. davidiana × P. bolleana* hybrid poplar clone Shanxin	1 More vigorous growth; 2 Greater biomass produced; 3 Less Na⁺ in the leaves; 4 Higher Na⁺/H⁺ exchange activity and Na⁺ efflux; 5 More scavenging of ROS.	[80]
PtSOS2 (See above)	*Populus tremula × Populus tremuloides* Michx clone T89	1 Improved PM Na⁺/H⁺ exchange activity, Na⁺ efflux; 2 Higher proline activity; 3 Higher RWC and sustained decrease of water loss; 4 Increased SOD, POD, CAT activity; 5 Decreased MDA concentration.	[105]
PtCBL10A and *PtCBL10B* (Calcineurin B-like from *P. trichocarpa*)	*P. davidiana × P. bolleana* hybrid poplar clone Shanxin	1 Less impairment by SS with higher stature and greater shoot biomass; 2 Lower Na⁺ in the leaves, more Na⁺ in the stem.	[84]
PeCBL6, PeCBL10 (Calcineurin B-like from *P. euphratica*)	triploid white poplar	1 Higher height growth rate; 2 Less wilted leaves; 3 Lower MDA content; 4 Higher chl content.	[106]
PtSOS3 (Salt overly sensitive 3 from *P. trichocarpa*)	*P. davidiana × P. bolleana* hybrid poplar clone Shanxin	1 Lower Na⁺ in the root; 2 Higher K⁺ content in the root; 3 More Na⁺ in the stem.	[84]
TaMnSOD (Mn-superoxide dismutases from *Tamarix Androssowii*)	*P. davidiana × P. bolleana* hybrid poplar clone Shanxin	1 Higher SOD activity; 2 Lower MDA contents; 3 Lower REC; 4 More weight gains.	[71]

Table 1. *Cont.*

Candidate Genes and Source	Transgenic Species	Effect of SS in Transgenic Species Compared with WT	Reference
TaLEA (Late embryogenesis abundant from *T. androssowii*)	*Populus simonii × Populus nigra* Xiaohei poplar	1 Decrease in MDA content; 2 Decrease in relative electrolyte leakage; 3 Improved salt and drought resistance.	[107]
	P. davidiana × P. bolleana	1 Higher Survival percentages; 2 Higher Seedling height and photosynthetic capabilities; 3 Lower Na$^+$ in young leaves but higher in yellow and withered leaves.	[108]
ERF76 (Ethylene response factor from di-haploid *P. simonii × P. nigra*)	*P. simonii × P. nigra* di-haploid	1 Higher plant height, root length, fresh weight; 2 Higher in ABA and GA concentration.	[91]
JERFs (Jasmonic ethylene responsive factor from the tomato)	*Populus alba × Populus berolinensis*	1 Lower reductions of height, basal diameter, and biomass; 2 Lower reduction in leaf water content and increase in root/crown ratio; 3 Greater increase of foliar proline concentration; 4 Higher foliar Na$^+$ concentration.	[109]
LbDREB (dehydration responsive element binding TF from *Limonium bicolor*)	*Populus ussuriensis* Kom. Chinese Daqing poplar	1 Higher SOD, POD activity; 2 Less MDA accumulation in the leaves; 3 More proline accumulation; 4 Increased root/shoot ratio; 5 Reduced decrease of RWC; 6 Lower increase of relative electrolytic leakage.	[93]
AhDREB1 (dehydration responsive element binding-like TF from the halophyte *Atriplex hortensis*)	*Populus tomentosa*	1 Higher survival rate; 2 High proline content.	[110]
AtSTO1 (Salt tolerant1 from *Arabidopsis thaliana*)	*P. tremula × P. alba* Poplar 717-1B4	1 Higher aboveground biomass; 2 Higher root biomass; 3 Higher shoot height; 4 Higher chl content.	[111]
AtPLDα (Phospholipase Dα from *A. thaliana*)	*P. tomentosa*	1 Higher root rate and root length; 2 Reduced decrease of total chl content; 3 Lower REC and MDA content 4 Higher SOD, POD, and CAT activities.	[112]
AtSRK2C, AtGolS2 (Stress responses, SNF1-related protein kinase 2C, galactinol synthase 2 from *A. thaliana*)	*P. tremula × tremuloides*	1 Reduced decrease of dry weight; 2 Reduced decrease of total adventitious root length.	[113]
PtRabE1b(Q74L) (Rab GTPase from *P. trichocarpa*)	*P. alba × P. glandulosa* clone 84 K	1 More adventitious roots; 2 Greater root growth status in seedlings.	[96]
PagSAP1 (stress-associated proteins from *P. alba × P. glandulosa*)	*P. alba × P. glandulosa*	RNAi plants accumulate more Ca^{2+}, and K$^+$ and less Na$^+$.	[99]

SS, salt stress; PM, plasma membrane; SOD, superoxide dismutase; POD, peroxidase; CAT, catalase; MDA, malondialdehyde; Chl, chlorophyll; Car, carotenoid; PSII, actual quantum yield of PSII; Fv/Fm, maximum photochemical efficiency; qP, photochemical quenching coefficient; Pn, net photosynthetic rate; Gs, stomatal conductance; Ci, internal CO_2; CE, carboxylation efficiency concentration; ROS, reactive oxygen species; REC, relative electrical conductivity; RWC, relative water content; ABA, abscisic acid; GA, gibberellic acid.

5. Conclusions and Outlook

Soil salinization is increasingly problematic and is now a dominant factor limiting *Populus* growth [7]. Therefore, it is important to improve the salt tolerance of poplar trees. Plant salt tolerance is a typical quantitative trait affected by many physiological and biochemical factors [15]. Different *Populus* species, such as the salt-resistant poplar species *P. euphratica* and the salt-sensitive *Populus* species *P. × canescens*, have different SS responses [114,115]. Furthermore, trees such as the hybrid poplar 107 display different responses when exposed to different salinity levels [15]. Overall, *Populus* adapt to SS by maintaining suitable Na$^+$/K$^+$ ratios, accumulating osmotic-adjustment substances, activating antioxidative enzymes and antioxidants, and activating stress response signaling networks to reduce the negative effects of high salinity. With the innovation of transcriptomics technologies, a substantial number of stress-responsive and/or stress-regulated genes have been identified—in addition to the signal regulatory networks in which they function—and transferred between *Populus* and *Arabidopsis* or other species with great success. This has greatly elucidated the molecular mechanisms of the poplar stress responses [113].

ROS/RNS and hormones were identified as signaling molecules involved in the response to SS avoiding high salinity damage. The crosstalk between ROS, RNS, ABA, ethylene, and/or other hormones in poplar salt stress will be further studied. Molecular chaperones, especially dehydrins and osmotin, which contribute to protect proteins, are supposed key factors for coping with SS [16]. Nutrient fertilization with N and P was reported to reduce the accumulation of ROS (e.g., O_3) and enhance membrane stability, thus protecting from oxidative stress by activating a cross-talk between antioxidant and osmotic mechanisms [116]. The role of mycorrhization and polymer amendment in enhancing mineral nutrition and improving salt tolerance is also a topic of future study. All of the above topics need to be more deeply studied in the future to improve salt tolerance in poplar species as well as in other tree species.

Determining the key transcription factors and molecular mechanisms underlying salt tolerance is an important goal for future research and will facilitate the enhancement of salt tolerance in *Populus*.

Author Contributions: All authors worked on manuscript preparation.

Acknowledgments: We are grateful to Li Liu in KIB for the kind discussion and guidance to the project.

Abbreviations

SOS	salt overly sensitive
ROS	reactive oxygen species
SOD	superoxide dismutase
POD	peroxidase
MDA	malondialdehyde
GSH-Px	glutathione peroxidases
GST	glutathione S-transferases
GR	glutathione reductase
GPX	glutathione peroxidases
GPD	glucose phosphate dehydrogenase
G6PD	glucose-6-phosphate dehydrogenase
APX	ascorbate peroxidase
IDH	Isocitrate dehydrogenase
GA	gibberellic acid
ABA	abscisic acid
REC	relative electrical conductivity
RWC	relative water content
TFs	transcription factors
Chl	chlorophyll
Pn	net photosynthetic rate
Gs	stomatal conductance
Ci	internal CO_2
CE	carboxylation efficiency concentration;
Car	carotenoid
PSII	actual quantum yield of PSII
qP	photochemical quenching coefficient
NPQ	non photochemical quenching
Fv/Fm	maximum photochemical efficiency
eATP	extracellular ATP
iATP	intracellular ATP
DAB	3,3'-diaminobenzidine
PM	plasma membrane

D A	depolarization activated
KORCs	K^+ outward rectifying channels
NSCCs	non-selective cation channels
SS	salt stress

References

1. Wu, Z.Y.; Raven, P.H. (Eds.) *Flora of China*; Beijing: Science Press: Beijing, China; Missouri Botanical Garden Press: St. Louis, MO, USA, 1999; Volume 4, pp. 139–162.

2. Sixto, H. Response to sodium chloride in different species and clones of genus *Populus* L. *Forestry* **2005**, *78*, 93–104. [CrossRef]

3. Jansson, S.; Douglas, C.J. *Populus*: A model system for plant biology. *Annu. Rev. Plant Biol.* **2007**, *58*, 435–458. [CrossRef]

4. Bradshaw, H.D.; Ceulemans, R.; Davis, J.; Stettler, R. Emerging Model Systems in Plant Biology: Poplar (*Populus*) as A Model Forest Tree. *J. Plant Growth Regul.* **2000**, *19*, 306–313. [CrossRef]

5. Polle, A.; Douglas, C. The molecular physiology of poplars: Paving the way for knowledge-based biomass production. *Plant Biol.* **2010**, *12*, 239–241. [CrossRef]

6. Taylor, G. *Populus*: Arabidopsis for Forestry. Do We Need a Model Tree? *Ann. Bot.* **2002**, *90*, 681–689. [CrossRef] [PubMed]

7. Wang, R.G.; Chen, S.L.; Deng, L.; Fritz, E.; Hüttermann, A.; Polle, A. Leaf photosynthesis, fluorescence response to salinity and the relevance to chloroplast salt compartmentation and anti-oxidative stress in two poplars. *Trees* **2007**, *21*, 581–591. [CrossRef]

8. Munns, R.; Tester, M. Mechanisms of salinity tolerance. *Annu. Rev. Plant Biol.* **2008**, *59*, 651–681. [CrossRef]

9. Hirayama, T.; Shinozaki, K. Research on plant abiotic stress responses in the post-genome era: Past, present and future. *Plant J.* **2010**, *61*, 1041–1052. [CrossRef]

10. Chen, S.L.; Hawighorst, P.; Sun, J.; Polle, A. Salt tolerance in *Populus*: Significance of stress signaling networks, mycorrhization, and soil amendments for cellular and whole-plant nutrition. *Environ. Exp. Bot.* **2014**, *107*, 113–124. [CrossRef]

11. Gu, R.S.; Fonseca, S.; PuskÁs, L.G.; Hackler, L.J.; Zvara, Á.; Dudits, D.; Pais, M.S. Transcript identification and profiling during salt stress and recovery of *Populus euphratica*. *Tree Physiol.* **2004**, *24*, 275–276. [CrossRef]

12. Wang, S.; Chen, B.; Li, H. *Euphrates Poplar Forest*; China Environmental Science Press: Beijing, China, 1996; pp. 43–52.

13. Fu, A.H.; Li, W.H.; Chen, Y.N. The threshold of soil moisture and salinity influencing the growth of *Populus euphratica* and *Tamarix ramosissima* in the extremely arid region. *Environ. Earth Sci.* **2012**, *66*, 2519–2529. [CrossRef]

14. Qiu, Q.; Ma, T.; Hu, Q.J.; Liu, B.B.; Wu, Y.X.; Zhou, H.H.; Wang, Q.; Wang, J.; Liu, J.Q. Genome-scale transcriptome analysis of the desert poplar, *Populus euphratica*. *Tree Physiol.* **2011**, *31*, 452–461. [CrossRef] [PubMed]

15. Chen, P.F.; Zuo, L.H.; Yu, X.Y.; Dong, Y.; Zhang, S.; Yang, M.S. Response mechanism in *Populus × euramericana* cv. '74/76' revealed by RNA-seq under salt stress. *Acta Physiol. Plant* **2018**, *40*. [CrossRef]

16. Brinker, M.; Brosche, M.; Vinocur, B.; Abo-Ogiala, A.; Fayyaz, P.; Janz, D.; Ottow, E.A.; Cullmann, A.D.; Saborowski, J.; Kangasjarvi, J.; et al. Linking the salt transcriptome with physiological responses of a salt-resistant *Populus* species as a strategy to identify genes important for stress acclimation. *Plant Physiol.* **2010**, *154*, 1697–1709. [CrossRef] [PubMed]

17. Yer, E.N.; Baloglu, M.C.; Ayan, S. Identification and expression profiling of all Hsp family member genes under salinity stress in different poplar clones. *Gene* **2018**, *678*, 324–336. [CrossRef] [PubMed]

18. Zhao, K.; Li, S.X.; Yao, W.J.; Zhou, B.R.; Li, R.H.; Jiang, T.B. Characterization of the basic helix-loop-helix gene family and its tissue-differential expression in response to salt stress in poplar. *PeerJ* **2018**, *6*, e4502. [CrossRef] [PubMed]

19. Zhao, K.; Zhang, X.M.; Cheng, Z.H.; Yao, W.J.; Li, R.H.; Jiang, T.B.; Zhou, B.R. Comprehensive analysis of the three-amino-acid-loop-extension gene family and its tissue-differential expression in response to salt stress in poplar. *Plant Physiol. Biochem.* **2019**, *136*, 1–12. [CrossRef]

20. Zhang, Y.N.; Wang, Y.; Sa, G.; Zhang, Y.H.; Deng, J.Y.; Deng, S.R.; Wang, M.J.; Zhang, H.L.; Yao, J.; Ma, X.Y.; et al. Populus euphratica J3 mediates root K^+/Na^+ homeostasis by activating plasma membrane H^+-ATPase in transgenic Arabidopsis under NaCl salinity. Plant Cell Tiss Organ Cult. **2017**, *131*, 75–88. [CrossRef]

21. Liu, J.P.; Li, Z.J.; He, L.R.; Zhou, Z.L.; Xu, Y.L. Salt-tolerance of Populus euphratica and P. pruinosa seed during germination. Sci. Silvae Sin. **2004**, *40*, 165–169. [CrossRef]

22. Abassi, M.; Mguis, K.; Béjaoui, Z.; Albouchi, A. Morphogenetic responses of Populus alba L. under salt stress. J. For. Res.-JPN **2014**, *25*, 155–161. [CrossRef]

23. Meilan, R.; Sabatti, M.; Ma, C.P.; Elena, K. An Early-Flowering Genotype of Populus. J. Plant Biol. **2004**, *47*, 52–56. [CrossRef]

24. Zhao, C.Y.; Si, J.H.; Feng, Q.; Deo, R.C.; Yu, T.F.; Li, P.D. Physiological response to salinity stress and tolerance mechanics of Populus euphratica. Environ. Monit. Assess. **2017**, *189*, 533. [CrossRef] [PubMed]

25. Abbruzzese, G.; Beritognolo, L.; Muleo, R.; Piazzai, M.; Sabatti, M.; Mugnozza, G.S.; Kuzminsky, E. Leaf morphological plasticity and stomatal conductance in three Populus alba L. genotypes subjected to salt stress. Environ. Exp. Bot. **2009**, *66*, 381–388. [CrossRef]

26. Rajput, V.D.; Chen, Y.; Ayup, M. Effects of high salinity on physiological and anatomical indices in the early stages of Populus euphratica growth. Russ. J. Plant Physiol. **2015**, *62*, 229–236. [CrossRef]

27. Rajput, V.D.; Chen, Y.N.; Ayup, M.; Minkina, T.; Sushkova, S.; Mandzhieva, S. Physiological and hydrological changes in Populus euphratica seedlings under salinity stress. Acta Ecol. Sin. **2017**, *37*, 229–235. [CrossRef]

28. Awad, H.; Barigah, T.; Badel, E.; Cochard, H.; Herbette, S. Poplar vulnerability to xylem cavitation acclimates to drier soil conditions. Physiol. Plant. **2010**, *139*, 280–288. [CrossRef] [PubMed]

29. Junghans, U.; Polle, A.; Duchting, P.; Weiler, E.; Kuhlman, B.; Gruber, F.; Teichmann, T. Adaptation to high salinity in poplar involves changes in xylem anatomy and auxin physiology. Plant Cell Environ. **2006**, *29*, 1519–1531. [CrossRef] [PubMed]

30. Strassert, R.J.; Srivastava, A. Polyphasic chlorophyll a fluorescence transient in plants and cyanobacteria. Photochem. Photobiol. **1995**, *61*, 32–42. [CrossRef]

31. Sixto, H.; Aranda, I.; Grau, J.M. Assessment of salt tolerance in Populus alba clones using chlorophyll fluorescence. Photosynthetica **2006**, *44*, 169–173. [CrossRef]

32. Wu, H.H. Plant salt tolerance and Na^+ sensing and transport. Crop J. **2018**, *6*, 215–225. [CrossRef]

33. Ward, J.M.; Hirschi, K.D.; Sze, H. Plants pass the salt. Trends Plant Sci. **2003**, *8*, 200–201. [CrossRef]

34. Olias, R.; Eljakaoui, Z.; Li, J.; de Morales, P.A.; Marin-Manzano, M.C.; Pardo, J.M.; Belver, A. The plasma membrane Na^+/H^+ antiporter SOS1 is essential for salt tolerance in tomato and affects the partitioning of Na^+ between plant organs. Plant Cell Environ. **2009**, *32*, 904–916. [CrossRef] [PubMed]

35. Shi, H.Z.; Quintero, F.J.; Pardo, J.M.; Zhu, J.K. The Putative Plasma Membrane Na^+/H^+ Antiporter SOS1 Controls Long-Distance Na^+ Transport in Plants. Plant Cell Online **2002**, *14*, 465–477. [CrossRef]

36. Zhao, N.; Wang, S.J.; Ma, X.J.; Zhu, H.P.; Sa, G.; Sun, J.; Li, N.F.; Zhao, C.J.; Zhao, R.; Chen, S.L. Extracellular ATP mediates cellular K^+/Na^+ homeostasis in two contrasting poplar species under NaCl stress. Trees **2015**, *30*, 825–837. [CrossRef]

37. Ma, X.Y.; Deng, L.; Li, J.K.; Zhou, X.Y.; Li, N.Y.; Zhang, D.C.; Lu, Y.J.; Wang, R.G.; Sun, J.; Lu, C.F.; et al. Effect of NaCl on leaf H^+-ATPase and the relevance to salt tolerance in two contrasting poplar species. Trees **2010**, *24*, 597–607. [CrossRef]

38. Li, J.; Bao, S.Q.; Zhang, Y.H.; Ma, X.J.; Mishra-Knyrim, M.; Sun, J.; Sa, G.; Shen, X.; Polle, A.; Chen, S.L. Paxillus involutus strains MAJ and NAU mediate K^+/Na^+ homeostasis in ectomycorrhizal Populus × canescens under sodium chloride stress. Plant Physiol. **2012**, *159*, 1771–1786. [CrossRef] [PubMed]

39. Chen, S.L.; Li, J.K.; Wang, S.S.; Fritz, E.; Hüttermann, A.; Altman, A. Effects of NaCl on shoot growth, transpiration, ion compartmentation, and transport in regenerated plants of Populus euphratica and Populus tomentosa. Can. J. For. Res. **2003**, *33*, 967–975. [CrossRef]

40. Sun, J.; Dai, S.X.; Wang, R.G.; Chen, S.L.; Li, N.Y.; Zhou, X.Y.; Lu, C.F.; Shen, X.; Zheng, X.J.; Hu, Z.M.; et al. Calcium mediates root K^+/Na^+ homeostasis in poplar species differing in salt tolerance. Tree Physiol. **2009**, *29*, 1175–1186. [CrossRef]

41. Mansour, M.M.F.; Salama, K.H.A.; Al Mutawa, M.M. Transport proteins and salt tolerance in plants. Plant Sci. **2003**, *164*, 891–900. [CrossRef]

42. Silva, P.; Façanha, A.R.; Tavares, R.M.; Gerós, H. Role of Tonoplast Proton Pumps and Na^+/H^+ Antiport System in Salt Tolerance of Populus euphratica Oliv. J. Plant Growth Regul. **2010**, *29*, 23–34. [CrossRef]

43. Jiang, C.Q.; Zheng, Q.S.; Liu, Z.P.; Liu, L.; Zhao, G.M.; Long, X.H.; Li, H.Y. Seawater-irrigation effects on growth, ion concentration, and photosynthesis of transgenic poplar overexpressing the Na$^+$/H$^+$ antiporter AtNHX1. *J. Plant Nutr. Soil Sci.* **2011**, *174*, 301–310. [CrossRef]

44. Jiang, C.Q.; Zheng, Q.S.; Liu, Z.P.; Xu, W.J.; Liu, L.; Zhao, G.M.; Long, X.H. Overexpression of *Arabidopsis thaliana* Na$^+$/H$^+$ antiporter gene enhanced salt resistance in transgenic poplar (*Populus* × *euramericana* 'Neva'). *Trees* **2012**, *26*, 685–694. [CrossRef]

45. Qiao, G.R.; Zhuo, R.Y.; Liu, M.Y.; Jiang, J.; Li, H.Y.; Qiu, W.M.; Pan, L.Y.; lin, S.; Zhang, X.G.; Sun, Z.X. Over-expression of the *Arabidopsis* Na$^+$/H$^+$ antiporter gene in *Populus deltoides* CL × *P. euramericana* CL "NL895" enhances its salt tolerance. *Acta Physiol. Plant* **2011**, *33*, 691–696. [CrossRef]

46. Liu, H.; Tang, R.J.; Zhang, Y.; Wang, C.T.; Lv, Q.D.; Gao, X.S.; Li, W.B.; Zhang, H.X. AtNHX3 is a vacuolar K$^+$/H$^+$ antiporter required for low-potassium tolerance in *Arabidopsis thaliana*. *Plant Cell Environ.* **2010**, *33*, 1989–1999. [CrossRef]

47. Yang, L.; Liu, H.; Fu, S.M.; Ge, H.M.; Tang, R.J.; Yang, Y.; Wang, H.H.; Zhang, H.X. Na$^+$/H$^+$ and K$^+$/H$^+$ antiporters AtNHX1 and AtNHX3 from *Arabidopsis* improve salt and drought tolerance in transgenic poplar. *Biol. Plant.* **2017**. [CrossRef]

48. Jaime-Perez, N.; Pineda, B.; Garcia-Sogo, B.; Atares, A.; Athman, A.; Byrt, C.S.; Olias, R.; Asins, M.J.; Gilliham, M.; Moreno, V.; et al. The sodium transporter encoded by the HKT1;2 gene modulates sodium/potassium homeostasis in tomato shoots under salinity. *Plant Cell Environ.* **2017**, *40*, 658–671. [CrossRef]

49. Kobayashi, N.I.; Yamaji, N.; Yamamoto, H.; Okubo, K.; Ueno, H.; Costa, A.; Tanoi, K.; Matsumura, H.; Fujii-Kashino, M.; Horiuchi, T.; et al. OsHKT1;5 mediates Na$^+$ exclusion in the vasculature to protect leaf blades and reproductive tissues from salt toxicity in rice. *Plant J.* **2017**, *91*, 657–670. [CrossRef]

50. Xu, M.; Sun, Z.M.; Liu, S.A.; Chen, C.H.; Xu, L.A.; Huang, M.R. Cloning and Expression Analysis of Salinity Stress Related Peu HKT1 Gene from *Populus euphratica*. *MPB* **2016**, in press.

51. Britto, D.T.; Kronzucker, H.J. Cellular mechanisms of potassium transport in plants. *Physiol. Plant.* **2008**, *133*, 637–650. [CrossRef]

52. Shabala, L.; Zhang, J.; Pottosin, I.; Bose, J.; Zhu, M.; Fuglsang, A.T.; Velarde-Buendia, A.; Massart, A.; Hill, C.B.; Roessner, U.; et al. Cell-Type-Specific H$^+$-ATPase Activity in Root Tissues Enables K$^+$ Retention and Mediates Acclimation of Barley (*Hordeum vulgare*) to Salinity Stress. *Plant Physiol.* **2016**, *172*, 2445–2458. [CrossRef]

53. Sun, J.; Li, L.S.; Liu, M.Q.; Wang, M.J.; Ding, M.Q.; Deng, S.R.; Lu, C.F.; Zhou, X.Y.; Shen, X.; Zheng, X.J.; et al. Hydrogen peroxide and nitric oxide mediate K$^+$/Na$^+$ homeostasis and antioxidant defense in NaCl-stressed callus cells of two contrasting poplars. *Plant Cell Tiss Organ Cult.* **2010**, *103*, 205–215. [CrossRef]

54. Sun, J.; Wang, M.J.; Ding, M.Q.; Deng, S.R.; Liu, M.Q.; Lu, C.F.; Zhou, X.Y.; Shen, X.; Zheng, X.J.; Zhang, Z.K.; et al. H$_2$O$_2$ and cytosolic Ca^{2+} signals triggered by the PM H$^+$-coupled transport system mediate K$^+$/Na$^+$ homeostasis in NaCl-stressed *Populus euphratica* cells. *Plant Cell Environ.* **2010**, *33*, 943–958. [CrossRef] [PubMed]

55. Ding, M.Q.; Hou, P.C.; Shen, X.; Wang, M.J.; Deng, S.R.; Sun, J.; Xiao, F.; Wang, R.G.; Zhou, X.Y.; Lu, C.F.; et al. Salt-induced expression of genes related to Na$^+$/K$^+$ and ROS homeostasis in leaves of salt-resistant and salt-sensitive poplar species. *Plant Mol. Biol.* **2010**, *73*, 251–269. [CrossRef] [PubMed]

56. Polle, A.; Chen, S. On the salty side of life: Molecular, physiological and anatomical adaptation and acclimation of trees to extreme habitats. *Plant Cell Environ.* **2015**, *38*, 1794–1816. [CrossRef] [PubMed]

57. Beritognolo, I.; Piazzai, M.; Benucci, S.; Kuzminsky, E.; Sabatti, M.; Mugnozza, G.S.; Muleo, R. Functional characterisation of three Italian *Populus alba* L. genotypes under salinity stress. *Trees* **2007**, *21*, 465–477. [CrossRef]

58. Ma, T.; Wang, J.Y.; Zhou, G.K.; Yue, Z.; Hu, Q.J.; Chen, Y.; Liu, B.B.; Qiu, Q.; Wang, Z.; Zhang, J.; et al. Genomic insights into salt adaptation in a desert poplar. *Nat. Commun.* **2013**, *4*, 2797. [CrossRef]

59. Zeng, F.J.; Yan, H.L.; Arndt, S.K. Leaf and whole tree adaptations to mild salinity in field grown *Populus euphratica*. *Tree Physiol.* **2009**, *29*, 1237–1246. [CrossRef]

60. Chen, S.L.; Li, J.K.; Fritzb, E.; Wang, S.S.; Huttermann, A. Sodium and chloride distribution in roots and transport in three poplar genotypes under increasing NaCl stress. *For. Ecol. Manag.* **2002**, *168*, 217–230. [CrossRef]

61. Verslues, P.E.; Agarwal, M.; Katiyar-Agarwal, S.; Zhu, J.; Zhu, J.K. Methods and concepts in quantifying resistance to drought, salt and freezing, abiotic stresses that affect plant water status. *Plant J.* **2006**, *45*, 523–539. [CrossRef]

62. Ottow, E.A.; Brinker, M.; Teichmann, T.; Fritz, E.; Kaiser, W.; Brosche, M.; Kangasjarvi, J.; Jiang, X.N.; Polle, A. *Populus euphratica* displays apoplastic sodium accumulation, osmotic adjustment by decreases in calcium and soluble carbohydrates, and develops leaf succulence under salt stress. *Plant Physiol.* **2005**, *139*, 1762–1772. [CrossRef]

63. Liang, W.J.; Ma, X.L.; Wan, P.; Liu, L.Y. Plant salt-tolerance mechanism: A review. *Biochem. Biophys. Res. Commun.* **2018**, *495*, 286–291. [CrossRef]

64. Hasanuzzaman, M.; Alam, M.M.; Rahman, A.; Hasanuzzaman, M.; Nahar, K.; Fujita, M. Exogenous proline and glycine betaine mediated upregulation of antioxidant defense and glyoxalase systems provides better protection against salt-induced oxidative stress in two rice (*Oryza sativa* L.) varieties. *Biomed. Res. Int.* **2014**, *2014*, 757219. [CrossRef]

65. Watanabe, S.; Kojima, K.; Ide, Y.; Sasaki, S. Effects of saline and osmotic stress on proline and sugar accumulation in *Populus euphratica* in vitro. *Plant Cell Tiss Organ Cult.* **2000**, *63*, 199–206. [CrossRef]

66. Janz, D.; Polle, A. Harnessing salt for woody biomass production. *Tree Physiol.* **2012**, *32*, 1–3. [CrossRef]

67. Apel, K.; Hirt, H. Reactive oxygen species: Metabolism, oxidative stress, and signal transduction. *Annu. Rev. Plant Biol.* **2004**, *55*, 373–399. [CrossRef]

68. Miller, G.; Suzuki, N.; Ciftci-Yilmaz, S.; Mittler, R. Reactive oxygen species homeostasis and signalling during drought and salinity stresses. *Plant Cell Environ.* **2010**, *33*, 453–467. [CrossRef]

69. Del Rio, L.A. ROS and RNS in plant physiology: An overview. *J. Exp. Bot.* **2015**, *66*, 2827–2837. [CrossRef]

70. Sies, H. Role of metabolic H_2O_2 generation: Redox signaling and oxidative stress. *J. Biol. Chem.* **2014**, *289*, 8735–8741. [CrossRef]

71. Wang, Y.C.; Qu, G.Z.; Li, H.Y.; Wu, Y.J.; Wang, C.; Liu, G.F.; Yang, C.P. Enhanced salt tolerance of transgenic poplar plants expressing a manganese superoxide dismutase from *Tamarix androssowii*. *Mol. Biol. Rep.* **2010**, *37*, 1119–1124. [CrossRef]

72. Zheng, L.Y.; Meng, Y.; Ma, J.; Zhao, X.L.; Cheng, T.L.; Ji, J.; Chang, E.M.; Meng, C.; Deng, N.; Chen, L.Z.; et al. Transcriptomic analysis reveals importance of ROS and phytohormones in response to short-term salinity stress in *Populus tomentosa*. *Front Plant Sci.* **2015**, *6*, 678. [CrossRef]

73. Gruszecki, W.I.; Strzalka, K. Carotenoids as modulators of lipid membrane physical properties. *Biochim. Biophys. Acta* **2005**, *1740*, 108–115. [CrossRef]

74. Bible, A.N.; Fletcher, S.J.; Pelletier, D.A.; Schadt, C.W.; Jawdy, S.S.; Weston, D.J.; Engle, N.L.; Tschaplinski, T.; Masyuko, R.; Polisetti, S.; et al. A Carotenoid-Deficient Mutant in *Pantoea* sp. YR343, a Bacteria Isolated from the Rhizosphere of *Populus deltoides*, Is Defective in Root Colonization. *Front. Microbiol.* **2016**, *7*, 491. [CrossRef]

75. Shen, Z.D.; Ding, M.Q.; Sun, J.; Deng, S.R.; Zhao, R.; Wang, M.J.; Ma, X.J.; Wang, F.F.; Zhang, H.L.; Qian, Z.Y.; et al. Overexpression of *PeHSF* mediates leaf ROS homeostasis in transgenic tobacco lines grown under salt stress conditions. *Plant Cell Tiss Organ Cult.* **2013**, *115*, 299–308. [CrossRef]

76. Li, J.B.; Zhang, J.; Jia, H.X.; Li, Y.; Xu, X.D.; Wang, L.J.; Lu, M.Z. The *Populus trichocarpa PtHSP17.8* involved in heat and salt stress tolerances. *Plant Cell Rep.* **2016**, *35*, 1587–1599. [CrossRef]

77. Poor, P.; Kovacs, J.; Borbely, P.; Takacs, Z.; Szepesi, A.; Tari, I. Salt stress-induced production of reactive oxygen-and nitrogen species and cell death in the ethylene receptor mutant Never ripe and wild type tomato roots. *Plant Physiol. Biochem.* **2015**, *97*, 313–322. [CrossRef]

78. Corpas, F.J.; Palma, J.M.; Del Rio, L.A.; Barroso, J.B. Protein tyrosine nitration in higher plants grown under natural and stress conditions. *Front Plant Sci.* **2013**, *4*, 29. [CrossRef]

79. Ji, H.T.; Pardo, J.M.; Batelli, G.; Van Oosten, M.J.; Bressan, R.A.; Li, X. The Salt Overly Sensitive (SOS) pathway: Established and emerging roles. *Mol. Plant* **2013**, *6*, 275–286. [CrossRef]

80. Yang, Y.; Tang, R.J.; Jiang, C.M.; Li, B.; Kang, T.; Liu, H.; Zhao, N.; Ma, X.J.; Yang, L.; Chen, S.L.; et al. Overexpression of the *PtSOS2* gene improves tolerance to salt stress in transgenic poplar plants. *Plant Biotechnol. J.* **2015**, *13*, 962–973. [CrossRef]

81. Zhu, J.K. Salt and drought stress signal transduction in plants. *Annu. Rev. Plant Biol.* **2002**, *53*, 247–273. [CrossRef]

82. Tang, R.J.; Liu, H.; Bao, Y.; Lv, Q.D.; Yang, L.; Zhang, H.X. The woody plant poplar has a functionally conserved salt overly sensitive pathway in response to salinity stress. *Plant Mol. Biol.* **2010**, *74*, 367–380. [CrossRef]

83. Wu, Y.X.; Ding, N.; Zhao, X.; Zhao, M.G.; Chang, Z.Q.; Liu, J.Q.; Zhang, L.X. Molecular characterization of PeSOS1: The putative Na^+/H^+ antiporter of *Populus euphratica*. *Plant Mol. Biol.* **2007**, *65*, 1–11. [CrossRef]

84. Tang, R.J.; Yang, Y.; Yang, L.; Liu, H.; Wang, C.T.; Yu, M.M.; Gao, X.S.; Zhang, H.X. Poplar calcineurin B-like proteins PtCBL10A and PtCBL10B regulate shoot salt tolerance through interaction with PtSOS2 in the vacuolar membrane. *Plant Cell Environ.* **2014**, *37*, 573–588. [CrossRef]

85. Zhu, J.K. Plant salt tolerance. *Trends Plant Sci.* **2001**, *6*, 66–71. [CrossRef]

86. Zhu, J.K. Regulation of ion homeostasis under salt stress. *Curr. Opin. Plant Biol.* **2003**, *6*, 441–445. [CrossRef]

87. Zhang, F.; Wang, Y.; Yang, Y.; Wu, H.; Wang, D.; Liu, J. Involvement of hydrogen peroxide and nitric oxide in salt resistance in the calluses from *Populus euphratica*. *Plant Cell Environ.* **2007**, *30*, 775–785. [CrossRef]

88. Choi, J.; Tanaka, K.; Cao, Y.R.; Qi, Y.; Qiu, J.; Liang, Y.; Lee, S.Y.; Stacey, G. Identification of a Plant Receptor for Extracellular ATP. *Science* **2014**, *343*. [CrossRef]

89. Sun, J.; Zhang, X.; Deng, S.R.; Zhang, C.L.; Wang, M.J.; Ding, M.Q.; Zhao, R.; Shen, X.; Zhou, X.Y.; Lu, C.F.; et al. Extracellular ATP signaling is mediated by H_2O_2 and cytosolic Ca^{2+} in the salt response of *Populus euphratica* cells. *PLoS ONE* **2012**, *7*, e53136. [CrossRef]

90. Wang, S.J.; Zhou, B.R.; Yao, W.J.; Jiang, T.B. PsnERF75 Transcription Factor from *Populus simonii* × *P. nigra* Confers Salt Tolerance in Transgenic Arabidopsis. *J. Plant Biol.* **2018**, *61*, 61–71. [CrossRef]

91. Yao, W.J.; Wang, S.J.; Zhou, B.R.; Jiang, T.B. Transgenic poplar overexpressing the endogenous transcription factor *ERF76* gene improves salinity tolerance. *Tree Physiol.* **2016**, *36*, 896–908. [CrossRef]

92. Wang, Z.L.; Liu, J.; Guo, H.Y.; He, X.; Wu, W.B.; Du, J.C.; Zhang, Z.Y.; An, X.M. Characterization of two highly similar CBF/DREB1-like genes, PhCBF4a and PhCBF4b, in *Populus hopeiensis*. *Plant Physiol. Biochem.* **2014**, *83*, 107–116. [CrossRef]

93. Zhao, H.; Zhao, X.Y.; Li, M.Y.; Jiang, Y.; Xu, J.Q.; Jin, J.J.; Li, K.L. Ectopic expression of *Limonium bicolor* (Bag.) Kuntze *DREB (LbDREB)* results in enhanced salt stress tolerance of transgenic *Populus ussuriensis* Kom. *Plant Cell Tiss Organ Cult.* **2017**, *132*, 123–136. [CrossRef]

94. Zhou, M.L.; Ma, J.T.; Zhao, Y.M.; Wei, Y.H.; Tang, Y.X.; Wu, Y.M. Improvement of drought and salt tolerance in Arabidopsis and Lotus corniculatus by overexpression of a novel DREB transcription factor from *Populus euphratica*. *Gene* **2012**, *506*, 10–17. [CrossRef]

95. Speth, E.B.; Imboden, L.; Hauck, P.; He, S.Y. Subcellular Localization and Functional Analysis of the Arabidopsis GTPase RabE. *Plant Physiol.* **2009**, *149*, 1824–1837. [CrossRef]

96. Zhang, J.; Li, Y.; Liu, B.L.; Wang, L.J.; Zhang, L.; Hu, J.J.; Chen, J.; Zheng, H.Q.; Lu, M.Z. Characterization of the *Populus Rab* family genes and the function of *PtRabE1b* in salt tolerance. *BMC Plant Biol.* **2018**, *18*, 124. [CrossRef]

97. Perez-Salamo, I.; Papdi, C.; Rigo, G.; Zsigmond, L.; Vilela, B.; Lumbreras, V.; Nagy, I.; Horvath, B.; Domoki, M.; Darula, Z.; et al. The heat shock factor A4A confers salt tolerance and is regulated by oxidative stress and the mitogen-activated protein kinases MPK3 and MPK6. *Plant Physiol.* **2014**, *165*, 319–334. [CrossRef]

98. Li, F.; Zhang, H.; Zhao, H.; Gao, T.; Song, A.; Jiang, J.; Chen, F.; Chen, S. Chrysanthemum *CmHSFA4* gene positively regulates salt stress tolerance in transgenic chrysanthemum. *Plant Biotechnol. J.* **2018**, *16*, 1311–1321. [CrossRef]

99. Yoon, S.K.; Bae, E.K.; Lee, H.; Choi, Y.L.; Han, M.; Choi, H.; Kang, K.S.; Park, E.J. Downregulation of stress-associated protein 1 (*PagSAP1*) increases salt stress tolerance in poplar (*Populus alba* × *P. glandulosa*). *Trees* **2018**, *32*, 823–833. [CrossRef]

100. Xue, S.W.; Yao, X.; Luo, W.; Jha, D.; Tester, M.; Horie, T.; Schroeder, J.I. AtHKT1;1 mediates nernstian sodium channel transport properties in *Arabidopsis* root stelar cells. *PLoS ONE* **2011**, *6*, e24725. [CrossRef]

101. Cheng, N.H.; Pittman, J.K.; Zhu, J.K.; Hirschi, K.D. The protein kinase SOS2 activates the Arabidopsis H^+/Ca^{2+} antiporter CAX1 to integrate calcium transport and salt tolerance. *J. Biol. Chem.* **2004**, *279*, 2922–2926. [CrossRef]

102. Baxter, A.; Mittler, R.; Suzuki, N. ROS as key players in plant stress signalling. *J. Exp. Bot.* **2014**, *65*, 1229–1240. [CrossRef]

103. Huertas, R.; Olias, R.; Eljakaoui, Z.; Galvez, F.J.; Li, J.; De Morales, P.A.; Belver, A.; Rodriguez-Rosales, M.P. Overexpression of *SlSOS2 (SlCIPK24)* confers salt tolerance to transgenic tomato. *Plant Cell Environ.* **2012**, *35*, 1467–1482. [CrossRef]

104. Qiu, Q.S.; Guo, Y.; Quintero, F.J.; Pardo, J.M.; Schumaker, K.S.; Zhu, J.K. Regulation of vacuolar Na^+/H^+ exchange in *Arabidopsis thaliana* by the salt-overly-sensitive (SOS) pathway. *J. Biol. Chem.* **2004**, *279*, 207–215. [CrossRef]

105. Zhou, J.; Wang, J.J.; Bi, Y.F.; Wang, L.K.; Tang, L.Z.; Yu, X.; Ohtani, M.; Demura, T.; Zhu Ge, Q. Overexpression of *PtSOS2* Enhances Salt Tolerance in Transgenic Poplars. *Plant Mol. Biol. Rep.* **2014**, *32*, 185–197. [CrossRef]

106. Li, D.D.; Song, S.Y.; Xia, X.L.; Yin, W.L. Two CBL genes from *Populus euphratica* confer multiple stress tolerance in transgenic triploid white poplar. *Plant Cell Tiss Organ Cult.* **2012**, *109*, 477–489. [CrossRef]

107. Gao, W.; Bai, S.; Li, Q.; Gao, C.; Liu, G.; Li, G.; Tan, F. Overexpression of *TaLEA* Gene from *Tamarix androssowii* improves salt and drought tolerance in transgenic Poplar (*Populus simonii* × *P. nigra*). *PLoS ONE* **2013**, *8*, e67462. [CrossRef]

108. Sun, Y.S.; Chen, S.; Huang, H.J.; Jiang, J.; Bai, S.; Liu, G.F. Improved salt tolerance of *Populus davidiana* × *P. bolleana* overexpressed LEA from *Tamarix androssowii*. *J. For. Res.-JPN* **2014**, *25*, 813–818. [CrossRef]

109. Li, Y.L.; Su, X.H.; Zhang, B.Y.; Huang, Q.J.; Zhang, X.H.; Huang, R.F. Expression of jasmonic ethylene responsive factor gene in transgenic poplar tree leads to increased salt tolerance. *Tree Physiol.* **2009**, *29*, 273–279. [CrossRef]

110. Du, N.X.; Liu, X.; Li, Y.; Chen, S.Y.; Zhang, J.S.; Ha, D.; Deng, W.G.; Sun, C.K.; Zhang, Y.Z.; Pijut, P.M. Genetic transformation of *Populus tomentosa* to improve salt tolerance. *Plant Cell Tiss Organ Cult.* **2011**, *108*, 181–189. [CrossRef]

111. Lawson, S.S.; Michler, C.H. Overexpression of *AtSTO1* leads to improved salt tolerance in *Populus tremula* × *P. alba*. *Transgenic Res.* **2014**, *23*, 817–826. [CrossRef]

112. Zhang, T.T.; Song, Y.Z.; Liu, Y.D.; Guo, X.Q.; Zhu, C.X.; Wen, F.J. Overexpression of phospholipase Dα gene enhances drought and salt tolerance of *Populus tomentosa*. *Chin. Sci. Bull.* **2008**, *53*, 3656–3665. [CrossRef]

113. Yu, X.; Ohtani, M.; Kusano, M.; Nishikubo, N.; Uenoyama, M.; Umezawa, T.; Saito, K.; Shinozaki, K.; Demura, T. Enhancement of abiotic stress tolerance in poplar by overexpression of key Arabidopsis stress response genes, *AtSRK2C* and *AtGolS2*. *Mol. Breed.* **2017**, *37*. [CrossRef]

114. Chen, S.; Polle, A. Salinity tolerance of *Populus*. *Plant Biol.* **2010**, *12*, 317–333. [CrossRef] [PubMed]

115. Janz, D.; Behnke, K.; Schnitzler, J.P.; Kanawati, B.; Schmitt Kopplin, P.; Polle, A. Pathway analysis of the transcriptome and metabolome of salt sensitive and tolerant poplar species reveals evolutionary adaption of stress tolerance mechanisms. *BMC Plant Biol.* **2010**, *10*, 150. [CrossRef] [PubMed]

116. Podda, A.; Pisuttu, C.; Hoshika, Y.; Pellegrini, E.; Carrari, E.; Lorenzini, G.; Nali, C.; Cotrozzi, L.; Zhang, L.; Baraldi, R.; et al. Can nutrient fertilization mitigate the effects of ozone exposure on an ozone-sensitive poplar clone? *Sci. Total Environ.* **2019**, *657*, 340–350. [CrossRef] [PubMed]

14

Overexpression of the Stress-Inducible *SsMAX2* Promotes Drought and Salt Resistance via the Regulation of Redox Homeostasis in *Arabidopsis*

Qiaojian Wang [1,2,†], Jun Ni [2,*,†], Faheem Shah [2], Wenbo Liu [2], Dongdong Wang [1,2], Yuanyuan Yao [2], Hao Hu [2], Shengwei Huang [2], Jinyan Hou [2], Songling Fu [1,*] and Lifang Wu [2,*]

[1] College of Forestry and Landscape Architecture, Anhui Agricultural University, Hefei 230000, Anhui, China; wangqj521@126.com (Q.W.); 15755059531@163.com (D.W.)

[2] Key Laboratory of High Magnetic Field and Ion Beam Physical Biology, Hefei Institutes of Physical Science, Chinese Academy of Sciences, Hefei 230000, Anhui, China; faheemhorticulturist@gmail.com (F.S.); liuwenbo9261@sina.com (W.L.); 17355356851@163.com (Y.Y.); huhaoasd@mail.ustc.edu.cn (H.H.); swhuang@ipp.ac.cn (S.H.); jyhou@ipp.ac.cn (J.H.)

* Correspondence: nijun@ipp.ac.cn (J.N.); fusongl001@outlook.com (S.F.); lfwu@ipp.ac.cn (L.W.)

† These authors contributed equally to this work.

Abstract: Recent studies have demonstrated that strigolactones (SLs) also participate in the regulation of stress adaptation; however, the regulatory mechanism remains elusive. In this study, the homolog of *More Axillary Branches 2*, which encodes a key component in SL signaling, in the perennial oil plant *Sapium sebiferum* was identified and functionally characterized in *Arabidopsis*. The results showed that the expression of *SsMAX2* in *S. sebiferum* seedlings was stress-responsive, and *SsMAX2* overexpression (OE) in *Arabidopsis* significantly promoted resistance to drought, osmotic, and salt stresses. Moreover, *SsMAX2* OE lines exhibited decreased chlorophyll degradation, increased soluble sugar and proline accumulation, and lower water loss ratio in response to the stresses. Importantly, anthocyanin biosynthesis and the activities of several antioxidant enzymes, such as superoxide dismutase (SOD), peroxidase (POD), and ascorbate peroxidase (APX), were enhanced in the *SsMAX2* OE lines, which further led to a significant reduction in hydrogen peroxide levels. Additionally, the *SsMAX2* OE lines exhibited higher expression level of several abscisic acid (ABA) biosynthesis genes, suggesting potential interactions between SL and ABA in the regulation of stress adaptation. Overall, we provide physiological and biochemical evidence demonstrating the pivotal role of *SsMAX2* in the regulation of osmotic, drought, and salt stress resistance and show that *MAX2* can be a genetic target to improve stress tolerance.

Keywords: *SsMAX2*; *Sapium sebiferum*; drought, osmotic stress; salt stress; redox homeostasis; strigolactones; ABA

1. Introduction

Abiotic stresses, such as drought, salt, cold, and flooding, significantly affect vegetative and reproductive growth and cause devastating yield losses each year. Plants have developed different coping mechanisms to deal with these stresses, mainly through the regulation of phytohormonal networks and dynamic changes of intracellular chemicals [1,2]. Phytohormones play a central role in the regulation of both vegetative and reproductive growth as well as adaptation to adverse growth conditions [3]. Hormones, such as abscisic acid (ABA), cytokinin, auxin, and salicylic acid (SA), have been proposed to be directly involved in the regulation of stress tolerance [4–7]. Strigolactones (SLs), which are a group of terpenoid compounds, play a key role in the regulation of shoot branching and the symbiosis with fungi in interactions with other hormones [8–11]. Recently, it was revealed that SLs also regulate plant adaptations to abiotic stresses [12,13]. In *Arabidopsis*, exogenous application of GR24, a SL analog, significantly improved salt and drought resistance, while the mutation of SL signaling gene *MAX2* made it sensitive to abiotic stresses [13,14]. However, the regulatory mechanism of SLs in stress tolerance still remains largely elusive.

In the perennial woody plant, the biological and molecular functions of strigolactones in the regulation of plant growth and stress adaptation have barely been studied. In the bioenergy plant *Jatropha curcas*, SLs antagonistically regulate the axillary bud outgrowth in interactions with cytokinin and gibberellin [15,16]. Several recent reports have also demonstrated that manipulation of the expression of SL biosynthesis genes can lead to significant change of the shoot branching phenotype in *Populus* and *Malus* [17,18], indicating that SLs have significant functions on the morphogenesis of woody plants. Plant growth and seed yield of woody plants are also threatened by abiotic stresses, such as salt, drought, and cold. Based on recent discoveries on the role of SLs in stress regulation in *Arabidopsis*, a functional study of the key genes in SL biosynthesis or signaling in woody plants would provide potential targets for genetic modifications to generate new cultivars with higher tolerance to abiotic stresses.

Sapium sebiferum, the seeds of which contain high level of fatty acids, has been considered as one of the most promising bioenergy plants. The oil from its seed coat and kernel can be manufactured into resources for lubricants, candles, cosmetics, and biodiesels [19,20]. It is widely distributed in most areas of China and even in the marginal land. However, while the plant adapts well to flooding and cold conditions, it is more sensitive to drought and salt stresses. The selection of high-yield cultivars with high resistance to drought and salt stresses is therefore the foremost goal in the molecular breeding of *Sapium sebiferum*. In this study, we identified that the expression of the SL signaling component *MAX2* was strongly responsive to abiotic stresses in *S. sebiferum* seedlings. Then, the biological functions of *SsMAX2* in the regulation of drought, osmotic, and salt tolerance were evaluated in *Arabidopsis*. The regulatory mechanism of *SsMAX2* to stress tolerance was further investigated at the physiological, molecular, and biochemical levels. This study not only reveals a pivotal role of *SsMAX2* in the regulation of drought, osmotic, and salt stress resistance but also provides evidence that *SsMAX2* can be a useful target for genetic engineering to produce stress-resistant plants.

2. Results

2.1. Gene Cloning of SsMAX2 from Sapium sebiferum Seedlings and the Gene Expression Profile in Response to Abiotic Stresses

The MAX2 homolog with 76% sequence similarity with AtMAX2 was identified from the *S. sebiferum* transcriptome database (Figure 1A). Then, the phylogenetic analysis of the MAX2 sequences from more than 20 plant species was carried out (Table S1). The results showed that SsMAX2 had the highest sequence identity with several perennial woody plants, such as *Jatropha curcas*, *Ricinus communis*, and Hevea brasiliensis, which also belong to the Euphorbiaceae family (Figure 1A).

MAX2 was the key component for SL signaling. Mutation of *MAX2* led to a significant increase in axillary branching and decrease in hypocotyl elongation. Our results are in accord with previous findings that constitutive expression of *SsMAX2* in *Arabidopsis* inhibits shoot branching and hypocotyl elongation, while *max2 Arabidopsis* mutant exhibits elongated hypocotyl growth (Figure 1B,C) and increased shoot branching (Figure 1D,E), demonstrating that *SsMAX2* has conserved functions with its homologs from *Arabidopsis*, rice, and pea.

To investigate whether *SsMAX2* is involved in the regulation of abiotic regulation, we first characterized the time-course expression profile of *SsMAX2* of *S. sebiferum* seedlings in response to osmotic and salt stresses. The results showed that osmotic treatment, which was mimicked by mannitol, significantly increased *SsMAX2* expression at 3 h after treatment (Figure 2A), whereas salt treatment induced a significant increase in *SsMAX2* expression at 12 h (Figure 2B). This demonstrated that *SsMAX2* is a stress-responsive gene, which might be involved in the regulation of adaptation to abiotic stresses.

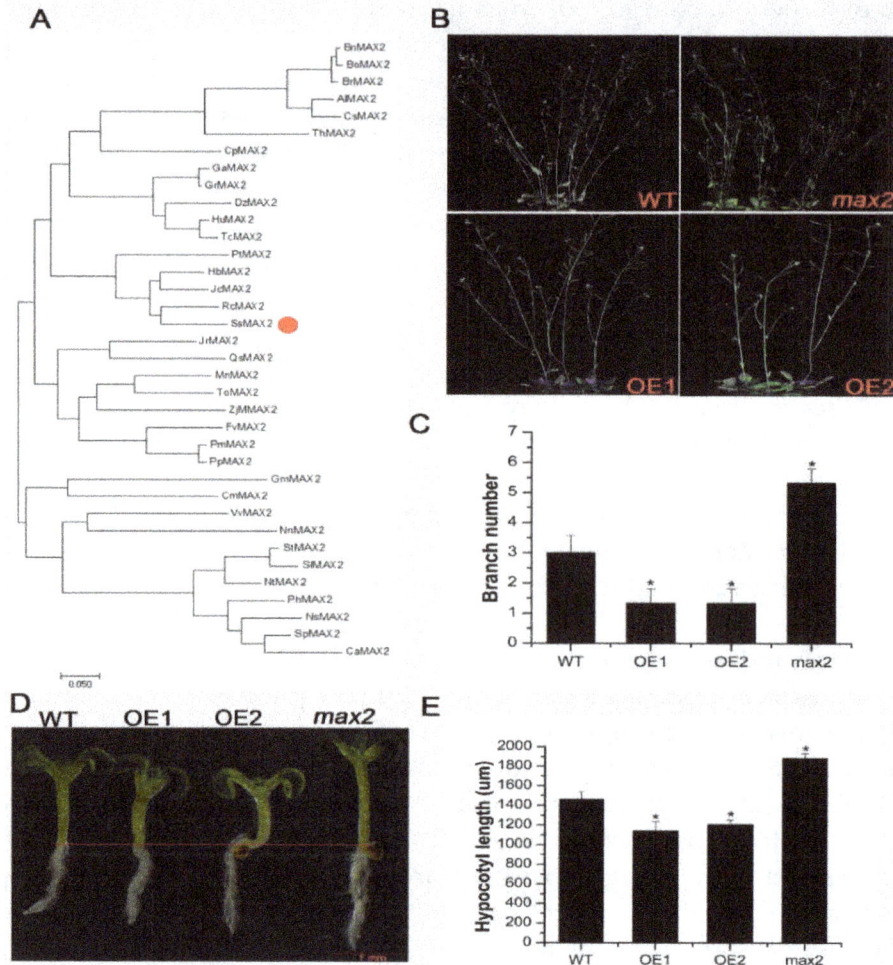

Figure 1. Phylogenetic analysis and functional characterization of *SsMAX2* in *Arabidopsis*. (**A**) Phylogenetic analysis of *SsMAX2* with its homologs from other species; (**B,C**) Rosette branching of 30-day-old plants of wild-type (WT), *SsMAX2* overexpression1 (OE1), OE2, and *max2*; (**D,E**) Hypocotyl length of 5-day-old seedlings of WT, *SsMAX2* OE1, OE2, and *max2* in half Murashige and Skoog (MS) medium. Data are presented as means ± SD of 20 replicates. Significant differences were determined by Student's *t*-test. Significance level: * $p < 0.05$.

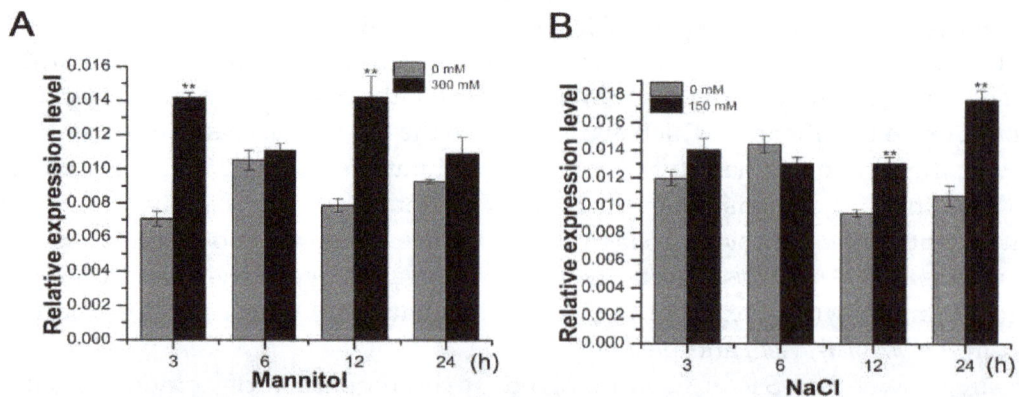

Figure 2. Expression profile of *SsMAX2* of 4-week-old *S. sebiferum* seedlings in response to (**A**) osmotic and (**B**) salt stresses. 300 mM mannitol and 150 mM NaCl were applied to 20-day-old *S. sebiferum* seedlings. *SsACT* was used as the internal control. Data are presented as means ± SD of three biological replicates. Significant differences were determined by Student's *t*-test. Significance level: ** $p < 0.01$.

2.2. SsMAX2 Conferred Drought and Osmotic Stress Tolerance in Arabidopsis

As the expression of *SsMAX2* in *S. sebiferum* seedlings was significantly induced by osmotic stress (Figure 2), we further investigated whether the constitutive expression of *SsMAX2* in *Arabidopsis* could confer drought and osmotic stress resistance. Results from the petri experiment showed that the *SsMAX2* OE lines exhibited significantly higher adaptation to osmotic stress, which was mimicked by the mannitol treatment (Figure 3A). Moreover, after withholding water for 11 days, all *Arabidopsis* lines exhibited significant dehydration, especially the wild-type (WT) and *max2* mutant seedlings. Seven days after re-watering, almost half of the *SsMAX2* OE lines survived (Figure 3B,C). Furthermore, chlorophyll fluorescence parameters, such as maximum photochemical efficiency of PSII (Fv/Fm), were investigated. The results showed that the *SsMAX2* OE lines exhibited much higher ratio of Fv/Fm under drought stress compared with the WT and *max2* mutant seedlings. Additionally, the water loss ratio in the leaf, which is an important characteristic of drought adaptation in plants, was much lower in the *SsMAX2* OE lines. These results suggest that *SsMAX2* positively regulates drought and osmotic stress adaptation in *Arabidopsis*. Interestingly, significant anthocyanin increase in the leaves of the *SsMAX2* OE lines was detected after drought treatment compared with the WT and *max2* mutant seedlings (Figure 4A). As previously reported, anthocyanin plays a key role in the regulation of the endogenous reactive oxygen species (ROS) level in response to abiotic stresses [21]. In this study, the leaves of two *SsMAX2* OE lines exhibited triple anthocyanin content compared to the WT and *max2* mutant seedlings (Figure 4B). Accordingly, the expression of the anthocyanin biosynthesis genes *chalcone synthase (CHS)*, *chalcone isomerase (CHI)*, *flavanone 3-hydroxylase (F3H)*, *flavanone 3'-hydroxylase (F3'H)*, *dihydroflavonol reductase (DFR)*, and *anthocyanin synthase (ANS)* was more significantly upregulated in the *SsMAX2* OE lines and downregulated in the *max2* mutant in response to drought stress (Figure 4C), suggesting SL may also regulate anthocyanin biosynthesis. These results demonstrate that overexpression (OE) of *SsMAX2* confer drought and osmotic stress tolerance, and the significant upregulation of anthocyanin accumulation in the *SsMAX2* OE lines may contribute to drought and osmotic stress resistance in the *SsMAX2* OE lines.

2.3. SsMAX2 Conferred Salt Tolerance in Arabidopsis

We further investigated the salt responses of different *Arabidopsis* lines (two *SsMAX2* OE lines, wild-type, and *max2* mutant). The results showed that seedlings of different lines exhibited no significant growth variations under normal conditions (Figure 5A). However, the WT and *max2* plants showed significant blushing phenotype after seven days of growth in half MS medium containing 100 mM NaCl, while the *SsMAX2* OE lines exhibited significantly higher tolerance to salt stress, even in the 150 mM NaCl medium (Figure 5A). The salt stress experiment was also conducted on different *Arabidopsis* lines growing in the soil. The results were in accord with that of the petri experiment, with the *SsMAX2* OE lines exhibiting robust salt tolerance (Figure 5B). The survival rate of the *SsMAX2* OE lines could reach as high as 65% at 7 d after 150 mM treatment (Figure 5C). It is worth noting here that the chlorophyll content of the *max2* mutant was significantly lower than that of *SsMAX2* OE and WT plants (Figure 5D). Stress can induce senescence and cause significant chlorophyll degradation in the leaf [22]. The results showed that the decrease in stress-induced chlorophyll in the leaves of the *SsMAX2* OE lines was significantly lower than that of the *max2* mutant and wild-type plants (Figure 5D). This suggests that *MAX2* may positively regulate chlorophyll synthesis and that *SsMAX2* overexpression in *Arabidopsis* can retard leaf senescence induced by salt stress.

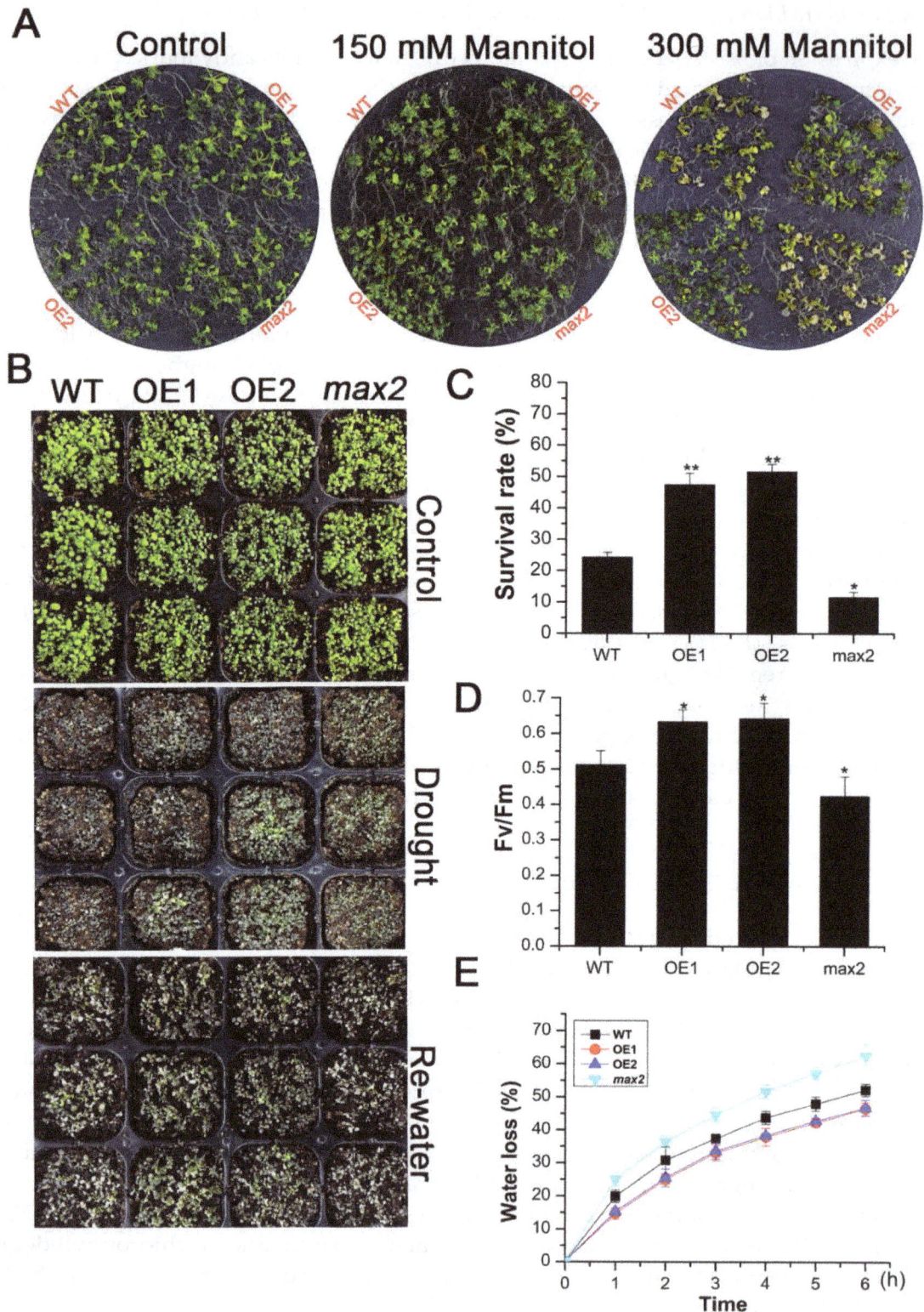

Figure 3. Constitutive expression of *SsMAX2* promoted osmotic and drought stress resistance in *Arabidopsis*. (**A**) Phenotype of *SsMAX2* OE lines, *max2*, and wild-type (WT) in half MS medium containing 150 and 300 mM mannitol; (**B**) Phenotype of different *Arabidopsis* lines after withholding water and rewatering treatment; (**C**) Survival rate of seedlings after drought treatment; (**D**) Maximum photochemical efficiency of PSII (Fv/Fm) of different lines under drought stress; (**E**) Water loss rate. Data are presented as means ± SD of three biological replicates. Significant differences were determined by Student's *t*-test. Significance level: * $p < 0.05$, ** $p < 0.01$.

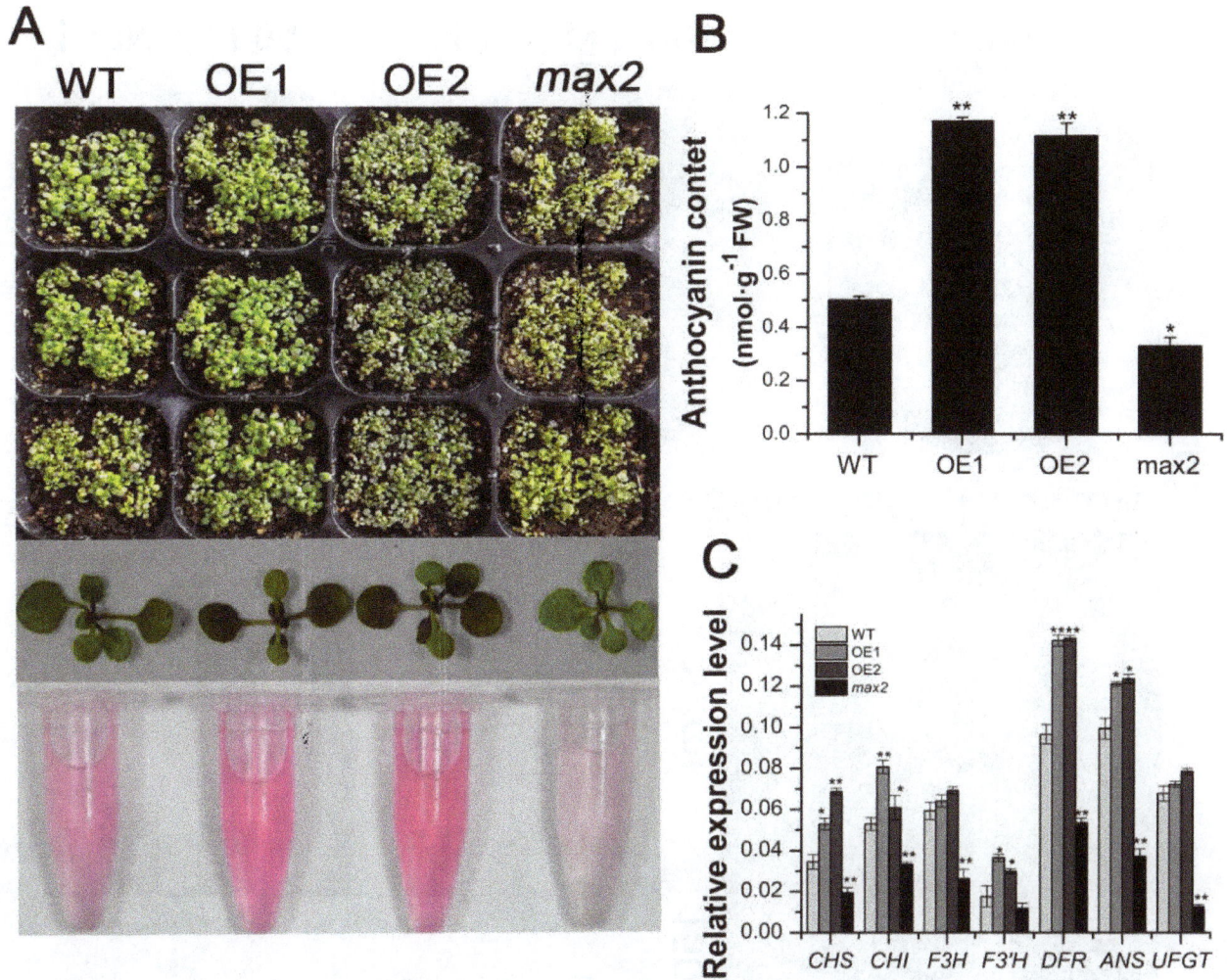

Figure 4. *SsMAX2* promoted anthocyanin accumulation in *Arabidopsis* leaves under drought stress. (**A**) Drought-induced anthocyanin accumulation in the leaves; (**B**) Anthocyanin level of different lines; (**C**) Relative expression of anthocyanin biosynthesis genes in the leaves of different lines in response to drought stress. Data are presented as means $\pm$ SD of three biological replicates. Significant differences between WT and the other groups were determined by Student's *t*-test. Significance level: * $p < 0.05$, ** $p < 0.01$.

2.4. SsMAX2 Promoted Seed Germination under Both Salt and Osmotic Stresses

We further investigated whether the *SsMAX2* OE lines could also improve stress resistance during the seed germination stage. We evaluated the effects of different concentrations of mannitol and NaCl on the seed germination of different *Arabidopsis* lines. The results showed that, under normal conditions, the germination rate of *SsMAX2* OE1 lines, *max2*, and WT showed no significant variations (Figure 6A,B). However, the seed germination of WT and *max2* mutant was more likely to be inhibited, whereas *SsMAX2* OE lines exhibited much higher germination ratio, with increasing concentrations of both mannitol and NaCl (Figure 6A,B). The results also showed that, even under 200 mM mannitol and 150 mM NaCl, the seed germination of *SsMAX2* OE lines was still over 50% (Figure 6A,B). Furthermore, the time-course assay of the seed germination showed that the seed germination of the *max2* mutant was significantly delayed compared with that of the *SsMAX2* OE lines under both salt and drought stress (Figure 6C–E). These results suggest that *SsMAX2* confer significant salt and osmotic stress resistance during seed germination.

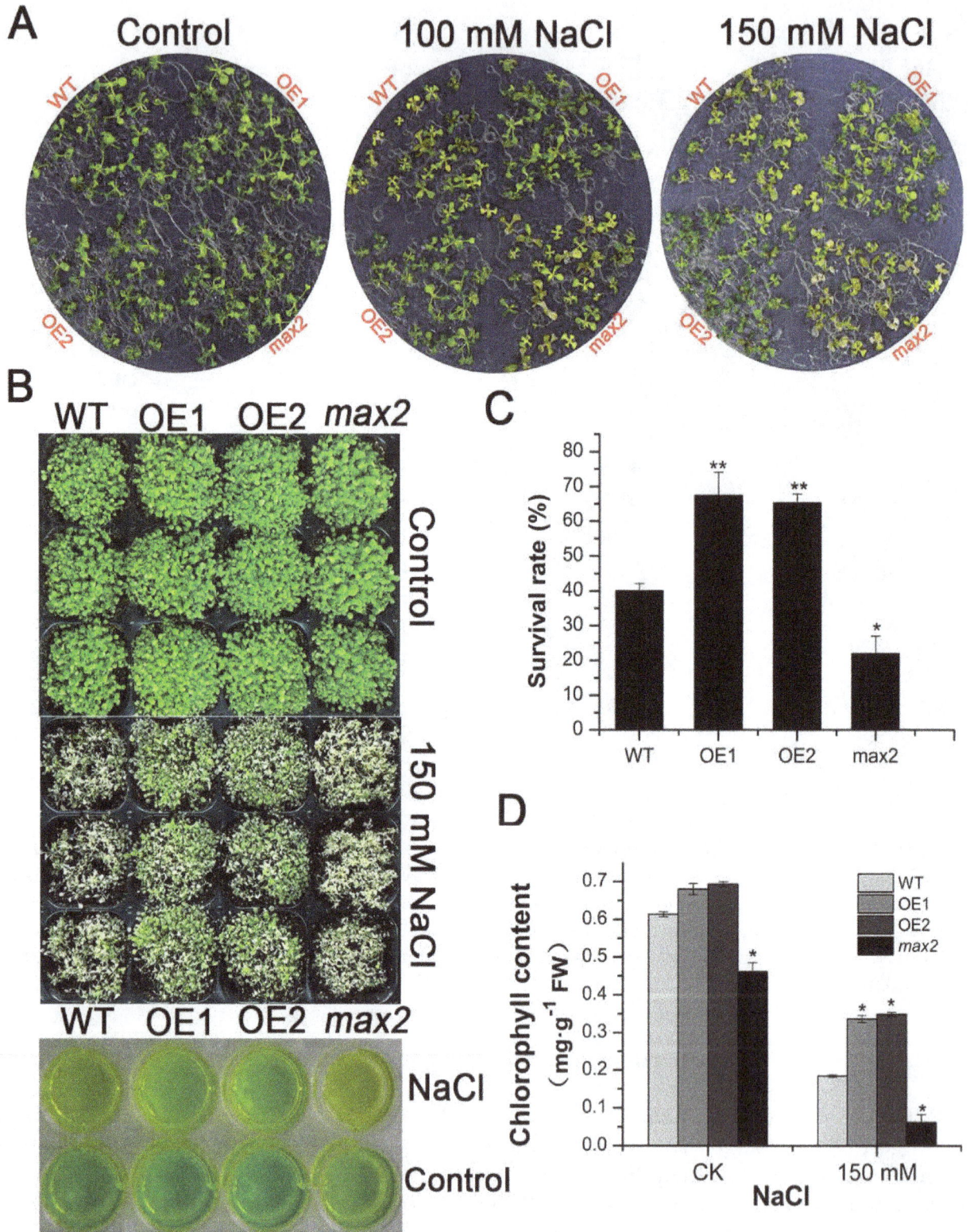

Figure 5. *SsMAX2* conferred salt resistance in *Arabidopsis*. (**A**) Growth phenotype of 5-day-old *SsMAX2* OE lines, *max2*, and WT seedlings under salt stress; (**B,C**) The growth and survival rate 7 days after 150 mM NaCl treatment on 15-day-old seedlings; (**D**) Chlorophyll content of the seedling leaves before or 7 days after salt stress treatment. Data are presented as means ± SD of three biological replicates. Significant differences between WT and the other groups were determined by Student's *t*-test. Significance level: * $p < 0.05$, ** $p < 0.01$.

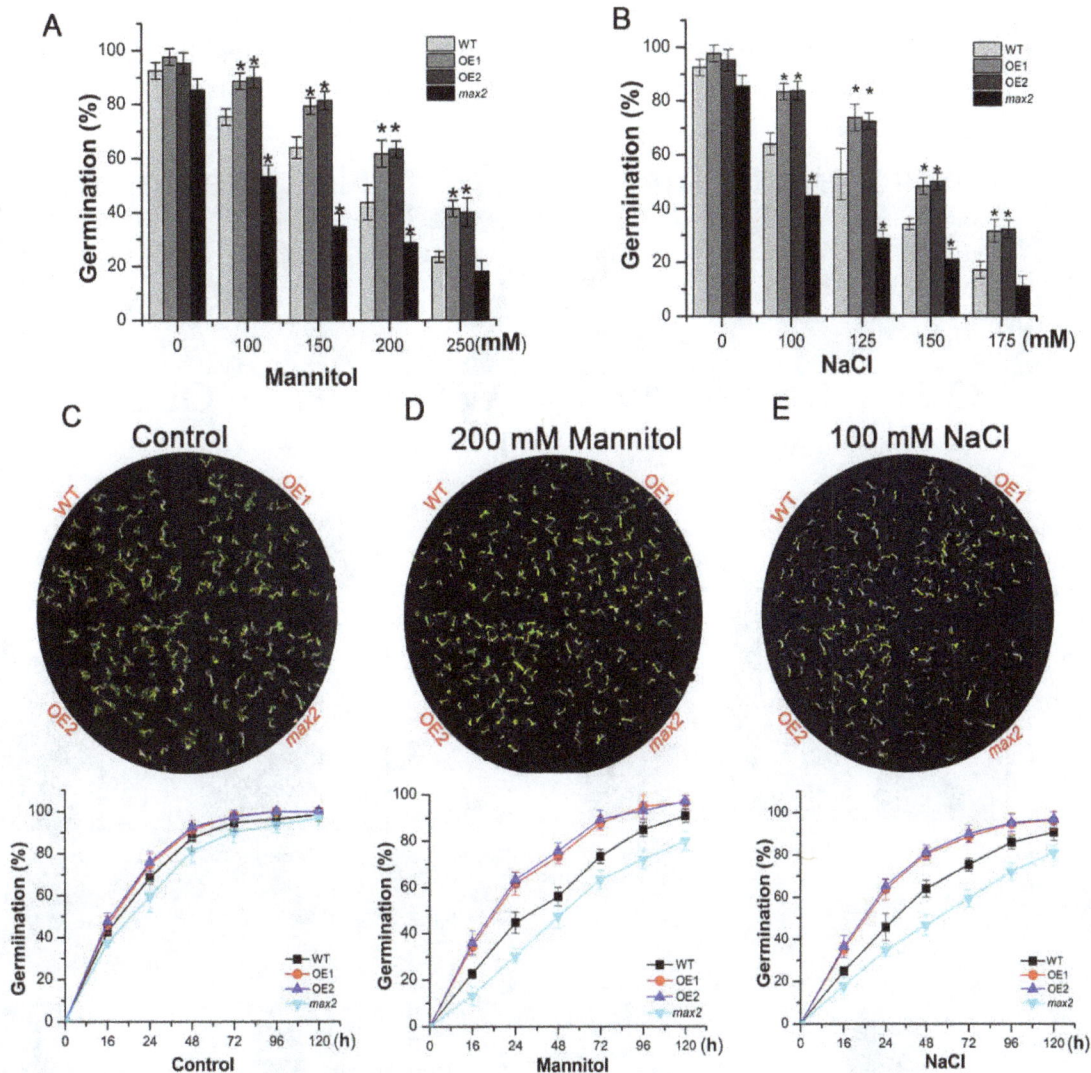

Figure 6. *SsMAX2* OE lines exhibited higher salt and osmotic stress tolerance during the seed germination stage. (**A,B**) Seed germination rate of the *SsMAX2* OE lines, *max2*, and WT under various concentrations of NaCl and mannitol treatment; (**C–E**) Time-course assay of the seed germination of different lines under 200 mM mannitol and 100 mM NaCl treatment. Data are presented as means ± SD of three biological replicates. Significant differences between WT and the other groups were determined by Student's *t*-test. Significance level: * $p < 0.05$.

2.5. SsMAX2 Regulated the Hydrogen Peroxide, Malondialdehyde (MDA), Proline, and Soluble Sugar Accumulation in the Seedlings in Response to the Stresses

The significant increase in endogenous peroxide or superoxide chemical levels induced by the abiotic stresses is responsible for the initiation of leaf senescence and death [23]. Malondialdehyde (MDA) is an important marker for lipid peroxidation due to overproduction of ROS in the cell [24]. Here, the results showed that both osmotic and salt stress could cause a significant increase in hydrogen peroxide and MDA in all lines, while *SsMAX2* OE lines had a significant lower level of both hydrogen peroxide and MDA (Figure 7A–D), suggesting a tightly regulated ROS and MDA homeostasis in the *SsMAX2* OE lines. This was also in accord with the physiological results, which showed that *SsMAX2* OE lines had better resistance and delayed leaf senescence to the stresses.

Proline and soluble sugars play an important role in maintaining osmotic homeostasis in plant cells [23]. Our results showed that the *SsMAX2* OE lines accumulated higher proline and soluble sugars in leaves than WT and *max2* mutant (Figure 7E,F). As both drought and salt stress can break the osmosis homeostasis in the plant, the enhanced accumulation of proline and soluble sugars can significantly prevent water loss from leaves under osmotic stresses. The results show that decreased water loss in the *SsMAX2* OE lines contribute to drought and salt stress resistance.

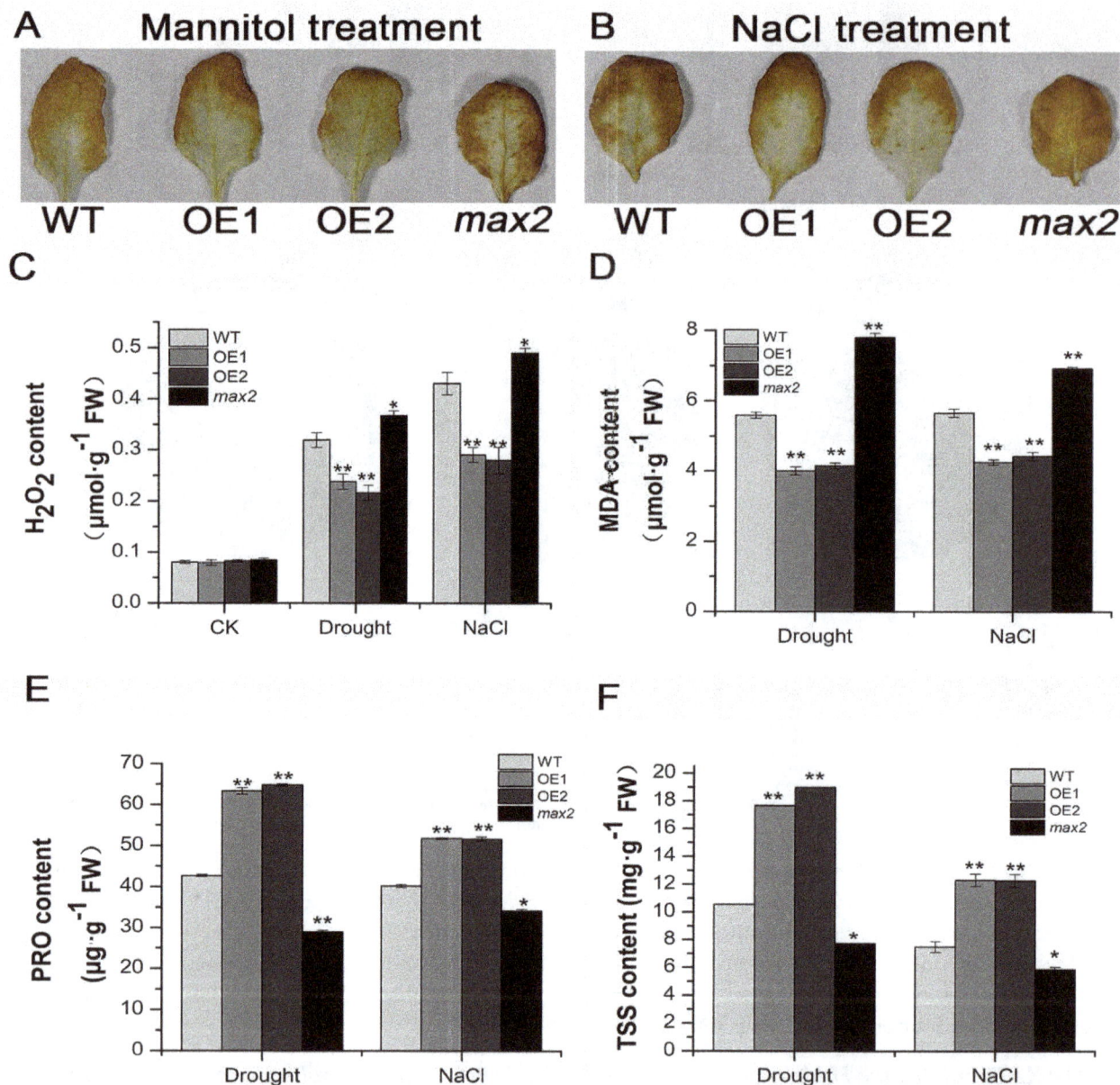

Figure 7. *SsMAX2* OE lines exhibited lower hydrogen peroxide and malondialdehyde (MDA) levels and increased proline and soluble sugar levels in response to osmotic and salt stresses. (**A,B**) DAB staining of the leaves of *SsMAX2* OE lines, *max2*, and WT after mannitol and NaCl treatment. The (**C**) hydrogen peroxide, (**D**) MDA, (**E**) proline, and (**F**) soluble sugar level of 15-day-old seedlings of the *SsMAX2* OE lines, *max2*, and WT were determined 5 days after withholding water or 150 mM NaCl treatment. Data are presented as means $\pm$ SD of three biological replicates. Significant differences between WT and the other groups were determined by Student's *t*-test. Significance level: * $p < 0.05$, ** $p < 0.01$.

2.6. SsMAX2 Increased the Enzyme Activity of Superoxide Dismutase (SOD), Peroxidase (POD), and Ascorbate Peroxidase (APX)

As the hydrogen level in the *SsMAX2* OE lines was significantly lower than that in WT and *max2* mutant, we further investigated whether the key enzymes involved in the regulation of ROS degradation were also affected in response to drought and salt treatment. POD, SOD, and CAT are the main oxidative enzymes involved in the regulation of ROS homeostasis in the cell [25,26]. The results showed that, under both drought and salt stress, the activity of POD, SOD, and CAT of the *SsMAX2* OE lines was significantly higher than that of *max2* and WT (Figure 8), whereas the *max2* mutant exhibited the lowest activity of the antioxidative enzymes, further demonstrating a tightly controlled ROS scavenge ability controlled by SL signaling. These results suggest that *MAX2* may be involved in the regulation of plant ROS homeostasis via controlling the activities of oxidative enzymes.

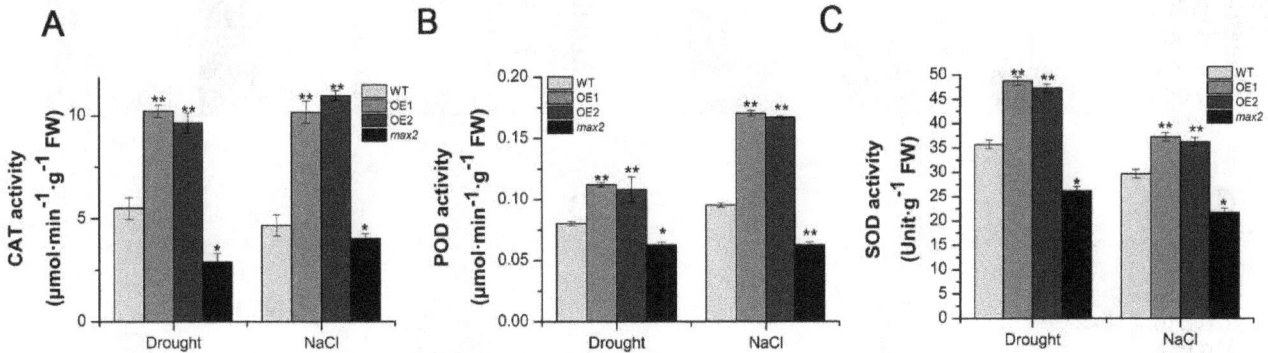

Figure 8. *SsMAX2* increased the activity of the antioxidant enzymes. The enzyme activities of (**A**) CAT, (**B**) POD, and (**C**) SOD of the extracts of 15-day-old seedlings were separately determined 5 days after withholding water or 150 mM NaCl treatment. Data are presented as means ± SD of three biological replicates. Significant differences between WT and the other groups were determined by Student's *t*-test. Significance level: * $p < 0.05$, ** $p < 0.01$.

2.7. Diverse Regulation of the Abscisic Acid (ABA) Biosynthesis Genes in SsMAX2 OE Lines and max2 in Response to Drought and Salt Stress

ABA is the key phytohormone that directly regulates abiotic stresses. Thus, in this study, we further investigated whether the expression of ABA biosynthesis genes (*CYP707-A1, -A2, -A3, NCED3,* and *OAA3*) was diversely regulated in response to drought and salt stress. After salt treatment, the significant upregulation of *NCED3, OAA3,* and *CYO707A1* in the *SsMAX2* OE lines could be detected at 6 h after treatment compared with the WT and *max2* mutant (Figure 9). The expression of *CYP707A3* and *NCED3* was relatively higher in the *SsMAX2* OE lines (Figure 9C,D). Specifically, the basic expression of *CYP707A2* in the *max2* mutant was higher than the WT and *SsMAX2* OE lines (Figure 9B), which is in accord with previously published results in *Arabidopsis* [27]. These results suggest a potential interaction between SL and ABA in the regulation of abiotic stress adaptation.

3. Discussion

Sapium sebiferum, which is one of the most important commercial woody plants in China, has received considerable attention due to its high oil content in the seed coat and kernel, excellent sightseeing value as a landscape plant, and its high adaptation to the adverse marginal land. Recent studies on the plant have mainly focused on flower sex determination, seed yield, oil extraction and production, and herb values [28–32]. However, only few reports have demonstrated its antistress abilities. Abiotic stresses, such as drought and salt, can significantly reduce the yield output in *S. sebiferum* [33], which significantly limits its industrial potential. As the transgenic approaches have been widely used and proven to be very effective in the regulation of abiotic stress tolerance in many species [34–36], the generation of high-stress-resistant cultivars via genetic modifications is the foremost mission in *S. sebiferum* breeding.

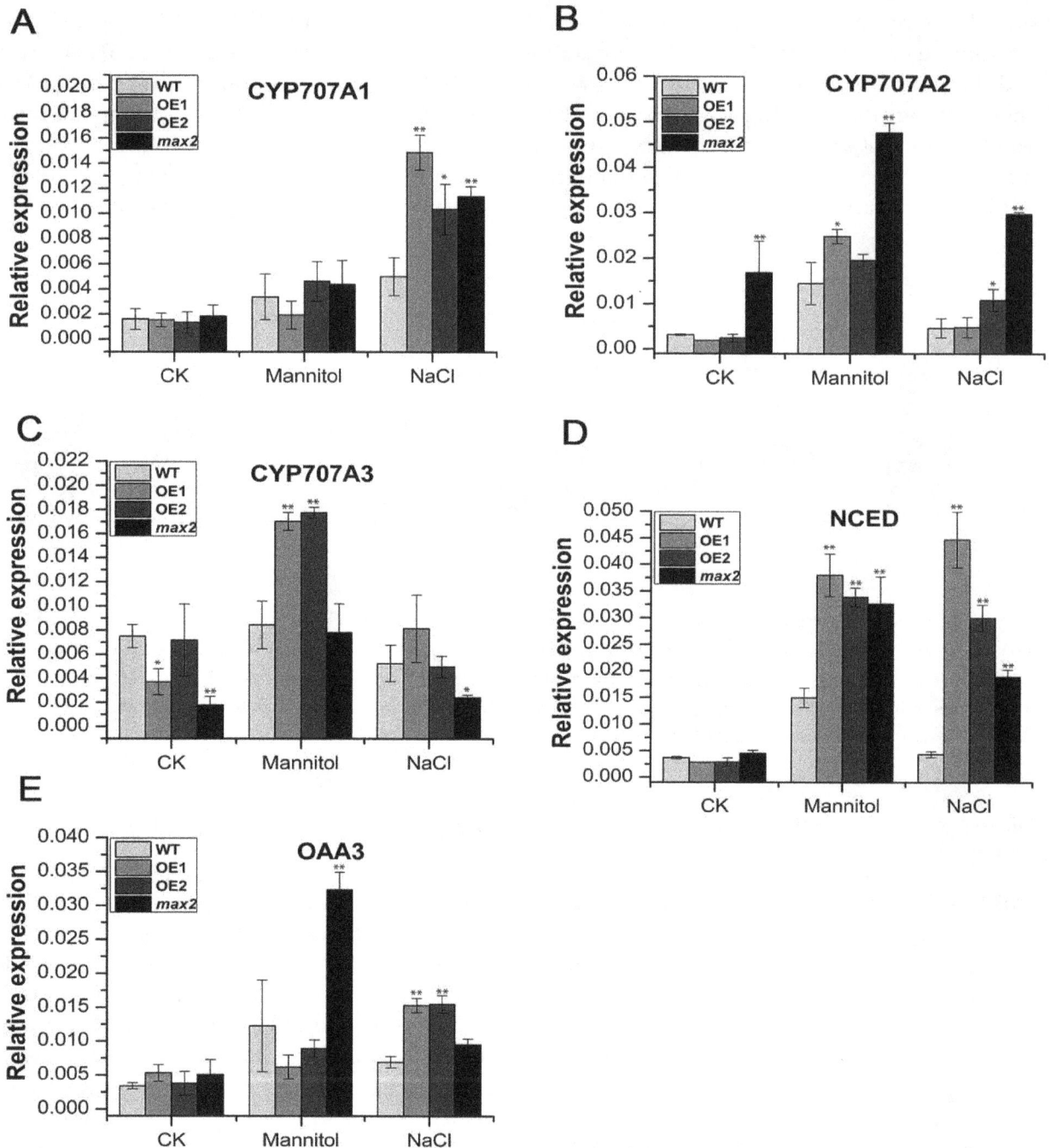

Figure 9. Expression of key ABA biosynthesis genes of the *SsMAX2* OE lines, *max2*, and WT was diversely regulated in response to salt and osmotic treatment. The gene expression of (**A**) *CYP707A1*, (**B**) *CYP707A2*, (**C**) *CYP707A3*, (**D**) *NCED3*, and (**E**) *OAA3* was determined at 6 h after mannitol and NaCl treatment by qPCR. *SsACT* was used as the internal control. Data are presented as means ± SD of three biological replicates. Student's *t*-test was used to determine the significant differences between WT and other *Arabidopsis* lines. Significance level: * $p < 0.05$, ** $p < 0.01$.

Previous studies have demonstrated that SLs are involved in the regulation of shoot branching, senescence, and photomorphogenesis [37,38]. Some recent researches have revealed the pivotal role of SLs in the regulation of stress adaptation [39]. Thus, the identification and characterization of SL biosynthesis and signaling genes in woody plants are important for generating stress-tolerant cultivators. *More Axillary Branches 2* (*MAX2*), which encodes a F-box E3 ligase in *Arabidopsis*, is the key component involved in SL signal transduction [40]. *MAX2* plays a key role in the regulation of shoot branching, photomorphogenesis, and stress adaptation [13,27,41,42]. In this work, we identified and functionally characterized the *MAX2* homolog (*SsMAX2*) from the oil plant *S. sebiferum*. Constitutive expression of *SsMAX2* in *Arabidopsis* inhibited the shoot branching and the hypocotyl elongation (Figure 1), further confirming the biological functions of *SsMAX2* as *MAX2* homolog in *Arabidopsis*. In *S. sebiferum* seedlings, it was interesting to find that the expression of *MAX2* was significantly upregulated in response to drought and salt treatment (Figure 2). Many abiotic stress-inducible genes, such as *NAC5* [43], *XTH3* [44], and *UGT87A2* [45], are important in controlling the adaptation to stresses. Thus, the significant upregulation of *SsMAX2* indicates that it may be correlated with stress adaptation. In comparison with the *max2 Arabidopsis* mutant, our results further demonstrated that *SsMAX2* OE lines exhibited significant drought and salt tolerance (Figures 3 and 5). The seed germination in the *SsMAX2* OE lines also had higher drought and salt tolerance (Figure 6), whereas the *max2* mutant was more sensitive to the stresses, as previously described [14,42]. These results demonstrate that *MAX2* participates in the regulation of stress adaptation. However, the regulatory mechanism of how *MAX2* controls the increased tolerance to abiotic stresses remains elusive.

Plants have developed many mechanisms to cope with biotic and abiotic stresses, such as accumulation of secondary metabolites, activated oxidative enzyme, or nonenzyme systems [36,46,47]. Anthocyanins, which consist of a group of phenolic compounds, act as important antioxidants in plants suffering from abiotic stresses [48]. Increased anthocyanin production has been shown to significantly enhance tolerance to abiotic stresses in *Arabidopsis*, grapevine, and bamboo [49–51]. Our results also showed that, under drought stress, significant accumulation of anthocyanins in the leaves of the *SsMAX2* OE lines was detected, while its content was relatively lower in the *max2* mutant and WT plants (Figure 4). The qPCR results further showed that the expression of the key genes in the anthocyanin biosynthesis pathway was significantly induced in the *SsMAX2* OE lines (Figure 4C). These results suggest that SLs may be involved in the regulation of anthocyanin biosynthesis in *Arabidopsis*, which is important for adaptation to salt stress. It is worth noting here that the *SsMAX2* OE lines had higher chlorophyll content in the leaves, while the *max2* mutant exhibited much lower chlorophyll content than WT plants under normal growth conditions (Figure 5). Furthermore, drought-stress-induced chlorophyll degradation in the leaves was much lower than that in WT and *max2* plants (Figure 5D), suggesting *MAX2* may positively regulate chlorophyll biosynthesis or accumulation, the level of which can be an important indicator of adaptation to abiotic stresses [52]. Many researchers have suggested that the accumulation of soluble sugars and amino acids is key for osmotic stress resistance [53,54] due to their direct role in the regulation of water uptake and loss. Our results also showed that the level of both proline and soluble sugars was significantly higher in the *SsMAX2* OE lines after drought and salt treatment in comparison with the WT and *max2* seedlings (Figure 7E,F), suggesting a pivotal role of *MAX2* in the regulation of cellular metabolite homeostasis.

The generation of oxidative chemicals induced by abiotic stresses in cells is the main cause of cell apoptosis and death [46]. Plants have developed a tightly regulated mechanism to maintain endogenous oxidative chemicals at a certain level. Many reports have demonstrated that exogenous application of ROS cleavage chemicals (e.g., melatonin) or overexpression of oxidative chemical cleavage enzymes (e.g., SOD, POD, and APX), can significantly promote tolerance to abiotic stresses due to the efficient cleavage of the stress-induced oxidative chemical level [55,56]. In this study, we presumed that *MAX2* could be involved in the regulation of oxidative chemical levels in plants. The results showed that both drought and salt treatment significantly induced hydrogen peroxide accumulation in *Arabidopsis* seedlings. However, the level of this was significantly lower in the *SsMAX2* OE lines compared with that of the *max2* mutant and wild-type (Figure 7A,B). Accordingly, the activity analysis of key oxidative enzymes, such as CAT, POD, and SOD, also proved that *SsMAX2* OE plants had higher capability in the cleavage of hydrogen peroxide induced by salt and drought stress, whereas *max2* mutant exhibited much lower enzyme activity (Figure 8). These results suggest that the SL signaling may be directly involved in the regulation of redox homeostasis, although the molecular mechanism still needs further investigation.

ABA is the key phytohormone that positively regulates abiotic stress adaptation. Many reports have demonstrated that ABA accumulation can happen immediately when plants are subjected to drought, salt, cadmium, or cold stresses [57]. Exogenous application with ABA or overexpression of ABA biosynthesis genes can significantly promote abiotic stress resistance in many species [58]. A recent study also demonstrated that ABA and SL coordinately regulated salt stress tolerance in *Sesbania cannabina* [59]. In this study, the *SsMAX2* OE lines exhibited higher expression level of key ABA biosynthesis-related genes, such as *CYP707A1, CYP707A3, NCED,* and *OAA3,* compared with the *max2* mutant or WT plants (Figure 9), further indicating that *SsMAX2*-induced stress tolerance may be partially ABA-dependent.

In this study, we isolated and functionally characterized the *MAX2* homolog in the oil plant *Sapium sebiferum.* We not only investigated the gene function in controlling shoot branching and hypocotyl elongation but, most importantly, we characterized the novel function of *SsMAX2* in the regulation of drought and salt adaptation. We showed that *MAX2* potentially controls chlorophyll biosynthesis and degradation, anthocyanin biosynthesis, soluble sugars, and proline accumulation (Figure 10). The physiological and biochemical results demonstrate that *SsMAX2* plays a pivotal role in the regulation of redox homeostasis via the regulation of antioxidative enzymes (Figure 10). The results also suggest that there may be potential interactions between SL and ABA in the regulation of abiotic stress adaptation (Figure 10). Further research will be focused on the identification of the molecular network of SL in the regulation of stress adaptation.

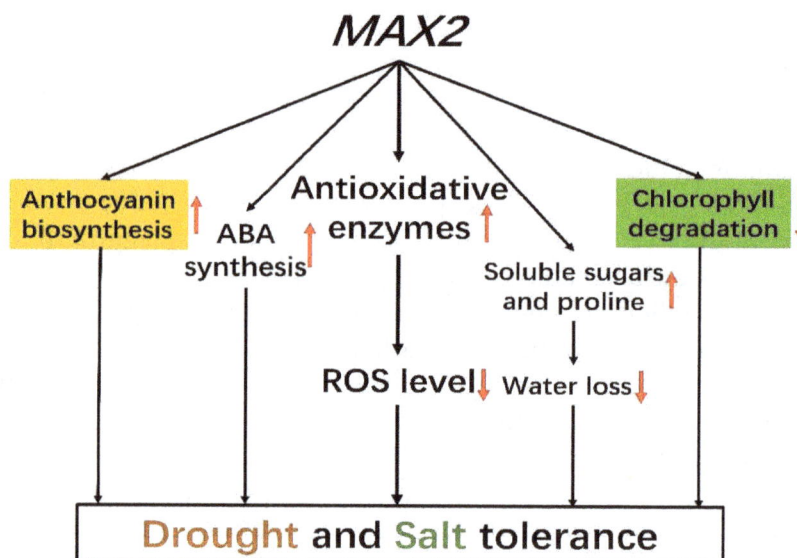

Figure 10. Model of *SsMAX2* in the regulation of drought and salt adaptation. The arrows indicate the downregulated or upregulated activities.

4. Materials and Methods

4.1. Plant Materials and Growth Conditions

The Col-0 ecotype *Arabidopsis* and *max2* mutant (CS9565) were ordered from the Arabidopsis Biological Resource Center (ABRC). *Arabidopsis* seeds were sterilized with 75% ethanol solution for one minute, followed by sterilization with 8% sodium hypochlorite for 15 min, and then germinated in half MS medium. *Arabidopsis* seedlings were grown in the growth chamber (22 °C; 16-h light/8-h dark photoperiod; 120 mol·m^{-2}·s^{-1} radiation strength; 75% humidity). All plants were fertilized with half Hoagland solution every other week.

Sapium sebiferum seeds were germinated in the peat soil as previously described [60]. Two-week-old *S. sebiferum* seedlings were transplanted into the garden pot (10 cm × 10 cm) and grown in the chamber as described above.

4.2. Gene Cloning, Vector Construction, and Arabidopsis Transformation

The protein sequence of AtMAX2 (AT2G42620.1) was used to blast against the local *S. sebiferum* transcriptome database using the NCBI local blast package: BLAST 2.7.1. The coding sequence (CDS) and amino acid sequence information is listed in Table S2. The complete CDS of *SsMAX2* was cloned into the pOCA30 expression vector and transformed into the *Agrobacterium* EHA105. The *Arabidopsis* transformation was carried out following a previously described method [61]. The candidate transgenic *Arabidopsis* plants were firstly screened on half MS containing 40 mg/L kanamycin and then further confirmed by RT-PCR. Two out of over 10 independent homozygous transgenic lines were used for further experiments.

4.3. Drought and Salt Treatment

In the petri experiment, different concentrations of mannitol and NaCl were separately used for drought and salt stress treatment. In the soil experiment, plants of different lines were grown in moss peat soil. To induce drought conditions, water was withheld for a number of days, as indicated in the figures. The plant growth was monitored after three days after rewatering. For salt stress, the pods of *Arabidopsis* were directly submerged in different concentrations of sodium chloride until the soil was completely saturated. Then, the plants were put under normal growth conditions.

4.4. RNA Extraction and Quantitative Real-Time PCR (qPCR)

Total RNA was extracted from *Arabidopsis* and *S. sebiferum* seedlings according to the instructions of HP Total Plant RNA kit (Omega, Shanghai, China). RNA concentration and integrity were further analyzed by a micro-analyzer and gel electrophoresis. For cDNA synthesis, 1–1.5 g of total RNA was used using the TransScript II One-Step gDNA removal and cDNA Synthesis SuperMix (Transgen, Beijing, China). qPCR was performed using the Premix Ex Taq^{TM} II (Transgen, Beijing, China) on the LightCycler 96 System (Roche, Basel, Switzerland). The qPCR program was set as follows: preheating, 95 °C, 10 min; amplification (45 cycles), 95 °C, 10 s, 60 °C, 20 s, and 72 °C, 20 s; melting curve: 95 °C, 2 min, 60 °C, 30 s, then continuously increased to 95 °C. The calculation of the relative gene expression was based on the $2^{-\Delta\Delta Cq}$ method as described previously [62]. The detailed primer information for each gene is listed in Table S3.

4.5. Total Chlorophyll and Anthocyanin Determination

The total chlorophyll content was analyzed based on a previously described method [63]. The leaf tissue was homogenized in liquid nitrogen and subsequently extracted in 80% acetone containing 1 M KOH overnight. After centrifugation at $12,000 \times g$ for 10 min, the supernatant was used for chlorophyll determination using a Scandrop spectrophotometer (Analytikjena, Jena, Germany). The anthocyanin was determined as previously described [64] with minor modifications. Briefly, the *Arabidopsis* leaves were ground into fine powders in liquid nitrogen. Then, the powders were transferred to methanol containing 1% HCl at 4 °C in the dark for 24 h. The aqueous phase was then used for anthocyanin determination in the spectrophotometer with the following formula: $OD = (A_{530} - A_{620}) - 0.1 (A_{650} - A_{620})$.

4.6. Determination of the Water Loss Rate

Approximately 0.5 g fresh leaves of 15-day-old *Arabidopsis* plants were collected and weighed immediately. The leaves were kept in a petri dish in open air. The water loss rate was calculated every hour based on the change in leaf weight, as previously described [65].

4.7. Diaminobenzidine (DAB) Staining of Hydrogen Peroxide in the Leaves

Diaminobenzidine (DAB) staining was used for in situ detection of hydrogen peroxide in *Arabidopsis* leaves as previously described [56]. The detached *Arabidopsis* leaves of different *Arabidopsis* lines were submerged in the DAB solution (1 mg·mL^{-1}, pH 3.8) overnight at room temperature. The leaves were submerged in ethanol until the chlorophyll was washed off. Then, the leaves were used for hydrogen peroxide detection.

4.8. Determination of Hydrogen Peroxide, MDA, Proline and Total Soluble Sugar Level, and Antioxidant Enzyme Activity

The plant samples were firstly ground into powder in liquid nitrogen and then suspended in ice-cold phosphate buffer (0.1 M, pH = 7). The sample was vortexed at maximum speed for 1 min and centrifuged at 12,000 rpm for 15 min. The supernatants were then used for further determination of the hydrogen peroxide level and antioxidant enzyme activity. The hydrogen peroxide level of different samples was determined using the Hydrogen Peroxide Assay Kit (Jiancheng Bioengineering Institute, Nanjing, China). The absorbance at 405 nm was determined by a Scandrop spectrophotometer (Analytikjena, Germany). The hydrogen peroxide level was calculated based on a previously described formula [56]. The MDA, proline, and total soluble sugar level were separately determined using the MDA Assay Kit, the Proline Assay Kit, and the Plant Soluble Sugar Content Test Kit (Jiancheng Bioengineering Institute, Nanjing, China), as previously described [24].

The antioxidant enzyme activities were also determined using the spectrophotometric method. SOD, POD, and CAT activities were separately determined using the Total Superoxide Dismutase (T-SOD) Assay Kit, Peroxidase Assay Kit, and Catalase (CAT) Assay Kit (Jiancheng Bioengineering Institute, Nanjing, China), as previously described [56].

4.9. Chlorophyll Fluorescence Measurement

The photosynthesis rate of the *Arabidopsis* plants of different lines after drought treatment was analyzed using the Portable Photosynthesis Rate Detector AGHJ-PPF (Anhui Institute of Optics and Fine Mechanics, Chinese Academy of Sciences) after a 30 min dark adaptation, as previously described [66]. Parameters including the maximum photochemical efficiency of PSII (Fv/Fm), minimal fluorescence (F0), maximal fluorescence (Fm), and PSII (as shown in Figure S1) were calculated according to a previous method [66].

4.10. Phylogenetic Analysis

The homologs of MAX2 of the other species were obtained by blasting against nucleic acid sequences in GenBank. Phylogenetic analysis was carried out using MEGA software Version 7.0 [67]. One thousand bootstrap replicates were performed for the phylogenetic tree construction.

4.11. Statistical Analysis

Multiple comparisons between different samples were carried out using Statistical Product and Service Solutions (SPSS, Chicago, IL, USA) with one-way ANOVA, followed by the Tukey's test ($p < 0.05$). Student's t-test was used to analyze the significant difference between the indicated groups and control.

Author Contributions: Conceptualization, J.N.; data curation, Q.W., J.N., F.S., W.L., D.W., Y.Y., H.H., S.H., J.H., S.F., and L.W.; investigation, Q.W., J.N., and F.S.; supervision, L.W.; validation, Q.W.; writing—original draft, J.N.; writing—review & editing, J.N.

Acknowledgments: We thank Kaiqin Ye for critically reading the manuscript.

Abbreviations

ABA	Abscisic acid
MDA	Malondialdehyde
MS	Murashige and Skoog
OE	Overexpression
PEG	polyethylene glycol
qPCR	Quantitative real-time PCR
ROS	Reactive oxygen species
rpm	Round per minute
RT-PCR	Reverse transcription PCR
SL	Strigolactone
WT	Wild-type

References

1. Zvi, P.; Eduardo, B. Hormone balance and abiotic stress tolerance in crop plants. *Curr. Opin. Plant Biol.* **2011**, *14*, 290–295.

2. Raja, V.; Majeed, U.; Kang, H.; Andrabi, K.I.; John, R. Abiotic stress: Interplay between ROS, hormones and MAPKs. *Environ. Exp. Bot.* **2017**, *137*, 142–157. [CrossRef]

3. Munne-Bosch, S.; Muller, M. Hormonal cross-talk in plant development and stress responses. *Front. Plant Sci.* **2013**, *4*, 529. [CrossRef] [PubMed]

4. Jones, A.M. A new look at stress: Abscisic acid patterns and dynamics at high-resolution. *New Phytol.* **2016**, *210*, 38–44. [CrossRef]

5. Zwack, P.J.; Rashotte, A.M. Interactions between cytokinin signalling and abiotic stress responses. *J. Exp. Bot.* **2015**, *66*, 4863–4871. [CrossRef]

6. Kang, G.Z.; Li, G.Z.; Guo, T.C. Molecular mechanism of salicylic acid-induced abiotic stress tolerance in higher plants. *Acta Physiol. Plant.* **2014**, *36*, 2287–2297. [CrossRef]

7. Korver, R.A.; Koevoets, I.T.; Testerink, C. Out of shape during stress: A key role for auxin. *Trends Plant Sci.* **2018**, *23*, 783–793. [CrossRef]

8. Van Zeijl, A.; Liu, W.; Xiao, T.T.; Kohlen, W.; Yang, W.C.; Bisseling, T.; Geurts, R. The strigolactone biosynthesis gene *DWARF27* is co-opted in rhizobium symbiosis. *BMC Plant Biol.* **2015**, *15*, 260. [CrossRef]

9. Gomez-Roldan, V.; Fermas, S.; Brewer, P.B.; Puech-Pages, V.; Dun, E.A.; Pillot, J.P.; Letisse, F.; Matusova, R.; Danoun, S.; Portais, J.C.; et al. Strigolactone inhibition of shoot branching. *Nature* **2008**, *455*, 189–194. [CrossRef]

10. Yamaguchi, S.; Kyozuka, J. Branching hormone is busy both underground and overground. *Plant Cell Physiol.* **2010**, *51*, 1091–1094. [CrossRef]

11. Foo, E. Auxin influences strigolactones in pea mycorrhizal symbiosis. *J. Plant Physiol.* **2013**, *170*, 523–528. [CrossRef] [PubMed]

12. Cardinale, F.; Krukowski, P.K.; Schubert, A.; Visentin, I. Strigolactones: Mediators of osmotic stress responses with a potential for agrochemical manipulation of crop resilience. *J. Exp. Bot.* **2018**, *69*, 2291–2303. [CrossRef] [PubMed]

13. Ha, C.V.; Leyva-Gonzalez, M.A.; Osakabe, Y.; Tran, U.T.; Nishiyama, R.; Watanabe, Y.; Tanaka, M.; Seki, M.; Yamaguchi, S.; Dong, N.V.; et al. Positive regulatory role of strigolactone in plant responses to drought and salt stress. *Proc. Natl. Acad. Sci. USA* **2014**, *111*, 851–856. [CrossRef] [PubMed]

14. Bu, Q.; Lv, T.; Shen, H.; Luong, P.; Wang, J.; Wang, Z.; Huang, Z.; Xiao, L.; Engineer, C.; Kim, T.H. Regulation of drought tolerance by the F-box protein MAX2 in *Arabidopsis*. *Plant Physiol.* **2014**, *164*, 424–439. [CrossRef] [PubMed]

15. Ni, J.; Gao, C.C.; Chen, M.S.; Pan, B.Z.; Ye, K.Q.; Xu, Z.F. Gibberellin promotes shoot branching in the perennial woody plant *Jatropha curcas*. *Plant Cell Physiol.* **2015**, *56*, 1655–1666. [CrossRef] [PubMed]

16. Ni, J.; Zhao, M.L.; Chen, M.S.; Pan, B.Z.; Tao, Y.B.; Xu, Z.F. Comparative transcriptome analysis of axillary buds in response to the shoot branching regulators gibberellin A3 and 6-benzyladenine in *Jatropha curcas*. *Sci. Rep.* **2017**, *7*, 11417. [CrossRef] [PubMed]

17. Muhr, M.; Prufer, N.; Paulat, M.; Teichmann, T. Knockdown of strigolactone biosynthesis genes in *Populus* affects *BRANCHED1* expression and shoot architecture. *New Phytol.* **2016**, *212*, 613–626. [CrossRef]

18. Foster, T.M.; Ledger, S.E.; Janssen, B.J.; Luo, Z.W.; Drummond, R.S.M.; Tomes, S.; Karunairetnam, S.; Waite, C.N.; Funnell, K.A.; van Hooijdonk, B.; et al. Expression of *MdCCD7* in the scion determines the extent of sylleptic branching and the primary shoot growth rate of apple trees. *J. Exp. Bot.* **2018**, *69*, 2379–2390. [CrossRef]

19. Wang, R.; Hanna, M.A.; Zhou, W.W.; Bhadury, P.S.; Chen, Q.; Song, B.A.; Yang, S. Production and selected fuel properties of biodiesel from promising non-edible oils: *Euphorbia lathyris* L., *Sapium sebiferum* L. and *Jatropha curcas* L. *Bioresour. Technol.* **2011**, *102*, 1194–1199. [CrossRef]

20. Xu, J.S.; Chikashige, T.; Meguro, S.; Kawachi, S. Effective utilization of stillingia or Chinese tallow-tree (*Sapium sebiferum*) fruits. *Mok. Gakk.* **1991**, *37*, 494–498.

21. Xu, Z.; Mahmood, K.; Rothstein, S.J. ROS induces anthocyanin production via late biosynthetic genes and anthocyanin deficiency confers the hypersensitivity to ROS-generating stresses in *Arabidopsis*. *Plant Cell Physiol.* **2017**, *58*, 1364–1377. [CrossRef] [PubMed]

22. Aarti, P.D.; Tanaka, R.; Tanaka, A. Effects of oxidative stress on chlorophyll biosynthesis in cucumber (*Cucumis sativus*) cotyledons. *Physiol. Plant.* **2010**, *128*, 186–197. [CrossRef]

23. Liu, C.; Xu, Y.; Feng, Y.; Long, D.; Cao, B.; Xiang, Z.; Zhao, A. Ectopic expression of mulberry G-Proteins alters drought and salt stress tolerance in tobacco. *Int. J. Mol. Sci.* **2018**, *20*, 89. [CrossRef] [PubMed]

24. Wang, X.; Gao, F.; Bing, J.; Sun, W.; Feng, X.; Ma, X.; Zhou, Y.; Zhang, G. Overexpression of the *Jojoba aquaporin* gene, *ScPIP1*, enhances drought and salt tolerance in transgenic *Arabidopsis*. *Int. J. Mol. Sci.* **2019**, *20*, 153. [CrossRef] [PubMed]

25. Wang, J.; Chen, G.; Zhang, C. The effects of water stress on soluble protein content, the activity of SOD, POD and CAT of two ecotypes of reeds (*Phragmites communis*). *Acta Bot. Boreal.-Occident. Sin.* **2002**, *22*, 561–565.

26. Wang, C.-T.; Ru, J.-N.; Liu, Y.-W.; Li, M.; Zhao, D.; Yang, J.-F.; Fu, J.D.; Xu, Z.-S. Maize *WRKY* transcription factor *ZmWRKY106* confers drought and heat tolerance in transgenic plants. *Int. J. Mol. Sci.* **2018**, *19*, 3046. [CrossRef] [PubMed]

27. Shen, H.; Zhu, L.; Bu, Q.Y.; Huq, E. MAX2 affects multiple hormones to promote photomorphogenesis. *Mol. Plant* **2012**, *5*, 750–762. [CrossRef]

28. Ni, J.; Shah, F.A.; Liu, W.; Wang, Q.; Wang, D.; Zhao, W.; Lu, W.; Huang, S.; Fu, S.; Wu, L. Comparative transcriptome analysis reveals the regulatory networks of cytokinin in promoting the floral feminization in the oil plant *Sapium sebiferum*. *BMC Plant Biol.* **2018**, *18*, 96. [CrossRef]

29. Wang, Y.Q.; Peng, D.; Zhang, L.; Tan, X.F.; Yuan, D.Y.; Liu, X.M.; Zhou, B. Overexpression of *SsDGAT2* from *Sapium sebiferum* (L.) roxb increases seed oleic acid level in *Arabidopsis*. *Plant Mol. Biol. Rep.* **2016**, *34*, 638–648.

30. Fu, R.; Zhang, Y.; Guo, Y.; Chen, F. Chemical composition, antioxidant and antimicrobial activity of Chinese tallow tree leaves. *Ind. Crop Prod.* **2015**, *76*, 374–377. [CrossRef]

31. Divi, U.K.; Zhou, X.R.; Wang, P.H.; Butlin, J.; Zhang, D.M.; Liu, Q.; Vanhercke, T.; Petrie, J.R.; Talbot, M.; White, R.G.; et al. Deep sequencing of the fruit transcriptome and lipid accumulation in a non-seed tissue of Chinese tallow, a potential biofuel crop. *Plant Cell Physiol.* **2016**, *57*, 125–137. [CrossRef] [PubMed]

32. Wang, X.; Luo, X.Y. Study on herbicidal activities of different organs of *Sapium sebiferum*. *Weed Sci.* **2011**, *2011*, 4.

33. Zhu, W.; Li, X. Stress resistance of *Sapium sebiferum* and its forestation at wind gap. *Prot. For. Sci. Technol.* **2017**, *2017*, 10.

34. El-Esawi, M.A.; Alayafi, A.A. Overexpression of rice *Rab7* gene improves drought and heat tolerance and increases grain yield in rice (*Oryza sativa* L.). *Genes* **2019**, *10*, 56. [CrossRef] [PubMed]

35. Polle, A.; Chen, S.L.; Eckert, C.; Harfouche, A. Engineering drought resistance in forest trees. *Front. Plant Sci.* **2019**, *9*, 18. [CrossRef]

36. Zwanenburg, B.; Blanco-Ania, D. Strigolactones: New plant hormones in the spotlight. *J. Exp. Bot.* **2018**, *69*, 2205–2218. [CrossRef] [PubMed]

37. Waters, M.T.; Gutjahr, C.; Bennett, T.; Nelson, D.C. Strigolactone signaling and evolution. *Annu. Rev. Plant Biol.* **2017**, *68*, 291–322. [CrossRef] [PubMed]

38. Mostofa, M.G.; Li, W.; Nguyen, K.H.; Fujita, M.; Lam-Son Phan, T. Strigolactones in plant adaptation to abiotic stresses: An emerging avenue of plant research. *Plant Cell Environ.* **2018**, *41*, 2227–2243. [CrossRef]

39. Stirnberg, P.; Furner, I.J.; Ottoline Leyser, H.M. MAX2 participates in an SCF complex which acts locally at the node to suppress shoot branching. *Plant J.* **2010**, *50*, 80–94. [CrossRef]

40. Stirnberg, P.; Van, D.S.K.; Leyser, H.M. MAX1 and MAX2 control shoot lateral branching in *Arabidopsis*. *Development* **2002**, *129*, 1131–1141.

41. An, J.-P.; Li, R.; Qu, F.-J.; You, C.-X.; Wang, X.-F.; Hao, Y.-J. Apple F-Box protein MdMAX2 regulates plant photomorphogenesis and stress response. *Front. Plant Sci.* **2016**, *7*, 1685. [CrossRef] [PubMed]

42. Takasaki, H.; Maruyama, K.; Kidokoro, S.; Ito, Y.; Fujita, Y.; Shinozaki, K.; Yamaguchsi-Shinozaki, K.; Nakashima, K. The abiotic stress-responsive NAC-type transcription factor OsNAC5 regulates stress-inducible genes and stress tolerance in rice. *Mol. Genet. Genom.* **2010**, *284*, 173–183. [CrossRef] [PubMed]

43. Cho, S.K.; Kim, J.E.; Park, J.A.; Eom, T.J.; Kim, W.T. Constitutive expression of abiotic stress-inducible hot pepper *CaXTH3*, which encodes a xyloglucan endotransglucosylase /hydrolase homolog, improves drought and salt tolerance in transgenic *Arabidopsis* plants. *FEBS Lett.* **2006**, *580*, 3136–3144. [CrossRef] [PubMed]

44. Li, P.; Li, Y.J.; Wang, B.; Yu, H.M.; Li, Q.; Hou, B.K. The *Arabidopsis UGT87A2*, a stress-inducible family 1 glycosyltransferase, is involved in the plant adaptation to abiotic stresses. *Physiol. Plant.* **2016**, *159*, 416–432. [CrossRef] [PubMed]

45. Keunen, E.; Remans, T.; Bohler, S.; Vangronsveld, J.; Cuypers, A. Metal-induced oxidative stress and plant mitochondria. *Int. J. Mol. Sci.* **2011**, *12*, 6894–6918. [CrossRef]

46. Roy, S.J.; Tucker, E.J.; Tester, M. Genetic analysis of abiotic stress tolerance in crops. *Curr. Opin. Plant Biol.* **2011**, *14*, 232–239. [CrossRef] [PubMed]

47. Nguyen, H.-C.; Lin, K.-H.; Ho, S.-L.; Chiang, C.-M.; Yang, C.-M. Enhancing the abiotic stress tolerance of plants: From chemical treatment to biotechnological approaches. *Physiol. Plant.* **2018**, *164*, 452–466. [CrossRef] [PubMed]

48. Eryılmaz, F. The relationships between salt stress and anthocyanin content in higher plants. *Biotechnol. Biotechnol. Equip.* **2006**, *20*, 47–52. [CrossRef]

49. Naing, A.H.; Il Park, K.; Ai, T.N.; Chung, M.Y.; Han, J.S.; Kang, Y.W.; Lim, K.B.; Kim, C.K. Overexpression of snapdragon *Delila* (*Del*) gene in tobacco enhances anthocyanin accumulation and abiotic stress tolerance. *BMC Plant Biol.* **2017**, *17*, 65. [CrossRef] [PubMed]

50. Lotkowska, M.E.; Tohge, T.; Fernie, A.R.; Xue, G.P.; Balazadeh, S.; Muellerroeber, B. The *Arabidopsis* transcription factor *MYB112* promotes anthocyanin formation during salinity and under high light stress. *Plant Physiol.* **2015**, *169*, 1862–1880. [CrossRef]

51. Castellarin, S.D.; Pfeiffer, A.; Sivilotti, P.; Degan, M.; Peterlunger, E.; Di Gaspero, G. Transcriptional regulation of anthocyanin biosynthesis in ripening fruits of grapevine under seasonal water deficit. *Plant Cell Environ.* **2007**, *30*, 1381–1399. [CrossRef] [PubMed]

52. Na, Y.W.; Jeong, H.J.; Lee, S.Y.; Choi, H.G.; Kim, S.H.; Rho, I.R. Chlorophyll fluorescence as a diagnostic tool for abiotic stress tolerance in wild and cultivated strawberry species. *Hort. Environ. Biotech.* **2014**, *55*, 280–286. [CrossRef]

53. Nuccio, M.L.; Rhodest, D.; McNeil, S.D.; Hanson, A.D. Metabolic engineering of plants for osmotic stress resistance. *Curr. Opin. Plant Biol.* **1999**, *2*, 128–134. [CrossRef]

54. Wani, S.H.; Gosal, S.S. Genetic engineering for osmotic stress tolerance in plants—Role of proline. *J. Genet. Evol.* **2011**, *3*, 14–25.

55. Shi, H.; Wang, X.; Tan, D.X.; Reiter, R.J.; Chan, Z. Comparative physiological and proteomic analyses reveal the actions of melatonin in the reduction of oxidative stress in Bermuda grass (*Cynodon dactylon* (L). Pers.). *J. Pineal Res.* **2015**, *59*, 120–131. [CrossRef] [PubMed]

56. Ni, J.; Wang, Q.; Shah, F.A.; Liu, W.; Wang, D.; Huang, S.; Fu, S.; Wu, L. Exogenous melatonin confers cadmium tolerance by counterbalancing the hydrogen peroxide homeostasis in wheat seedlings. *Molecules* **2018**, *23*, 799. [CrossRef] [PubMed]

57. Seiler, C.; Rajesh, K.; Reddy, P.S.; Strickert, M.; Rolletschek, H.; Scholz, U.; Wobus, U.; Sreenivasulu, N. ABA biosynthesis and degradation contributing to ABA homeostasis during barley seed development under control and terminal drought-stress conditions. *J. Exp. Bot.* **2011**, *62*, 2615–2632. [CrossRef]

58. Vishwakarma, K.; Upadhyay, N.; Kumar, N.; Yadav, G.; Singh, J.; Mishra, R.K.; Kumar, V.; Verma, R.; Upadhyay, R.G.; Pandey, M. Abscisic acid signaling and abiotic stress tolerance in plants: A review on current knowledge and future prospects. *Front. Plant Sci.* **2017**, *8*, 161. [CrossRef] [PubMed]

59. Ren, C.G.; Kong, C.C.; Xie, Z.H. Role of abscisic acid in strigolactone-induced salt stress tolerance in arbuscular mycorrhizal *Sesbania cannabina* seedlings. *BMC Plant Biol.* **2018**, *18*, 74. [CrossRef] [PubMed]

60. Shah, F.A.; Ni, J.; Chen, J.; Wang, Q.; Liu, W.; Chen, X.; Tang, C.; Fu, S.; Wu, L. Proanthocyanidins in seed coat tegmen and endospermic cap inhibit seed germination in *Sapium sebiferum*. *Peer J.* **2018**, *6*, 10. [CrossRef] [PubMed]

61. Zhang, X.; Henriques, R.; Lin, S.S.; Niu, Q.W.; Chua, N.H. Agrobacterium-mediated transformation of *Arabidopsis thaliana* using the floral dip method. *Nat. Protoc.* **2006**, *1*, 641–646. [CrossRef] [PubMed]

62. Livak, K.J.; Schmittgen, T.D. Analysis of relative gene expression data using real-time quantitative PCR and the $2^{-\Delta\Delta Ct}$ method. *Methods* **2001**, *25*, 402–408. [CrossRef] [PubMed]

63. Adriana, P.; Gaby, T.; Sylvain, A.; Iwona, A.; Simone, M.; Thomas, M.; Karl-Hans, O.; Bernhard, K.U.; Ji-Young, Y.; Liljegren, S.J. Chlorophyll breakdown in senescent *Arabidopsis* leaves. Characterization of chlorophyll catabolites and of chlorophyll catabolic enzymes involved in the degreening reaction. *Plant Physiol.* **2005**, *139*, 52–63.

64. Cinzia, S.; Alessandra, P.; Elena, L.; Amedeo, A.; Pierdomenico, P. Sucrose-specific induction of the anthocyanin biosynthetic pathway in *Arabidopsis*. *Plant Physiol.* **2006**, *140*, 637–646.

65. Zhang, K.W.; Xia, X.Y.; Zhang, Y.Y.; Gan, S.S. An ABA-regulated and Golgi-localized protein phosphatase controls water loss during leaf senescence in *Arabidopsis*. *Plant J.* **2012**, *69*, 667–678. [CrossRef] [PubMed]

66. Yin, G.F.; Zhao, N.J.; Shi, C.Y.; Chen, S.; Qin, Z.S.; Zhang, X.L.; Yan, R.F.; Gan, T.T.; Liu, J.G.; Liu, W.Q. Phytoplankton photosynthetic rate measurement using tunable pulsed light induced fluorescence kinetics. *Opt. Express* **2018**, *26*, A293–A300. [CrossRef] [PubMed]

67. Kumar, S.; Stecher, G.; Tamura, K. MEGA7: Molecular evolutionary genetics analysis version 7.0 for bigger datasets. *Mol. Biol. Evol.* **2016**, *33*, 1870–1874. [CrossRef]

PERMISSIONS

The contributors of this book come from diverse backgrounds, making this book a truly international effort. This book will bring forth new frontiers with its revolutionizing research information and detailed analysis of the nascent developments around the world.

We would like to thank all the contributing authors for lending their expertise to make the book truly unique. They have played a crucial role in the development of this book. Without their invaluable contributions this book wouldn't have been possible. They have made vital efforts to compile up to date information on the varied aspects of this subject to make this book a valuable addition to the collection of many professionals and students.

This book was conceptualized with the vision of imparting up-to-date information and advanced data in this field. To ensure the same, a matchless editorial board was set up. Every individual on the board went through rigorous rounds of assessment to prove their worth. After which they invested a large part of their time researching and compiling the most relevant data for our readers.

The editorial board has been involved in producing this book since its inception. They have spent rigorous hours researching and exploring the diverse topics which have resulted in the successful publishing of this book. They have passed on their knowledge of decades through this book. To expedite this challenging task, the publisher supported the team at every step. A small team of assistant editors was also appointed to further simplify the editing procedure and attain best results for the readers.

Apart from the editorial board, the designing team has also invested a significant amount of their time in understanding the subject and creating the most relevant covers. They scrutinized every image to scout for the most suitable representation of the subject and create an appropriate cover for the book.

The publishing team has been an ardent support to the editorial, designing and production team. Their endless efforts to recruit the best for this project, has resulted in the accomplishment of this book. They are a veteran in the field of academics and their pool of knowledge is as vast as their experience in printing. Their expertise and guidance has proved useful at every step. Their uncompromising quality standards have made this book an exceptional effort. Their encouragement from time to time has been an inspiration for everyone.

The publisher and the editorial board hope that this book will prove to be a valuable piece of knowledge for researchers, students, practitioners and scholars across the globe.

LIST OF CONTRIBUTORS

Muhammad Zeeshan and Shafaque Sehar
Institute of Crop Science, Department of Agronomy, College of Agriculture and Biotechnology, Zijingang Campus, Zhejiang University, Hangzhou 310058, China

Feibo Wu
Institute of Crop Science, Department of Agronomy, College of Agriculture and Biotechnology, Zijingang Campus, Zhejiang University, Hangzhou 310058, China
Jiangsu Co-Innovation Center for Modern Production Technology of Grain Crops, Yangzhou University, Yangzhou 225009, China

Meiqin Lu
Australian Grain Technologies, Narrabri, NSW 2390, Australia

Paul Holford
School of Science and Health, Western Sydney University, Penrith, NSW 2751, Australia

Kadir Uçgun
Department of Plant and Animal Production, Technical Sciences Vocational School, Karamanoğlu Mehmetbey University, Karaman 70200, Turkey

Jorge F. S. Ferreira, Xuan Liu, Donald L. Suarez and Devinder Sandhu
US Salinity Laboratory (USDA-ARS), 450W. Big Springs Rd., Riverside, CA 92507, USA

Jaime Barros da Silva Filho
Departments of Microbiology and Plant Pathology, University of California Riverside, 900 University Ave., Riverside, CA 92521, USA

Claudivan F. de Lacerda
Department of Agricultural Engineering, Federal University of Ceará, Fortaleza-CE 60450-760, Brazil

Jie Yu and Wei Tong
Department of Plant Resources, College of Industrial Sciences, Kongju National University, Yesan 32439, Korea
State Key Laboratory of Tea Plant Biology and Utilization, Anhui Agricultural University, Hefei 230036, China

Weiguo Zhao
Department of Plant Resources, College of Industrial Sciences, Kongju National University, Yesan 32439, Korea
School of Biotechnology, Jiangsu University of Science and Technology, Sibaidu, Zhenjiang, Jiangsu 212018, China

Qiang He
Department of Plant Resources, College of Industrial Sciences, Kongju National University, Yesan 32439, Korea
National Key Facility for Crop Resources and Genetic Improvement, Institute of Crop Science, Chinese Academy of Agricultural Sciences, Beijing 100081, China

Min-Young Yoon
Department of Plant Resources, College of Industrial Sciences, Kongju National University, Yesan 32439, Korea
Leader of Eco. Energy & Bio (LEEBCOR), 190-26 Hwangyeonggongwon-ro, Asan-si, Chungcheongnam-do 31529, Korea

Feng-Peng Li
Department of Plant Resources, College of Industrial Sciences, Kongju National University, Yesan 32439, Korea
Suzhou GENEWIZ Biotechnology Co. LTD, C3 218 Xinghu Road Suzhou Industrial Park, Suzhou 215123, China

Buung Choi
Department of Plant Resources, College of Industrial Sciences, Kongju National University, Yesan 32439, Korea
Chemical Safety Division, National Institute of Agricultural Sciences (NIAS), Wanju 55365, Korea

Eun-Beom Heo
Department of Plant Resources, College of Industrial Sciences, Kongju National University, Yesan 32439, Korea
Breeding & Research Institute, Koregon Co. LTD, Anseong Center 60-34, Gokcheon-gil, Bogae-Myeon, Anseong-Si, Gyeonggi-Do 17509, Korea

Kyu-Won Kim
Center of Crop Breeding on Omics and Artificial Intelligence, Kongju National University, Yesan 32439, Korea

Yong-Jin Park
Department of Plant Resources, College of Industrial Sciences, Kongju National University, Yesan 32439, Korea
Center of Crop Breeding on Omics and Artificial Intelligence, Kongju National University, Yesan 32439, Korea

Hongxia Miao, Juhua Liu, Jingyi Wang and Biyu Xu
Key Laboratory of Biology and Genetic Resources of Tropical Crops, Ministry of Agriculture, Institute of Tropical Bioscience and Biotechnology, Chinese Academy of Tropical Agricultural Sciences, Xueyuan Road 4, Haikou 571101, China

Peiguang Sun
Key Laboratory of Genetic Improvement of Bananas, Hainan Province, Haikou Experimental Station, Chinese Academy of Tropical Agricultural Sciences, Xueyuan Road 4, Haikou 570102, China

Zhiqiang Jin
Key Laboratory of Biology and Genetic Resources of Tropical Crops, Ministry of Agriculture, Institute of Tropical Bioscience and Biotechnology, Chinese Academy of Tropical Agricultural Sciences, Xueyuan Road 4, Haikou 571101, China
Key Laboratory of Genetic Improvement of Bananas, Hainan Province, Haikou Experimental Station, Chinese Academy of Tropical Agricultural Sciences, Xueyuan Road 4, Haikou 570102, China

Gan Zhao, Yingying Zhao, Xiuli Yu, Felix Kiprotich, Han Han and Wenbiao Shen
College of Life Sciences, Laboratory Center of Life Sciences, Nanjing Agricultural University, Nanjing 210095, China

Rongzhan Guan
National Key Laboratory of Crop Genetics and Germplasm Enhancement, Jiangsu Collaborative Innovation Center for Modern Crop Production, Nanjing Agricultural University, Nanjing 210095, China

Ren Wang
Institute of Botany, Jiangsu Province and Chinese Academy of Sciences, Nanjing 210014, China

Dan Luo, Xiaoming Hou, Yumeng Zhang, Yuancheng Meng, Huafeng Zhang, Suya Liu, Xinke Wang and Rugang Chen
College of Horticulture, Northwest A&F University, Yangling 712100, China

Kun Yan, Wenjun He, Guangxuan Han and Mengxue Lv
Key Laboratory of Coastal Environmental Processes and Ecological Remediation, Yantai Institute of Coastal Zone Research, Chinese Academy of Sciences, Yantai 264003, China

Tiantian Bian
Key Laboratory of Coastal Environmental Processes and Ecological Remediation, Yantai Institute of Coastal Zone Research, Chinese Academy of Sciences, Yantai 264003, China
School of Life Sciences, Ludong University, Yantai 264025, China

Mingzhu Guo and Ming Lu
College of Life Sciences, Yantai University, Yantai 264005, China

Ning Wang, Zhixin Qian, Manwei Luo, Shoujin Fan, Xuejie Zhang and Luoyan Zhang
Key Lab of Plant Stress Research, College of Life Science, Shandong Normal University, No. 88 Wenhuadong Road, Jinan 250014, China

Agustina Bernal-Vicente and José Antonio Hernández
Biotechnology of Fruit Trees Group, Department Plant Breeding, CEBAS-CSIC, Campus Universitario de Espinardo, 25, 30100 Murcia, Spain

Daniel Cantabella
Biotechnology of Fruit Trees Group, Department Plant Breeding, CEBAS-CSIC, Campus Universitario de Espinardo, 25, 30100 Murcia, Spain
IRTA, XaRTA-Postharvest, Edifici Fruitcentre, Parc Científic i Tecnològic Agroalimentari de Lleida, 25003 Lleida, Catalonia, Spain

Cesar Petri
Departamento de Producción Vegetal, Universidad Politécnica de Cartagena, Paseo Alfonso XIII, 48, 30203 Cartagena, Spain

Pedro Diaz-Vivancos
Biotechnology of Fruit Trees Group, Department Plant Breeding, CEBAS-CSIC, Campus Universitario de Espinardo, 25, 30100 Murcia, Spain
Department of Plant Biology, Faculty of Biology, University of Murcia, Campus de Espinardo, E-30100 Murcia, Spain

Huilong Zhang, Yinan Zhang, Chen Deng, Shurong Deng, Nianfei Li, Chenjing Zhao, Rui Zhao and Shaoliang Chen
Beijing Advanced Innovation Center for Tree Breeding by Molecular Design, College of Biological Sciences and Technology, Beijing Forestry University, Beijing 100083, China

Shan Liang
Beijing Advanced Innovation Center for Food Nutrition and Human Health, School of Food and Chemical Engineering, Beijing Technology and Business University, Beijing 100048, China

Jianzhong Wu, Yanhua Ma, Hong Lin, Liyan Pan, Suiyan Li and Dequan Sun
Institute of Forage and Grassland Sciences, Heilongjiang Academy of Agricultural Sciences, Harbin 150086, China

Qian Zhao, Guangwen Wu and Hongmei Yuan
Institute of Industrial Crop, Heilongjiang Academy of Agricultural Sciences, Harbin 150086, China

Akhtar Ali and Dae-Jin Yun
Department of Biomedical Science & Engineering, Konkuk University, Seoul 05029, Korea

Albino Maggio
Department of Agriculture, University of Naples Federico II, Via Universita 100, I-80055 Portici, Italy

Ray A. Bressan
Department of Horticulture and Landscape Architecture, Purdue University, West Lafayette, IN 47907-2010, USA

Xiaoning Zhang, Bowen Chen, Zihai Qin, Yufei Xiao, Ye Zhang, Ruiling Yao and Hailong Liu
Guangxi Key Laboratory of Superior Timber Trees Resource Cultivation, Guangxi Forestry Research Institute, 23 Yongwu Road, Nanning 530002, China

Lijun Liu
Key Laboratory of State Forestry Administration for Silviculture of the lower Yellow River, College of Forestry, Shandong Agricultural University, Taian 271018, Shandong, China

Hong Yang
Key Laboratory of Economic Plants and Biotechnology, Kunming Institute of Botany, Academy of Sciences, Yunnan Key Laboratory for Wild Plant Resources, Kunming 650201, China

Songling Fu
College of Forestry and Landscape Architecture, Anhui Agricultural University, Hefei 230000, Anhui, China

Jun Ni, Faheem Shah, Wenbo Liu, Yuanyuan Yao, Hao Hu, Shengwei Huang, Jinyan Hou and Lifang Wu
Key Laboratory of High Magnetic Field and Ion Beam Physical Biology, Hefei Institutes of Physical Science, Chinese Academy of Sciences, Hefei 230000, Anhui, China

Qiaojian Wang and Dongdong Wang
College of Forestry and Landscape Architecture, Anhui Agricultural University, Hefei 230000, Anhui, China
Key Laboratory of High Magnetic Field and Ion Beam Physical Biology, Hefei Institutes of Physical Science, Chinese Academy of Sciences, Hefei 230000, Anhui, China

Index

9 781647 404192